Coevolution of Black Holes and Galaxies

Black holes are among the most mysterious objects in the Universe. Weighing up to several billion Suns, massive black holes have long been suspected to be the central powerhouses of energetic phenomena such as quasars. Recent advances in astronomy have not only provided spectacular proof of this long-standing paradigm, but also revealed the unexpected result that, far from being rare, exotic beasts, they inhabit the center of virtually all large galaxies. Candidate black holes have been identified in increasingly large numbers of galaxies, both inactive and active, to the point where statistical studies are now possible. Recent work has highlighted the close connection between the formation, growth, and evolution of supermassive black holes and their host galaxies. This volume contains the invited lectures from an international symposium that was held to explore this exciting theme. With contributions from leading authorities in the field with diverse but interrelated observational and theoretical expertise, this is a valuable review for professional astronomers and graduate students.

LUIS C. HO received his undergraduate education at Harvard University and his Ph.D. in astronomy from the University of California at Berkeley. He is currently a staff astronomer at the Carnegie Observatories, where he conducts research on black holes, accretion physics in galactic nuclei, and star formation processes. He is the editor for this series.

This series of four books celebrates the Centennial of the Carnegie Institution of Washington, and is based on a set of four special symposia held by the Observatories in Pasadena. Each symposium explored an astronomical topic of major historical and current interest at the Observatories, and each resulting book contains a set of comprehensive, authoritative review articles by leading experts in the field.

T0192674

Carnegie Observatories Astrophysics Series
Volume 1

COEVOLUTION OF BLACK HOLES AND GALAXIES

Edited by

LUIS C. HO

CAMBRIDGE
UNIVERSITY PRESS

CAMBRIDGE UNIVERSITY PRESS
Cambridge, New York, Melbourne, Madrid, Cape Town, Singapore,
São Paulo, Delhi, Dubai, Tokyo

Cambridge University Press
The Edinburgh Building, Cambridge CB2 8RU, UK

Published in the United States of America by Cambridge University Press, New York

www.cambridge.org
Information on this title: www.cambridge.org/9780521141567

© The Observatories of the Carnegie Institution of Washington 2004

This publication is in copyright. Subject to statutory exception
and to the provisions of relevant collective licensing agreements,
no reproduction of any part may take place without the written
permission of Cambridge University Press.

First published 2004
This digitally printed version 2010

A catalogue record for this publication is available from the British Library

ISBN 978-0-521-82449-1 Hardback
ISBN 978-0-521-14156-7 Paperback

Cambridge University Press has no responsibility for the persistence or accuracy of
URLs for external or third-party internet websites referred to in this publication, and
does not guarantee that any content on such websites is, or will remain, accurate or
appropriate.

Contents

Preface

In 1902, Andrew Carnegie, a man of uncommon vision and philanthropy, bequeathed a sizable sum to establish a scientific research organization whose purpose was "to encourage, in the broadest and most liberal manner, investigation, research, and discovery, and the application of knowledge to the improvement of mankind." For the past century, the Carnegie Institution of Washington has been a haven for many influential and creative scientists, covering disciplines ranging from geophysics, earth and planetary science, cellular and genetic biology, plant science, global ecology, and last, but not least, astronomy.

Astronomy has been a major part of the Institution almost from its inception, thanks to persistence and courage of another visionary, George Ellery Hale. Convinced that southern California had conditions favorable for astronomical observations, Hale persuaded Carnegie in 1904 to establish an observatory at Mount Wilson, located near Pasadena, California. There, Hale consecutively built the world's next two largest telescopes, the 60-inch in 1908 and the renowned 100-inch Hooker telescope in 1917. Arguably no other telescope since Galileo's has had a more profound impact on astronomy—indeed in shaping our view of humankind's footing in the cosmos—than those on Mount Wilson. Using the 60-inch to map the distribution of globular clusters in the Milky Way, Harlow Shapley concluded that our Galaxy was significantly larger than previously thought, and deduced that the Sun lies not at the center of the Galaxy but in its remote outskirts. But the Universe proved to be far greater still. With the 100-inch Edwin Hubble established the extragalactic nature of "nebulae," and therefore the existence of a multitude of galaxies beyond our own, followed by the discovery that the Universe is expanding. This was the birth of modern cosmology. Among the many other significant, if less sensational, advances attributable to the 100-inch includes Walter Baade's recognition of two distinct stellar populations, a concept central to the subsequent development of stellar and galactic evolution.

The supremacy of the 100-inch was not eclipsed until the completion in 1948 of Hale's last and most ambitious feat—the mighty 200-inch reflector at Mount Palomar. Although the "Big Eye" was not finished before he died, Hale was chiefly responsible for securing the funding for the project, and he was the main driving force behind its long, difficult construction. The accomplishments stemming from the 200-inch are too numerous and varied to be recounted here. It suffices to say that the Palomar 200-inch, which was operated in partnership with Caltech until 1980, has played a major role in ground-based optical astronomy for the latter half of the twentieth century.

In search of more pristine skies and to gain access to the southern hemisphere, Carnegie astronomers in 1969 established the Las Campanas Observatory in Chile's Atacama Desert, where they operate the Swope 1-meter and the du Pont 2.5-meter telescopes. However, in the era of ever-increasing large telescopes, this was not enough. In the mid-1980's, plans were under way for the design and construction of a pair of large optical telescopes at Las Campanas. The outcome—the twin 6.5-meter Magellan telescopes—is a collaboration between Carnegie, University of Arizona, Harvard University, University of Michigan, and Massachusetts Institute of Technology. Both the Baade and Clay telescopes, each equipped with state-of-the-art instrumentation, are now fully functional. Though nominally smaller than the current generation of 8–10 meter-class telescopes, Magellan is every bit as competitive thanks to its superb image quality and wide field-of-view.

Headquartered at 813 Santa Barbara Street, Carnegie Observatories presently supports a small, but distinguished group of about two dozen scientific staff members and postdoctoral fellows, along with a sizable group of engineers and instrument scientists who are responsible for technical developments. The research interests at the Observatories are diverse, ranging from observational cosmology to galaxy formation and evolution, large-scale structure, the intergalactic medium, stellar populations, stellar chemical composition, supernovae, star clusters, black holes, and accretion processes in galactic nuclei. In keeping with the tradition of the Observatories, some of staff devote considerable effort building innovative instruments for the telescopes.

This year marks the 100th anniversary of the founding of Carnegie Observatories. We stand at an important crossroad. To be sure, we glance back at our accomplishments of the past century with significant pride. But we are also confronted with many challenges for the years ahead, for our discipline is constantly driven by larger and more ambitious telescope enterprises, which carry sky-rocketing price tags and daunting technological hurdles. While the future success of a research institution, no matter how distinguished its past, should not be taken for granted, in reflecting on Carnegie's legacy in astronomy we cannot help but draw from it a measure of inspiration and optimism.

To commemorate our Centennial, we thought it would be fitting to host a series of scientific meetings, organized by Carnegie astronomers, on a range of topics that both celebrates Carnegie's past astronomical contributions and recognizes its current, diverse research interests. In the end, we organized four international-level meetings, held in Pasadena, from Fall 2002 to Winter 2003. The Carnegie Observatories Centennial Symposia covered the following topics: (1) *Coevolution of Black Holes and Galaxies* (hosted by Luis Ho; 20–25 October 2002), (2) *Measuring and Modeling the Universe* (hosted by Wendy Freedman; 17–22 November 2002), (3) *Clusters of Galaxies: Probes of Cosmological Structure and Galaxy Evolution* (hosted by John Mulchaey, Alan Dressler, and Gus Oemler; 27–31 January 2003), and (4) *Origin and Evolution of the Elements* (hosted by Andy McWilliam and Michael Rauch; 16–21 February 2003). The meetings were very well attended, and, by most measures, highly successful.

To complement the Symposia, we have planned from the outset to use the invited papers to compile a set of volumes of sufficiently high standards to have lasting value as an authoritative reference, one that potentially can be used for graduate-level course work. To achieve this goal, we have subjected each contribution to a battery of quality controls atypical for conventional conference proceedings; this includes a formal peer-review process, careful editing by the scientific organizers, and final scrutiny and copy-editing by the series editor. The product of this exercise is the first four volumes of the *Carnegie Observatories Astrophysics Series*.

An undertaking of this scope would not have been possible without the help of many people. First and foremost, I would like to thank the organizers of the Symposia and editors of this *Series*, my colleagues Wendy Freedman, John Mulchaey, Alan Dressler, Gus Oemler, Andy McWilliam, and Michael Rauch, for allowing me to twist their arms into this zany venture. Paul Martini helped me through many queries on how to set up HTML pages and troubleshoot Latex class files. There were myriad details associated with the local organization, from seemingly trivial items like how many cookies to order for coffee breaks to major ones like securing a venue, all necessary for the successful execution of the meetings. They were handled patiently and efficiently by Karen Gross during the initial phase, and later by Silvia Hutchison and Becky Lynn. I am most grateful for their assistance. I also appreciate the help of the facilities staff, especially Steve Wilson, Scott Rubel, Earl Harris, and Greg Ortiz, who worked hard to set up the technical logistics and to ensure the smooth operation of the audio-visual equipment. Lastly, I thank P Street for their financial support to help cover the cost overrun incurred for the meetings.

Luis C. Ho
Carnegie Observatories
January 2004

Introduction

Few subjects in astronomy capture the popular imagination like black holes. Black holes and their varied manifestations as active galaxies certainly occupy the attention of a large segment of the current astronomical community. While Carnegie's involvement in this subject may not be widely known, the fact is that it has had a long historical connection to this field. Following the initial work by Edward Fath at Lick Observatory in 1908 and by Vesto Slipher at Lowell Observatory in 1917, Edwin Hubble himself noted in 1926 the unusual nature of the emission-line spectrum in NGC 1068, NGC 4051, and NGC 4151. But it was really the 1943 paper by Carl Seyfert, based on observations obtained at Mount Wilson, which first systematically studied the class of active galaxies that today bear his name, although the significance of this work remained unrecognized for some time to come.

Carnegie's role in the early development of AGN research was most pronounced after World War II, when advances in radio astronomy led to the discovery of extragalactic radio sources. In the ensuing period, much of the community with access to large optical telescopes was keen on obtaining optical identifications of these mysterious sources. At Carnegie, the early effort was led by Rudolf Minkowski and Walter Baade, and subsequently by Allan Sandage and his colleagues. Sandage's extensive work on optical identification and spectroscopy of radio sources led to the discovery of a large population of radio-quiet objects that show an ultraviolet excess.

Once the redshift puzzle of quasars was solved in 1963 by Maarten Schmidt and Jesse Greenstein, it was thereafter quickly realized that the quasar phenomenon most likely draws its power from the gravitational energy of a massive collapsed object—a massive black hole. There is just too much energy coming out of too tiny a volume for anything else to be viable. In the 1970s, Jerome Kristian, working with Peter Young and others at Caltech, was one of the first to search for supermassive objects in the centers of giant elliptical galaxies. Kristian's work on quasar host galaxies was also very much a forerunner to what has become a lively pursuit. In more recent times, a number of Carnegie astronomers have also been involved in various aspects of black hole and AGN research. Some of the notable examples include Ray Weymann's seminal contributions on quasar absorption-line systems and quasar outflows, Alan Dressler's mass determination for the nuclei of M31 and M32, Pat McCarthy's multifaceted work on radio galaxies, and John Mulchaey's investigations on Seyfert galaxies and AGN fueling.

While it has long been suspected that massive black holes and nuclear activity are somehow related to galaxy formation and evolution, it is fair to say that until recently very few people realized the depth of the interconnection. Most of the practitioners in these subfields belong to communities that rarely overlapped. Though AGN research commands a strong following, it has had a somewhat checkered reputation in the broader community as a largely phenomenological endeavor, too often preoccupied with taxonomy. AGNs are useful for exploring some aspects of relativistic and high-energy astrophysics, and they make good background probes for absorption-line work, but beyond that they are really not that relevant to mainstream work on "normal" galaxies. Such is often the prejudice.

There has been a refreshing change of attitude in the last few years, which is largely triggered, I think, by the persuasive evidence that massive black holes are not only common but evidently tightly coupled to the life-cycle of galaxies. The term "normal galaxy" is woefully inadequate. Most respectable-sized galaxies, we now suspect, come naturally endowed with a massive central black hole, whose mass—somehow— has an uncanny familiarity with the large-scale properties of its host galaxy. A symbiotic relationship between black hole growth and galaxy assembly seems inescapable. As black holes grow through accretion, they ignite briefly as AGNs of many flavors, dumping radiation and kinetic energy into their host galaxies, and perhaps beyond. The cumulative deposition of accretion energy lights up the sky in X-rays. Black holes inspiral during galaxy mergers; some may coalesce, generating gravitational radiation. The landscape for observational and theoretical astrophysics has never been so rich.

Given these healthy developments, it seemed opportune to convene a meeting to bring together specialists working on different but interrelated subjects concerning black holes and galaxies. I had tried such an experiment before, in a 1998 meeting in Nagoya, Japan, entitled *The AGN-Galaxy Connection*, which I co-organized with Anne Kinney and Henrique Schmitt. The justification was strong then, and it is even stronger now.

On the occasion of the Centennial of the Carnegie Institution of Washington, Carnegie Observatories hosted a series of four astrophysics symposia in Pasadena, from Fall 2002 to Winter 2003. The first of these symposia, *Coevolution of Black Holes and Galaxies*, was held on 20–25 October 2002. By most accounts, it was a somewhat unusual, but highly effective gathering, which brought together people with very different backgrounds, in an intense but lively atmosphere. A total of 28 invited speakers covered topics ranging from black hole searches to formation and fueling mechanisms of black holes, gas-dynamical processes, dynamical evolution of dense stellar systems, the central and global structures of galaxies, binary black holes, gravitational radiation, AGN statistics, galaxy formation, AGN feedback, reionization, and the X-ray background. At least 100 other participants gave contributed talks or presented posters.

This book contains the review papers based on the presentations of the invited speakers, which forms the first volume of the *Carnegie Observatories Astrophysics Series*. (The contributed papers are published separately in electronic form at the Carnegie web site.) I am happy to say that these papers are of exceptionally high quality. As explained in the Preface, it has been my intention from the outset that the *Series* should aim for a high standard of scholarship, to ensure that the contributions contained therein would have a lasting impact. I am most grateful to all the authors for the enormous effort they have invested in conscientiously preparing the manuscripts, and for agreeing to have them subjected to a peer-review process and to entrust them to my editorial oversight. I can only hope that they agree that their efforts have been worthwhile.

Luis C. Ho
Carnegie Observatories
January 2004

List of Participants

Agol, Eric	Caltech, USA
Alexander, Tal	The Weizmann Institute of Science, Israel
Amaro-Seoane, Pau	Astronomisches Rechen-Institut, Germany
Armitage, Philip	University of Colorado, USA
Asada, Keiichi	The Graduate University for Advanced Studies, Japan
Axon, Dave	University of Hertfordshire, UK
Backer, Donald	U. C. Berkeley, USA
Barth, Aaron	Caltech, USA
Barthel, Peter	Kapteyn Institute, Netherlands
Begelman, Mitch	JILA/University of Colorado, USA
Bender, Peter	JILA/University of Colorado, USA
Blandford, Roger	Caltech, USA
Burbidge, Geoffrey	U.C. San Diego, USA
Burkert, Andreas	Max-Planck-Institute for Astronomy Heidelberg, Germany
Cappellari, Michele	Leiden Observatory, Leiden, Netherlands
Cappi, Massimo	TeSRE-CNR, Bologna, Italy
Carollo, Marcella	ETH-Zurich, Zurich, Switzerland
Cavaliere, Alfonso	Univ. Roma Tor Vergata, Italy
Celotti, Annalisa	S.I.S.S.A., Italy
Chornock, Ryan	U. C. Berkeley, USA
Clarke, Cathie	Cambridge University, UK
Colpi, Monica	University of Milano Bicocca, Italy
Cretton, Nicolas	European Southern Observatory, Germany
Cruz, Fidel	UNAM, Mexico
de Zeeuw, Tim	Leiden Observatory, Leiden, Netherlands
Dressel, Linda	Space Telescope Science Institute, USA
Dunlop, James	University of Edinburgh, UK
Emsellem, Eric	Observatoire de Lyon, France
Erwin, Peter	Instituto de Astrofisica de Canarias, Spain
Escala, Andres	Yale University, USA
Fabian, Andy	Cambridge University, UK
Fall, S. Mike	Space Telescope Science Institute, USA

Falomo, Renato	Osservatorio Astronomico di Padova, Italy
Fan, Xiaohui	Institute of Advanced Studies, USA
Filho, Mercedes	Kapteyn Institute, Netherlands
Floyd, David	University of Edinburgh, UK
Freitag, Marc	Caltech, USA
Gallagher, Sarah	Penn State University, USA
Gebhardt, Karl	University of Texas, Austin, USA
Gerhard, Ortwin	University of Basel, Switzerland
Gerssen, Jouris	Space Telescope Science Institute, USA
Gezari, Suvi	Columbia University, USA
Ghez, Andrea	UCLA, USA
Gorjian, Varoujan	JPL, Pasadena, USA
Graham, Alister	University of Florida, USA
Granato, Gian Luigi	Osservatorio Astronomico di Padova, Italy
Green, Richard	NOAO, USA
Greene, Jenny	Harvard University, USA
Haehnelt, Martin	Institute of Astronomy, Cambridge, UK
Haiman, Zoltan	Princeton University, USA
Hao, Lei	Princeton University, USA
Hayashida, Kiyoshi	Osaka University, Japan
Heckman, Timothy	Space Telescope Science Institute, USA
Heidt, Jochen	Landessternwarte Heidelberg, Germany
Ho, Luis	Carnegie Observatories, USA
Horiuchi, Shinji	JPL, Pasadena, USA
Hosokawa, Takashi	Kyoto University, Japan
Huang, JieHao	Nanjing University, China
Hughes, Mark	University of Hertfordshire, UK
Hutchings, John	HIA, Canada
Jarvis, Matt	Leiden University, Netherlands
Jian, Hung-Yu	National Taiwan University, Taiwan
Jones, Dayton	JPL, Pasadena, USA
Kalogera, Vicky	Northwestern University, USA
Kauffmann, Guinevere	MPA, Garching, Germany
Kawakatu, Nozomu	University of Tsukuba, Japan
Kollatschny, Wolfram	University Goettingen, Germany
Komossa, Stefanie	Max-Planck-Institut fuer extraterrestrische Physik, Germany
Kormendy, John	University of Texas, Austin, USA
Kukula, Marek	University of Edinburgh, UK
Lacy, Mark	IPAC, Pasadena, USA
Laine, Seppo	Space Telescope Science Institute, USA
Lauer, Tod	NOAO, USA
Lu, Youjun	Princeton University, USA
Maciejewski, Wiltold	Osservatorio Astrofisico di Arcetri, Italy
MacMillan, Joseph	Queen's University, Canada
Malkan, Matt	UCLA, USA
Maoz, Dani	Tel-Aviv University, Israel
Marchesini, Danilo	S.I.S.S.A., Italy
Marconi, Alessandro	Osservatorio Astrofisico di Arcetri, Italy

Markowitz, Alex	UCLA, Los Angeles, USA
Martini, Paul	Carnegie Observatories, USA
Matsumoto, Hironori	Kyoto University, Japan
McLure, Ross	Oxford University, UK
Meier, David	JPL, Pasadena, USA
Merritt, David	Rutgers University, USA
Miller, Mark	JPL, Pasadena, USA
Milosavljevic, Milos	Rutgers University, USA
Nakamura, Masanori	JPL, Pasadena, USA
Nelson, Charles	Drake University, USA
Newman, Peter	Apache Point Observatory/NMSU, USA
Noel-Storr, Jacob	Columbia University, USA
Novak, Gregory	U. C., Santa Cruz, USA
Ohsuga, Ken	Kyoto University, Japan
Oshlack, Alicia	University of Melbourne, Australia
Osmer, Patrick	Ohio State University, USA
Panessa, Francesca	TeSRE-CNR, Bologna, Italy
Peng, Chien	Steward Observatory, USA
Peterson, Bradley	Ohio State University, USA
Petric, Andreea	Columbia University, USA
Phinney, Sterl	Caltech, USA
Rasio, Fred	Northwestern University, USA
Ravindranath, Swara	Carnegie Observatories, USA
Rector, Travis	NRAO, USA
Rich, R. Michael	UCLA, USA
Richstone, Douglas	University of Michigan, USA
Sadler, Elaine	University of Sydney, Australia
Sarzi, Marc	University Durham, UK
Schinnerer, Eva	NRAO, USA
Scoville, Nick	Caltech, USA
Sellwood, Jerry	Rutgers University, USA
Shankar, Francesco	S.I.S.S.A., Italy
Shapiro, Stuart	University of Illinois, USA
Sheinis, Andrew	U. C., Santa Cruz, USA
Shen, Juntai	Rutgers University, USA
Shields, Joe	Ohio University, USA
Sigurdsson, Steinn	Penn State University, USA
Somerville, Rachel	University of Michigan, USA
Strateva, Iskra	Princeton University, USA
Szuszkiewicz, Ewa	University of Szczecin, Poland
Umemura, Masayuki	University of Tsukuba, Japan
Ulvestad, James	NRAO, USA
van Breugel, Wil	LLNL, USA
van der Marel, Roeland	Space Telescope Science Institute, USA
Verdoes Kleijn, Gijs	Space Telescope Science Institute, USA
Vestergaard, Marianne	Ohio State University, USA
Viollier, Raoul	Inst. Theoretical Physics and Astrophysics, South Africa
Wada, Keiichi	National Astronomical Observatory, Japan

Walcher, Jakob   Max-Planck-Institute for Astronomy Heidelberg, Germany
Wandel, Amri   The Hebrew University of Jerusalem, Israel
Wang, Yiping   Purple Mountain Observatory, China
Yu, Qingjuan   CITA, Canada
Yuan, Chi   ASIAA, Taiwan

1

The stellar-dynamical search for supermassive black holes in galactic nuclei

JOHN KORMENDY

Department of Astronomy, University of Texas at Austin

Abstract

The robustness of stellar-dynamical black hole (BH) mass measurements is illustrated using six galaxies that have results from independent research groups. Derived BH masses have remained constant to a factor of ~ 2 as spatial resolution has improved by a factor of $2 - 330$, as velocity distributions have been measured in increasing detail, and as the analysis has improved from spherical, isotropic models to axisymmetric, three-integral models. This gives us confidence that the masses are reliable and that the galaxies do not indulge in a wide variety of perverse orbital structures. Another successful test is the agreement between a preliminary stellar-dynamical BH mass for NGC 4258 and the accurate mass provided by the maser disk. Constraints on BH alternatives are also improving. In M 31, *Hubble Space Telescope* (*HST*) spectroscopy shows that the central massive dark object (MDO) is in a tiny cluster of blue stars embedded in the P2 nucleus of the galaxy. The MDO must have a radius $r \lesssim 0\rlap{.}''06$. M 31 becomes the third galaxy in which dark clusters of brown dwarf stars or stellar remnants can be excluded. In our Galaxy, spectacular proper motion observations of almost-complete stellar orbits show that the central dark object has radius $r \lesssim 0.0006$ pc. Among BH alternatives, this excludes even neutrino balls. Therefore, measurements of central dark masses and the conclusion that these are BHs have both stood the test of time. Confidence in the BH paradigm for active galactic nuclei (AGNs) is correspondingly high.

Compared to the radius of the BH sphere of influence, BHs are being discovered at similar spatial resolution with *HST* as in ground-based work. The reason is that *HST* is used to observe more distant galaxies. Typical BHs are detectable in the Virgo cluster, and the most massive ones are detectable $3 - 6$ times farther away. Large, unbiased samples are accessible. As a result, *HST* has revolutionized the study of BH demographics.

1.1   Introduction

The supermassive black hole paradigm for AGNs was launched by Zel'dovich (1964), Salpeter (1964), and Lynden-Bell (1969, 1978), who argued that the high energy production efficiencies required to make quasars are provided by gravity power. Eddington-limited accretion suggested that BH engines have masses of 10^6 to $10^9\ M_\odot$. Confidence grew rapidly with the amazing progress in AGN observations and with the paradigm's success in weaving these results into a coherent theoretical picture. Unlike the normal course of scientific research, acceptance of the AGN paradigm came long before there was any dynamical evidence that BHs exist.

© The Observatories of the Carnegie Institution of Washington 2004.

Table 1.1 *Black Hole Mass Measurements*

Galaxy	D (Mpc)	σ_e (km/s)	$M_\bullet$ (M_{low}, M_{high}) ($M_\odot$)	r_{cusp} (arcsec)	σ_* (arcsec)	r_{cusp}/σ_*	Reference
Galaxy	0.008	103	3.7 (3.3−4.1) e6	38.8	0.0159	2438.	Ghez 2004
Galaxy			3.7 (2.2−5.2) e6		0.0159	2438.	Schödel + 2002
Galaxy			2.0 (1.3−2.7) e6		0.113	343.	Chakrabarty + 2001
Galaxy			3.0 (2.6−3.3) e6		0.26	150.	Genzel + 2000
Galaxy			2.6 (2.4−2.8) e6		0.39	100.	Ghez + 1998
Galaxy			2.6 (2.3−3.0) e6		0.39	100.	Genzel + 1997
Galaxy			2.5 (2.1−2.9) e6		0.39	100.	Eckart + 1997
Galaxy			2.7 (2.4−3.0) e6		2.60	14.9	Genzel + 1996
Galaxy			1.8 (1.3−2.3) e6		3.6	10.8	Haller + 1996
Galaxy			2.8 (1.9−3.8) e6		3.4	11.4	Krabbe + 1995
Galaxy			2. e6		5.2	7.5	Evans + 1994
Galaxy			3. e6		5.2	7.5	Kent 1992
Galaxy			5.2 (3.8−6.6) e6		5.2	7.5	Sellgren + 1990
M 31			1.0 e8		0.297	10.8	Peiris + 2004
M 31	0.76	160	7.0 (3.0–20.0) e7	3.20	0.039	81.	Bender + 2004
M 31			7.0 (3.5−8.5) e7		0.052	61.	Bacon + 2001
M 31			3.3 (1.5−4.5) e7		0.297	10.8	Kormendy + 1999
M 31			5.9 (5.7−6.1) e7		0.297	10.8	Magorrian + 1998
M 31			7.4 e7		≈ 0.57	≈ 5.6	Tremaine 1995
M 31			7.8 e7		0.39	8.2	Bacon + 1994
M 31			5.0 (4.4−5.5) e7		0.60	5.3	Richstone + 1990
M 31			3.6 (1.1–10.9) e7		0.57	5.6	Kormendy 1988a
M 31			7.7 (3.3−7.7) e7		0.60	5.3	Dressler + 1988
M 32	0.81	75	2.9 (2.3−3.5) e6	0.56	0.052	10.83	Verolme + 2002
M 32			3.7 (2.6−5.0) e6		0.052	10.83	Joseph + 2001
M 32			2.4 (2.2−2.6) e6		0.23	2.41	Magorrian + 1998
M 32			4.0 (3.1−4.8) e6		0.050	11.39	van der Marel + 1998b
M 32			4.0 (2.1−5.8) e6		0.050	11.39	van der Marel + 1997ab
M 32			3.2 (2.6−3.7) e6		0.23	2.41	Bender + 1996
M 32			2.1 (1.9−2.3) e6		0.34	1.66	Dehnen 1995
M 32			2.1 e6		0.34	1.66	Qian + 1995
M 32			2.1 (1.7−2.4) e6		0.34	1.66	van der Marel + 1994b
M 32			2.2 (0.8−3.5) e6		0.59	0.95	Richstone + 1990
M 32			9.4 (4.7–18.9) e6		0.59	0.95	Dressler + 1988
M 32			7.6 (3.5–11.6) e6		0.76	0.75	Tonry 1987
M 32			5.9 e6		1.49	0.38	Tonry 1984
M 81	3.9	143	6.8 (5.5−7.5) e7	0.76	0.068	11.08	Bower + 2000
NGC 821	24.1	209	3.7 (2.9−6.1) e7	0.031	0.052	0.60	Gebhardt + 2003
NGC 1023	11.4	205	4.4 (3.9−4.8) e7	0.081	0.068	1.18	Bower + 2001
NGC 2778	22.9	175	1.4 (0.5−2.2) e7	0.018	0.052	0.34	Gebhardt + 2003
NGC 3115	9.7	182	1.0 (0.4−2.0) e9	2.77	0.047	59.	Tremaine + 2002
NGC 3115			6.3 (2.9−9.7) e8		0.111	24.9	Emsellem + 1999
NGC 3115			4.7 (4.4−4.9) e8		0.26	10.6	Magorrian + 1998
NGC 3115			1.5 e9		0.047	59.	Kormendy + 1996a
NGC 3115			1.6 (1.1−2.1) e9		0.50	5.5	Kormendy + 1992
NGC 3377			5.7 (3.4–11.) e7		0.29	1.3	Cretton + 2004
NGC 3377	11.2	145	1.0 (0.9−1.9) e8	0.38	0.111	3.4	Gebhardt + 2003
NGC 3377			6.9 (6.3−7.7) e7		0.24	1.57	Magorrian + 1998
NGC 3377			2.0 (1.1−2.9) e8		0.24	1.57	Kormendy + 1998
NGC 3379	10.6	206	1.0 (0.6−2.0) e8	0.201	0.111	1.81	Gebhardt + 2000a
NGC 3384	11.6	143	1.6 (1.4−1.7) e7	0.060	0.052	1.15	Gebhardt + 2003
NGC 3608	22.9	182	1.9 (1.3−2.9) e8	0.223	0.052	4.3	Gebhardt + 2003
NGC 4258	7.2	105	2.0 (1.0−3.0) e7	0.44	0.052	8.4	Siopis + 2004

Table 1.1 　*Black Hole Mass Measurements*

Galaxy	D (Mpc)	σ_e (km/s)	$M_\bullet$ (M_{low}, M_{high}) ($M_\odot$)	r_{cusp} (arcsec)	σ_* (arcsec)	r_{cusp}/σ_*	Reference
NGC 4291	26.2	242	3.1 (0.8−3.9) e8	0.180	0.052	3.45	Gebhardt + 2003
NGC 4342	15.3	225	3.1 (2.0−4.8) e8	0.351	0.135	2.60	Cretton + 1999a
NGC 4473	15.7	190	1.1 (0.3−1.5) e8	0.173	0.052	3.31	Gebhardt + 2003
NGC 4486B	16.1	185	6.0 (4.0−9.0) e8	0.97	0.258	3.75	Kormendy + 1997
NGC 4564	15.0	162	5.6 (4.8−5.9) e7	0.127	0.052	2.43	Gebhardt + 2003
NGC 4594			6.9 (6.7−7.0) e8		0.46	3.78	Magorrian + 1998
NGC 4594	9.8	240	1.1 (0.3−3.4) e9	1.73	0.111	15.61	Kormendy + 1996b
NGC 4594			5.4 (4.9−6.0) e8		0.46	3.78	Emsellem + 1994
NGC 4594			5.4 (1.7−17.2) e8		0.46	3.78	Kormendy 1988b
NGC 4649	16.8	385	2.0 (1.4−2.4) e9	0.71	0.052	13.71	Gebhardt + 2003
NGC 4697	11.7	177	1.7 (1.6−1.9) e8	0.41	0.052	7.9	Gebhardt + 2003
NGC 4742	15.5	90	1.4 (0.9−1.8) e7	0.099	0.068	1.45	Kaiser + 2004
NGC 5845	25.9	234	2.4 (1.0−2.8) e8	0.150	0.111	1.36	Gebhardt + 2003
NGC 7457	13.2	67	3.5 (2.1−4.6) e6	0.053	0.052	1.01	Gebhardt + 2003
IC 1459	29.2	340	2.5 (2.1−3.0) e9	0.661	0.052	12.69	Cappellari + 2002
NGC 2787	7.5	140	4.1 (3.6−4.5) e7	0.248	0.068	3.63	Sarzi + 2001
M 81	3.9	143	7.5 (6.4−9.7) e7	0.76	0.052	14.6	Devereux + 2003
NGC 3245	20.9	205	2.1 (1.6−2.6) e8	0.213	0.068	3.11	Barth + 2001
NGC 4261	31.6	315	5.2 (4.1−6.2) e8	0.146	0.058	2.54	Ferrarese + 1996
NGC 4374	18.4	296	1.6 (0.4−2.8) e9	0.89	0.068	13.1	Bower + 1998
NGC 4459	16.1	186	7.0 (5.7−8.3) e7	0.112	0.068	1.63	Sarzi + 2001
M 87	16.1	375	3.4 (2.5−4.4) e9	1.35	0.043	31.3	Macchetto + 1997
M 87			2.6 (1.8−3.3) e9		0.135	9.98	Harms + 1994
NGC 4596	16.8	152	7.8 (4.5−11.6) e7	0.179	0.068	2.61	Sarzi + 2001
NGC 5128	4.2	150	2.4 (0.7−6.0) e8	2.26	0.205	11.03	Marconi + 2001
NGC 6251	93	290	5.3 (3.7−6.8) e8	0.060	0.050	1.21	Ferrarese + 1999
NGC 7052	58.7	266	3.3 (2.0−5.6) e8	0.071	0.135	0.52	van der Marel + 1998a
NGC 1068	15	151	1.5 　e7	0.039	0.008	4.8	Greenhill + 1997a
NGC 4258	7.2	105	3.9 (3.8−4.0) e7	0.44	0.0047	93.	Herrnstein + 1999
NGC 4945	3.7		1.4 　e6				Greenhill + 1997b

Parameters – Column 2 is the distance (Tonry et al. 2001). Column 3 is the galaxy's velocity dispersion outside the sphere of influence of the BH. Column 4 is the BH mass $M_\bullet$ with error bars (M_{low}, M_{high}) from the sources in Column 8 corrected to the adopted distance. The line with all columns filled in contains the adopted BH mass. Column 5 is the radius of the sphere of influence of the BH, $r_{cusp} = GM_\bullet/\sigma_e^2$. Column 6 is the effective spatial resolution of the spectroscopy (see § 1.3.1). Column 7 is the measure of spatial resolution that shows how much leverage the observations have on the BH detection and mass measurement. Parameters not credited are from Tremaine et al. (2002) or from Kormendy & Gebhardt (2001). Notes on individual objects:

　Galaxy: For Ghez (2004) and Schödel et al. (2002), σ_* is the pericenter orbital radius of star S2. Otherwise, it is the radius for the centermost radial bin of stars used in the mass analysis.

　M 81 and NGC 4258: $M_\bullet$ is adopted from Bower et al. (2000) and Herrnstein et al. (1999).

　NGC 3115: Kormendy & Richstone (1992) provide σ_e. The resolution σ_* for Kormendy et al. (1996a) is based on the size of the nuclear star cluster, not on the *HST* spectroscopy. The corresponding BH mass is given by the virial theorem applied to this nucleus (see their § 6). Anders et al. (2001) modeled published data and their ground-based, integral field spectroscopy. Isotropic models implied $M_\bullet \simeq 10^9$ $M_\odot$, consistent with previous results. However, they find that "anisotropic models reduce this to ca. 2×10^7 $M_\odot$." This is inconsistent with our conclusion from the escape velocity argument that $M_\bullet \approx 10^9$ $M_\odot$, independent of anisotropy. Therefore, pending publication of the details of the the Anders et al. (2001) preliminary work, I omit this result.

　NGC 4374: I adopted $M_\bullet$ from Bower et al. (1998), but the low-$M_\bullet$ error bar includes the value suggested by Maciejewski & Binney (2001).

　For the maser galaxies, σ_* is the radius of the innermost maser source used in the analysis.

The stellar-dynamical BH search began with two papers on M 87 by Young et al. (1978) and by Sargent et al. (1978). Based on the non-isothermal (cuspy) surface brightness profile of its core and an observed rise in velocity dispersion toward the center, they showed that M 87 contains an $M_\bullet \simeq 4 \times 10^9 \, M_\odot$ MDO if the stellar velocity distribution is isotropic. At about the same time, it became clear that almost no giant ellipticals like M 87 are isotropic (e.g., Illingworth 1977; Binney 1978) and that anisotropic models can explain the cuspy core and the dispersion gradient without a BH (Duncan & Wheeler 1980; Binney & Mamon 1982; Richstone & Tremaine 1985; Dressler & Richstone 1990). Nevertheless, the Young and Sargent papers were seminal. They set the field in motion.

The dynamical detection of dark objects in galaxy centers began with the discovery of an $M_\bullet \approx 10^{6.5} \, M_\odot$ mass in M 32 (Tonry 1984, 1987; Dressler & Richstone 1988), a $10^{7.5} \, M_\odot$ object in M 31 (Dressler & Richstone 1988; Kormendy 1988a), and $10^9 \, M_\odot$ objects in NGC 4594 (Kormendy 1988b) and NGC 3115 (Kormendy & Richstone 1992). The observations were ground-based with resolution FWHM $\approx 1''$. The BH case in our Galaxy developed slowly (see Genzel, Hollenbach, & Townes 1994; Kormendy & Richstone 1995 for reviews), for two reasons. Dust extinction made it necessary to use infrared techniques that were just being developed in the early 1990s. And the $M_\bullet$ measurement in our Galaxy requires the study of a relatively small number of stars that are bright enough to be observed individually. As a result, graininess in the light and velocity distributions becomes a problem. On the other hand, the Galactic Center is very close, so progress in the past decade has been spectacular. Now the Galaxy is by far the best supermassive BH case (§ 1.3.2).

The BH search speeded up dramatically once *HST* provided spatial resolution a factor of 3 to 10 better than ground-based telescopes (see Kormendy & Gebhardt 2001 for a review). By now, almost all galaxies in which BHs were discovered from the ground have undergone several iterations of improved spatial resolution. Analysis machinery has improved just as dramatically. This is an opportune time to take stock of the past 15 years of progress. Are the detections of central dark objects reliable? Are the derived masses robust? And are the dark objects really BHs? The BH search is starting to look like a solved problem; assuming this, emphasis has shifted to demographic studies of BHs and their relation to galaxy evolution (see Richstone et al. 1998; Ho 1999; Kormendy & Gebhardt 2001; Richstone 2004 for reviews). Is this a reasonable attitude? Sanity checks are the purpose of this paper.

1.2   The History of BH Mass Measurements

The history of supermassive BH mass measurements is summarized in Table 1.1. In focusing on this history, I will be concerned with whether we achieve approximately the accuracies that we believe. That is, I concentrate on errors of $\gtrsim 0.2$ dex. To what extent hard work can further squeeze the measurement errors is discussed by Gebhardt (2004).

In Table 1.1, horizontal lines separate BH detections based on stellar dynamics (first group), ionized gas dynamics (middle), and maser dynamics (last group). All multiple stellar-dynamical $M_\bullet$ estimates for the same galaxy are listed. Our Galaxy, M 31, M 32, NGC 3115, NGC 3377, and NGC 4594 have all been measured by at least two competing groups. M 81 has been observed independently in stars and ionized gas; both measurements are listed and they agree. However, consistency checks of $M_\bullet$ values based on ionized gas dynamics have revealed some problems in other galaxies; these are discussed by Maciejewski & Binney (2001), Barth et al. (2001), Verdoes Kleijn et al. (2002), Barth (2004), and Sarzi (2004). I have not included all multiple measurements based on ionized gas dynamics.

1.3   How Robust Are Stellar-Dynamical BH Mass Estimates?

1.3.1   The History of the BH Search As Seen Through Work on M 32

M 32 was the first application of many improvements in spatial resolution, in kinematic analysis techniques, and in dynamical modeling machinery. It provides an excellent case study for a review of these developments. Figure 1.1 illustrates the remarkable result that BH mass estimates for M 32 have remained stable for more than 15 years while a variety of competing groups have improved the observations and analysis*.

The BH in M 32 was discovered as early as possible, when the spatial resolution was so poor that $r_{cusp}/\sigma_* < 1$. This is not surprising, given the importance of the problem. In astronomy as in other sciences, if you wait for a 5 σ result, someone else is likely to make the discovery when it is still a 2 σ result. The trick is to be careful enough to get the right answer even when the result is uncertain. Tonry (1984, 1987) got within a factor of 2.5 of the current best BH mass even though he made serious simplifying assumptions. His spectra did not resolve the intrinsic velocity dispersion gradient near the center; rotational line broadening accounted for the apparent dispersion gradient. Without an intrinsic dispersion gradient, his models were guaranteed not to be self-consistent, because there was no dynamical support in the axial direction. Despite this approximation, Tonry derived $M_\bullet \simeq (6 \text{ to } 8) \times 10^6 \, M_\odot$, close to the modern value. Poor spatial resolution allowed considerable freedom to interpret dispersion gradients as unresolved rotation; since V and σ contribute comparably to the dynamical support, trading one for the other results in no large change in $M_\bullet$.

The spatial resolution of the spectroscopy improved by a factor of 30 from the discovery observations (Tonry 1984) to the Space Telescope Imaging Spectrograph (STIS) data from *HST*. In Column 6 of Table 1.1, the Gaussian dispersion radius of the PSF is estimated as follows. First, I estimate the resolution in the directions parallel and perpendicular to the slit as $\sigma_{*\|}$, the sum in quadrature of the radius σ_{*tel} of the telescope PSF and of 1/2 pixel, and $\sigma_{*\perp}$, the sum in quadrature of the radius of the telescope PSF and half of the slit width. The *HST* PSF was modeled in van der Marel, de Zeeuw, & Rix (1997b) as the sum of three Gaussians; for all *HST* observations, I use $\sigma_{*tel} \simeq 0''036$, the best single Gaussian dispersion radius that fits this sum. Finally, the effective σ_* is the geometric mean of $\sigma_{*\|}$ and $\sigma_{*\perp}$. I do not take into account slit centering errors; for some observations, these are larger than σ_*.

---

* The referee suggests that this result is caused by two effects that accidentally cancel because spatial resolution and dynamical models have improved in parallel. He suggests (1) that $M_\bullet$ estimates increase with improving spatial resolution because we reach farther into the BH sphere of influence and (2) that $M_\bullet$ estimates decrease as dynamical models get more sophisticated because the models have more freedom to tinker the orbital structure to fit the data without a BH. I disagree. (1) Reaching farther into the BH sphere of influence should not change $M_\bullet$ if we model the stellar dynamics adequately well. Instead, we should get more "leverage" and smaller mass error bars. Of course, if we model the physics incorrectly, then more leverage may result in a systematic change in $M_\bullet$. But the change could go either way, depending on how the models err in approximating the true velocity anisotropy. In fact, Figure 1.4 shows that improving the spatial resolution does not increase the $M_\bullet$ values given by the Gebhardt et al. (2003) three-integral models, although it does, as expected, improve the error bars. For the Magorrian et al. (1998) models, improving the resolution decreases $M_\bullet$, an effect opposite to that predicted by the referee. (2) Improving modeling techniques provides more degrees of freedom on the orbital structure, but modeling programs do not have any built-in desire to decrease the BH mass. Instead, they have instructions to fit the data. Again, if the real orbital structure is sufficiently well approximated by simple models, then making the models more complicated will not change the BH mass. And if the orbital structure is not well approximated by the simple models, then better models could just as easily increase $M_\bullet$ as decrease it. However, the low-mass error bar on $M_\bullet$ will decrease, for the reason the referee suggests. The high-mass error bar will increase. As a result, the error bars become larger and more realistic. This effect is evident in Table 1.1. I conclude that the consistency of $M_\bullet$ estimates in Figures 1.1 and 1.2 tells us something important, namely that we have been modeling the stellar dynamics of power-law galaxies well enough to derive robust BH masses.

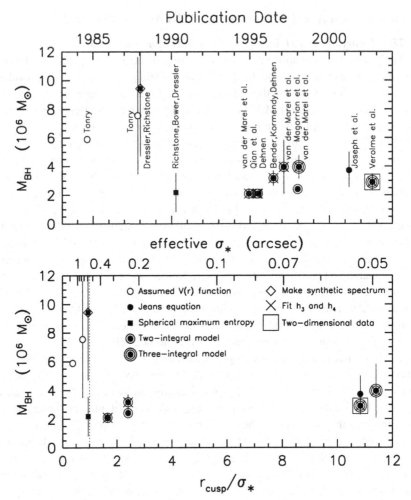

Fig. 1.1. History of the stellar-dynamical BH search as seen through work on M 32: derived BH mass as a function of (top) publication date and (bottom) spatial resolution. Resolution is measured along the top axis by the Gaussian dispersion radius σ_* of the effective PSF (see text). More relevant physically (bottom axis) is the ratio of the radius of the sphere of influence of the BH, $r_{cusp} = GM_\bullet/\sigma^2$, to σ_*. If $r_{cusp}/\sigma_* \lesssim 1$, then the measurements are dominated by the mass distribution of the stars rather than by the BH. If $r_{cusp}/\sigma_* \gg 1$, then we reach well into the part of the galaxy where velocities are dominated by the BH. Symbols shapes encode improvements in observations or kinematic measurements (right key) and in dynamical modeling techniques (left key). The data are listed in Table 1.1.

Dressler & Richstone (1988) and Richstone, Bower, & Dressler (1990) followed with better observations and analysis. They fitted spherical maximum entropy models including velocity anisotropy. By this time, it was well known that unknown velocity dispersion anisotropy was the biggest uncertainty in $M_\bullet$ measurements based on stellar dynamics. They were unable to explain the central kinematic gradients in M 32 without a BH. Rapid confirmation of Tonry's BH detection contributed to the early acceptance of this subject.

Since then, dynamical modeling machinery has improved remarkably. The next major step defined the state of the art from 1995 through 1997. This was the use of two-integral models that included flattening and velocity dispersion anisotropy. Essentially simultaneous work by van der Marel et al. (1994b), Qian et al. (1995), and Dehnen (1995) all derived $M_{\bullet} = 2.1 \times 10^6 \, M_{\odot}$ from van der Marel's data. Soon thereafter, Bender, Kormendy, & Dehnen (1996) got $3.2 \times 10^6 \, M_{\odot}$ using the same machinery on CFHT data of slightly higher resolution. The limitation of these models, as the authors realized, was the fact that two-integral models are approximations. They work best for cuspy and relatively rapidly rotating galaxies like M 32, but they are not fully general. Still, by this time, it was routine to measure not just the first two moments of the line-of-sight velocity distributions (LOSVDs)—that is, V and σ—but also the next two coefficients h_3 and h_4 in a Gauss-Hermite expansion of the LOSVDs. These measure asymmetric and symmetric departures from Gaussian line profiles. In a transparent galaxy that rotates differentially, projection guarantees that $h_3 \neq 0$. In general, h_3 is antisymmetric with V. A galaxy containing a BH is likely to have $h_4 > 0$; that is, an LOSVD that is more centrally peaked than a Gaussian. The reason is that stars close to the BH move very rapidly and give the LOSVD broader symmetric wings than they would otherwise have (van der Marel 1994). Thus, as emphasized especially by van der Marel et al. (1994a), measuring and fitting h_3 and h_4 adds important new constraints both to the stellar distribution function and to the BH detection and mass determination.

HST Faint Object Spectrograph (FOS) observations of M 32 were obtained by van der Marel et al. (1998b). These authors further "raised the bar" on BH mass measurements by fitting their data with three-integral dynamical models constructed using Schwarzschild's (1979) method. Such models now define the state of the art (see Cretton et al. 1999b; Gebhardt et al. 2000a, 2003; Richstone et al. 2004 for more detail).

Finally, the most thorough data set and modeling analysis for M 32 is provided by Verolme et al. (2002). They use the SAURON two-dimensional spectrograph to measure V, σ, h_3, and h_4 in the central $9'' \times 11''$. Also, *HST* STIS spectroscopy (Joseph et al. 2001) provides improved data near the BH. These observations fitted with three-integral models for the first time break the near-degeneracy between the stellar mass-to-light ratio, M/L, and the unknown inclination of the galaxy. Because the mass in stars is better known, the BH mass is more reliable. Again, the derived BH mass is similar to that given in previous analyses, $M_{\bullet} = (2.9 \pm 0.6) \times 10^6 \, M_{\odot}$.

So the BH mass derived for M 32 has remained almost unchanged while the observations and analysis have improved dramatically. It was exceedingly important to our confidence in the BH detection to test whether the apparent kinematic gradients near the center could be explained without a BH. Asked to do this, a dynamical modeling code attempts to fine-tune the stellar velocity dispersion anisotropy. In general, it tries to add more radial orbits near the center, because doing so implies less mass for the same σ. Nowadays, its freedom to tinker is severely restricted by the need to match the full LOSVDs. However, even simple approximations to the dynamical structure gave essentially the correct BH mass. *That is, M 32 does not use its freedom to indulge in perverse orbit structure.* The following sections show that this is also true in our Galaxy, M 31, NGC 3117, NGC 3377, and NGC 4594. Dynamical mass modeling is relatively benign in such galaxies that have power-law profiles (for more details, see Kormendy et al. 1994; Lauer et al. 1995; Gebhardt et al. 1996; Faber et al. 1997; Lauer 2004). It would not be safe to assume that this result applies equally well to galaxies with cuspy cores.

1.3.2 *The Best Case of a Supermassive Black Hole: Our Galaxy*

Figure 1.2 summarizes the history of BH mass measurements in galaxies with observations or stellar-dynamical mass analyses by different research groups. The BH case that has improved the most is the one in our Galaxy. Both the evidence for a central dark object and the arguments that this is a BH and not something less exotic like a cluster of dark stars are better in our Galaxy than anywhere else.

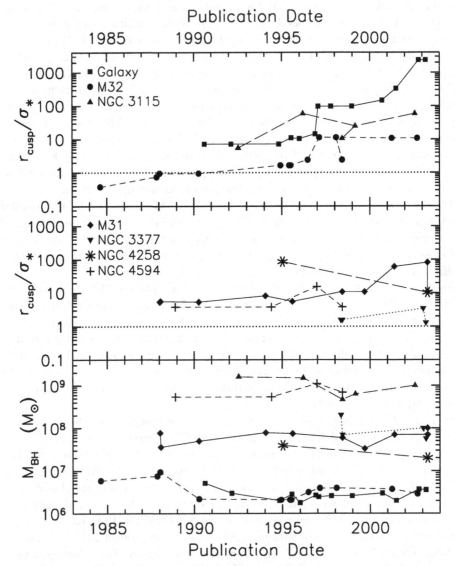

Fig. 1.2. Effective resolution of the best spectroscopy (top two panels) and resulting BH mass estimates (bottom) versus publication date. The data are listed in Table 1.1. For M 31 and M 32, steep rises in r_{cusp}/σ_* occur when *HST* was first used to observe the galaxies. For our Galaxy, two jumps in r_{cusp}/σ_* occur when the kinematic work switched from radial velocities to proper motions in the Sgr A* star cluster and when the first nearly complete stellar orbit in that cluster was observed.

A complete review of the BH search in our Galaxy is beyond the scope of this paper. Early work is discussed in Genzel & Townes (1987); Genzel et al. (1994); Kormendy & Richstone (1995), and in conference proceedings such as Backer (1987), Morris (1989), and Genzel & Harris (1994). Observations of our Galactic Center benefit from the fact that it is 100 times closer than the next nearest good BH cases, M 31 and M 32. For a distance of 8 kpc, the scale is $25\rlap{.}''8$ pc^{-1}. Early gas- and stellar-dynamical studies suggested the presence of a several-million-solar-mass dark object. In Table 1.1 and Figure 1.2, I date the convincing case for a BH to Sellgren et al. (1990) and to Kent (1992). Since then, two dramatic improvements in spatial resolution have taken place.

Research groups led by Reinhard Genzel and Andrea Ghez have pioneered the use of speckle interferometry and, more recently, adaptive optics imaging and spectroscopy to achieve spatial resolutions good enough to resolve a tiny cluster of stars (radius $\sim 1''$) that surrounds the compact radio source Sgr A* at the Galactic Center. The Sgr A* cluster is so tiny that stars move fast enough to allow us to observe proper motions. This provides a direct measure of the velocity dispersion anisotropy. It is not large. The derived central mass is about $2.5 \times 10^6 \, M_\odot$. And, even though the number density of stars is higher than we observe anywhere else, the volume is so small that the stellar mass is negligible. The advent of proper motion measurements accounts for the jump in $r_{\rm cusp}/\sigma_*$ at the start of 1997.

A second jump in $r_{\rm cusp}/\sigma_*$ has just occurred as a result of an even more remarkable observational coup. As reviewed in this volume by Ghez (2004), Schödel et al. (2002), Ghez et al. (2003), and Ghez (1994) have independently measured several individual stellar orbits through pericenter passage. In the case of star S2, more than half of an orbit has been observed (period = 15.78 ± 0.82 years). The orbit is closed, so the controlling mass resides inside $r_{\rm peri} \simeq 0\rlap{.}''0159 \simeq 0.00062$ pc $\simeq 127$ AU $\simeq 1790$ Schwarzschild radii. This accounts for the current jump in spatial resolution. As measurement accuracies improve, the observation of individual closed orbits will rapidly obsolete the complicated analysis of stellar distribution functions that describe ensembles of stars at larger radii. Rather, the analysis will acquire the much greater rigor inherent in the two-body problem. Arguably the orbit of S2 already contributes as much to our confidence in the BH detection as all stars at larger radii combined. The best-fitting BH mass, $M_\bullet = (3.7 \pm 0.4) \times 10^6 \, M_\odot$, is in good agreement with, but slightly larger than, the value derived from the stellar-dynamical modeling. This leads to an important point: The above comparison in our Galaxy and a similar one in NGC 4258 (see the next section) are currently the only reliable external checks on our stellar-dynamical modeling machinery. The measurement accuracies are not good enough yet to show whether the models achieve the accuracies that we expect for the best data ($\pm 30\%$: Gebhardt 2004). But neither test points to modeling errors that range over a factor of ~ 6 as feared by Valluri, Merritt, & Emsellem (2004).

Finally, these new observations have an implication that is actually more fundamental than the mass measurement. They restrict the dark mass to live inside such a small radius that even neutrino balls (Tsiklauri & Viollier 1998, 1999; Munyaneza, Tsiklauri, & Viollier 1998, 1999; Munyaneza & Viollier 2002) with astrophysically allowable neutrino masses are excluded. The exclusion principle forces them to be too fluffy to be consistent with the radius constraints. Dark clusters of brown dwarf stars or stellar remnants were already excluded (Maoz 1995, 1998)—brown dwarfs would collide, merge, and become visible stars, and stellar remnants would evaporate via relaxation processes. The maximum lifetime of dark cluster alternatives to a BH is now a few times 10^5 yr (Schödel et al. 2002).

1.3.3   The Best Test of Stellar-Dynamical $M_\bullet$ Estimates: NGC 4258

The galaxy that stands out as having the most reliable BH mass measurement is NGC 4258. Very Long Baseline Array measurements of its nuclear water maser disk reach to within $0\rlap{.}''0047 = 0.16$ pc of the BH (Miyoshi et al. 1995). The rotation curve, $V(r) = 2180\,(r/0\rlap{.}''001)^{-1/2}$ km s^{-1}, is Keplerian to high precision. Proper motion and acceleration observations of the masers in front of the Seyfert nucleus are consistent with the radial velocity measurements along the orbital tangent points (Herrnstein et al. 1999). All indications are that the rotation is circular. Therefore $M_\bullet = (3.9 \pm 0.1) \times 10^7\ M_\odot$ is generally regarded as bomb-proof.

This provides a unique opportunity to test the three-integral dynamical modeling machinery used by the Nuker team (Gebhardt et al. 2000a, b, 2003; Richstone et al. 2004). NGC 4258 contains a normal bulge much like the one in M 31 (Kormendy et al. 2004a). Siopis et al. (2004) have obtained *HST* STIS spectra and WFPC2 images of NGC 4258. The STIS spectroscopy has spatial resolution $r_{\rm cusp}/\sigma_* \simeq 8.4$ well within the range of the BH discoveries in Table 1.1. The kinematic gradients are steep, consistent with the presence of a BH. Three-integral models are being calculated as I write this; the preliminary result is that $M_\bullet = (2 \pm 1) \times 10^7\ M_\odot$. The agreement with the maser $M_\bullet$ is fair. The problem is the brightness profile, which involves more complications than in most BH galaxies. A color gradient near the center may be a sign of dust obscuration, and correction for the bright AGN (Chary et al. 2000) is nontrivial. Both problems get magnified by deprojection.

1.3.4   A Case History of Improving Spatial Resolution: NGC 3115

One sanity check on BH detections is that apparent kinematic gradients should get steeper as the spectroscopic resolution improves. We have seen this test work in M 32 and in our Galaxy. This section is a brief discussion of NGC 3115. At $r_{\rm cusp}/\sigma_* = 59$, NGC 3115 is surpassed in spectroscopic resolution only by our Galaxy, NGC 4258, and M 31.

Exploiting the good seeing on Mauna Kea, Kormendy & Richstone (1992) found a central dark object of $10^9\ M_\odot$ in NGC 3115 using the Canada-France-Hawaii Telescope (CFHT). The resolution was not marginal; $r_{\rm cusp}/\sigma_* \simeq 5.5$. This is higher than the median for *HST* BH discoveries in Figure 1.3 (§ 1.3.7). Since then, there have been two iterations in improved spectroscopic resolution (Kormendy et al. 1996a). The apparent central velocity dispersion increased correspondingly: it was $\sigma = 295 \pm 9$ km s^{-1} at $r_{\rm cusp}/\sigma_* \simeq 5.5$, $\sigma = 343 \pm 19$ km s^{-1} at $r_{\rm cusp}/\sigma_* \simeq 10.6$ (CFHT plus Subarcsecond Imaging Spectrograph), and $\sigma = 443 \pm 18$ km s^{-1} at $r_{\rm cusp}/\sigma_* \simeq 59$ (*HST* FOS). These are projected velocity dispersions: they include the contribution of foreground and background stars that are far from the BH and so have relatively small velocity dispersions. However, NGC 3115 has a tiny nuclear star cluster that is very distinct from the rest of the bulge. It is just the sort of high-density concentration of stars that we always expected to find around a BH. From a practical point of view, it is a great convenience, because it is easy to subtract the foreground and background light as estimated from the spectra immediately adjacent to the nucleus. This procedure is analogous to sky subtraction. It provides the velocity dispersion of the nuclear cluster by itself and is, in effect, another way to increase the spatial resolution. The result is that the nuclear cluster has a velocity dispersion of $\sigma = 600 \pm 37$ km s^{-1}. The effective spatial resolution of this measurement is not determined by the spectrograph but rather by the half-radius $r_h = 0\rlap{.}''052 \pm 0\rlap{.}''010$ of the nuclear cluster. This is smaller than the entrance aperture of the FOS. It implies that $r_{\rm cusp}/\sigma_* \simeq 59$, as quoted in Table 1.1.

The nucleus allows us to estimate the BH mass independent of any velocity anisotropy. If the nucleus consisted only of old stars with the mass-to-light ratio measured for the bulge, then its mass would be $\sim 4 \times 10^7 \, M_\odot$ and its escape velocity would be ~ 352 km s^{-1}. This is much smaller than the observed velocities of the stars. The nucleus would fly apart in a few crossing times T_{cross}. But $T_{\mathrm{cross}} \simeq 16{,}000$ yr is very short. Therefore, a dark object of $10^9 \, M_\odot$ must be present to confine the stars within the nucleus.

1.3.5 A Comparison of Ground-based and HST Studies of NGC 3377 and NGC 4594

Besides M 31 (§ 1.4), *HST* has confirmed ground-based BH detections in two more galaxies (Fig. 1.2).

NGC 4594, the Sombrero galaxy, was observed with the CFHT by Kormendy (1988b), yielding a BH mass of $M_\bullet \approx 10^{8.7} \, M_\odot$. Resolution was average for BH detections; $r_{\mathrm{cusp}}/\sigma_* = 3.8$. The galaxy was reobserved with *HST* by Kormendy et al. (1996b) using the FOS at $r_{\mathrm{cusp}}/\sigma_* \approx 15.6$. They confirmed the BH detection and quoted a slightly higher mass of $10^9 \, M_\odot$. This test is weaker than those quoted above because the same research group was involved and because three-integral models were not constructed. However, independent dynamical models by Emsellem et al. (1994) agree very well with the results in Kormendy (1988b).

NGC 3377 also has a CFHT BH detection; $r_{\mathrm{cusp}}/\sigma_* = 1.57$ (Kormendy et al. 1998). The BH mass was $M_\bullet = (2 \pm 1) \times 10^8 \, M_\odot$. Gebhardt et al. (2003) reobserved the galaxy with *HST* at $r_{\mathrm{cusp}}/\sigma_* = 3.4$. The improvement in resolution is smaller than normal because the CFHT seeing was very good and because the *HST* FOS aperture size was 0.″2. Nevertheless, the improvement is substantial. Also, the analysis machinery was updated; Kormendy et al. (1998) fitted analytic approximations to V and σ and, independently, spherical maximum entropy models with post-hoc flattening corrections. Gebhardt et al. (2003) fitted three-integral models. They obtained $M_\bullet = 1.0^{+0.9}_{-0.1} \times 10^8 \, M_\odot$, confirming the earlier result. Also, Cretton et al. (2004) report two-dimensional spectroscopy in the inner $6'' \times 3''$ of NGC 3377. Three-integral models give $M_\bullet = 5.7^{+5.6}_{-2.3} \times 10^7 \, M_\odot$, corrected to our adopted distance. Again, the published results are consistent.

1.3.6 Robustness of Stellar-Dynamical $M_\bullet$ Values. I. Conclusion from §§ 1.3.1 – 1.3.5

All of the ground-based, stellar-dynamical BH detections discussed in Kormendy & Richstone (1995) have now been confirmed at higher spatial resolution and with more sophisticated modeling machinery. All of the original mass estimates agree with the best current values to factors of $2 - 3$ or better.

Given the above tests, given the agreement between the BH parameter correlations implied by the dynamics of stars, of ionized gas, and of maser gas, and especially given the tightness of the scatter in the $M_\bullet - \sigma$ correlation, it seems unlikely that $M_\bullet$ values are still uncertain to factors of several, as suggested by Valluri et al. (2004). Nevertheless, so much is at stake that we must continue to test the stellar-dynamical modeling codes. For example, triaxiality is not yet included. It is unlikely that triaxiality provides enough new degrees of freedom to greatly change the results; very triaxial configurations would have been seen with *HST*. But checking the consequences of triaxiality is under way by the SAURON team.

All papers contain simplifying assumptions. Science is the art of getting the right answer using approximate analysis of imperfect data. We should not get complacent, but we appear to be doing reasonably well.

1.3.7   Application: HST BH Discoveries

Having shown from repeat observations at better spatial resolution how well we do when $r_{cusp}/\sigma_* \simeq 1-10$, we now apply these results to *HST* BH discoveries that do not have repeat measurements.

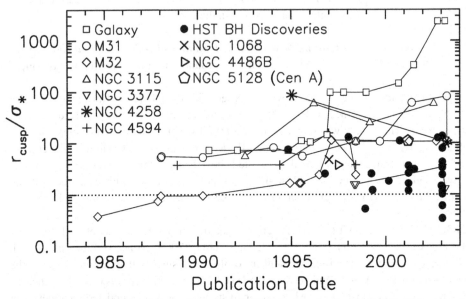

Fig. 1.3. Spectroscopic spatial resolution for all BH discoveries in Table 1.1. Galaxies with repeat measurements are from Figures 1.1 and 1.2. Note that *HST* and ground-based BH discoveries have similar distributions of r_{cusp}/σ_*. However, *HST* has $5-10$ times better spatial resolution in arcsec than ground-based observations (absent adaptive optics). This means that *HST* is being used to discover lower-mass BHs in more distant galaxies.

Figure 1.3 shows the distribution of r_{cusp}/σ_* values for all BH detections made with *HST*. It contains a number of surprises. Contrary to popular belief, *HST BH discoveries are not being made with much better spatial resolution than ground-based BH discoveries.* Only a few of the best *HST* cases have $r_{cusp}/\sigma_* \simeq 10$ comparable to the ground-based BH detections in our Galaxy, in M 31, and in NGC 3115. On average, *HST* BH discoveries are being made at lower r_{cusp}/σ_* values than those made from the ground. Several have $r_{cusp}/\sigma_* < 1$, similar to the early measurements of M 32. I am not suggesting that *HST* and ground-based spatial resolutions are similar *in arcsec*. *HST* is better by a factor of 10 (if a 0″.1 slit is used) or at least 5 (for measurements with the 0″.2 aperture or slit). What is really going on is this: The ground-based observations "used up" the best galaxies. For example, our Galaxy, M 31, and M 32 are unusually close, and NGC 3115 has an unusually large BH mass fraction. So *HST* is necessarily being used on more distant galaxies or ones that have smaller BH mass fractions. This puts the exceedingly important contributions of *HST* into perspective:

(1) *HST* did not find the strongest BH cases. NGC 4258 and our Galaxy were observed from the ground. *HST* observations of NGC 4258 serve to test the stellar-dynamical models.

(2) *HST* has confirmed and greatly strengthened the BH cases for BH discoveries made from the ground. The spectroscopic resolutions r_{cusp}/σ_* for M 31 and for NGC 3115 are now essentially as good as that for the famous maser case, NGC 4258.

(3) *HST* did not revolutionize BH detections by finding them at higher resolution.

(4) *HST* has revolutionized the BH search by allowing us to find smaller BHs and ones in more distant galaxies. This has two important implications.

(5a) There has always been a danger that ground-based observations would be biased in favor of BHs that are unusually massive. Any such bias is rapidly being diluted away. In fact, it was not large. Kormendy & Richstone (1995) found from ground-based observations that the mean ratio of BH mass to bulge mass was $\langle M_\bullet/M_{\text{bulge}}\rangle = 0.0022^{+0.0016}_{-0.0009}$ (they averaged $\log M_\bullet/M_{\text{bulge}}$ for eight BH detections, six made with stellar dynamics and one each with masers and ionized gas disks). Now, the data in Table 1.1 give $\langle M_\bullet/M_{\text{bulge}}\rangle = 0.0013$ (Kormendy & Gebhardt 2001; Merritt & Ferrarese 2001).

(5b) *HST* has made it possible to detect canonical BHs (ones within the scatter of the $M_\bullet$ correlations) out to the distance of the Virgo cluster. The largest BHs can be detected several times farther away. This has revolutionized the subject of BH demographics. We now have enough detections to address the question of how BH growth is related to galaxy formation.

(6) As Figure 1.3 emphasizes, this subject has speeded up enormously because of *HST*.

1.3.8 Caveat: Cuspy Core Galaxies

The caveat to this rosy story is that the above tests were carried out for galaxies with "power-law profiles" (Lauer et al. 1995). The physical distinction between such galaxies and ones with cuspy cores is discussed by Kormendy et al. (1994), Lauer et al. (1995), Gebhardt et al. (1996), and especially Faber et al. (1997). The observations imply that cuspy core galaxies have more anisotropic velocity distributions than do power-law galaxies (Kormendy & Bender 1996). They are fundamentally more difficult for the BH search (Kormendy 1993). The shallower volume brightness profile $\rho(r)$ gives, in projection, less luminosity weight to the stars in the sphere of influence of the BH. The $d\ln\rho/d\ln r$ term in the mass derivation is smaller and more easily cancelled by the effects of velocity anisotropy, which is larger than in power-law galaxies. Stellar dynamical BH detections in cuspy core galaxies are few and not well tested. Comparisons between stellar-dynamical and gas-dynamical $M_\bullet$ measurements do not show universally good agreement. BH masses in cuspy core galaxies are more uncertain than those in power-law galaxies, and the above conclusions cannot confidently be applied to them. We need better tests of BH detections in core galaxies.

1.3.9 Robustness of Stellar-Dynamical $M_\bullet$ Values. II. What Resolution Do We Need?

BH mass estimates made with the spatial resolution shown in Figures 1.2 – 1.3 appear to be reliable. So how good does the spatial resolution have to be? We can now answer this question for two $M_\bullet$ analysis machines, the two-integral models of Magorrian et al. (1998) and the three-integral models of Gebhardt et al. (2003).

Gebhardt et al. (2003) investigate, for their objects with *HST* spectra and BH detections, how the BH mass would be affected if only the supporting ground-based observations were used in the modeling. The *HST* data are higher in resolution than the ground-based data by a factor of 11.2 ± 1.2. If the ground-based observations are comparable in quality to *HST* observations made with the same effective spatial resolution r_{cusp}/σ_*, then Gebhardt's exercise distills a clean test of the effects of spatial resolution. Modeling uncertainties are minimized because the same analysis machinery is used on both sets of data. Gebhardt et al. (2003) conclude that, when the *HST* data are omitted, the error bars on $M_\bullet$ are larger but the systematic errors in $M_\bullet$ are small. Here we ask how these results depend on r_{cusp}/σ_*.

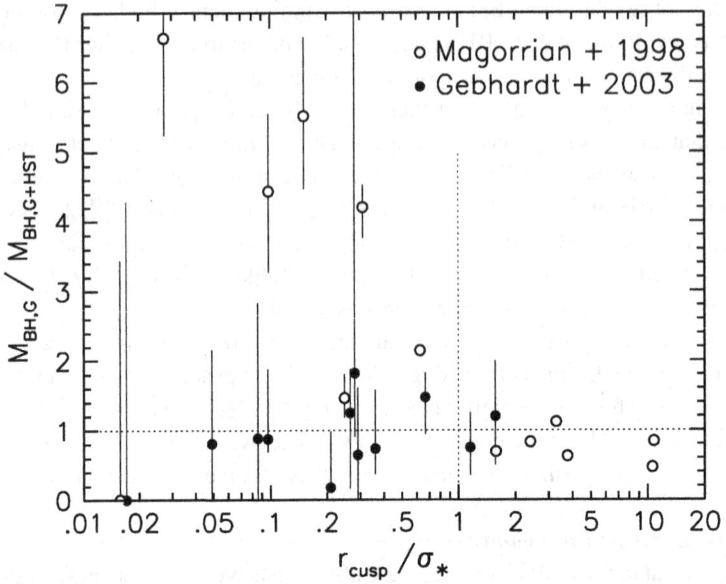

Fig. 1.4. Reliability and precision of BH masses as a function of the spatial resolution of the observations. The ordinate is the ratio of the BH mass as obtained from ground-based data to that obtained with *HST* kinematic data included. The error bars are from the ground-based data only, because I want to illustrate how estimated errors grow as resolution deteriorates.

Figure 1.4 shows no systematic errors in the $M_\bullet$ values given by three-integral models, even at low resolution. BH masses are accurate to a factor of 1.5 or better provided that $r_{cusp}/\sigma_* \gtrsim 0.3$. All BH detections in Table 1.1 satisfy this criterion. At lower resolution, $M_\bullet$ can be wrong by a factor of 2 or more, but the error bars remain realistic.

The $M_\bullet$ measurements in Magorrian et al. (1998) have two main limitations; they are based on two-integral models, and they are derived from low-resolution, ground-based spectroscopy. They can be tested with *HST* spectroscopy and (mostly) three-integral models for 13 galaxies (open circles). When $r_{cusp}/\sigma_* \gtrsim 1$, the two-integral models work well; they underestimate the best current BH masses by a factor of 0.76 ± 0.09. But when $r_{cusp}/\sigma_* < 1$, they overestimate the BH mass by larger factors at lower resolution. The reason for the systematic error is unclear. At $r_{cusp}/\sigma_* \lesssim 0.1$, $M_\bullet$ is overestimated by a factor of ~ 5. The majority of the Magorrian galaxies that have not been reobserved with *HST* are more distant than the ones represented in Figure 1.4. Therefore poor resolution plus the assumption of two-integral models appear to be the reasons why the ratio of BH mass to bulge mass found by Magorrian et al. (1998) is larger than the current value of 0.0013 by a factor of 4.

All ground-based BH discoveries in Kormendy & Richstone (1995) had $r_{cusp}/\sigma_* > 1$ except in the earliest papers on M 32. These papers also overestimated $M_\bullet$ (Fig. 1.1). *HST* BH discoveries made with $r_{cusp}/\sigma_* \simeq 0.3$ to 1 are more secure than the M 32 results derived at the same resolution because we now fit full LOSVDs and because three-integral models are more reliable than simpler models.

Given these tests and the ones in Gebhardt (2004), it seems entirely appropriate that the emphasis in current work has shifted from the reliability of BH discovery to the use of BH demographics to study the relationship between BH growth and galaxy formation.

1.4 Are They Really Black Holes?

Astrophysical arguments that the dark objects detected in galaxy centers are not clusters of underluminous stars (Maoz 1995, 1998) are well known. Dark clusters made of brown dwarf stars become luminous when the stars collide, merge, and become massive enough for nuclear energy generation. Clusters of stellar remnants (white dwarf stars, neutron stars, or stellar-mass black holes) evaporate as a result of two-body relaxation. The time scales for these processes are compellingly short (i.e., $\lesssim 10^9$ yr) only for the Milky Way and for NGC 4258. The next best case has been M 32 (van der Marel et al. 1998b), although Maoz argued that it is not conclusive. News in this subject involves our Galaxy and M 31.

As discussed in § 1.3.2, the observation of an almost-complete, closed orbit for star S2 in the Sgr A* cluster restricts the central dark mass to live inside the orbit's pericenter radius, $r_{\mathrm{peri}} = 1790$ Schwarzschild radii. Demise time scales for dark star clusters are now $< 10^6$ yr. Even neutrino balls are excluded (Schödel et al. 2002; Ghez et al. 2003; Ghez 2004).

Second, M 31 becomes the third galaxy in which astrophysical arguments make a strong case against dark star clusters. Bender et al. (2004) have used the *HST* STIS to measure the velocity dispersion of the tiny cluster of blue stars (King, Stanford, & Crane 1995; Lauer et al. 1998; Kormendy & Bender 1999) embedded in the fainter of the two nuclei ("P2") of the galaxy (Lauer et al. 1993). Kormendy & Bender (1999) already suggested that the central dark object in M 31 is embedded in this blue cluster. The STIS spectra now show that the velocity dispersion of the blue cluster is $\sigma = 940 \pm 100$ km s^{-1} (Fig. 1.5). This is remarkably high; the red stars along the same line of sight have a velocity dispersion of only 300 to 400 km s^{-1}. We can now be sure that the dark object is in the blue cluster.

From WFPC2 photometry in Lauer et al. (1998), the half-light radius of the blue cluster is $r_h \simeq 0\!\!''\!06$. Since all of the light of the A-type stars comes from this cluster, r_h and not the *HST* PSF or slit defines the effective spatial resolution of the spectroscopy (Table 1.1). To confine the stars within the blue cluster, the dark object must have a radius $r_\bullet \lesssim r_h$. Also, $M_\bullet$ is larger than we thought: the virial theorem gives $M_\bullet \approx 2 \times 10^8\ M_\odot$. This approximation is an overestimate if the light in the blue cluster is very centrally concentrated. However, it is likely that $M_\bullet$ is at least $7 \times 10^7\ M_\odot$. This is the value adopted in Table 1.1.

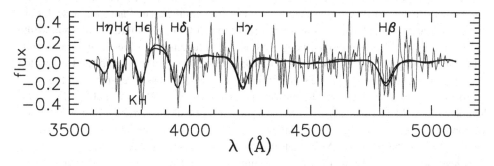

Fig. 1.5. Spectrum (thin line) of the central $0\!\!''\!2$ of the blue cluster. The adjacent spectrum of the stars in the bulge and nucleus has been subtracted. The spectrum is continuum-subtracted and normalized for the Fourier correlation quotient program (Bender 1990). Flux is in arbitrary linear units. The blue cluster has an A-type spectrum. Heavy lines show the spectra of an A0 V star and an A0 III star broadened to the line-of-sight velocity distribution that best fits the cluster spectrum. This figure is from Bender et al. (2004).

It is important to note that $M_\bullet$ is more uncertain in M 31 than in other galaxies even though this is the second-nearest BH case. The reason is the double nucleus. Three-integral models are not available and would omit important physics. Four techniques have been used. (1) Axisymmetric models of P1 and P2 give $M_\bullet \simeq (4 \text{ to } 8) \times 10^7\ M_\odot$. (2) Models of the double nucleus as an eccentric disk of stars give $M_\bullet \sim 7 \times 10^7\ M_\odot$ (Tremaine 1995). (3) The requirement that the center of mass of the BH and the asymmetric distribution of stars be at the center of the bulge gives $M_\bullet \sim 3 \times 10^7\ M_\odot$ (Kormendy & Bender 1999). (4) The virial theorem applied to the blue cluster gives $M_\bullet \approx 2 \times 10^8\ M_\odot$. These masses range over a factor of 7. However, all four techniques are more uncertain than three-integral models applied to nearly axisymmetric galaxies. An improved eccentric disk model has just become available; it gives $M_\bullet \simeq 1 \times 10^8\ M_\odot$ (Peiris & Tremaine 2004). The most accurate BH mass is likely to come from such detailed analysis of the asymmetric nucleus. Here, I adopt a BH mass in the middle of the above range; it should be accurate to a factor of ~ 2.

We can now ask: Can we stuff $10^8\ M_\odot$ of brown dwarfs or stellar remnants into the central $0\overset{''}{.}06$ without getting into trouble? The answer is "no" (Kormendy et al. 2004b). Following Maoz (1995, 1998), brown dwarfs are strongly excluded. The collision time for even the most massive brown dwarf (which becomes a luminous star after only one merger) is less than 10^9 yr. Less massive brown dwarfs collide more quickly. Dark clusters made of stellar-mass BHs or neutron stars evaporate in several billion years and are at least weakly excluded. The most viable dark cluster would be made of $0.6\ M_\odot$ white dwarfs. Such a cluster would have an evaporation time of 10^{10} yr and is not excluded by the arguments made so far.

However, we can add a new argument. An MDO made of stellar remnants is viable only if its progenitor stars can safely live their lives and deliver their remnants at suitable radii. Progenitors get into more trouble than their remnants. They are so close together that they collide too quickly, as follows. The progenitor cluster must be as small as the dark cluster, because dynamical friction is too slow to deliver remnants from much larger radii. We get into less trouble with collisions if fewer progenitors are resident at one time. That is, if the dark cluster was made in time T, and if the progenitor star lifetime is T_*, the safest strategy is to have T/T_* successive generations, each with an equal number of progenitors. For $T = 10^{10}$ yr, we then calculate the time scale on which any one progenitor star collides with another as a function of the stellar mass and generation number. The longest time scales are 10^8 yr for black hole and neutron star progenitors and shorter for the more troublesome white dwarf progenitors. Colliding stars merge and become progenitors of higher-mass remnants. Also, stellar mergers decrease the number of stars and increase the mass range and so shorten the dynamical evolution time. The result is a dark cluster with a short evaporation time.

So astrophysically plausible alternatives to a supermassive BH are likely to fail. Our leverage on the M 31 BH, $r_{\mathrm{cusp}}/\sigma_* \simeq 81$, is almost as good as $r_{\mathrm{cusp}}/\sigma_* \simeq 93$ for NGC 4258. Astrophysical arguments against BH alternatives are stronger in the latter case because we know in NGC 4258 but not in M 31 that the rotation curve is accurately Keplerian at $r \gtrsim \sigma_*$. This leads to a factor-of-ten stronger constraint on the half-mass radius of the dark object in NGC 4258 (Maoz 1995, 1998). Nevertheless, M 31 becomes the third galaxy in which astrophysical arguments favor the conclusion that a dynamically detected dark object is a BH. This increases our confidence that all of them are BHs.

Finally, a great variety of AGN observations, including relativistic jets and X-ray Fe Kα line widths as large as 1/3 of the speed of light, argue forcefully that the engines for nuclear activity in galaxies are BHs.

1.5 Conclusion

The sanity checks that were the purpose of this paper have succeeded. Progress on a broad front is on the agenda for this meeting. It is gratifying to see the developing connection between the dynamical BH search and the AGN work that motivated it. The limited contact between these subjects was a complaint in Kormendy & Richstone (1995). Now, reverberation mapping (Blandford & McKee 1982; Netzer & Peterson 1997) has become a reliable tool to measure BH masses (Gebhardt et al. 2000c; Barth 2004). Ionization models of AGNs (Netzer 1990; Rokaki, Boisson, & Collin-Souffrin 1992) are consistent with other techniques (McLure & Dunlop 2001; Wandel 2002; Shields et al. 2003). The growing connection between BH dynamical searches, AGN physics, and the study of galaxy formation is a sign of the developing maturity of this subject (e.g., Kormendy 2000). The emphasis on BH discovery has given way to the richer field of BH astrophysics.

Kormendy & Richstone (1995) was entitled "Inward Bound: The Search for Supermassive Black Holes in Galaxy Nuclei" because the BH search is an iterative process. "We make incremental improvements in spatial resolution, each expensive in ingenuity and money. [The above] paper reviews the first order of magnitude of the inward journey in radius." At that time, the best BH candidate, NGC 4258, was observed with a resolution $\sigma_* \simeq 44,000$ Schwarzschild radii (Miyoshi et al. 1995). Now $\sigma_* \simeq 23,000$ Schwarzschild radii for M 31 and for NGC 3115. The best BH case, our own Galaxy, has $\sigma_* \simeq 1790$ Schwarzschild radii. The gap between the smallest radii reached by dynamical studies and the radii studied by the well-developed industry on accretion disk physics is shrinking. This, too, is a sign of the growing maturity of the subject.

But it is too early to declare the problem solved. Loren Eiseley (1975) wrote:

"The universe [may be] too frighteningly queer to be understood by minds like ours. It's not a popular view. One is supposed to flourish Occam's razor and reduce hypotheses about a complex world to human proportions. Certainly I try. Mostly I come out feeling that whatever else the universe might be, its so-called simplicity is a trick. I know that we have learned a lot, but the scope is too vast for us. Every now and then if we look behind us, everything has changed. It isn't precisely that nature tricks us. We trick ourselves with our own ingenuity."

However reassured we may be by the tests reviewed here, it is worth remembering that even star S2 in the Galaxy's Sgr A* cluster, which approaches to within 1790 Schwarzschild radii of the central engine, lives well outside the region of strong gravity. Surprises are not out of the question. Further tests of the BH paradigm are worthwhile to make sure that we do not suddenly find ourselves in an unfamiliar landscape.

Acknowledgements. It is a pleasure to thank my collaborators, R. Bender, G. Bower, the Nuker team (D. Richstone, PI), and the STIS GTO team (R. Green, PI) for many helpful discussions and for permission to discuss our results before publication. I am grateful to Luis Ho for his invitation to present this review and for his patience and meticulous editing. Careful reading by Scott Tremaine and the referee resulted in important improvements to this paper. My *HST* work on BHs is supported by grants GO-06587.07, GO-07388.07, GO-08591.09-A, GO-08687.01-A, and GO-09107.07-A.

References

Anders, S. W., Thatte, N., & Genzel, R. 2001, in Black Holes in Binaries and Galactic Nuclei, ed. L. Kaper, E. P. J. van den Heuvel, & P. A. Woudt (New York: Springer), 88

Bacon, R., Emsellem, E., Combes, F., Copin, Y., Monnet, G., & Martin, P. 2001, A&A, 371, 409

Bacon, R., Emsellem, E., Monnet, G., & Nieto, J.-L. 1994, A&A, 281, 691

Backer, D. C., ed. 1987, The Galactic Center (New York: Amer. Inst. Phys.)

Barth, A. J. 2004, in Carnegie Observatories Astrophysics Series, Vol. 1: Coevolution of Black Holes and Galaxies, ed. L. C. Ho (Cambridge: Cambridge Univ. Press), in press

Barth, A. J., Sarzi, M., Rix, H.-W., Ho, L. C., Filippenko, A. V., & Sargent, W. L. W. 2001, ApJ, 555, 685

Bender, R. 1990, A&A, 229, 441

Bender, R., et al. 2004, ApJ, submitted

Bender, R., Kormendy, J., & Dehnen, W. 1996, ApJ, 464, L123

Binney, J. 1978, MNRAS, 183, 501

Binney, J., & Mamon, G. A. 1982, MNRAS, 200, 361

Blandford, R. D., & McKee, C. F. 1982, ApJ, 255, 419

Bower, G. A., et al. 1998, ApJ, 492, L111

——. 2001, ApJ, 550, 75

Bower, G. A., Wilson, A. S., Heckman, T. M., Magorrian, J., Gebhardt, K., Richstone, D. O., Peterson, B. M., & Green, R. F. 2000, BAAS, 32, 1566

Cappellari, M., Verolme, E. K., van der Marel, R. P., Verdoes Kleijn, G. A., Illingworth, G. D., Franx, M., Carollo, C. M., & de Zeeuw, P. T. 2002, ApJ, 578, 787

Chakrabarty, D., & Saha, P. 2001, AJ, 122, 232

Chary, R., et al. 2000, ApJ, 531, 756

Cretton, N., Copin, Y., Emsellem, E., & de Zeeuw, T. 2004, in Carnegie Observatories Astrophysics Series, Vol. 1: Coevolution of Black Holes and Galaxies, ed. L. C. Ho (Pasadena: Carnegie Observatories, http://www.ociw.edu/ociw/symposia/series/symposium1/proceedings.html)

Cretton, N., de Zeeuw, P. T., van der Marel, R. P., & Rix, H.-W. 1999b, ApJS, 124, 383

Cretton, N., & van den Bosch, F. C. 1999a, ApJ, 514, 704

Dehnen, W. 1995, MNRAS, 274, 919

Devereux, N., Ford, H., Tsvetanov, Z., & Jacoby, G. 2003, AJ, 125, 1226

Dressler, A., & Richstone, D. O. 1988, ApJ, 324, 701

——. 1990, ApJ, 348, 120

Duncan, M. J., & Wheeler, J. C. 1980, ApJ, 237, L27

Eckart, A., & Genzel, R. 1997, MNRAS, 284, 576

Eiseley, L. 1975, All the Strange Hours (New York: Scribner)

Emsellem, E., Dejonghe, H., & Bacon, R. 1999, MNRAS, 303, 495

Emsellem, E., Monnet, G., Bacon, R., & Nieto, J.-L. 1994, A&A, 285, 739

Evans, N. W., & de Zeeuw, P. T. 1994, MNRAS, 271, 202

Faber, S. M., et al. 1997, AJ, 114, 1771

Ferrarese, L., & Ford, H. C. 1999, ApJ, 515, 583

Ferrarese, L., Ford, H. C., & Jaffe, W. 1996, ApJ, 470, 444

Gebhardt, K. 2004, in Carnegie Observatories Astrophysics Series, Vol. 1: Coevolution of Black Holes and Galaxies, ed. L. C. Ho (Cambridge: Cambridge Univ. Press), in press

Gebhardt, K. et al. 2000a, AJ, 119, 1157

——. 2000b, ApJ, 539, L13

——. 2000c, ApJ, 543, L5

——. 2003, ApJ, 583, 92

Gebhardt, K., Richstone, D., Ajhar, E. A., Kormendy, J., Dressler, A., Faber, S. M., Grillmair, C., & Tremaine, S. 1996, AJ, 112, 105

Genzel, R., Eckart, A., Ott, T., & Eisenhauer, F. 1997, MNRAS, 291, 219

Genzel, R., & Harris, A. I., ed. 1994, The Nuclei of Normal Galaxies: Lessons From The Galactic Center (Dordrecht: Kluwer)

Genzel, R., Hollenbach, D., & Townes, C. H. 1994, Rep. Prog. Phys., 57, 417

Genzel, R., Pichon, C., Eckart, A., Gerhard, O. E., & Ott, T. 2000, MNRAS, 317, 348

Genzel, R., Thatte, N., Krabbe, A., Kroker, H., & Tacconi-Garman, L. E. 1996, ApJ, 472, 153

Genzel, R., & Townes, C. H. 1987, ARA&A, 25, 377

Ghez, A. M. 2004, in Carnegie Observatories Astrophysics Series, Vol. 1: Coevolution of Black Holes and Galaxies, ed. L. C. Ho (Cambridge: Cambridge Univ. Press), in press

Ghez, A. M., et al. 2003, ApJ, 586, L127

Ghez, A. M., Klein, B. L., Morris, M., & Becklin, E. E. 1998, ApJ, 509, 678

Greenhill, L. J., & Gwinn, C. R. 1997a, Ap&SS, 248, 261

Greenhill, L. J, Moran, J. M., & Herrnstein, J. R. 1997b, ApJ, 481, L23

Haller, J. W., Rieke, M. J., Rieke, G. H., Tamblyn, P., Close, L., & Melia, F. 1996, ApJ, 456, 194

Harms, R. J., et al. 1994, ApJ, 435, L35

Herrnstein, J. R., Moran, J. M., Greenhill, L. J., Diamond, P. J., Inoue, M., Nakai, N., Miyoshi, M., Henkel, C., & Riess, A. 1999, Nature, 400, 539

Ho, L. C. 1999, in Observational Evidence for Black Holes in the Universe, ed. S. K. Chakrabarti (Dordrecht: Kluwer), 157

Illingworth, G. 1977, ApJ, 218, L43

Joseph, C. L., et al. 2001, ApJ, 550, 668

Kaiser, M. E., et al. 2004, in preparation

Kent, S. M. 1992, ApJ, 387, 181

King, I. R., Stanford, S. A., & Crane, P. 1995, AJ, 109, 164

Kormendy, J. 1988a, ApJ, 325, 128

——. 1988b, ApJ, 335, 40

——. 1993, in The Nearest Active Galaxies, eds. J. Beckman, L. Colina, & H. Netzer (Madrid: Consejo Superior de Investigaciones Científicas), 197

——. 2000, Science, 289, 1484

Kormendy, J., & Bender, R. 1996, ApJ, 464, L119

——. 1999, ApJ, 522, 772

Kormendy, J., & Bender, R., Evans, A. S., & Richstone, D. 1998, AJ, 115, 1823

Kormendy, J., Bender, R., Ambrose, E., Tonry, J. L., & Freeman, K. C. 2004a, in preparation

Kormendy, J., Dressler, A., Byun, Y.-I., Faber, S. M., Grillmair, C., Lauer, T. R., Richstone, D., & Tremaine, S. 1994, in ESO/OHP Workshop on Dwarf Galaxies, ed. G. Meylan & P. Prugniel (Garching: ESO), 147

Kormendy, J., & Gebhardt, K. 2001, in 20th Texas Symposium on Relativistic Astrophysics, ed. J. C. Wheeler & H. Martel (Melville: AIP), 363

Kormendy, J., & Richstone, D. 1992, ApJ, 393, 559

——. 1995, ARA&A, 33, 581

Kormendy, J., et al. 1996a, ApJ, 459, L57

——. 1996b, ApJ, 473, L91

——. 1997, ApJ, 482, L139

——. 2004b, in preparation

Krabbe, A., et al. 1995, ApJ, 447, L95

Lauer, T. R. 2004, in Carnegie Observatories Astrophysics Series, Vol. 1: Coevolution of Black Holes and Galaxies, ed. L. C. Ho (Cambridge: Cambridge Univ. Press), in press

Lauer, T. R., et al. 1993, AJ, 106, 1436

——. 1995, AJ, 110, 2622

Lauer, T. R., Faber, S. M., Ajhar, E. A., Grillmair, C. J., & Scowen, P. A. 1998, AJ, 116, 2263

Lynden-Bell, D. 1969, Nature, 223, 690

——. 1978, Physica Scripta, 17, 185

Macchetto, F., Marconi, A., Axon, D. J., Capetti, A., Sparks, W., & Crane, P. 1997, ApJ, 489, 579

Maciejewski, W., & Binney, J. 2001, MNRAS, 323, 831

Magorrian, J., et al. 1998, AJ, 115, 2285

Maoz, E. 1995, ApJ, 447, L91

——. 1998, ApJ, 494, L181

Marconi, A., Capetti, A., Axon, D. J., Koekemoer, A., Macchetto, D., & Schreier, E. J. 2001, ApJ, 549, 915

McLure, R. J., & Dunlop, J. S. 2001, MNRAS, 327, 199

Merritt, D., & Ferrarese, L. 2001, MNRAS, 320, L30

Miyoshi, M., Moran, J., Herrnstein, J., Greenhill, L., Nakai, N., Diamond, P., & Inoue, M. 1995, Nature, 373, 127

Morris, M., ed. 1989, IAU Symp. 136, The Center of Our Galaxy (Dordrecht: Kluwer)

Munyaneza F., Tsiklauri D., & Viollier R. D. 1998, ApJ, 509, L105

——. 1999, ApJ, 526, 744

Munyaneza F., & Viollier R. D. 2002, ApJ, 564, 274

Netzer, H. 1990, in Active Galactic Nuclei, Saas-Fee Advanced Course 20, ed. T. J.-L. Courvoisier & M. Mayor (Berlin: Springer), 57

Netzer, H., & Peterson, B. M. 1997, in Astronomical Time Series, ed. D. Maoz, A. Sternberg, & E. M. Leibowitz (Dordrecht: Kluwer), 85

Peiris, H. V., & Tremaine, S. 2004, ApJ, submitted

Qian, E. E., de Zeeuw, P. T., van der Marel, R. P., & Hunter, C. 1995, MNRAS, 274, 602

Richstone, D. 2004, in Carnegie Observatories Astrophysics Series, Vol. 1: Coevolution of Black Holes and Galaxies, ed. L. C. Ho (Cambridge: Cambridge Univ. Press), in press

Richstone, D., et al. 1998, Nature, 395, A14

——. 2004, in preparation

Richstone, D., Bower, G., & Dressler, A. 1990, ApJ, 353, 118

Richstone, D. O., & Tremaine, S. 1985, ApJ, 296, 370

Rokaki, E., Boisson, C., & Collin-Souffrin, S. 1992, A&A, 253, 57

Salpeter, E. E. 1964, ApJ, 140, 796

Sargent, W. L. W., Young, P. J., Boksenberg, A., Shortridge, K., Lynds, C. R., & Hartwick, F. D. A. 1978, ApJ, 221, 731

Sarzi, M. 2004, in Carnegie Observatories Astrophysics Series, Vol. 1: Coevolution of Black Holes and Galaxies, ed. L. C. Ho (Pasadena: Carnegie Observatories, http://www.ociw.edu/ociw/symposia/series/symposium1/proceedings.html)

Sarzi, M., Rix, H.-W., Shields, J. C., Rudnick, G., Ho, L. C., McIntosh, D. H., Filippenko, A. V., & Sargent, W. L. W. 2001, ApJ, 550, 65

Schödel, R., et al. 2002, Nature, 419, 694

Schwarzschild, M. 1979, ApJ, 232, 236

Sellgren, K., McGinn, M. T., Becklin, E. E., & Hall, D. N. B. 1990, ApJ, 359, 112

Shields, G. A., Gebhardt, K., Salviander, S., Wills, B. J., Xie, B., Brotherton, M. S., Yuan, J., & Dietrich, M. 2003, ApJ, 583, 124

Siopis, C., et al. 2004, in preparation

Tonry, J. L. 1984, ApJ, 283, L27

——. 1987, ApJ, 322, 632

Tonry, J. L., et al. 2001, ApJ, 546, 681

Tremaine, S. 1995, AJ, 110, 628

Tremaine, S., et al. 2002, ApJ, 574, 740

Tsiklauri D., & Viollier R. D. 1998, ApJ, 500, 591

——. 1999, Astroparticle Phys., 12, 199

Valluri, M., Merritt, D., & Emsellem, E. 2004, ApJ, submitted (astro-ph/0210379)

van der Marel, R. P. 1994, ApJ, 432, L91

van der Marel, R. P., Cretton, N., de Zeeuw, P. T., & Rix, H.-W. 1998b, ApJ, 493, 613

van der Marel, R. P., de Zeeuw, P. T., Rix, H.-W. 1997b, ApJ, 488, 119

van der Marel, R. P., de Zeeuw, P. T., Rix, H.-W., & Quinlan, G. D. 1997a, Nature, 385, 610

van der Marel, R. P., Evans, N. W., Rix, H.-W., White, S. D. M., & de Zeeuw, T. 1994b, MNRAS, 271, 99

van der Marel, R. P., Rix, H.-W., Carter, D., Franx, M., White, S. D. M., & de Zeeuw, P. T. 1994a, MNRAS, 268, 521

van der Marel, R. P., & van den Bosch, F. C. 1998a, AJ, 116, 2220

Verdoes Kleijn, G. A., van der Marel, R. P., de Zeeuw, P. T., Noel-Storr, J., & Baum, S. A. 2002, AJ, 124, 2524

Verolme, E. K., et al. 2002, MNRAS, 335, 517

Wandel, A. 2002, ApJ, 565, 762

Young, P. J., Westphal, J. A., Kristian, J., Wilson, C. P., & Landauer, F. P. 1978, ApJ, 221, 721

Zel'dovich, Ya. B. 1964, Soviet Physics – Doklady, 9, 195

2

Black holes in active galaxies

AARON J. BARTH

California Institute of Technology

Abstract

Recent years have seen tremendous progress in the quest to detect supermassive black holes in the centers of nearby galaxies, and gas-dynamical measurements of the central masses of active galaxies have been valuable contributions to the local black hole census. This review summarizes measurement techniques and results from observations of spatially resolved gas disks in active galaxies, and reverberation mapping of the broad-line regions of Seyfert galaxies and quasars. Future prospects for the study of black hole masses in active galaxies, both locally and at high redshift, are discussed.

2.1   Introduction

The detection of supermassive black holes in the nuclei of many nearby galaxies has been one of the most exciting discoveries in extragalactic astronomy during the past decade. Accretion onto black holes has long been understood as the best explanation for the enormous luminosities of quasars (Salpeter 1964; Zel'dovich & Novikov 1964; Rees 1984), and the luminosity generated by quasars over the history of the Universe implies that most large galaxies must contain a black hole as a relic of an earlier quasar phase (Sołtan 1982; Chokshi & Turner 1992; Small & Blandford 1992). While the search for evidence of black holes in nearby galaxies began 25 years ago with the seminal studies of M87 by Sargent et al. (1978) and Young et al. (1978), only a handful of galaxies were accessible to such measurements until the repair of the *Hubble Space Telescope* (*HST*) in 1993 made it possible to study the central dynamics of galaxies routinely at $0\rlap{.}''1$ resolution. In addition to the recent dynamical searches for black holes in active and inactive galaxies with *HST*, the existence of black holes has been further confirmed by ground-based observations of the Galactic Center (see Ghez, this volume) and by radio observations of the H_2O maser disk in the Seyfert 2 galaxy NGC 4258 (Miyoshi et al. 1995). As the evidence for supermassive black holes in galaxy centers has strengthened, it has become clear that nuclear activity and the growth of black holes must be integral components of the galaxy formation process.

With measurements of black hole masses in several galaxies, it became possible for the first time to study the demographics of the black hole population and the connection of the black holes with their host galaxies. Kormendy & Richstone (1995) showed that $M_\bullet$ was correlated with L_{bulge}, the luminosity of the spheroidal "bulge" component of the host galaxy, albeit with substantial scatter. More intriguing was the discovery that $M_\bullet$ is very tightly correlated with $\sigma_\star$, the stellar velocity dispersion in the host galaxy (Ferrarese & Merritt 2000; Gebhardt et al. 2000a). The scatter in this relation is surprisingly small;

© The Observatories of the Carnegie Institution of Washington 2004.

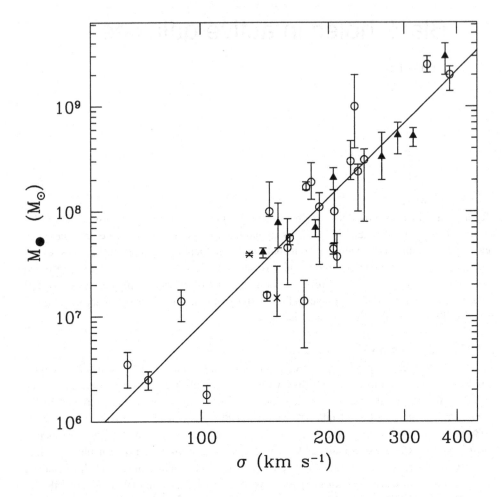

Fig. 2.1. The correlation between black hole mass and stellar velocity dispersion. Triangles denote galaxies measured with *HST* observations of gas dynamics, crosses are H_2O maser galaxies, and circles denote stellar-dynamical detections. The diagonal line is the best fit to the data as determined by Tremaine et al. (2002).

Tremaine et al. (2002) estimate the dispersion to be < 0.3 dex in log $M_\bullet$ at a given value of $\sigma_\star$. This is a remarkable finding: it implies that the masses of black holes, objects that inhabit scales of $\lesssim 10^{-4}$ pc in galaxy nuclei, are almost *completely* determined by the bulk properties of their host galaxies on scales of hundreds or thousands of parsecs. Although the $M_\bullet - \sigma_\star$ correlation is well established, its slope, and the amount of intrinsic scatter, remain somewhat controversial. The currently available sample of galaxies with accurate determinations of $M_\bullet$ is still modest. More measurements of black hole masses in nearby galaxies are needed, over the widest possible range of host galaxy types and velocity dispersions, in order to obtain a definitive present-day black hole census.

Gas-dynamical measurements of black hole masses in active galactic nuclei (AGNs) are an essential contribution to this pursuit, as illustrated in Figure 2.1. *HST* observations of ionized gas disks are vitally important for tracing the upper end of the black hole mass function, where stellar-dynamical measurements are hampered both by the low stellar surface brightness of the most massive elliptical galaxies and by the possibility of velocity anisotropy in nonrotating ellipticals. Observations of maser emission from molecular disks in active galaxies have provided the most solid black hole detection outside of our own Galaxy, strengthening the case that the massive dark objects discovered in *HST* surveys are indeed likely to be supermassive black holes. Reverberation mapping, and secondary methods that are calibrated by comparison with the reverberation technique, offer the most promising methods to determine black hole masses at high redshift.

The topic of black holes in active galaxies is vast, and this review will only concentrate on gas-dynamical measurements of black hole masses in AGNs. Before discussing the methods and results, a few general comments are in order. As Kormendy & Richstone (1995) have pointed out, there is a potentially serious drawback to any measurement technique based on gas dynamics: unlike stars, gas can respond to nongravitational forces, and the motions of gas clouds do not always reflect the underlying gravitational potential. For all methods based on gas dynamics, it is absolutely crucial to verify that the gas is actually in gravitational orbits about the central mass. If, for example, AGN-driven outflows or other nongravitational motions dominate, then black hole masses derived under the assumption of gravitational dynamics will be seriously compromised or completely erroneous. With that said, there are now numerous examples of ionized gas disks, and at least one maser disk, that clearly show orderly circular rotation. For reverberation mapping, the dynamical state of the broad-line emitting gas is more difficult to ascertain, but as discussed in §2.4 below, recent observations have provided some encouragement.

It must also be emphasized that, while these measurement techniques are capable of detecting dark mass concentrations in the centers of galaxies and determining their masses with varying degrees of accuracy, the observations do not actually prove that the dark mass is in the form of a supermassive black hole. The spatial resolution of gas-dynamical observations with *HST* typically corresponds to $\sim 10^{5-6}$ Schwarzschild radii. This is often sufficient to resolve the region over which the black hole dominates the gravitational potential of its host galaxy, but optical techniques are incapable of resolving the region in which relativistic motion occurs in the strong gravitational field near the black hole's event horizon. The conclusion that the massive dark objects detected in nearby galaxies are actually black holes is supported by the two most convincing dynamical detections, in our own Galaxy and in NGC 4258; in both objects the density of the central dark mass is inferred to be so large that reasonable alternatives to a black hole can be ruled out (Maoz 1998; Ghez, this volume). The best evidence for highly relativistic motion in the inner accretion disks of AGNs comes from X-ray spectra showing extremely broadened ($\sim 0.3c$), gravitationally redshifted Fe K line emission in Seyfert nuclei (Tanaka et al. 1995; Nandra et al. 1997). While this signature has only been convincingly detected in a handful of objects, it offers a powerful confirmation of the AGN paradigm, and analysis of the relativistically broadened line profiles may even reveal evidence for the black hole's spin (e.g., Iwasawa et al. 1996).

2.2   Black Hole Masses from Dynamics of Ionized Gas Disks

A striking discovery from the first years of *HST* observations was the presence of round, flattened disks of ionized gas and dust in the centers of some nearby radio galaxies

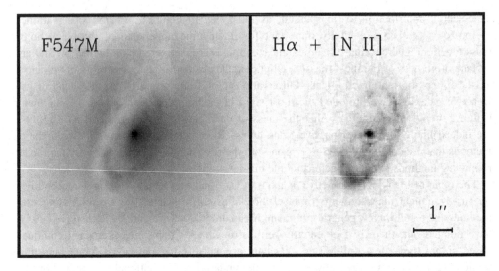

Fig. 2.2. An example of a dusty emission-line disk: *HST* images of the S0 galaxy NGC 3245. The left panel is a continuum image taken with the *HST* F547M filter (equivalent to the *V* band), and the right panel shows a continuum-subtracted, narrow-band image isolating the Hα and [N II] emission lines. At $D = 21$ Mpc, $1''$ corresponds to 100 pc.

(Jaffe et al. 1993; Ford et al. 1994). It was previously known from ground-based imaging that many ellipticals contained nuclear patches of dust (Kotanyi & Ekers 1979; Sadler & Gerhard 1985; Ebneter, Davis, & Djorgovski 1988), but *HST* revealed that the dust was often arranged in well-defined disks too small to be resolved from the ground. Imaging surveys with *HST* have found such disks, with typical radii of $100 - 1000$ pc, in $\sim 20\%$ of giant elliptical galaxies (e.g., van Dokkum & Franx 1995; Verdoes Kleijn et al. 1999; Capetti et al. 2000; de Koff et al. 2000; Tomita et al. 2000; Tran et al. 2001; Laine et al. 2003).

Jaffe et al. (1999) show that nuclear gas disks in early-type galaxies fall into two general categories. The most common type are dusty disks, which are easily detected by their obscuration in broad-band optical *HST* images. The dust is usually accompanied by an ionized component. Figure 2.2 shows an example, the disk in the S0 galaxy NGC 3245. The second class consists of ionized gas without associated dust disks. M87 (Ford et al. 1994) is the prototype of this category. Ionized disks are sometimes found to have filamentary or spiral structure, and patches of dust may be present as well. A comprehensive study of disk orientations in radio galaxies by Schmitt et al. (2002) finds that the radio jets are not preferentially aligned along the disk rotation axis, although jets tend not to be oriented close to the disk plane.

Soon after the first *HST* servicing mission, the first spectroscopic investigations of the kinematics of these disks were performed with the Faint Object Spectrograph (FOS). The first target was M87, for which Harms et al. (1994) detected a steep velocity gradient across the nucleus in the Hα, [N II], and [O III] emission lines, consistent with Keplerian rotation in an inclined disk. The central mass was found to be $(2.4 \pm 0.7) \times 10^9 \, M_\odot$, remarkably close to the values first determined by Young et al. (1978) and Sargent et al. (1978). The second gas-dynamical study with *HST* found a central dark mass of $(4.9 \pm 1.0) \times 10^8$

$M_\odot$ in the radio galaxy NGC 4261 (Ferrarese, Ford, & Jaffe 1996). These dramatic results opened a new chapter in the search for supermassive black holes, demonstrating that spatially resolved gas disks could indeed be used to measure the central masses of galaxies. In contrast to stellar dynamics, the gas-dynamical method is extremely appealing in its simplicity. Modeling the kinematics of a thin, rotating disk is conceptually straightforward. Furthermore, observations of emission-line velocity fields require less telescope time than absorption-line spectroscopy. Since the FOS was a single-aperture spectrograph, however, it was not well-suited to the task of mapping out emission-line velocity fields in detail, and FOS gas-dynamics data were only obtained for a few additional galaxies (van der Marel & van den Bosch 1998; Ferrarese & Ford 1999; Verdoes Kleijn et al. 2000).

After the initial FOS detections, progress was made on two fronts. The installation of the Space Telescope Imaging Spectrograph (STIS), a long-slit instrument, greatly expanded the capabilities of *HST* for dynamical measurements. In addition, the development of techniques to model the kinematic data in detail led to more robust measurements. At the center of a disk, the large spatial gradients in rotation velocity and emission-line surface brightness are smeared out by the telescope point-spread function (PSF) and by the nonzero size of the spectroscopic aperture. Macchetto et al. (1997) and van der Marel & van den Bosch (1998) were the first to model the effects of instrumental blurring on *HST* gas-kinematic data, and detailed descriptions of modeling techniques have been given by Barth et al. (2001), Maciejewski & Binney (2001), and Marconi et al. (2003).

The feasibility of performing a black hole detection in any given galaxy can be roughly quantified in terms of r_G, the radius of the "sphere of influence" over which the black hole dominates the gravitational potential of its host galaxy. This quantity is given by $r_G = GM_\bullet / \sigma_\star^2$. Projected onto the sky, and scaled to typical parameters for an *HST* measurement, this corresponds to

$$ r_G = 0.11 \left(\frac{M_\bullet}{10^8 M_\odot} \right) \left(\frac{200 \text{ km s}^{-1}}{\sigma_\star} \right)^2 \left(\frac{20 \text{ Mpc}}{D} \right) \text{ arcsec.} \qquad (2.1) $$

Detection of black holes via their influence on the motions of stars or gas is most readily accomplished when observations are able to probe spatial scales smaller than r_G, but it should be borne in mind that this is at best an approximate criterion. The stellar velocity dispersion is an aperture-dependent quantity, so there is no uniquely determined value of r_G for a given galaxy. Even when r_G is unresolved the black hole will still influence the motions of stars and gas at larger radii. In stellar-dynamical measurements, it is possible to obtain information from spatial scales smaller than the instrumental resolution by measuring higher-order moments of the central line-of-sight velocity profile, since extended wings on the velocity profile are the signature of high-velocity stars orbiting close to the black hole (e.g., van der Marel 1994). With sufficiently high signal-to-noise ratio, the information contained in the full line-of-sight velocity profile could be exploited in gas-dynamical measurements as well, although measurements to date have generally been performed by fitting models to the first and second moments of the velocity distribution function (i.e., the mean velocity and line width at each observed position), or in some cases only to the mean velocities.

The analysis of a gas-dynamical dataset consists of the following basic steps. The galaxy's stellar light profile must be measured, corrected for dust absorption if necessary, and converted to a three-dimensional luminosity density. The stellar mass density is generally

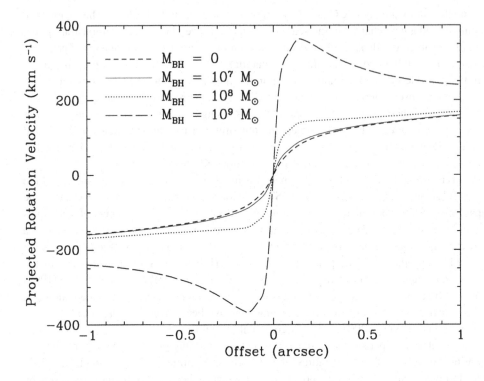

Fig. 2.3. Projected radial velocity curves for the major axis of a model disk with $i = 60°$ at $D = 20$ Mpc, for $M_\bullet = 0$, 10^7, 10^8, and 10^9 $M_\odot$. The models are convolved with the *HST* PSF and sampled over a slit width of $0.''1$. The curves illustrate the mean velocity observed at each position along the spectrograph slit.

assumed to be axisymmetric or spherically symmetric. A model velocity field is computed for the combined potential of the black hole and the galaxy mass distribution, usually assuming a spatially constant stellar mass-to-light ratio (M/L). After projecting the velocity field to a given distance and inclination angle, the model is synthetically "observed" by simulating the passage of light through the spectrograph optics and measuring the resulting model line profiles. Finally, the model fit to the measured emission-line velocity field is optimized to obtain the best-fitting value of $M_\bullet$. In addition to $M_\bullet$, the free parameters in the kinematic model fit include the disk inclination and major axis position angle, and the stellar M/L; these can be determined from the kinematic data if observations are obtained at three or more parallel positions of the spectrograph slit. Maciejewski & Binney (2001) have shown that when the slit is wider than the PSF core, there is an additional signature of the black hole: at one particular location in the velocity field, the rotational and instrumental broadening will be oppositely directed, and they will very nearly cancel, giving a very narrow line profile. The location of this feature can be used as an additional diagnostic of $M_\bullet$. With up-to-date analysis techniques and high-quality STIS data, it is possible to achieve formal measurement uncertainties on $M_\bullet$ of order $\sim 25\%$ or better for galaxies with well-behaved disks (e.g., Barth et al. 2001). This makes the gas-dynamical method very competitive with the precision that can be achieved by stellar-dynamical measurements.

Figure 2.3 illustrates model calculations for the projected radial velocities along the major axis of an inclined gas disk. The models have been calculated for a disk inclined at $60°$ to the line of sight, in a galaxy at a distance of 20 Mpc with $M_\bullet = 0$, 10^7, 10^8, and 10^9 $M_\odot$. To demonstrate the effects of varying $M_\bullet$, the same stellar mass profile has been used for all four models. The model velocity fields have been convolved with the STIS PSF and sampled over an aperture corresponding to an $0.''1$-wide slit, and the curves represent the mean velocity that would be observed as a function of position along the slit. At one extreme, the sphere of influence of the 10^7 $M_\odot$ black hole is unresolved, and the 10^7 $M_\odot$ model is barely distinguishable from the model with no black hole, although the rapidly rotating gas near the black hole will give rise to high-velocity wings on the central emission-line profile. The opposite extreme is the 10^9 $M_\odot$ black hole, for which the Keplerian region is extremely well resolved and the black hole would be readily detected. Instrumental blurring causes a turnover in the velocity curves at about $0.''15$ in this case, and Keplerian rotation can only be clearly detected at larger radii. Thus, Keplerian rotation can only be verified in detail (in the sense of having several independent data points to trace the $v \propto r^{-1/2}$ dependence) for galaxies with exceptionally well-resolved r_G. To date, the only published *HST* gas-dynamical measurements that have unambiguously detected the Keplerian region are those of M87 (Macchetto et al. 1997) and M84 (Bower et al. 1998).

The middle ground between these two extremes is illustrated by the 10^8 $M_\odot$ black hole in Figure 2.3. In this case, the Keplerian rise in velocity is not detected; instead, a steep but smooth velocity gradient across the nucleus is observed. The mass of the black hole can still be determined from the steepness of this central gradient, provided that the observations give sufficient information to distinguish the velocity curves from those of the $M_\bullet = 0$ case. It is possible in principle to measure $M_\bullet$ even in galaxies for which r_G is formally unresolved, if it can be shown that the disk has an excess rotation velocity relative to the best-fitting model without a black hole. However, in such cases there are practical complications that may limit the measurement accuracy: it is critical to determine the stellar luminosity density profile and M/L accurately, and spatial gradients in M/L are an added complication. In general, the most confident detections of black holes will be in those relatively rare objects for which the Keplerian region can be clearly traced, but the majority of gas-dynamical mass measurements will come from objects for which the mass is determined primarily from the steepness of the central velocity gradient rather than from fitting models to a well-resolved Keplerian velocity field.

Since the gas can respond to nongravitational forces, it is essential for the observations to map the disk structure in sufficient detail that the assumption of gravitational motion can be tested. This is a nontrivial concern, as there are examples of galaxies in which the gas does not rotate at the circular velocity (e.g., Fillmore, Boroson, & Dressler 1986; Kormendy & Westpfahl 1989). The case of IC 1459 serves as a cautionary tale. FOS observations at 6 positions within $0.''3$ of the nucleus revealed a steep central velocity gradient in the ionized gas, and disk model fits implied the presence of a central dark mass of $(1-4) \times 10^8$ $M_\odot$ (Verdoes Kleijn et al. 2000). More recently, STIS observations have mapped out the circumnuclear kinematics in much greater detail and found an irregular and asymmetric velocity field that cannot be interpreted in terms of flat disk models; a stellar-dynamical analysis finds $M_\bullet = (2.6 \pm 1.1) \times 10^9$ $M_\odot$ (Cappellari et al. 2002). Thus, the ionized gas fails to be a useful probe of the gravitational potential in this galaxy.

Most of the nuclear disks observed spectroscopically with *HST* have revealed evidence for substantial internal velocity dispersions, even when the velocity field is clearly dominated by rotation. That is, the gas disks are not in purely quiescent circular rotation, and the disks have some degree of internal velocity structure or turbulence. The intrinsic emission-line velocity dispersion σ_{gas} tends to be greatest at the nucleus, and observed values span a wide range, exceeding 500 km s^{-1} in extreme cases (e.g., van der Marel & van den Bosch 1998).

The origin of this internal velocity dispersion is unknown, and the interpretation of σ_{gas} remains the single most important unresolved problem for gas-dynamical measurements of black hole masses. In a study of the radio galaxy NGC 7052, van der Marel & van den Bosch (1998) argued that σ_{gas} was due to local turbulence in gas that remained in bulk motion on circular orbits at the local circular velocity. Another possibility that has been considered is that the disks are composed of a large number of clouds with small filling factor, and that the individual clouds are on noncircular orbits that are nevertheless dominated by gravity (e.g., Verdoes Kleijn et al. 2000). This is analogous to the effect of "asymmetric drift" that is well known in stellar dynamics (e.g., Binney & Tremaine 1987). In this situation, the gas disk would be supported against gravity by both rotation and random motions, and models based on pure circular rotation would underestimate the true black hole masses. The asymmetric drift correction to $M_\bullet$ is expected to be small when $(\sigma_{gas}/v_{rot})^2 \ll 1$, but this must be tested on a case-by-case basis by computing models for the emission-line widths. Various approaches to calculating the asymmetric drift in gas disks have been presented by Cretton, Rix, & de Zeeuw (2000), Verdoes Kleijn et al. (2000, 2002), and Barth et al. (2001).

Since different groups have used a variety of methods to treat this problem in dynamical analyses (sometimes ignoring it altogether), and since σ_{gas} varies widely among different galaxies, the influence of the intrinsic velocity dispersion could be responsible for some of the apparent scatter in the $M_\bullet - \sigma_\star$ relation. Now that large samples of galaxies have been observed with STIS, it may be possible to discern whether there are any correlations of σ_{gas} with the level of nuclear activity or with the Hubble type or any other property of the host galaxy; detection of any clear trends might help to elucidate the origin of the intrinsic dispersion. Three-dimensional hydrodynamical simulations have evolved to the point where it is now becoming feasible to model the turbulent structure of gas disks in galaxies (e.g., Wada, Meurer, & Norman 2002); this work may lead to new insights on how best to interpret σ_{gas} in gas-dynamical measurements. Further discussion of the intrinsic velocity dispersion problem is presented by Verdoes Kleijn, van der Marel, & Noel-Storr (2004).

Most gas-dynamical studies to date have concentrated on elliptical and S0 galaxies, but STIS data have now been obtained for dozens of spiral galaxies as well (e.g., Marconi et al. 2003). In principle, the gas-dynamical method can work equally well for spirals, but there are some additional complications. One is the presence of nuclear star clusters, which are nearly ubiquitous in late-type spirals (Carollo, Stiavelli, & Mack 1998; Böker et al. 2002). The nuclear star clusters are typically young (Walcher et al. 2004), and the dynamical modeling must take into account the possible radial gradient in M/L. Another caveat is that spiral arms or bar structure can lead to departures from circular rotation in the ionized gas (Koda & Wada 2002; Maciejewski 2004). Only $\sim 10\%-20\%$ of spirals appear to have orderly emission-line velocity fields suitable for disk-model fitting to derive $M_\bullet$ (Sarzi et al. 2001), so some care is needed when selecting targets for *HST* gas-dynamical surveys. Ho et al. (2002) have shown that orderly rotation in the emission-line gas almost exclusively occurs in galaxies having orderly, symmetric circumnuclear dust-lane morphology that can be detected in *HST* imaging data, and this offers a promising way to maximize the rate of successful $M_\bullet$ measurements in future programs.

Even when gas-dynamical data cannot provide a direct measurement of black hole mass, either due to insufficient resolution or irregular kinematics, they can still be used to derive upper limits to $M_\bullet$ from the central amplitude of the rotation curve or the central emission-line width. For galaxies with $\sigma_\star \lesssim 100$ km s^{-1}, where direct measurements of $M_\bullet$ are scarce, upper limits are valuable additions to the black hole census. For example, from a large sample of ground-based rotation curves, Salucci et al. (2000) demonstrated that late-type spirals must generally have $M_\bullet \lesssim 10^{6-7} M_\odot$. Sarzi et al. (2002) measured upper limits to $M_\bullet$ from the central emission-line widths observed in STIS spectra of 16 galaxies having $\sigma_\star$ between 80 and 270 km s^{-1}. The derived upper limits are consistent with the $M_\bullet - \sigma_\star$ relation, further confirming that few galaxies (perhaps none) are strong outliers with very overmassive black holes relative to their bulge luminosity or velocity dispersion.

Gas-dynamical measurements using *HST* data have now been published for about a dozen galaxies (Harms et al. 1994; Ferrarese et al. 1996; Macchetto et al. 1997; Bower et al. 1998; van der Marel & van den Bosch 1998; Ferrarese & Ford 1999; Barth et al. 2001; Sarzi et al. 2001; Verdoes Kleijn et al. 2002; Marconi et al. 2003). The results, together with stellar-dynamical measurements, have been compiled by Kormendy & Gebhardt (2001), Merritt & Ferrarese (2001), and Tremaine et al. (2002). This number will continue to increase over the next several years, since STIS gas-kinematic observations have been obtained or scheduled for over 100 galaxies. However, it remains to be seen how many of these datasets will lead to accurate measurements of black hole masses. A substantial fraction of the galaxies will probably not have well-behaved gas disks suitable for modeling, and in many of the target galaxies r_G may be much smaller than the resolution limit of *HST* (Merritt & Ferrarese 2002). Thus, it remains worthwhile to search for additional nearby galaxies having morphologically regular disks that will make promising targets for future spectroscopic observations.

With the small projected sizes of r_G in most nearby galaxies, ground-based observations cannot generally be used to perform gas-dynamical measurements of $M_\bullet$. One exception is the nearby ($D = 3.5$ Mpc) radio galaxy Cen A. Marconi et al. (2001) obtained near-infrared spectra of this galaxy in 0''5 seeing, and measured the velocity field of the Paβ and [Fe II] emission lines. They detected a steep central velocity gradient with a turnover to Keplerian rotation outside the region dominated by atmospheric seeing, and derived a central dark mass of $2^{+3.0}_{-1.4} \times 10^8 M_\odot$. Although the black hole mass was not determined with very high precision, this measurement serves as a proof of concept that ground-based observations of near-infrared emission lines can be used for gas-dynamical analysis. With future laser guide star systems, such measurements can be performed using adaptive optics on 8–10 meter telescopes, although the possibility of a spatially and temporally variable PSF will make the data analysis a formidable challenge. Also, the *Atacama Large Millimeter Array* will provide new high-resolution views of circumnuclear disks and new probes of disk dynamics with molecular emission lines.

2.3 Black Hole Masses from Observations of H_2O Masers

Water vapor maser emission at $\lambda = 1.35$ cm from the nucleus of the low-luminosity Seyfert 2 galaxy NGC 4258 was detected by Claussen, Heiligman, & Lo (1984). The maser emission consists of a bright core with several distinct components near the systemic velocity arranged in an elongated region (Greenhill et al. 1995), as well as "satellite" lines separated by $\pm 800-1000$ km s^{-1} from the systemic features (Nakai, Inoue, & Miyoshi 1993). The presence of high-velocity emission was suggestive of a disk rotating about a central mass of $\sim 10^7 \, M_\odot$ (Watson & Wallin 1994), and the breakthrough observation came when Miyoshi et al. (1995) used the Very Long Baseline Array to map out the positions of the high-velocity features. They found that the satellite masers traced out a near-perfect Keplerian velocity curve on either side of the nucleus, allowing a precise determination of the enclosed mass within the inner radius of the disk (3.9 milliarcsec, corresponding to a radius of only 0.14 pc). For a distance of 7.2 Mpc, the central mass is found to be $3.9 \times 10^7 \, M_\odot$. It is very unlikely that this mass could be composed of a cluster of dark objects such as stellar remnants or brown dwarfs, since the cluster density would be so large that its lifetime against evaporation or collisions would be short compared to the age of the galaxy (Maoz 1995, 1998). This makes NGC 4258 the most compelling dynamical case for the existence of a supermassive black hole outside of our own Galaxy. Recently, a preliminary analysis of *HST* stellar-dynamical data for NGC 4258 has yielded a value of $M_\bullet$ consistent with the maser measurement (Siopis et al. 2002); this is a reassuring confirmation of the stellar-dynamical technique.

A detailed review of the properties of the NGC 4258 maser system and a listing of other known H$_2$O maser galaxies is given by Moran, Greenhill, & Herrnstein (1999). Hundreds of galaxies have been surveyed for H$_2$O masers (e.g., Braatz, Wilson, & Henkel 1996; Greenhill et al. 2002, 2003), and there are about 30 known sources to date. Braatz et al. (1997) find that powerful H$_2$O masers are only detected in Seyfert 2 and LINER nuclei, and not in Seyfert 1 galaxies. This is consistent with the geometrical picture of AGN unification models, which posit the existence of an edge-on torus or disk in the Type 2 objects. The large path length along our line of sight in an edge-on disk permits maser amplification, while a more face-on disk in a Type 1 AGN would not emit maser lines in our direction. Further confirmation of this geometric structure comes from detection of large X-ray obscuring columns (e.g., Makishima et al. 1994; Iwasawa, Maloney, & Fabian 2002) and polarized emission lines from obscured nuclei (e.g., Antonucci & Miller 1985; Wilkes et al. 1995) in some H$_2$O maser galaxies.

Disklike structures have been detected in several other maser galaxies, but NGC 4258 remains the only one for which the central mass has been derived with high precision. The maser clouds in NGC 1068 appear to trace the surface of a geometrically thick torus rather than a thin disk, and their velocities fall below a Keplerian curve, suggesting a central mass of $\sim 10^7 \, M_\odot$ (Greenhill et al. 1996). NGC 4945 contains high-velocity maser clumps in a roughly disklike arrangement, but with larger uncertainties than for NGC 4258; the central mass is $\sim 10^6 \, M_\odot$ within a radius of 0.3 pc (Greenhill, Moran, & Herrnstein 1997). Recently, high-velocity maser features have been detected in IC 2560 (Ishihara et al. 2001), NGC 5793 (Hagiwara et al. 2001), NGC 2960 (Henkel et al. 2002), and the Circinus galaxy (Greenhill et al. 2003), and VLBI measurements can determine whether the masers in these galaxies trace out Keplerian rotation curves. Additional maser disks will continue to be found in future surveys, although the nearby AGN population has now been surveyed so thoroughly that it is unlikely that many more new examples will be found at distances comparable to that of NGC 4258.

2.4   Reverberation Mapping

Reverberation mapping uses temporal variability of active nuclei to probe the size and structure of the broad-line region (BLR) of Seyfert 1 galaxies and quasars. The basic principle is that variability in the ionizing photon output of the central engine will be followed by corresponding variations in the emission-line luminosity, after a time delay dependent on the light-travel time between the ionizing source and the emission-line clouds (Blandford & McKee 1982). By monitoring the continuum and emission-line brightness of a broad-lined AGN over a sufficient time period, the lag between continuum variations and emission-line response can be derived, giving a size scale for the region emitting the line; typical sizes range from a few light-days up to ~ 1 light-year. Thus, the method makes use of the time domain to resolve structures that cannot be resolved spatially, and the measurement accuracy depends on temporal sampling rather than spatial resolution. Peterson (2001) gives a thorough discussion of the observational and analysis techniques used in reverberation-mapping campaigns. The literature on reverberation mapping is extensive, and this section reviews only a few recent results on the determination of black hole masses from reverberation data.

If the motions in the BLR are dominated by gravity (rather than, for example, radiatively driven outflows), then the central mass can be derived from the BLR radius combined with a characteristic velocity. Essentially, the black hole mass is derived as $M_\bullet = fv^2r/G$, where v is some measure of the broad-line velocity width (typically the full width at half maximum), r is the BLR radius derived from the measured time delay, and f is an order-unity factor that depends on the geometry of the BLR (i.e., disklike or spherical). Reverberation masses for 34 Seyfert 1 galaxies and low-redshift quasars have been determined by Wandel, Peterson, & Malkan (1999) and Kaspi et al. (2000). The method is subject to some potentially serious systematic errors, however, as emphasized by Krolik (2001); the unknown geometry and emissivity distribution of the BLR clouds can lead to biases in the derived masses, and it is critically important to verify that the BLR velocity field is in fact dominated by gravitational motion.

Recent observations have yielded some encouraging results. Since the BLR is radially stratified in ionization level, highly ionized clouds emitting He II and C IV respond most quickly to continuum variations, with response times of a few days, while lines such as Hβ and C III] λ1909 have longer lag times as well as narrower widths. Thus, if lag times t_{lag} can be measured for multiple broad emission lines in a given galaxy, it is possible to trace out the velocity structure of the BLR as a function of radius. Peterson & Wandel (1999, 2000) and Onken & Peterson (2002) have shown that for four of the best-observed reverberation targets, NGC 3783, NGC 5548, NGC 7469, and 3C 390.3, the relation between t_{lag} and emission-line width shows exactly the dependence expected for Keplerian motion. Within the measurement uncertainties, the emission lines in each galaxy yield a correlation consistent with FWHM $\propto t_{\text{lag}}^{0.5}$, and for each galaxy the time lags and line widths of the different emission lines give consistent results for the central mass.

Another key question is whether reverberation mapping yields black hole masses that are consistent with the host galaxy properties of the AGNs. Gebhardt et al. (2000b) and Ferrarese et al. (2001) demonstrated that the reverberation masses and velocity dispersions of several Seyfert nuclei are in good agreement with the $M_\bullet - \sigma_\star$ relation of inactive galaxies. Nelson (2000) performed a similar analysis for a larger sample, using [O III] line widths as a substitute for $\sigma_\star$, and found similar results, albeit with additional scatter that could be attributed to some nongravitational motion in the narrow-line region. Some previous studies had found the puzzling result that the AGNs seemed to have systematically lower $M_\bullet/M_{\text{bulge}}$ ratios than inactive galaxies (Ho 1999; Wandel 1999). Since the AGNs do not appear discrepant in the $M_\bullet - \sigma_\star$ relation, a likely conclusion (e.g., Wandel 2002) is that the bulge luminosities of the Seyferts had been systematically overestimated, due to a combination of their relatively large distances and the dominance of the central point sources; starburst activity in the Seyferts might further bias the photometric decompositions toward anomalously high bulge luminosities.

Overall, the agreement between the $M_\bullet - \sigma_\star$ relation of Seyferts with that of inactive galaxies suggests that the reverberation masses are probably accurate to a factor of ~ 3 on average (Peterson 2003). Direct measurements of $M_\bullet$ in reverberation-mapped Seyferts with *HST* would be a valuable cross-check, but unfortunately only a few bright Seyfert 1 galaxies are near enough for r_G to be resolved, and attempts to perform stellar-dynamical measurements of the nearest objects with *HST* have been thwarted by the dominance of the bright nonstellar continuum. Gas-dynamical observations are not affected by this problem, and there have been attempts to detect cleanly rotating kinematic components in Seyfert narrow-line regions that could be used for $M_\bullet$ measurements with *HST* (Winge et al. 1999). Unfortunately, the presence of outflows or other kinematic disturbances typically precludes the use of emission-line velocity fields as probes of of $M_\bullet$ in Seyfert 1 galaxies (e.g., Crenshaw et al. 2000). Since there is no way to perform direct stellar- or gas-dynamical measurements of $M_\bullet$ in most reverberation-mapped AGNs, the comparison with the $M_\bullet - \sigma_\star$ relation remains the main consistency check that can currently be applied to the reverberation-based masses.

Reverberation mapping can be extended to higher redshifts, since it is not dependent on spatial resolution, but the longer variability time scales for higher-mass black holes in luminous quasars, combined with cosmological time dilation, can require monitoring campaigns with durations that are a significant fraction of an individual astronomer's career! Kaspi et al. (2004) present preliminary results from an ongoing, eight-year campaign to monitor 11 luminous quasars at $2.1 < z < 3.2$. Continuum variations have been detected, but the corresponding emission-line variability has not yet been seen. Reverberation observations of quasars are a fundamental probe of black hole masses at high redshift, and efforts to monitor additional quasars over a wide redshift range should be encouraged, despite the long time scales involved.

One important consequence of the reverberation campaigns has been the detection of a correlation between the BLR radius and the continuum luminosity, a result that is expected on the basis of simple photoionization considerations (e.g., Wandel 1997). With a sample of 34 reverberation-mapped AGNs, Kaspi et al. (2000) find $r_{\text{BLR}} \propto L^{0.7}$ using continuum luminosity at 5100 Å. This correlation offers an extremely valuable shortcut to estimate the BLR size, and black hole mass, in distant quasars. While reverberation campaigns require years of intensive observations, $L(5100 \text{ Å})$ and FWHM(Hβ) can be measured from a single spectrum, and then combined to yield an estimate of $M_\bullet$ under the assumption of virial motion of the BLR clouds. This is the only technique that can routinely be applied to derive $M_\bullet$ in distant AGNs, and it has recently been the subject of intense interest, with numerous studies focused on topics such as the $M_\bullet/M_{\text{bulge}}$ ratio in quasars and the search for possible differences between radio-loud and radio-quiet objects (e.g., Laor 1998, 2001; Lacy et al. 2001; McLure & Dunlop 2001, 2002; Jarvis & McLure 2002; Oshlack, Webster,

& Whiting 2002; Shields et al. 2003). The technique has also been extended to make use of continuum luminosity in the rest-frame ultraviolet combined with the velocity width of either C IV (Vestergaard 2002) or Mg II (McLure & Jarvis 2002), so that ground-based, optical spectra can be used to derive black hole masses for high-redshift quasars.

These secondary methods are extremely valuable since they offer the most straightforward estimates of black hole masses at high redshift, although there are potential biases that must be kept in mind. The derived BLR size and $M_\bullet$ depend on the observed continuum luminosity, but this can be affected by dust extinction (Baker & Hunstead 1995) or by relativistic beaming of synchrotron emission in radio-loud objects (Whiting, Webster, & Francis 2001). If the BLR has a flattened, disklike geometry, then the effects of source orientation on the observed line width must be accounted for (e.g., McLure & Dunlop 2002). An additional concern is that the r_{BLR}-L relation has only been calibrated against the Kaspi et al. reverberation sample, which covers a somewhat limited range both in black hole mass ($M_\bullet \lesssim 5 \times 10^8\ M_\odot$) and in luminosity ($\lambda L_\lambda \lesssim 7 \times 10^{45}$ erg s^{-1} at 5100 Å). Application of this method to high-luminosity quasars with $M_\bullet > 10^9\ M_\odot$ necessarily involves a large extrapolation (see Netzer 2003 for further discussion). These issues can be overcome if reverberation masses can be derived for larger samples of quasars, extending to high intrinsic luminosities and $M_\bullet > 10^9\ M_\odot$, so that the relations between $M_\bullet$, line width, and luminosity can be calibrated over a broader parameter space.

2.5   Future Work and Some Open Questions

I conclude with a very incomplete list of a few important and tractable problems that can be addressed in the foreseeable future by new observations.

1. More dynamical measurements of black hole masses in nearby galaxies are needed, over the widest possible range of host galaxy masses and velocity dispersions, so that the slope of the $M_\bullet - \sigma_\star$ and $M_\bullet$-L_{bulge} correlations can be determined definitively. To constrain the amount of intrinsic scatter in these correlations, realistic estimates of the measurement uncertainties are crucial. Gas-dynamical measurements with *HST* and, in the future, from ground-based telescopes with adaptive optics, will be a key component of this pursuit. Additional measurements of black hole masses from maser dynamics will be extremely valuable as well, if more galaxies with Keplerian maser disks can be found.

2. Direct comparisons of stellar- and gas-dynamical measurements for the same galaxies are a needed consistency check that should be performed for galaxies over a wide range of Hubble types and velocity dispersions.

3. What causes the intrinsic velocity dispersion observed in nuclear gas disks? Is it possible to determine central masses accurately for disks having $(\sigma_{\mathrm{gas}}/\upsilon_{\mathrm{rot}})^2 \approx 1$? Again, direct comparisons with stellar-dynamical observations would be very useful.

4. What can we learn about black hole demographics from AGNs at the extremes of the Hubble sequence? The broad-line widths and continuum luminosities of high-redshift quasars imply masses of up to $\sim 10^{10}\ M_\odot$ for some objects (e.g., Shields et al. 2003), but these measurements involve extrapolating the known correlation between r_{BLR} and L far beyond the mass and luminosity ranges over which it has been calibrated locally. At the other extreme, is there a lower limit to L_{bulge} or $\sigma_\star$ below which galaxies have no central black hole at all? Dynamical searches for black holes in the nuclei of dwarf ellipticals or very late-type spirals become extremely difficult for distances beyond the Local Group (see van der Marel, this volume). On the other hand, searches for accretion-powered nuclear

activity in dwarf galaxies can offer some constraints on the population of black holes with $M < 10^6\ M_\odot$. The case of NGC 4395, a dwarf Magellanic spiral hosting a full-fledged Seyfert 1 nucleus (Filippenko & Sargent 1989) with a black hole of $\lesssim 10^5\ M_\odot$ (Iwasawa et al. 2000; Filippenko & Ho 2003), demonstrates that at least some dwarf galaxies can host black holes that would be undetectable by dynamical means.

5. How do the $M_\bullet - \sigma_\star$ and $M_\bullet - L_{\text{bulge}}$ correlations evolve with redshift, and how early did the black holes in the highest-redshift quasars build up most of their mass? Reverberation mapping of high-redshift quasars, and further calibration and testing of the r_{BLR}-luminosity relationship in luminous quasars, will be of fundamental importance in answering these questions. Measurement of the masses of black holes at high redshift, as well as the luminosities and/or velocity dispersions of their host galaxies, will be a major observational step toward understanding the coevolution of black holes and their host galaxies.

References

Antonucci, R. R. J., & Miller, J. S. 1985, ApJ, 297, 621

Baker, J. C., & Hunstead, R. W. 1995, ApJ, 452, L95

Barth, A. J., Sarzi, M., Rix, H.-W., Ho, L. C., Filippenko, A. V., & Sargent, W. L. W. 2001, ApJ, 555, 685

Binney, J., & Tremaine, S. 1987, Galactic Dynamics (Princeton: Princeton Univ. Press)

Blandford, R. D., & McKee, C. F. 1982, ApJ, 255, 419

Böker, T., Laine, S., van der Marel, R. P., Sarzi, M., Rix, H.-W., Ho, L. C., & Shields, J. C. 2002, AJ, 123, 1389

Bower, G. A., et al. 1998, ApJ, 492, L111

Braatz, J. A., Wilson, A. S., & Henkel, C. 1996, ApJS, 106, 51

——. 1997, ApJS, 110, 321

Capetti, A., de Ruiter, H. R., Fanti, R., Morganti, R., Parma, P., & Ulrich, M.-H. 2000, A&A, 362, 871

Cappellari, M., Verolme, E. K., van der Marel, R. P., Verdoes Kleijn, G. A., Illingworth, G. D., Franx, M., Carollo, C. M., & de Zeeuw, P. T. 2002, ApJ, 578, 787

Carollo, C. M., Stiavelli, M., & Mack, J. 1998, AJ, 116, 68

Chokshi, A., & Turner, E. L. 1992, MNRAS, 259, 421

Claussen, M. J., Heiligman, G. M., & Lo, K.-Y. 1984, Nature, 310, 298

Crenshaw, D. M., et al. 2000, AJ, 120, 1731

Cretton, N., Rix, H.-W., & de Zeeuw, P. T. 2000, ApJ, 536, 319

de Koff, S., et al. 2000, ApJS, 129, 33

Ebneter, K., Davis, M., & Djorgovski, S. 1988, AJ, 95, 422

Ferrarese, L., & Ford, H. C. 1999, ApJ, 515, 583

Ferrarese, L., Ford, H. C., & Jaffe, W. 1996, ApJ, 470, 444

Ferrarese, L., & Merritt, D. 2000, ApJ, 539, L9

Ferrarese, L., Pogge, R. W., Peterson, B. M., Merritt, D., Wandel, A., & Joseph, C. L. 2001, ApJ, 555, L79

Filippenko, A. V., & Ho, L. C. 2003, ApJ, 588, L13

Filippenko, A. V., & Sargent, W. L. W. 1989, ApJ, 342, L11

Fillmore, J. A., Boroson, T. A., & Dressler, A. 1986, ApJ, 302, 208

Ford, H. C., et al. 1994, ApJ, 435, L27

Gebhardt, K., et al. 2000a, ApJ, 539, L13

——. 2000b, ApJ, 543, L5

Greenhill, L. J., et al. 2002, ApJ, 565, 836

Greenhill, L. J., Gwinn, C. R., Antonucci, R., & Barvainis, R. 1996, ApJ, 472, L21

Greenhill, L. J., Jiang, D. R., Moran, J. M., Reid, M. J., Lo, K.-Y., & Claussen, M. J. 1995, ApJ, 440, 619

Greenhill, L. J., Kondratko, P. T., Lovell, J. E. J., Kuiper, T. B. H., Moran, J. M., Jauncey, D. L., & Baines, G. P. 2003, ApJ, 582, L11

Greenhill, L. J., Moran, J. M., & Herrnstein, J. R. 1997, ApJ, 481, L23

Hagiwara, Y., Diamond, P. J., Nakai, N., & Kawabe, R. 2001, ApJ, 560, 119

Harms, R. J., et al. 1994, ApJ, 435, L35

Henkel, C., Braatz, J. A., Greenhill, L. J., & Wilson, A. S. 2002, A&A, 394, L23

Ho, L. C. 1999, in Observational Evidence for Black Holes in the Universe, ed. S. K. Chakrabarti (Dordrecht: Kluwer), 157

Ho, L. C., Sarzi, M., Rix, H.-W., Shields, J. C., Rudnick, G., Filippenko, A. V., & Barth, A. J. 2002, PASP, 114, 137

Ishihara, Y., Nakai, N., Iyomoto, N., Makishima, K., Diamond, P., & Hall, P. 2001, PASJ, 53, 215

Iwasawa, K., et al. 1996, MNRAS, 282, 1038

Iwasawa, K., Fabian, A. C., Almaini, O., Lira, P., Lawrence, A., Hayashida, K., & Inoue, H. 2000, MNRAS, 318, 879

Iwasawa, K., Maloney, P. R., & Fabian, A. C. 2002, MNRAS, 336, L71

Jaffe, W., Ford, H. C., Ferrarese, L., van den Bosch, F., & O'Connell, R. W. 1993, Nature, 364, 213

Jaffe, W., Ford, H. C., Tsvetanov, Z., Ferrarese, L., & Dressel, L. 1999, in Galaxy Dynamics, ed. D. Merritt, J. A. Sellwood, & M. Valluri (San Francisco: ASP), 13

Jarvis, M. J., & McLure, R. J. 2002, MNRAS, 336, L38

Kaspi, S., Netzer, H., Maoz, D., Shemmer, O., Brandt, W. N., & Schneider, D. P. 2004, in Carnegie Observatories Astrophysics Series, Vol. 1: Coevolution of Black Holes and Galaxies, ed. L. C. Ho (Pasadena: Carnegie Observatories, http://www.ociw.edu/ociw/symposia/series/symposium1/proceedings.html)

Kaspi, S., Smith, P. S., Netzer, H., Maoz, D., Jannuzi, B. T., & Giveon, U. 2000, ApJ, 533, 631

Koda, J., & Wada, K. 2002, A&A, 396, 867

Kormendy, J., & Gebhardt, K. 2001, in The 20th Texas Symposium on Relativistic Astrophysics, ed. H. Martel & J. C. Wheeler (Melville: AIP), 363

Kormendy, J., & Richstone, D. 1995, ARA&A, 33, 581

Kormendy, J., & Westpfahl, D. J. 1989, ApJ, 338, 752

Kotanyi, C. G., & Ekers, R. D. 1979, A&A, 73, L1

Krolik, J. H. 2001, ApJ, 551, 72

Lacy, M., Laurent-Muehleisen, S. A., Ridgway, S. E., Becker, R. H., & White, R. L. 2001, ApJ, 551, L17

Laine, S., van der Marel, R. P., Lauer, T. R., Postman, M., O'Dea, C. P., & Owen, F. N. 2003, AJ, 125, 478

Laor, A. 1998, ApJ, 505, L83

——. 2001, ApJ, 553, 677

Macchetto, F., Marconi, A., Axon, D. J., Capetti, A., Sparks, W., & Crane, P. 1997, ApJ, 489, 579

Maciejewski, W. 2004, in Carnegie Observatories Astrophysics Series, Vol. 1: Coevolution of Black Holes and Galaxies, ed. L. C. Ho (Pasadena: Carnegie Observatories, http://www.ociw.edu/ociw/symposia/series/symposium1/proceedings.html)

Maciejewski, W., & Binney, J. 2001, MNRAS, 323, 831

Makishima, K., et al. 1994, PASJ, 46, L77

Maoz, E. 1995, ApJ, 447, L91

——. 1998, ApJ, 494, L181

Marconi, A., et al. 2003, ApJ, 586, 868

Marconi, A., Capetti, A., Axon, D. J., Koekemoer, A., Macchetto, D., & Schreier, E. J. 2001, ApJ, 549, 915

McLure, R. J., & Dunlop, J. S. 2001, MNRAS, 327, 199

——. 2002, MNRAS, 331, 795

McLure, R. J., & Jarvis, M. J. 2002, MNRAS, 337, 109

Merritt, D., & Ferrarese, L. 2001, in The Central Kpc of Starbursts and AGN: The La Palma Connection, ed. J. H. Knapen et al. (San Francisco: ASP), 335

Miyoshi, M., Moran, J., Herrnstein, J., Greenhill, L., Nakai, N., Diamond, P., & Inoue, M. 1995, Nature, 373, 127

Moran, J. M., Greenhill, L. J., & Herrnstein, J. R. 1999, Jour. Astrophys. and Astron., 20, 165

Nakai, N., Inoue, M., & Miyoshi, M. 1993, Nature, 361, 6407

Nandra, K., George, I. M., Mushotzky, R. F., Turner, T. J., & Yaqoob, T. 1997, ApJ, 477, 602

Nelson, C. H. 2000, ApJ, 544, L91, 199

Netzer, H. 2003, ApJ, 583, L5

Onken, C. A., & Peterson, B. M. 2002, ApJ, 572, 746

Oshlack, A. Y. K. N., Webster, R. L., & Whiting, M. T. 2002, ApJ, 576, 81

Peterson, B. M. 2001, in Advanced Lectures on the Starburst-AGN Connection, ed. I. Aretxaga, D. Kunth, & R. Mújica (Singapore: World Scientific), 3

——. 2003, in Active Galactic Nuclei: from Central Engine to Host Galaxy, ed. S. Collin, F. Combes, & I. Shlosman (San Francisco: ASP), 43

Peterson, B. M., & Wandel, A. 1999, ApJ, 521, L95

——. 2000, ApJ, 540, L13

Rees, M. J. 1984, ARA&A, 22, 471

Sadler, E. M., & Gerhard, O. E. 1985, MNRAS, 214, 177

Salpeter, E. E. 1964, ApJ, 140, 796

Salucci, P., Ratnam, C., Monaco, P., & Danese, L. 2000, MNRAS, 317, 488

Sargent, W. L. W., Young, P. J., Boksenberg, A., Shortridge, K., Lynds, C. R., & Hartwick, F. D. A. 1978, ApJ, 221, 731

Sarzi, M., et al. 2002, ApJ, 567, 237

Sarzi, M., Rix, H.-W., Shields, J. C., Rudnick, G., Ho, L. C., McIntosh, D. H., Filippenko, A. V., & Sargent, W. L. W. 2001, ApJ, 550, 65

Schmitt, H. R., Pringle, J. E., Clarke, C. J., & Kinney, A. L. 2002, ApJ, 575, 150

Shields, G. A., Gebhardt, K., Salviander, S., Wills, B., Xie, B., Brotherton, M. S., Yuan, J., Dietrich, M. 2002, ApJ, 583, 124

Siopis, C., et al. 2002, BAAS, 201, 6802

Small, T. A., & Blandford, R. D. 1992, MNRAS, 259, 725

Sołtan, A. 1982, MNRAS, 200, 115

Tanaka, Y., et al. 1995, Nature, 375, 659

Tomita, A., Aoki, K., Watanabe, M., Takata, T., & Ichikawa, S. 2000, AJ, 120, 123

Tran, H. D., Tsvetanov, Z., Ford, H. C., Davies, J., Jaffe, W., van den Bosch, F. C., & Rest, A. 2001, AJ, 121, 2928

Tremaine, S., et al. 2002, ApJ, 574, 740

van der Marel, R. P. 1994, ApJ, 432, L91

van der Marel, R. P., & van den Bosch, F. C. 1998, AJ, 116, 2220

van Dokkum, P. G., & Franx, M. 1995, AJ, 110, 2027

Verdoes Kleijn, G. A., Baum, S. A., de Zeeuw, P. T., & O'Dea, C. P. 1999, AJ, 118, 2592

Verdoes Kleijn, G. A., van der Marel, R. P., Carollo, C. M., & de Zeeuw, P. T. 2000, AJ, 120, 1221

Verdoes Kleijn, G. A., van der Marel, R. P., de Zeeuw, P. T., Noel-Storr, J., & Baum, S. A. 2002, AJ, 124, 2524

Verdoes Kleijn, G. A., van der Marel, R. P., & Noel-Storr, J. 2004, in Carnegie Observatories Astrophysics Series, Vol. 1: Coevolution of Black Holes and Galaxies, ed. L. C. Ho (Pasadena: Carnegie Observatories, http://www.ociw.edu/ociw/symposia/series/symposium1/proceedings.html)

Vestergaard, M. 2002, ApJ, 571, 733

Wada, K., Meurer, G., & Norman, C. A. 2002, ApJ, 577, 197

Walcher, C. J., Häring, N., Böker, T., Rix, H.-W., van der Marel, R. P., Gerssen, J., Ho, L. C., & Shields, J. C. 2004, in Carnegie Observatories Astrophysics Series, Vol. 1: Coevolution of Black Holes and Galaxies, ed. L. C. Ho (Pasadena: Carnegie Observatories, http://www.ociw.edu/ociw/symposia/series/symposium1/proceedings.html)

Wandel, A. 1997, ApJ, 490, L131

——. 1999, ApJ, 519, L39

——. 2002, ApJ, 565, 762

Wandel, A., Peterson, B. M., & Malkan, M. A. 1999, ApJ, 526, 579

Watson, W. D. & Wallin, B. K. 1994, ApJ, 432, L35

Whiting, M. T., Webster, R. L., & Francis, P. J. 2001, MNRAS, 323, 718

Wilkes, B. J., Schmidt, G. D., Smith, P. S., Mathur, S., & McLeod, K. K. 1995, ApJ, 455, L13

Winge, C., Axon, D. J., Macchetto, F. D., Capetti, A., & Marconi, A. 1999, ApJ, 519, 134

Young, P. J., Westphal, J. A., Kristian, J., Wilson, C. P., & Landauer, F. P. 1978, ApJ, 221, 721

Zel'dovich, Ya. B., & Novikov, I. D. 1964, Sov. Phys. Dokl., 158, 811

3

Intermediate-mass black holes in the Universe: a review of formation theories and observational constraints

ROELAND P. VAN DER MAREL
Space Telescope Science Institute

Abstract

This paper reviews the subject of intermediate-mass black holes (IMBHs) with masses between those of "stellar-mass" and "supermassive" black holes (BHs). The existence of IMBHs is a real possibility: they might plausibly have formed as remnants of the first generation of stars (Population III), as the result of dense star cluster evolution, or as part of the formation process of supermassive BHs. Their cosmic mass density could exceed that of supermassive BHs ($\Omega \approx 10^{-5.7}$) and observations do not even rule out that they may account for all of the baryonic dark matter in the Universe ($\Omega \approx 10^{-1.7}$). Unambiguous detections of individual IMBHs currently do not exist, but there are observational hints from studies of microlensing events, "ultra-luminous" X-ray sources, and centers of nearby galaxies and globular clusters. Gravitational wave experiments will soon provide another method to probe their existence. IMBHs have potential importance for several fields of astrophysics and are likely to grow as a focus of research attention.

3.1 Introduction

BHs were long considered a mathematical curiosity, but it is now clear that they are an important and indisputable part of the astronomical landscape (e.g., Begelman & Rees 1998). In particular, there is unambiguous evidence for "stellar-mass" BHs and "supermassive" BHs.

Stellar-mass BHs form in a reasonably well-understood manner through stellar evolution (Fryer 1999). A BH with a companion star might accrete matter from it to produce an X-ray binary (XRB). It is sometimes possible to determine the mass of the accreting object in an XRB from detailed modeling. Neutron stars cannot be more massive than $2-3 M_\odot$ and compact objects with larger masses are therefore assumed to be BHs. Some dozen such objects are know, mostly with masses in the range $5-15 M_\odot$ (Charles 2001). The fraction of stellar-mass BHs that manages to remain in a close binary throughout its evolution and is also currently accreting is very small, so most stellar-mass BHs exist singly and go unnoticed (§ 3.4.1). The Milky Way hosts $\sim 10^{7-9}$ stellar-mass BHs (Brown & Bethe 1994).

The paradigm that there are also supermassive BHs in the Universe is based on the existence of active galactic nuclei (AGNs) in the centers of some galaxies. Their properties can only be plausibly explained by assuming that a BH of $10^6-10^9 M_\odot$ acts as the central engine (Rees 1984). The proper motions of stars around our Galactic Center (Sgr A*) provide direct evidence for this (Schödel et al. 2002; Ghez 2004). A variety of techniques now exist to detect and weigh supermassive BHs using stellar or gaseous kinematics

(Kormendy & Gebhardt 2001). The BH mass is always of the order 0.1% of the galaxy bulge/spheroid mass. An even better correlation exists with the velocity dispersion, $M_\bullet \propto \sigma^4$ (Tremaine et al. 2002). The origin of these correlations, the triggers of AGN activity, and the exact formation mechanisms of supermassive BHs remain poorly understood (§ 3.2.3).

Stellar-mass and supermassive BHs can be studied because they (sometimes) exist in environments favorable for the production of observable signatures through accretion or gravitational influence. This need not to be true for all BHs in the Universe, which may therefore also exist in other mass ranges. BHs in the intermediate-mass range of, say, 15–$10^6 M_\odot$ (i.e, between the familiar classes of BHs), are of particular interest. Such IMBHs might plausibly have formed in different ways (§ 3.2). They have been suggested as an important component of the missing baryonic dark matter in the Universe (§ 3.3) and recent observational studies have provided hints of IMBHs in various environments (§ 3.4). It can be concluded that IMBHs are an important topic for additional research (§ 3.5).

3.2   Formation Theories

IMBHs may plausibly have formed as the remnants of Population III stars (§ 3.2.1), through dynamical processes in dense star clusters (§ 3.2.2), or as an essential ingredient or occasional by-product of the formation of supermassive BHs (§ 3.2.3). Primordial formation of IMBHs is unlikely (§ 3.2.4).

3.2.1   *Formation from Population III Stellar Evolution*

The present-day stellar initial mass function (IMF) extends to $\sim 200 M_\odot$ (Larson 2003). Massive stars shed most of their mass through radiatively driven stellar winds. Above $\sim 100 M_\odot$ a nuclear pulsational instability sets in that generates additional mass loss. Evolutionary calculations indicate that massive stars leave compact remnants with masses below $\sim 15 M_\odot$, consistent with observations of X-ray binaries (Fryer & Kalogera 2001). The minimum initial mass for a star to become a stellar-mass BH (rather than a neutron star) is ~ 20–$25 M_\odot$ (Fryer 1999).

For the first generation of zero-metallicity stars in the Universe (Population III) the initial conditions and evolutionary path were quite different. There is now a growing body of evidence that suggests a top-heavy IMF in the early Universe (Schneider et al. 2002), although this issue continues to be debated. In the absence of metals, primordial molecular clouds cool through rotational-vibrational lines of H_2. Simulations of the collapse and fragmentation of such clouds (Abel, Bryan, & Norman 2000) suggest that the first generation of stars had typical masses of $\sim 100 M_\odot$, compared to the $\sim 1 M_\odot$ characteristic of stars at the present epoch. In addition, radiative mass losses are negligible at zero metallicity, and mass losses due to nuclear pulsational instability are greatly reduced (Fryer, Woosley, & Heger 2001).

The evolution of massive Population III stars depends on the initial mass (Bond, Arnett, & Carr 1984; Heger & Woosley 2001). Stars below $140 M_\odot$ probably evolve into BHs in similar fashion as do stars of normal metallicity (Fryer 1999), although the remnant BHs will be more massive than today's stellar-mass BHs due to the more limited mass loss. Stars that are initially more massive than $\sim 140 M_\odot$ encounter the electron-positron pair-instability during oxygen burning. In the range ~ 140–$260 M_\odot$ this yields an explosion that leaves no remnant and is considerably more energetic than a normal supernova. Above $\sim 260 M_\odot$ there is direct collapse into a BH because nuclear burning is unable to halt the collapse and generate an explosion. The remnant mass exceeds half of the initial stellar mass, thus constituting an IMBH. Objects that are initially more massive than $\sim 10^5 M_\odot$ cannot have stable hydrogen burning to begin with, due to a post-Newtonian instability. As a result, such objects quickly collapse into a BH (Baumgarte & Shapiro 1999; Shibata & Shapiro 2002).

It is thus possible, and maybe even likely, that a population of IMBHs was produced from Population III stars. The size of this population was recently estimated by Madau & Rees (2001) and Schneider et al. (2002). Many details of these calculations are uncertain, but both papers find that a population could easily have been produced with a global mass density similar to that of the supermassive BHs in the Universe, and possibly more. The IMBHs would presumably have formed at redshifts $z \approx 10$–20 in peaks of the mass distribution.

3.2.2 *Formation in Dense Star Clusters*

Star clusters have long been suspected as possible sites for the formation of IMBHs. The self-gravity of a cluster gives it a negative heat capacity that makes it vulnerable to the so-called "gravothermal catastrophe": the core collapses on a timescale proportional to the two-body relaxation time (Binney & Tremaine 1987). The resulting high central density may lead to BH formation in various ways. The crucial issue is whether realistic initial conditions ever lead to densities that are high enough for this to occur, or whether core-collapse is halted and reversed at lower densities. This question has been addressed theoretically using semi-analytic arguments, Fokker-Planck calculations, and direct N-body codes.

Lee (1987) and Quinlan & Shapiro (1990) studied the importance of stellar mergers during core collapse. These can give rise to the runaway growth of a supermassive star, which at the end of its lifetime collapses to a BH (§ 3.2.1). These studies found that runaway merging occurs naturally in very dense clusters ($\rho > 10^6 M_\odot\,\mathrm{pc}^{-3}$) of many stars ($N > 10^7$). These initial conditions correspond to velocity dispersions of hundreds of km s^{-1}, and may be relevant for (early) galactic nuclei. This provides a scenario for the formation of an IMBH, which through accretion might subsequently grow to become a supermassive BH. By contrast, Quinlan & Shapiro (1987, 1989) and Lee (1993) studied the fate of a cluster of compact objects (neutron stars and stellar-mass BHs) instead of normal stars. In this situation they found, also for initial conditions appropriate for galactic nuclei, that the core collapses all the way to a relativistic state. When the redshift reaches values $z > 0.5$ (velocities in excess of 10^5 km s^{-1}), a relativistic instability sets in that results in catastrophic collapse to a BH (Shapiro & Teukolsky 1985). However, this relativistic path to a BH may not be the most natural evolutionary scenario. Starting from a cluster of normal stars, runaway merging would likely produce a single IMBH before a cluster of compact objects could form (Quinlan & Shapiro 1990).

The aforementioned studies agreed that formation of an IMBH would not occur in star clusters with fewer than 10^6–10^7 stars, such as globular clusters. In such clusters core collapse is halted by binary heating (Hut et al. 1992) before the densities become high enough for runaway stellar merging. Three-body interactions between "hard" binaries and single stars add energy to the cluster (at the expenses of the binaries, which become harder; Heggie 1975). The binaries form primarily through tidal capture. This process is much more efficient at the low velocity dispersions characteristic of globular clusters than at the higher velocity dispersions of galactic nuclei.

Although it has long been thought that core collapse in globular clusters is generically halted by binary heating, it was realized recently that this is not always true. Stars of different masses are not always able to reach energy equipartition (Spitzer 1969), and in fact, the Salpeter IMF is unstable in this sense (Vishniac 1978). This causes the heaviest stars to undergo core collapse more or less independently of the other cluster stars, on a timescale that is much less than the core collapse time for the cluster as a whole. Portegies Zwart & McMillan (2002) used N-body simulations to show that a runaway merger among these massive stars leads to the formation of an IMBH, provided that the core collapse proceeds faster than their main-sequence lifetime. This implies an initial half-mass relaxation time < 25 Myr. For a globular cluster that evolves in the Galactic tidal field the corresponding present-day half-mass relaxation time would have to be $< 10^8$ yr. It has been proposed that this scenario might be important for young compact star clusters, such as those often observed in star-forming galaxies (Ebisuzaki et al. 2001). Many of the Milky Way's globular clusters have half-mass relaxation times in the range 10^8–10^9 yr, and some have half-mass relaxation times below 10^8 yr (Harris 1996). So this scenario may well be relevant for Milky Way globular clusters as well, in particular because the physical conditions during their formation are only poorly understood.

A more unlikely route for the formation of IMBHs in globular clusters is through the repeated merging of compact objects, such as stellar-mass BHs (Lee 1995; Taniguchi et al. 2000; Mouri & Taniguchi 2002). Such objects get caught in binaries through dynamical effects. After hardening by interactions with single stars, they eventually merge after losing energy by gravitational radiation. However, the interactions that produce hardening also provide recoils that tend to eject the binaries from the cluster (Kulkarni, Hut, & McMillan 1993; Sigurdsson & Hernquist 1993; Portegies Zwart & McMillan 2000). This limits the scope for considerable growth through repeated merging, although in some situations four-body interactions may boost the probability (Miller & Hamilton 2002a). One way to avoid the recoil problem is to assume that there is a single BH somewhere in the cluster that starts out at $\sim 50 M_\odot$. After sinking to the cluster center through dynamical friction, the BH could slowly grow in mass through merging with stellar-mass BHs. The mass of the BH would be large enough to prevent ejection through recoil (Miller & Hamilton 2002b).

3.2.3   *Relation to Supermassive Black Hole Formation*

The formation of supermassive BHs in the centers of galaxies is poorly understood, but there are many plausible scenarios (Begelman & Rees 1978; Rees 1984). Many scenarios are extensions of those discussed in the preceding sections, and involve IMBHs at some time in their evolution. It is therefore possible that supermassive BHs and IMBHs in the Universe are intimately linked. Also, not all BHs in galaxy centers may have had the opportunity to become supermassive, so some galaxies may have a central BH of intermediate mass.

Scenarios that evolve IMBHs into supermassive BHs usually invoke merging and/or accretion. Schneider et al. (2002) and Volonteri, Haardt, & Madau (2003) considered the case of IMBHs formed from Population III stars (§ 3.2.1). They envisaged that while galaxies are assembled hierarchically from smaller units, the IMBHs in these units sink to the center through dynamical friction. There they merge to form supermassive BHs. Haiman and Loeb (2001) found that this is a plausible scenario for building some $10^9 M_\odot$ BHs at very early times, as required observationally by the detection of bright quasars at redshifts as large as 6 (Fan et al. 2001). However, Islam, Taylor, & Silk (2003) argued that this scenario may not be able to account for all the mass observed today in supermassive BHs. Also, Hughes & Blandford (2003) showed that supermassive BHs that grow through mergers generally have little spin, which makes it unlikely that such BHs could power radio jets. As an alternative to merging, a single intermediate-mass "seed" BH might have grown supermassive through accretion. The growth may happen quickly through collapse of a surrounding protogalaxy onto the BH (Adams, Graff, & Richstone 2001) or it may happen slowly by accretion of material shed by surrounding stars (Murphy, Cohn, & Durisen 1991). Feedback from the energy release near the center may limit both the growth of the BH (Haehnelt, Natarajan, & Rees 1998) and the growth of the galaxy (Silk & Rees 1998). Feedback from star formation may also limit the BH growth (Burkert & Silk 2001), while fresh gas supply provided during the merging of galactic subunits (Haehnelt & Kauffmann 2000; Kauffmann & Haehnelt 2000) may provide increased growth. These scenarios can reproduce observed correlations such as those between BH mass and bulge mass or bulge velocity dispersion (§ 3.1).

Not all scenarios for supermassive BH formation proceed through an IMBH stage. If a collapsing gas cloud can loose its angular momentum and avoid fragmentation into stars, it may collapse to a BH directly. Haehnelt & Rees (1993) sketched a route by which this may have occurred. Bromm & Loeb (2003) investigated this possibility quantitatively by studying the collapse of metal-free primordial clouds of $10^8 M_\odot$ using hydrodynamical simulations. To avoid fragmentation into stars, they assumed that the presence of H_2 (which would otherwise be responsible for cooling) is suppressed by an intergalactic UV background. With this assumption, condensations of $\sim 5 \times 10^6 M_\odot$ form that can collapse to a BH through the post-Newtonian instability (§ 3.2.1). It is unclear whether feedback from a growing BH may limit the attainable mass, so it is possible that the result would be an IMBH.

3.2.4 Primordial Formation

BHs might have formed primordially in the early Universe. The mass of such BHs is generally of order the horizon mass at its formation time, $M \approx 10^5 (t/\text{sec}) M_\odot$ (Barrow & Carr 1996), although smaller values are not impossible (Hawke & Stewart 2002). At the Planck time ($\sim 10^{-43}$ sec) the horizon mass is the Planck mass ($\sim 10^{-38} M_\odot$) and at 1 sec it is $10^5 M_\odot$. Primorial BHs less massive than $\sim 10^{-18} M_\odot$ would by now have evaporated through the process of Hawking radiation. Primordial BHs around this mass would currently be evaporating, and the observed γ-ray background places useful limits on their existence (MacGibbon & Carr 1991). However, Hawking radiation becomes progressively less relevant for more massive BHs, and is negligible for the mass range of interest in the present context. So it places no useful observational limits on the existence of primordial BHs in the intermediate-mass regime, and our thinking must be guided by theoretical considerations.

One possible mechanism for primordial BH formation is through collapse of density fluctuations (Carr 1975; Carr & Lidsey 1993). However, in standard cold dark matter (CDM) cosmologies the early Universe is characterized by a very high degree of homogeneity and isotropy. The associated Gaussian density fluctuations are much too small to collapse to a BH (Begelman & Rees 1998). Another mechanism for the formation of primordial BHs does not require density fluctuations, but invokes collisions of bubbles of broken symmetry during phase transitions in the early Universe (Hawking, Moss & Stewart 1982; Rubin, Khlopov, & Sakharov 2000). For example, the quantum chromodynamic cosmic phase transition at $t = 10^{-5}$ sec might have produced BHs of order $\sim 1 M_\odot$ (Jedamzik 1997). Primordial BHs could also have formed spontaneously through the collapse of cosmic strings (MacGibbon, Brandenburger & Wichowski 1998) or through inflationary reheating (Garcia-Bellido & Linde 1998). Even if primordial BH formation is possible in these scenarios, it is by no means guaranteed. Also, these scenarios do not naturally lead to BHs in the intermediate-mass range. Afshordi, McDonald & Spergel (2003) recently addressed some cosmological implications of a potential large population of primoridial IMBHs. However, more conventional cosmological thinking suggests that such a population is not particularly likely.

3.3   IMBHs: The Missing Baryonic Dark Matter?

There is now considerable evidence that the matter density of the Universe is $\Omega_m \equiv \rho_m/\rho_{crit} \approx 0.3$, with an additional $\Omega_\Lambda \approx 0.7$ in a cosmological constant or "dark energy." Comparison of Big Bang nucleosynthesis calculations with the observed abundances of light elements yields the baryon density: $\Omega_b = 0.041 \pm 0.004$ (this number scales as H_0^{-2}, and $H_0 = 70$ km s^{-1} Mpc^{-1} was assumed; Burles, Nollett, & Turner 2001). The non-zero value of $\Omega_m - \Omega_b$ indicates the presence of non-baryonic dark matter, with some form of CDM being the most popular candidate. However, this is probably not the only missing matter in the Universe. A detailed inventory of the visible baryonic matter adds up to a best guess of only $\Omega_v = 0.021$ (Persic & Salucci 1992; Fukugita, Hogan, & Peebles 1998). Although this number can be stretched with various assumptions, it does appear that half of the baryons in the Universe are in some dark form.

Carr (1994) provided a general review of the candidates for, and constraints on, the baryonic dark matter. The hypothesis that it could be a population of IMBHs in the halos of galaxies (Lacey & Ostriker 1985) is of particular interest in the present context. Such a population is constrained observationally by the dynamical effects it would have on its environment (§ 3.3.1) and by its gravitational lensing properties (§ 3.3.2). Additional constraints exist if the IMBHs are assumed to have formed from Population III stars (§ 3.3.3).

3.3.1   Dynamical Constraints on IMBHs in Dark Halos

The gravitational interactions that one would expect IMBHs to have with other objects provide important constraints on their possible contribution to the baryonic dark matter (Carr & Sakellariadou 1999). For example, IMBHs in dark halos would heat (increase the stellar velocity dispersion) of galaxy disks, the more so for larger BH masses. The observed velocity dispersions of stellar disks therefore limit the masses of BHs in galactic halos. If the Milky Way dark halo were composed entirely of BHs, then their mass would have to be less than $\sim 3 \times 10^6 M_\odot$ (Carr & Sakellariadou 1999). This limit becomes more stringent for small, dark matter dominated galaxies. Rix & Lake (1993) find an upper limit of $\sim 6 \times 10^3 M_\odot$ for the Local Group galaxy GR8, although this is open to debate (Tremaine & Ostriker 1999). Halo IMBHs also tend to disrupt stellar systems in galaxies, in particular globular clusters (Carr & Sakellariadou 1999). Klessen & Burkert (1995) found that this excludes the possibility that dark halos are made up entirely of BHs more massive than $\sim 5 \times 10^4 M_\odot$. If there are more massive BHs, they would have to make up a smaller fraction of the halo: no more than 2.5%–5% for BHs of $\sim 10^6 M_\odot$ (Murali, Arras, & Wasserman 2000). However, these constraints are quite uncertain because it is unknown what the properties of globular cluster systems were when they formed. It could even be that disruption of clusters by IMBHs may have played an essential role in the shaping of the present-day number and mass distribution of globular clusters (Ostriker, Binney, & Saha 1989).

IMBHs in a galactic halo sink to the center through dynamical friction. If they merge and accumulate there, then the observed masses of supermassive BHs in galaxy centers constrain the mass and number of halo IMBHs (Carr & Sakellariadou 1999). Xu & Ostriker (1994) modeled this in detail, taking into account the timescale for merging through emission of gravitational radiation and the possibility of slingshot ejection of BHs in three-body interactions. They found that unacceptable build-up of a central object occurs only for a halo made up of BHs more massive than $\sim 3 \times 10^6 M_\odot$. Consistent with this, Islam et al. (2003) found in a study of a cosmologically motivated population of IMBHs (remnants of Population III stars) that the build-up of central objects remains within observationally acceptable limits.

3.3.2   Lensing Constraints on IMBHs in Dark Halos

Massive compact halo objects (MACHOs) can produce gravitational microlensing amplification of the intensity of background stars (Paczyński 1986). Several teams have monitored stars in the Large Magellanic Cloud (LMC) for a number of years to search for such signatures. For the specific case of the LMC, the average characteristic timescale for microlensing events is $\sim 130(M/M_\odot)^{1/2}$ days. Hence, MACHOs are progressively more difficult to detect for increasing masses. Still, Alcock et al. (2001) calculated that they should have been able detect ~ 1 event of multi-year duration toward the LMC if the Milky Way dark halo were made entirely of $100 M_\odot$ MACHOs. By contrast, no LMC events were detected with durations in excess of ~ 130 days. This rules out that the entire halo is made of MACHOs in the range 0.15–$30 M_\odot$ (95% confidence), although contribution of a fraction below $\sim 25\%$ is allowed. Alcock et al. (2000) did detect many shorter duration events, from which it was concluded that $\sim 20\%$ of the Milky Way halo may be composed of compact objects with masses in the range 0.15–$0.9 M_\odot$.

The possible contributions of compact objects to the dark halos of other (more distant) galaxies are constrained by various gravitational lensing effects as well. However, these do not yet place limits in the intermediate-mass regime that improve upon what is already known from Big Bang nucleosynthesis (Carr 1994).

3.3.3   Constraints on IMBHs Formed from Population III Stars

The cosmic density of IMBHs that have formed from Population III stars (§ 3.2.1) is limited by additional constraints (Carr 1994). Stars with initial masses below $260 M_\odot$ expel much of their metals at the end of their lifetime, enriching the interstellar medium (ISM). The fact that the enrichment must have been less than the lowest metallicities observed in Population I stars ($Z \approx 10^{-3}$) limits the cosmic mass density Ω in such Population III stars to no more than 10^{-4}. The cosmic density of stars more massive than $260 M_\odot$ (and their IMBH remnants) is not constrained by metallicity considerations because they do not end their life in a supernova explosion. However, they might shed helium before their ultimate collapse. This places some constraints on their potential contribution to Ω, but these are no more stringent than what is already known from Big Bang nucleosynthesis. Objects more massive than $10^5 M_\odot$ shed neither metals nor helium because they collapse to a BH without reaching stable hydrogen burning.

All Population III stars below $10^5 M_\odot$ shine brightly during their main-sequence phase. Their numbers are therefore constrained by observations of the extragalactic background light, particularly in the infrared. The constraints depend on the formation redshift and on whether or not allowance is made for possible reprocessing by dust. Depending on the exact assumptions, a cosmic density of Population III stars sufficient to explain all the baryonic dark matter may just barely be consistent with the available extragalactic background light data (Carr 1994; Schneider et al. 2002). There are also limits from the accretion that one would expect onto a cosmologically important population of BHs, but these do not place strong constraints in the intermediate-mass range (Ipser & Price 1977; Carr 1994).

3.4   Searches for Individual IMBHs

IMBHs may contribute as much as $\Omega \approx 0.02$ to the cosmic baryon budget (§ 3.3). However, even if they existed in far smaller numbers, they would be of great importance for astrophysics. For comparison, the cosmic mass density of supermassive BHs in galaxy centers is only $10^{-5.7}$ (Yu & Tremaine 2002). Cosmologically motivated scenarios of Population III evolution easily predict densities that rival or exceed this (Madau & Rees 2001; Schneider et al. 2002). Hence, it is important to search for evidence of individual IMBHs. Such IMBHs may exist in the main luminous bodies of galaxies, where they could reveal themselves through their microlensing properties (§ 3.4.1), or through accretion-powered X-ray emission (§ 3.4.2). Alternatively, they could exist in the centers of galaxies (§ 3.4.3) or globular star clusters (§ 3.4.4). In the near future it may be possible to search for gravitational-wave signatures of IMBHs (§ 3.4.5).

3.4.1   Bulge Microlensing

Compact objects can be detected through microlensing. The Einstein ring crossing time scales as $M^{1/2}$ and long-duration events are therefore of particular interest. No long-duration events were detected toward the LMC (§ 3.3.2) but the situation is different for the Galactic bulge: $\sim 10\%$ of the few hundred detected events have timescales exceeding ~ 140 days (Bennett et al. 2002).

The lensing timescale depends not only on the lens mass, but also on the unknown transverse velocity of the lens and the ratio of lens and source distances. The latter quantities can be constrained statistically from the fact that they are drawn from the known phase-space distribution function of the Galaxy. When many events are modeled as a statistical ensemble this yields an estimate of the mass distribution of the lenses (Han & Gould 1996). Additional information is needed to constrain the masses of individual lenses. This is often possible for long-duration events from the "microlensing parallax" effect, which produces a signature in the light curve due to the fact that the Earth moves around the Sun as the event progresses. Modeling yields the transverse velocity of the lens as projected to the solar position. Bennett et al. (2002) identified six events with sufficiently accurate parallax data to yield an estimate of the lens mass. The largest masses are $6^{+10}_{-3} M_{\odot}$ (MACHO-96-BLG-5) and $6^{+7}_{-3} M_{\odot}$ (MACHO-98-BLG-6). The observational limits on the lens brightness make these events excellent candidates for stellar-mass BHs, the first tentative detections outside of XRBs. The long-duration event MACHO-99-BLG-22/OGLE-1999-BUL-32 is even more interesting (Agol et al. 2002; Mao et al. 2002). The lens-mass likelihood function is bimodal, with maximum likelihood at mass $130^{+42}_{-114} M_{\odot}$ and with a secondary peak of lower likelihood at $4.0^{+1.5}_{-1.8} M_{\odot}$ (Bennett et al. 2004). So this lens could be an IMBH.

The bulge microlensing events suggest that stellar-mass BHs may contribute more than 1% of the Milky Way mass (Bennett et al. 2002, 2004), more than is traditionally believed (Brown & Bethe 1994; Fryer 1999). An important caveat in the analysis is that the dynamics of the lenses is assumed to follow that of the known stars. This would be violated if BHs are born with large kick velocities, as are neutron stars. There is conflicting observational evidence on this issue (Nelemans, Tauris, & van den Heuvel 1999; Mirabel et al. 2002).

3.4.2 Ultra-luminous X-ray Sources

The Eddington luminosity for an accreting compact object of mass M is $1.3 \times 10^{38} (M/M_\odot)$ erg s^{-1}. This is $\sim 2 \times 10^{38}$ erg s^{-1} for a neutron star and $(0.4–2) \times 10^{39}$ erg s^{-1} for a stellar-mass BH. Surprisingly, X-ray observations with *Einstein* (Fabbiano 1989), *ROSAT* (Colbert & Mushotzky 1999; Roberts & Warwick 2000; Colbert & Ptak 2002) and *Chandra* have shown that more luminous sources appear to exist in $\sim 30\%$ of nearby galaxies. If the emission of these sources is assumed to be isotropic, then the observed fluxes indicate X-ray luminosities $2 \times 10^{39} \leq L_X \leq 10^{41}$ erg s^{-1}. These sources are generally referred to as ultra-luminous X-ray sources (ULXs); their isotropic luminosities are less that those of bright Seyfert galaxies ($10^{42}–10^{44}$ erg s^{-1}), so they are also sometimes referred to as "intermediate-luminosity X-ray objects."

ULXs do not generally reside in the centers of galaxies, so they are unrelated to low-level AGN activity. They are generally unresolved at the high spatial resolution ($\sim 0\rlap{.}''5$) of *Chandra*. Combined with the fact that many show variability (Fabbiano et al. 2003), this rules out the hypothesis that ULXs are closely spaced aggregates of lower-luminosity sources. A detailed study of the "Antennae" galaxies (Zezas et al. 2002; Zezas & Fabbiano 2002) shows that the large majority do not have radio counterparts. Combined with the observed variability, this rules out that they are young supernovae. Hence, ULXs are believed to be powered by accretion onto a compact object. Bondi accretion from a dense ISM is insufficient to explain the observed luminosities (King et al. 2001), so the accretion is believed to be from a companion star in a binary system. This interpretation is supported by the variability seen in ULXs (in one case there is even evidence for periodicity; Liu et al. 2002) and the fact that some show transitions between hard and soft states (Kubota et al. 2001). These characteristics are commonly seen in Galactic XRBs.

If ULXs are emitting isotropically at the Eddington luminosity, then the accreting objects must be IMBHs with masses in the range $15–1000 M_\odot$. Sub-Eddington accretion or partial emission outside the X-ray band would imply even higher masses. However, the mass cannot be more than $\sim 10^6 M_\odot$, or else the BH would have sunk to the galaxy center through dynamical friction (§ 3.3.1; Kaaret et al. 2001). Either way, the IMBH interpretation of ULXs has several problems (King et al. 2001; Zezas & Fabbiano 2002). There is no known path of double-star evolution that produces a binary of the required characteristics (King et al. 2001). One would need to assume that the IMBH was born isolated, and subsequently acquired a binary companion through tidal capture in a dense environment. This predicts a one-to-one correspondence between ULXs and star clusters, which is not observed. In the Antennae, ULXs are often observed close to, but not coincident with, star clusters. This suggests a scenario in which ULXs are XRBs that have been ejected out of clusters through recoil (Portegies Zwart & McMillan 2000). This precludes an IMBH because the mass would be too large for the binary to be ejected (Miller & Hamilton 2002b).

An alternative to the IMBH interpretation is that ULXs are an unusual class of XRBs. One possibility is that radiation is emitted anisotropically, so that the luminosity is overestimated when assumed isotropic. Mild beaming (King et al. 2001) and a relativistic jet (Körding, Falcke, & Markoff 2002; Kaaret et al. 2003) have both been proposed. It is also possible that ULXs are in fact emitting at super-Eddington rates (Begelman 2002; Grimm, Gilfanov,

& Sunyaev 2002). Observations show that ULXs are often associated with actively star-forming regions or galaxies. The brightest known source resides in the starburst galaxy M82 (Matsushita et al. 2000; Kaaret et al. 2001; Matsumoto et al. 2001), and the merging Antennae galaxy pair has the most known sources in a single system (18 above 10^{39} erg s^{-1}). The association with young stellar populations suggests that ULXs might be related to high-mass XRBs (where "high-mass" refers to the companion). Indeed, optical counterparts reported for ULXs suggest a young star cluster in one case (Goad et al. 2002) and a single O-star in another case (Liu, Bregman, & Seitzer 2002). However, ULXs have also been identified in elliptical galaxies (Colbert & Ptak 2002) and globular clusters (Angelini, Loewenstein, & Mushotzky 2001; Wu et al. 2002), which suggests an association with low-mass XRBs. So it may be that the ULX population encompasses different types of objects, possibly related to Milky Way sources like SS433 and microquasars (King 2002). Outbursts with luminosities similar to those of ULXs have indeed been reported for some Milky Way sources (Revnivtsev et al. 2001; Grimm et al. 2002).

X-ray luminosity functions provide additional information. In both the Antennae and the interacting pair NGC 4485/4490 (Roberts et al. 2002) the luminosity function has constant slope across the luminosity boundary that separates normal XRBs from ULXs. This is not expected if the two classes formed through different evolutionary paths, and hence does not support the IMBH interpretation. However, if normal XRBs and ULXs differ only in beaming fraction then one would also have expected a break in the luminosity function (Zezas & Fabbiano 2002).

The X-ray spectra of ULXs are important as well. Most tend to have hard spectra that are well fit by a so-called multi-color disk black-body model (Makishima et al. 2000). Others are equally well fit by a single power law (Foschini et al. 2002; Roberts et al. 2002). These results are consistent with an association with accreting binaries. The good fits of accretion disk models suggest that the bulk of the emission is not relativistically beamed (Zezas et al. 2002). The inner-disk temperature in the models is of the order of $kT = 1$–2 keV. This is similar to values observed in Galactic microquasars, and is larger than what would naturally be expected for an IMBH (Makishima et al. 2000). On the other hand, Miller et al. (2003) recently found strong evidence for soft components in *XMM-Newton* spectra of the two ULXs in NGC 1313. These soft components are well fit with inner-disk temperatures of ~ 150 eV. Temperature scales with mass as $T \propto M^{-1/4}$, so this was interpreted as spectroscopic evidence that at least in these ULXs the accreting object is an IMBH of $\sim 10^3 M_\odot$.

3.4.3   Galaxy Centers

Dynamical studies of galaxies indicate that they generally have central supermassive BHs and that the BH mass scales with the velocity dispersion of the host spheroid as $M_\bullet \propto \sigma^4$ (§ 3.1). This result is based on data for galaxies with Hubble types earlier than Sbc, $\sigma > 70$ km s^{-1}, and $M_\bullet > 2 \times 10^6 M_\odot$. It is unknown whether the same $M_\bullet$–σ relation holds for later-type and/or dwarf galaxies. If so, then one would expect such galaxies to host IMBHs (owing to their less massive spheroids and correspondingly smaller velocity dispersions). However, no firm detections and mass measurements exist for such galaxies. In fact, it is not guaranteed that such galaxies have central BHs at all. This would provide a natural explanation for the scarcity of AGNs among late-type galaxies (Ho, Filippenko, & Sargent 1997; Ulvestad & Ho 2002). On the other hand, we do know that at least some late-type galaxies host AGNs. The most famous example is NGC 4395, a dwarf galaxy of type Sm, which has the nearest and lowest-luminosity Seyfert 1 nucleus yet found (Filippenko & Sargent 1989). The conventional explanation of Seyfert activity suggests that at least this galaxy must have a central BH. Filippenko & Ho (2003) argue that the BH mass lies in the range $10^4 - 10^5 M_\odot$, which puts it firmly in the intermediate-mass regime.

Dynamical measurements of BH masses in late-type and dwarf galaxies are complicated by the fact that such galaxies generally host a nuclear star cluster of mass 10^6–$10^7 M_\odot$ (Böker et al. 2002). The cluster is often barely resolved at *Hubble Space Telescope (HST)* resolution so that its gravitational influence resembles that of a point mass. This masks the dynamical effect of any BH, unless the BH is at least as massive as the cluster. This is not expected in view of the $M_\bullet$–σ relation, and is indeed generally ruled out by detailed modeling. Böker, van der Marel & Vacca (1999) inferred a BH mass upper limit of $5 \times 10^5 M_\odot$ for the nearby Scd spiral IC 342. Geha, Guhathakurta, & van der Marel (2002) inferred upper limits in the range 10^6–$10^7 M_\odot$ for six dwarf elliptical galaxies in Virgo. The only way to obtain more stringent limits is to study galaxies in the Local Group, for which it is possible to obtain spectroscopic observations that resolve the central star cluster itself. This was done for M33 ($\sigma = 24$ km s^{-1}), with no resulting BH detection. Two independent groups analyzed the same *HST* spectra, and obtained upper limits of $1500 M_\odot$ (Gebhardt et al. 2001) and $3000 M_\odot$ (Merritt, Ferrarese, & Joseph 2001). This is a factor ~ 10 below the value predicted by extrapolation of the $M_\bullet$–σ relation.

3.4.4   *Globular Clusters*

The existence of theoretical scenarios for IMBH formation in dense star clusters (§ 3.2.2) makes it natural to search for IMBHs in globular clusters. This search splits into two questions: does the mass-to-light ratio (M/L) increase toward the center in globular clusters? (§ 3.4.4.1); and can this be explained as a result of normal mass segregation, or must an IMBH be invoked? (§ 3.4.4.2).

3.4.4.1   *Centrally Peaked M/L Profiles in Globular Clusters*

The radial M/L profile of globular clusters is constrained by the observed profile of the line-of-sight velocity dispersion σ (through the equations of hydrostatic equilibrium). For distant clusters one can use integrated light techniques similar to those used for galaxy centers. Gebhardt, Rich, & Ho (2002) performed such a study for the globular cluster G1, the most massive cluster of M31. A constant M/L model cannot fit their *HST* data, and they inferred the presence of $M_d = 2.0^{+1.4}_{-0.8} \times 10^4 M_\odot$ of dark material near the center. The corresponding "sphere of influence" $r_d = GM_d/\sigma^2$ is only $0.''035$, which is less than the *HST* FWHM. Nonetheless, it is plausible that $M_d = 2.0 \times 10^4 M_\odot$ can indeed be detected in G1: it is similar in distance and physical properties to the central star cluster of M33, for which *HST* data yielded an upper limit as small as 1500–$3000 M_\odot$ (§ 3.4.3).

For Milky Way globular clusters, velocity determinations of individual stars are better than integrated light techniques. The cluster M15 has been observed from the ground by many groups (most recently by Gebhardt et al. 2000) and has long been a focus of discussions on IMBHs in globular clusters (as reviewed by van der Marel 2001). A recent *HST* study (van der Marel et al. 2002) added important stars in the central few arcsec of the cluster, yielding a combined sample of ~ 1800 stars with known velocities. The inferred velocity dispersion increases radially inward and cannot be fit with a constant M/L model.

Gerssen et al. (2002) modeled the data and inferred the presence of $M_d = 3.2^{+2.2}_{-2.2} \times 10^3 M_\odot$ of dark material near the center.

There are ~ 70 pulsars known in globular clusters and some of these have a negative period derivative $\dot{P}$. These can be used to constrain the cluster mass distribution. Pulsars are expected to be spinning down intrinsically (positive $\dot{P}$), so negative $\dot{P}$ must be due to acceleration by the mean gravitational field. This places a lower limit on the mass enclosed inside the projected radius R of the pulsar. In M15 there are two pulsars at $R \approx 1''$ (Phinney 1993) whose negative $\dot{P}$ values are consistent with the mass distribution implied by the stellar kinematics (Gerssen et al. 2002).

D'Amico et al. (2002) recently reported two pulsars with negative $\dot{P}$ at $6''$ and $7''$ from the center of the cluster NGC 6752. These suggest a large enclosed mass and a considerable central increase in M/L. However, the inferred masses may be inconsistent with the stellar kinematics of this cluster (Gebhardt 2002, priv. comm.). NGC 6752 is interesting also because it hosts a pulsar at an unusually large distance from the cluster center. It has been suggested that this pulsar may have been kicked there through interaction with an IMBH in the cluster core (Colpi, Possenti, & Gualandris 2002; Colpi, Mapelli, & Possenti 2004).

3.4.4.2   *IMBH versus Mass Segregation*

A natural consequence of two-body relaxation in globular clusters is mass segregation. In an attempt to reach equipartition of energy, heavy stars and dark remnants sink to the center of the cluster, which causes a central increase in M/L. One must model the time evolution of the cluster in considerable detail to determine the theoretically predicted M/L increase. M15 is one of the few clusters for which this has been done. The most recent and sophisticated Fokker-Planck models constructed for M15 are those of Dull et al. (1997). Gerssen et al. (2002) found that the M/L profile published by Dull et al. (1997) did not contain enough dark remnants near the cluster center to fit their *HST* data, which suggested the presence of an IMBH. However, it was subsequently reported that the M/L figure of Dull et al. (1997) contained an error in the labeling of the axes (Dull et al. 2003). A corrected data-model comparison shows that the Fokker-Planck models can provide a statistically acceptable fit to the *HST* data (Gerssen et al. 2003). Baumgardt et al. (2003a) performed direct N-body calculations and reached a similar conclusion.

Although models without an IMBH can fit the kinematical data for M15, this does not necessarily mean that such models are the correct interpretation. The dark remnants that segregate to the cluster center evolve from stars with initial masses $M \geq 3M_\odot$. The evolutionary end-products of such stars are only understood with limited accuracy (Fryer 1999; Claver et al. 2001), and the same is true for their IMF (especially at the low metallicities of globular clusters). Depending on the assumptions that are made on these issues, it is possible to create models that either do or do not fit the M15 data. Most of the neutron stars that form are expected to escape because of kicks received at birth (Pfahl, Rappaport, & Podsiadlowski 2002). The M/L increase from mass segregation is therefore due mostly to white dwarfs with masses $> 1M_\odot$. Such white dwarfs have cooled for too long to be observable in globular clusters, which makes this prediction hard to test. Another caveat is that M15 is known to have considerable rotation near its center. This is not naturally explained by evolutionary models and may hold important new clues to the structure of M15 (Gebhardt et al. 2000). Hence, an IMBH of mass $\leq 2 \times 10^3 M_\odot$ is certainly not ruled out in M15 (Gerssen et al. 2003; Baumgardt et al. 2003a).

There are no X-rays observed from the center of M15. Ho, Terashima, & Okajima (2003) find $L_X/L_{Edd} \leq 4 \times 10^{-9}$. This does not imply that there cannot be an IMBH. In globular clusters there is only a limited gas supply available for accretion (Miller & Hamilton 2002b) and an advection-dominated accretion flow (Narayan, Mahadevan & Quataert 1998) can naturally lead to very low values of L_X/L_{Edd}. The galaxy M32, which has a well-established supermassive BH, has an upper limit $L_X/L_{Edd} < 10^{-7}$ (van der Marel et al. 1998).

Scaling of the Dull et al. (1997, 2003) M15 models to the mass, size and distance of G1 does not yield a sufficient concentration of dark remnants to fit its *HST* data (Gebhardt 2002, priv. comm.). Hence, G1 may well contain an IMBH in its center, as suggested by Gebhardt et al. (2002). If true, this need not necessarily be representative for globular clusters in general. G1 is unusually massive, and it has been suggested to be the nucleus of a disrupted dwarf galaxy (Meylan et al. 2001). Either way, a simple scaling of the Dull et al. models to the case of G1 is likely to be an oversimplification. Baumgardt et al. (2003b) performed *N*-body calculations and argued that the G1 data can be explained without an IMBH. However, the proper scaling of these calculations with $N \approx 7 \times 10^4$ particles to the case of G1 (with $N \approx 10^7$ stars) is uncertain. The same argument applies to the Baumgardt et al. (2003a) models for M15; improved modeling of both clusters remains highly desirable.

It is intriguing that the BH mass detections/upper limits suggested for G1 and M15 fall right on the $M_\bullet$–σ relation for supermassive BHs. This leaves open the possibility that there may be some previously unrecognized connection between the formation and evolution of globular clusters, galaxies and central BHs.

3.4.5 *Gravitational Waves*

In the near future, gravitational wave detection experiments such as LIGO and LISA will provide a new way to probe the possible existence of IMBHs. Binary systems of compact objects and mergers of supermassive BHs are already well known as possible sources of gravitational radiation. Miller (2002) recently emphasized that a population of IMBHs could also be observable, especially if they reside in dense stars clusters. With optimistic assumptions, LIGO could see the coalescence of a stellar-mass BH with an IMBH up to several tens of times per year.

3.5 Concluding Remarks

The main conclusion to emerge from this review is that the existence of IMBHs in the Universe is not merely a remote possibility. IMBHs have been predicted theoretically as a natural result of several realistic scenarios. In addition, it has been shown that IMBHs might plausibly explain a variety of recent observational findings. Much progress has been made in the last few years, but certainly, even more work remains to be done. None of the theoretical arguments for IMBH formation are unique. Many alternative theoretical scenarios exist that do not lead to IMBHs. Similarly, none of the observational suggestions for IMBHs are clear cut. Alternative interpretations of the data exist that invoke known classes of objects and many would argue that such conservative interpretations are more plausible. Either way, these issues can only be addressed and resolved with additional research. IMBHs are therefore likely to grow as a focus of research attention.

References

Abel, T., Bryan, G., & Norman, M. 2000, ApJ, 540, 39

Adams, F. C., Graff, D. S., & Richstone, D. O. 2001, ApJ, 551, L31

Afshordi, N., McDonald, P., & Spergel, D. N. 2003, ApJ, 594, L71

Agol, E., Kamionkowski, M., Koopmans, L. V. E., & Blandford, R. D. 2002, ApJ, 576, L131

Alcock, C., et al. 2000, ApJ, 542, 281

——. 2001, ApJ, 550, L169

Angelini, L., Loewenstein, M., & Mushotzky, R. F. 2001, ApJ, 557, L35

Barrow, J. D., & Carr, B. J. 1996, Phys. Rev. D, 54, 3920

Baumgardt, H., Hut, P., Makino, J., McMillan, S., & Portegies Zwart, S. 2003a, ApJ, 582, L21

Baumgardt, H., Makino, J., Hut, P., McMillan, S., & Portegies Zwart, S. 2003b, ApJ, 589, L25

Baumgarte, T. W., & Shapiro, S. L. 1999, ApJ, 526, 941

Begelman, M. C. 2002, ApJ, 568, L97

Begelman, M. C., & Rees, M. J. 1978, MNRAS, 185, 847

——. 1998, Gravity's Fatal Attraction (New York: Scientific American Lib.)

Bennett, D. P., et al. 2002, ApJ, 579, 639

Bennett, D. P., Becker, A. C., Calitz, J. J., Johnson, B. R., Laws, C., Quinn, J. L., Rhie, S. H., & Sutherland, W. 2004, ApJ, submitted (astro-ph/0207006)

Binney, J., & Tremaine, S. 1987, Galactic Dynamics (Princeton: Princeton Univ. Press)

Böker, T., Laine, S., van der Marel, R. P., Sarzi, M., Rix, H.-W., Ho, L. C., & Shields, J. C. 2002, AJ, 123, 1389

Böker, T., van der Marel, R. P., & Vacca, W. D. 1999, AJ, 118, 831

Bond, J. R., Arnett, W. D., & Carr, B. J. 1984, ApJ, 280, 825

Bromm, V., & Loeb, A. 2003, ApJ, in press (astro-ph/0212400)

Brown, G. E., & Bethe, H. A. 1994, ApJ, 423, 659

Burkert, A., & Silk, J. 2001, ApJ, 554, L151

Burles, S., Nollett, K., & Turner, M. S. 2001, ApJ, 552, L1

Carr, B. J. 1975, ApJ, 201, 1

Carr, B. J. 1994, ARA&A, 1994, 32, 531

Carr, B. J., & Lidsey, J. E. 1993, Phys. Rev. D, 48, 543

Carr, B. J., & Sakellariadou, M. 1999, ApJ, 516, 195

Charles, P., 2001, in Black Holes in Binaries and Galactic Nuclei, ed. L. Kaper, E. P. J. van den Heuvel, & P. A. Woudt (New York: Springer), 27

Claver, C. F., Liebert, J., Bergeron, P., & Koester, D. 2001, ApJ, 563, 987

Colbert, E. J. M., & Mushotzky, R. F. 1999, ApJ, 519, 89

Colbert, E. J. M., & Ptak, A. F. 2002, ApJS, 143, 25

Colpi, M., Mapelli, M., & Possenti, A. 2004, Carnegie Obs. Astrophysics Series, Vol. 1: Coevolution of Black Holes and Galaxies, ed. L. C. Ho (Pasadena: Carnegie Observatories, http://www.ociw.edu/ociw/symposia/series/symposium1/proceedings.html)

Colpi, M., Possenti, A., & Gualandris, A. 2002, ApJ, 570, L85

D'Amico, N., Possenti, A., Fici, L., Manchester, R. N., Lyne, A. G., Camilo, F., & Sarkissian, J. 2002, ApJ, 570, L89

Dull, J. D., Cohn, H. N., Lugger, P. M., Murphy, B. W., Seitzer, P. O., Callanan, P. J., Rutten, R. G. M., & Charles, P. A. 1997, ApJ, 481, 267

——. 2003, ApJ, 585, 598 (Addendum to Dull et al. 1997)

Ebisuzaki, T., et al. 2001, ApJ, 562, L19

Fabbiano, G. 1989, ARA&A, 27, 87

Fabbiano, G., Zezas, A., King, A. R., Ponman, T. J., Rots, A., & Schweizer, F. 2003, ApJ, 584, L5

Fan, X., et al. 2001, AJ, 122, 2833

Filippenko, A. V., & Ho, L. C. 2003, ApJ, 588, L13

Filippenko, A. V., & Sargent, W. L. W. 1989, ApJ, 342, L11

Foschini, L., et al. 2002, A&A, 392, 817

Fryer, C. L. 1999, ApJ, 522, 413

Fryer, C. L., & Kalogera, V. 2001, ApJ, 554, 548

Fryer, C. L., Woosley, S. E., & Heger, A. 2001, ApJ, 550, 372

Fukugita, M., Hogan, C. J., & Peebles, P. J. E. 1998, ApJ, 503, 518

Garcia-Bellido, J., & Linde, A. 1998, Phys. Rev. D, 57, 6075

Gebhardt, K., et al. 2001, AJ, 122, 2469

Gebhardt, K., Pryor, C., O'Connell, R. D., Williams, T. B., & Hesser, J. E. 2000, AJ, 119, 1268

Gebhardt, K., Rich, R. M., & Ho, L. 2002, ApJ, 578, L41

Geha, M., Guhathakurta, P., & van der Marel, R. P. 2002, AJ, 124, 3073

Gerssen, J., van der Marel, R. P., Gebhardt, K. Guhathakurta, P., Peterson, R. C., & Pryor, C. 2002, AJ, 124, 3270

——. 2003, AJ, 125, 376 (Addendum to Gerssen et al. 2002)

Ghez, A. M. 2004, in Carnegie Observatories Astrophysics Series, Vol. 1: Coevolution of Black Holes and Galaxies, ed. L. C. Ho (Cambridge: Cambridge Univ. Press), in press

Goad, M. R., Roberts, T. P., Knigge, C., & Lira, P. 2002, MNRAS, 335, L67

Grimm, H.-J., Gilfanov, M., & Sunyaev, R. 2002, A&A, 391, 923

Han, C., & Gould, A. 1996, ApJ, 467, 540

Haehnelt, M., & Kauffmann, G. 2000, MNRAS, 318, L35

Haehnelt, M., Natarajan, P., & Rees, M. J. 1998, MNRAS, 300, 817

Haehnelt, M., & Rees, M. J. 1993, MNRAS, 263, 168

Haiman, Z., & Loeb, A. 2001, 552, 459

Harris, W. E. 1996, AJ, 112, 1487

Hawke, I., & Stewart, J. M. 2002, Class. Quant. Grav., 19, 3687

Hawking, S. W., Moss I. G., & Stewart, J. M. 1982, Phys. Rev. D, 26, 2681

Heger, A., & Woosley, S. E. 2001, ApJ, 567, 532

Heggie, D. C. 1975, MNRAS, 173, 729

Ho, L. C., Filippenko, A. V., & Sargent, W. L. W. 1997, ApJ, 487, 568

Ho, L. C, Terashima, Y., & Okajima, T. 2003, ApJ, L35

Hughes, S. A., & Blandford, R. D. 2003, ApJ, 585, L101

Hut, P., et al. 1992, PASP, 104, 981

Ipser, J. R., & Price, R. H. 1977, ApJ, 216, 578

Islam, R. R., Taylor, J. E., & Silk, J. 2003, MNRAS, 340, 647

Jedamzik, K. 1997, Phys. Rev. D., 55, R5871

Kaaret, P., Corbel, S., Prestwich, A. H., & Zezas, A. 2003, Science, 299, 365

Kaaret, P., Prestwich, A. H., Zezas, A. L., Murray, S. S., Kim, D.-W., Kilgard, R. E., Schlegel, E. M., & Ward, M. J. 2001, MNRAS, 321, L29

Kauffmann, G., & Haehnelt, M. 2000, MNRAS, 311, 576

King, A. R. 2002, MNRAS, 335, L13

King, A. R., Davies, M. B., Ward, M. J., Fabbiano, G., & Elvis, M. 2001, ApJ, 552, L109

Klessen, R., & Burkert, A. 1995, MNRAS, 280, 735

Körding, E., Falcke, H., & Markoff, S. 2002, A&A, 382, L13

Kormendy, J., & Gebhardt, K. 2001, in The 20th Texas Symposium on Relativistic Astrophysics, ed. H. Martel & J. C. Wheeler (New York: AIP), 363

Kubota, A., Mizuno, T., Makishima, K., Fukazawa, Y., Kotoku, J., Ohnishi, T., & Tashiro, M. 2001, ApJ, 547, L119

Kulkarni, S. R., Hut, P., & McMillan, S. 1993, Nature, 364, 421

Lacey, C. G., & Ostriker, J. P. 1985, ApJ, 299, 633

Larson, R. 2003, in Galactic Star Formation Across the Stellar Mass Spectrum, ed. J. M. De Buizer & N. S. van der Bliek (San Francisco: ASP), 65

Lee, H. M. 1987, ApJ, 319, 801

——. 1995, MNRAS, 272, 605

Lee, M. H. 1993, ApJ, 418, 147

Liu, J.-F., Bregman, J. N., Irwin, J., & Seitzer, P. 2002, ApJ, 581, L93

Liu, J.-F., Bregman, J. N., & Seitzer, P. 2002, ApJ, 580, L31

MacGibbon, J. H., Brandenburger, R. H., & Wichowski, U. F. 1998, Phys. Rev. D, 57, 2158

MacGibbon, J. H., & Carr, B. J. 1991, ApJ, 371, 447

Madau, P., & Rees, M. J. 2001, ApJ, 551, L27

Makishima, K., et al. 2000, ApJ, 535, 632

Mao, S., et al. 2002, MNRAS, 329, 349

Matsumoto, H., Tsuru, T., Koyama, K., Awaki, H., Canizares, C. R., Kawai, N., Matsushita, S., & Kawabe, R. 2001, ApJ, 547, L25

Matsushita, K., Kawabe, R., Matsumoto, H., Tsuru, T. G., Kohno, K., Morita, K.-I., Okumura, S. K., & Vila-Vilaro, B. 2000, ApJ, 545, L107

Merritt, D., Ferrarese, L., & Joseph, C. 2001, Science, 293, 1116

Meylan, G., Sarajedini, A., Jablonka, P., Djorgovski, S. G., Bridges, T., & Rich, R. M. 2001, AJ, 122, 830

Miller, J. M., Fabbiano, G., Miller, M. C., & Fabian, A. C. 2003, ApJ, 585, L37

Miller, M. C. 2002, ApJ, 581, 438

Miller, M. C., & Hamilton, D. P. 2002a, ApJ, 576, 894

——. 2002b, MNRAS, 330, 232

Mirabel, I. F., Mignani, R., Rodrigues, I., Combi, J. A., Rodriguez, L. F., & Guglielmetti, F. 2002, A&A, 395, 595

Mouri, H., & Taniguchi, Y. 2002, ApJ, 566, L17

Murali, C., Arras, P., & Wasserman, I. 2000, MNRAS, 313, 87

Murphy, B. W., Cohn, H. N., & Durisen, R. H. 1991, ApJ, 370, 60

Narayan, R., Mahadevan, R., & Quataert, E. 1998, in The Theory of Black Hole Accretion Disks, ed. M. Abramowicz, G. Bjö;rnsson, & J. E. Pringle (Cambridge: Cambridge Univ. Press), 148

Nelemans, G., Tauris, T. M., & van den Heuvel, E. P. J. 1999, A&A, 352, L87

Ostriker, J. P., Binney, J., & Saha, P. 1989, MNRAS, 241, 849

Paczyński, B. 1986, ApJ, 304, 1

Persic, M., & Salucci, P. 1992, MNRAS, 258, 14p

Pfahl, E., Rappaport, S., & Podsiadlowski, P. 2002, ApJ, 573, 283

Phinney, E. S. 1993, in Structure and Dynamics of Globular Clusters, ed. G. Djorgovski & G. Meylan (San Francisco: ASP), 141

Portegies Zwart, S. F., & McMillan, S. L. W. 2000, ApJ, 528, L17

——. 2002, ApJ, 576, 899

Quinlan, G. D., & Shapiro, S. L. 1987, ApJ, 321, 199

——. 1989, ApJ, 343, 725

——. 1990, ApJ, 356, 483

Rees, M. J. 1984, ARA&A, 22, 471

Revnivtsev, M., Sunyaev, R., Gilfanov, M., & Churazov, E. 2002, A&A, 385, 904

Rix, H.-W., & Lake, G. 1993, ApJ, 417, L1

Roberts, T. P., & Warwick, R. S., 2000, MNRAS, 315, 98

Roberts, T. P., Warwick, R. S., Ward, M. J., & Murray, S. S. 2002, MNRAS, 337, 677

Rubin, S. G., Khlopov, M. Yu., & Sakharov, A. S. 2000, Grav. Cosmol., S6, 1

Schneider, R., Ferrara, A., Natarajan, P., & Omukai, K. 2002, ApJ, 571, 30

Schödel, R., et al. 2002, Nature, 419, 694

Shapiro, S. L., & Teukolsky, S. A. 1985, ApJ, 292, L41

Shibata, M., & Shapiro, S. L. 2002, ApJ, 572, L39

Sigurdsson, S., & Hernquist, L. 1993, Nature, 364, 423

Silk, J., & Rees, M. J. 1998, A&A, 331, L1

Spitzer, L., Jr. 1969, ApJ, 158, L139

Taniguchi, Y., Shioya, Y., Tsuru, T. G., & Ikeuchi, S. 2000, PASJ, 52, 533

Tremaine, S., et al. 2002, ApJ, 574, 740

Tremaine, S., & Ostriker, J. P. 1999, MNRAS, 306, 662

Ulvestad, J. S., & Ho, L. C. 2002, ApJ, 581, 925

van der Marel, R. P. 2001, in Black Holes in Binaries and Galactic Nuclei, ed. L. Kaper, E. P. J. van den Heuvel, & P. A. Woudt (New York: Springer), 246

van der Marel, R. P., Cretton, N., de Zeeuw, P. T., & Rix, H.-W. 1998, ApJ, 493, 613

van der Marel, R. P., Gerssen, J., Guhathakurta, P., Peterson, R. C., & Gebhardt, K. 2002, AJ, 124, 3255

Vishniac, E. T. 1978, ApJ, 223, 986

Volonteri, M., Haardt, F., & Madau, P. 2003, ApJ, 582, 559

Wu, H., Xue, S. J., Xia, X. Y., Deng, Z. G., & Mao, S. 2002, ApJ, 576, 738

Xu, G., & Ostriker, J. P. 1994, ApJ, 437, 184

Yu, Q., & Tremaine, S. 2002, MNRAS, 335, 965

Zezas, A., & Fabbiano, G. 2002, ApJ, 577, 726

Zezas, A., Fabbiano, G., Rots, A. H., & Murray, S. S. 2002, ApJ, 577, 710

4

The supermassive black hole at the center of the Milky Way

ANDREA M. GHEZ
University of California, Los Angeles

Abstract

Within the last year two results dramatically strengthened the case for a supermassive black hole at the center of the Milky Way and our understanding the black hole's effect on its environment. First, orbital solutions for multiple stars have been obtained, using adaptive optics and speckle imaging to study the motions of stars in the plane of sky over the past decade. The three most remarkable orbits are those of the newly identified stars S0-16, which passed a mere 60 AU from the central dark mass at a velocity of 9,000 km s^{-1} in 2000, and S0-19, which at one time was posited to be the near-infrared counterpart to Sgr A*, and the previously recognized S0-2, which has an orbital period of a mere 16 years. A simultaneous orbital analysis yields the most accurate and precise values of the location and mass of the Galaxy's central dark matter. The location is consistent with the inferred infrared position of Sgr A* but with an uncertainty that is a factor of 7 smaller ($\pm$ 1.5 milliarcsec). The estimated mass from orbital motion is $3.7(\pm0.4) \times 10^6 (\frac{R_0}{8\,\mathrm{kpc}})^3 M_\odot$ and is a more direct measure of mass than that obtained from velocity dispersion measurements, which are as much as a factor of 2 times smaller; this brings the Milky Way into better agreement with the $M_\bullet - \sigma$ relationship. Furthermore, by confining the mass to within a radius of a mere 0.0003 pc or 1,000 R_{Sch}, this orbital analysis increases the inferred dark mass density by four orders of magnitude compared to earlier analyses based on velocity and acceleration vectors, making the Milky Way the strongest existing case for a supermassive black hole at the center of any normal type galaxy.

Second, with the introduction of an adaptive-optics-fed spectrometer, the first detection of spectral absorption lines in one of the high-velocity stars, S0-2, was obtained one month after its closest approach to the Galaxy's central supermassive black hole. Both Br γ $\lambda 2.1661$ μm and He I $\lambda 2.1126$ μm are seen in absorption with equivalent widths and an inferred stellar rotational velocity that are consistent with that of an O8-B0 dwarf, which suggests that S0-2 is a massive ($\sim$15 $M_\odot$), young (<10 Myr) main-sequence star. Similarly, the lack of CO detected in several of the other high-velocity stars suggests that they are also young. This presents a major challenge to star formation theories, given the strong tidal forces that prevail over all distances reached by these stars in their current orbits and the difficulty in migrating these stars inward during their lifetime from farther out where tidal forces should no longer preclude star formation.

© The Observatories of the Carnegie Institution of Washington 2004.

4.1   Introduction

While the Milky Way was neither the first nor the most obvious place to search for a supermassive black hole, the case for one at the center of the Galaxy is quickly becoming the most iron clad. The first hint of a central concentration of dark matter came from radial velocity measurements of ionized gas located in a three-armed structure known as the mini-spiral, which extends from the center out to about 1–2 pc (Lacy et al. 1980). Concerns that the gases' motions were not tracing the gravitational potential were quickly allayed by radial velocity measurements of stars, which are not susceptible to nongravitational forces (McGinn et al. 1989; Haller et al. 1996; Genzel et al. 1997). These early dynamical measurements of the gas and stars suggested the presence of $3 \times 10^6 M_\odot$ of dark matter and confined it to within a radius of ~0.1 pc. The implied dark matter density was not sufficiently high to definitively claim this as evidence for a single supermassive black hole, since the measurements imposed a lifetime for clusters of dark objects that was not significantly shorter than the age of the Galaxy (Maoz 1998). To make further progress in understanding the underlying source of dark matter at the center of the Galaxy, it was necessary to use techniques that compensated for the distorting effects of the Earth's atmosphere, which had restricted the earlier studies of the dark matter distribution to radii of 0.1 pc or larger.

In the early- to mid-1990's, two independent groups initiated 2 μm high spatial resolution imaging studies of the central stellar cluster to measure the motions of stars in the plane of the sky. In 1992, the ESO team began their program using speckle imaging at the 3.6 m NTT and in 2002 have moved to the adaptive optics system on the 8 m VLT. The Keck team initiated their program using speckle imaging on the 10 m Keck I telescope in 1995 and began using adaptive optics on the 10 m Keck II telescope in 1999. The first phase of these experiments yielded proper motion velocities, which increased the implied dark matter density by 3 orders of magnitude to $10^{12} M_\odot/pc^3$ (Eckart & Genzel 1997; Ghez et al. 1998). This eliminated a cluster of dark objects, such as neutron stars or stellar mass black holes, as a possible explanation of the Galaxy's central dark mass concentration (Maoz 1998) and left only the fermion ball hypothesis (e.g., Tsiklauri & Viollier 1998; Munyaneza & Viollier 2002) as an alternative to a single supermassive black hole. The velocity dispersion measurements also localized the dark matter to ± 100 mas (4 milli-pc) at a position consistent with the nominal location of the unusual radio source Sgr A* (Ghez et al. 1998), whose emission is posited to arise from accretion onto a central supermassive black hole (e.g., Lo et al. 1985). The proper motion experiments proceeded to strengthen both the case for a supermassive black hole and its association with Sgr A* with the detection of acceleration for three stars — S0-1, S0-2, and S0-4 — which increased the dark matter density to $10^{13} M_\odot/pc^3$ and positional accuracy to ± 30 mas (Ghez et al. 2000; Eckart et al. 2002). These experiments also revealed that the orbital periods for S0-2 and S0-1 could be as short as 15 and 35 years, respectively, which would open a new arena for dynamical studies of the central stellar cluster.

This paper summarizes the recent progress that has been made in this field on two fronts. First, with approximately a decade of proper motion measurements, is the derivation of complete 3-dimensional orbits for multiple stars that are making close approaches to the supermassive black hole at the center of the Milky Way. Second is the measurement of spectral lines in one of these high-velocity stars. These two steps forward make the strongest case yet for the presence of a supermassive black hole at the center of the Galaxy and, for the first time, allow us to take an in-depth look at the question of where these stars formed.

4.2   Proper Motion Measurements

Diffraction limited near-infrared imaging programs have published significant proper motions (S/N > 5) for 18 stars, which are located within 3″ of Sgr A* (Eckart & Genzel 1997; Ghez et al. 1998; Genzel et al. 2000). An additional ~80 stars have reported proper motion values, although ~50 of these have S/N < 3 and thus should be considered upper limits. Since the last reporting of stellar proper motions, significant progress has been made. The velocity estimates have been improved as more data have been taken, which has increased the time baseline of these studies by factors of 3.5 and 1.4 for the Keck and ESO programs, respectively. This has allowed proper motion velocity uncertainties of ~15 km s^{-1} to be obtained for the Keck data set. Furthermore, the sensitivity to additional proper motion sources also has been improved, due, in the case of the Keck program, to three separate factors. First, the original approach was conservative in an effort to avoid falsely identifying high-velocity stars on the basis of the three maps, each of which is separated by a full year (obtained in 1995, 1996, and 1997). With multiple observations each year beginning in 1998, it was clear that many real sources were not being identified, leading to a decrease in the choice of threshold used to identify sources. Second, in 1998 it became possible to calibrate the alignment of the Keck telescope's mirror segments on NIRC, the Keck I facility infrared camera, allowing the telescope to be calibrated during the observing run at the elevation of the Galactic Center, thereby improving the image quality of the speckle maps. Lastly, beginning in 1999 adaptive optics images have been obtained, which increased the sensitivity to fainter sources. Within the central 1″×1″ (R = 0.″5)* alone, the number of proper motion sources has quadrupled (Ghez et al. 2004). In this review, we focus on six of the most dynamically important proper motion stars, of which three are new (S0-16, S0-19, and S0-20) and three were previously known (S0-1, S0-2, and S0-4). Figure 4.1 identifies these sources in a cleaned image from 2001, and Figure 4.2 shows their proper motions.

In terms of source confusion, S0-19 is of particular note. Its large proper motion has caused it to be misidentified in earlier papers. While it was detected by Ghez et al. (1998) in 1995 as a K = 14.0 mag source, two possible counterparts were identified in 1996. With limited time coverage, it was not possible to definitively identify either as the correct counterpart, and it was not included in the proper motion sample. The same source was reported as S3 (K = 15 mag) moving westward by Eckart & Genzel (1997) and Genzel et al. (2000). Genzel et al. (1997) report the detection of an apparently new source, S12 (K ≈ 15 mag), in their 1996.43 image, located ~0.″1 W from S3 and proposed it as the best candidate for the infrared counterpart of the compact radio source Sgr A*. This has been used in several recent papers to constrain models of Sgr A*'s flared state (e.g., Narayan 2002). In Ghez et al. (2003b,c), it is clear that S12 is simply a high velocity star that was coincident with Sgr A* in 1996, which we label S0-19. The discrepancy in magnitudes arises from the difficulties of carrying out accurate photometry in such a crowded region. This source illustrates the challenges associated with making a definitive detection of infrared emission associated with Sgr A*, given the high stellar densities and velocities and modest stellar intensity variations in this region (see Hornstein et al. 2002 for a detailed discussion of near-infrared limits on Sgr A*).

The naming convention used here was introduced by Ghez et al. (1998) and was

---

* Both proper motion studies cover regions that are significantly larger than 1″×1″.

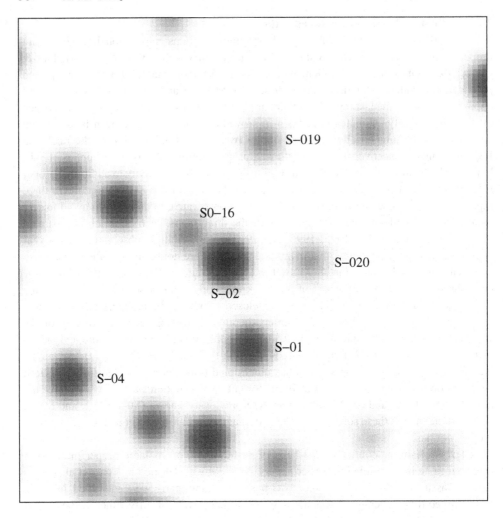

Fig. 4.1. A $1'' \times 1''$ cleaned image centered on the nominal position of Sgr A* showing the 2001 positions of some of the stars that have been followed over the course of the Keck proper motion study. Three of the newly identified stars are S0-16, S0-19, and S0-20.

designed to directly convey relevant information about the location of the source relative to the position of Sgr A*. Originally the Sgr A* position given by Menten et al. (1997) was adopted, and the surrounding field was divided into concentric arcsecond-wide annuli centered on Sgr A*. Stars lying within the central circle, which has radius $1''$, were given names S0-1, S0-2, S0-3, etc. Stars lying in the annulus between radii of $1''$ to $2''$ were given the names S1-1, S1-2, and so on. The number immediately following "S" thus refers to the inner radius of the annulus in which the star lies. The number following the hyphen was ordered in the sense of increasing distance from Sgr A* within each annulus. In this scheme, newly identified sources are named by incrementing the number following the hyphen within each annulus and ordered in the sense of increasing distance from Sgr A*. Since the original list within $1''$ ended at 15, the newly identified stars begin with 16. S0-16, S0-17, and S0-18

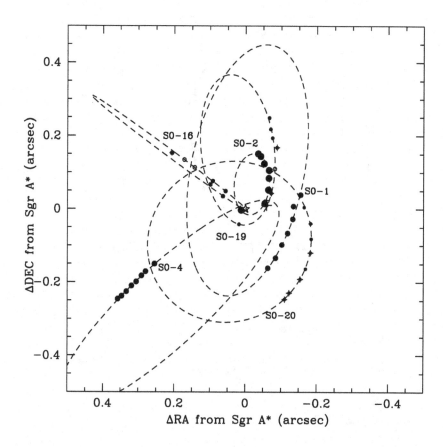

Fig. 4.2. The annual positions of the stars labeled in Fig. 4.1, which have orbital solutions. While several measurements are obtained each year and are critical to the orbital fits at the time of closest approach, the use of annual averages helps to depict the passage of time in this figure. Each star is labeled near its first measurement. (Adapted from Ghez et al. 2004.)

were labeled by Gezari et al. (2002), and S0-19 and S0-20 were first presented at the Rees Symposium "Making Light of Gravity," held in Cambridge, England (July 2002). Due to the motions of stars, the current distance rank does not necessarily match the one at the time of discovery.

4.3 Spectral Line Measurements

In 2002, the high proper motion star, S0-2, was observed with NIRC2, the W. M. Keck II 10 m telescope facility near-infrared adaptive-optics instrument (Matthews et al. 2004) in a mode that achieved a spectral resolution of $R \approx 4{,}000$ ($\sim$75 km s^{-1}). The resulting spectrum of S0-2, shown in Figure 4.3, has two identifiable spectral lines. These are both seen in absorption and are identified as the H I (4-7) or Brγ line at 2.166 μm and the He I triplet at 2.1126 μm ($3p\,^3P^0 - 4s\,^3S$), which is a blend of three transitions at 2.11274,

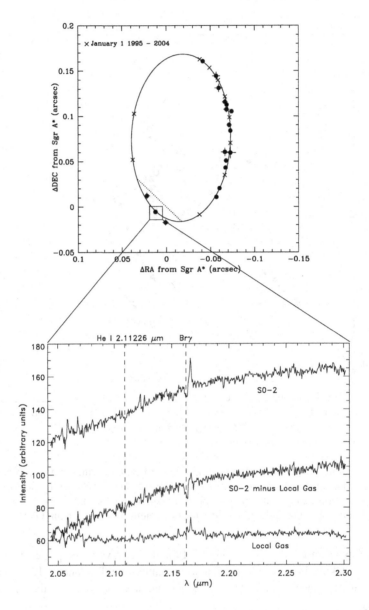

Fig. 4.3. In the lower panel is the first spectrum of S0-2 to show detectable photospheric absorption lines (Brγ and He I λ2.1126 μm). The final spectrum (middle) is the raw spectrum (top; with only an instrumental background removed) minus a local sky (bottom). The horizontal dimension has been re-binned by a factor of two for display purposes only. The vertical lines are drawn at 2.10878 and 2.16230 μm, which correspond to the locations of Brγ and He I for a V_{LSR} of −513 km s^{-1}. This spectrum was obtained in 2000 June at the same time as one of the proper motion measurements shown in the upper panel (filled circles). The crosses in the upper panel mark January 1 of each year between 1995 and 2004 for the best-fit orbit solution (solid line), which is based on both the radial velocity and proper motions. The dotted line is the line of nodes, which reveals S0-2 to be behind the black hole for a mere ∼0.5 years out of its 15-year orbit. (Adapted from Ghez et al. 2003a.)

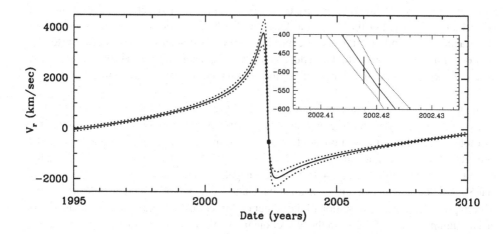

Fig. 4.4. Measured radial velocities along with predicted radial velocities. The solid curve comes from the best-fit orbit and the dotted curves display the range for the orbital solutions allowed with the present data sets, which include two radial velocity measurements separated by one night. The inset panel shows a detailed view of the plot centered on the two observations, which differ by 40 km s^{-1}. (Adapted from Ghez et al. 2003a.)

2.11267, and 2.11258 μm. The detailed properties of these two lines, which are obtained by fitting the background continuum over the whole spectrum with a low-order polynomial and fitting the lines with a Gaussian profile, are reported in Ghez et al. (2003a).

4.4   Discussion

4.4.1   Dynamics

Strong deviations from linear motions on the plane of the sky are observed for several stars that come within 0."2 of Sgr A*, allowing detailed orbital solutions to be carried out. The first orbital solutions reported by Ghez et al. (2000) and Eckart et al. (2002) for S0-1, S0-2, and S0-4, have the following assumptions made about the central dark mass: (1) it is confined to a point source, (2) its distance is 8 kpc (Reid 1993), (3) it has no significant velocity with respect to the Galaxy, which is supported by the lack of motion detected for Sgr A* by Reid et al. (1999) and Reid (2003) in the plane of the sky (<7 km s^{-1}), (4) its position (x_0 and y_0; the center of attraction) is coincident with Sgr A*, and (5) it has a mass $M = 2.6 \times 10^6 M_\odot$, based on velocity dispersion measurements reported by Genzel et al. (1997) and Ghez et al. (1998). This leaves six classical binary star parameters to be solved for from the proper motion data: period P^*, eccentricity e, time of periapse passage T_0, angle of nodes to periapse ω, angle of the line of nodes Ω, and inclination i. This is a well-determined problem as soon as acceleration is well measured.

Since these initial orbital estimates, more significant curvature in the proper motions of the stars has been detected, allowing less constrained orbital solutions to be obtained. For the

---

* With M fixed, semi-major axis A is not an independent variable.

brightest star in the central stellar cluster, S0-2[†], Schödel et al. (2002) add one additional parameter to the orbit fit, by dropping the mass assumption, and derive a central mass of $3.7(\pm 1.5) \times 10^6 M_\odot$. For the same star, Ghez et al. (2003a) drop both the mass and center of attraction assumptions and find a central mass that is consistent with that reported by Schödel et al. and, despite the two additional free parameters introduced by fitting for the center of attraction, obtain uncertainties of the orbital parameters that are reduced by a factor of 2–3[‡].

Inclusion of radial velocity measurements for S0-2 presented by Ghez et al. (2003a) breaks the ambiguity in the inclination angle. With proper motion data alone, only the absolute value of the inclination angle can be determined, leaving the questions of the direction of revolution and when the star is located behind the black hole unresolved. Radial velocity measurements (see Fig. 4.4) indicate a negative inclination angle and consequently that S0-2 is both counter-revolving against the Galaxy and behind the black hole at the time of periapse. The improved location of the center of attraction from the orbital analysis results in a minimum offset of S0-2 from the black hole in the plane of the sky of 11 ± 2 milliarcsec, which is significantly larger than the expected Einstein radius ($\theta_E = 0.42$ milliarcsec, for the S0-2 distance behind the black hole of ~ 100 AU) and therefore makes gravitational lensing an unlikely event (Wardle & Yusef-Zadeh 1992; Jaroszyński 1998a; Alexander & Loeb 2001).

In addition to S0-2, both S0-16 and S0-19 have sufficient curvature to be fit independently with a nine-parameter binary model, with the requirement that the total mass be determined with a S/N greater than 3 (Ghez et al. 2003b,c). The existence of multiple orbits is important for testing the initial assumption that the central dark mass is well modeled by a point source. Measurement of significant curvature in only one star may be equivalently well modeled by a point source and an extended mass distribution. This ambiguity can be easily understood by considering the case of a constant mass density distribution, which has closed elliptical orbit solutions but has the center of attraction located at the ellipse's center rather than at one of its focii (Binney & Tremaine 1987). The agreement between the center of attractions (Fig. 4.5) and total mass (Table 4.1 and Fig. 4.6) for these three independent orbit fits supports the use of a Keplerian orbit model, suggesting that it would be valid to assume a common center of attraction and central mass.

Fitting the orbital motions of these stars simultaneously with a common center of attraction and central mass yields more precise values than those obtained from the independent orbits. Compared to the nominal infrared position of Sgr A*, the new estimate of the center of attraction is a factor of 7 more precise (± 1.5 milliarcsec) and located within the uncertainties of the former. The central dark mass is estimated to be $3.7(\pm 0.4) \times 10^6 (\frac{R_0}{8\,\mathrm{kpc}})^3 M_\odot$. This mass is somewhat higher than that inferred from the velocity dispersion measurements; in particular, it is a factor of 2 higher than that estimated by a nonparametric approach presented by Chakrabarty & Saha (2001) and somewhat less discrepant with the parametric approaches (e.g., Ghez et al. 1998; Genzel et al. 2000). In principle, the difference could provide a handle on the level of anisotropy and the number density distribution of the central cluster (Genzel et al. 2000). However, it is worth recalling that the majority of velocity estimates are in fact limits (see earlier discussion of proper

---

† S0-2 is referred to as S2 by Schödel et al. (2002).
‡ The smaller orbital parameter uncertainties are primarily due to the higher astrometric accuracy of the Keck data set, rather than the inclusion of a radial velocity measurement.

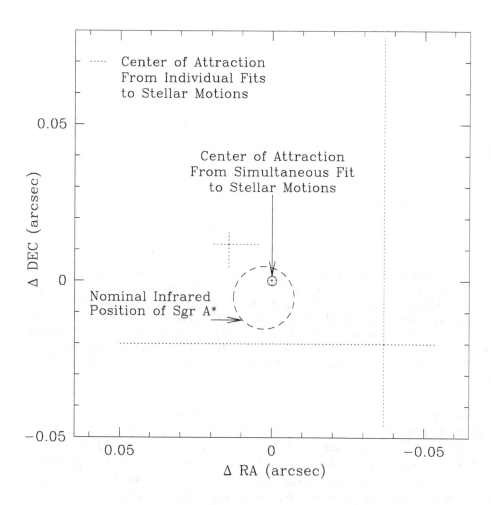

Fig. 4.5. The locations of the center of attraction from the orbital analysis of the proper motion data sets. The three independent orbits for S0-2, S0-16, and S0-19 yield consistent locations, which suggests that the central mass can indeed be modeled as a point source. A simultaneous fit to all three stars pinpoints the Galaxy's dynamical center to a location that is consistent with the nominal infrared position of Sgr A*, but with an uncertainty of a mere 1.5 milliarcsec. All positions are shown with 1 σ uncertainties.

motion measurements), which means that the intrinsic velocity dispersion can be altered significantly if the uncertainties are not well estimated (this affects the bias term that has to be removed) and are a strong function of how the data set is weighted. This discrepancy should therefore not be used to infer properties of the central stellar cluster and the more robust dynamical mass should be used in future characterizations of the Galaxy's central dark mass.

Orbital analysis dramatically raises estimates of the dark matter density. Since this mass is confined to radii less than the minimum periapse passage of any of the stars included in the orbital fit (set by S0-16 at 60 AU = 0.0003 pc), the lower limit on the central dark

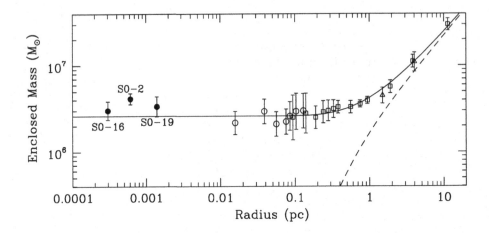

Fig. 4.6. Enclosed mass as a function of radius. The masses from the individual stars' orbital motion agree with one another. The solid line shows the best-fit black hole plus luminous cluster model based on the earlier measurements. While the new mass estimates from the orbital analysis are systematically higher than earlier estimates from velocity dispersion measurements, this difference is most likely primarily due to the way velocity dispersion uncertainties are treated. The new orbital masses increase the central dark mass density by 4 orders of magnitude, dramatically strengthening the case for a central supermassive black hole.

mass density is increased to $3 \times 10^{16} M_\odot/\text{pc}^3$, 3 orders of magnitude larger than previous estimates from measurements of acceleration vectors (Ghez et al. 1998). and 4 orders of magnitude larger than that based on velocity vectors (Genzel et al. 1997, 2000; Ghez et al. 1998). Although previous density estimates were sufficient to rule out the possibility of a cluster of dark objects based on evaporation and collisional lifetimes (Maoz 1998), the present observations dramatically strengthen this argument. At these densities, such a cluster of dark objects would have a lifetime of less than 10^5 years, which is significantly shorter than the age of the Galaxy, making this a highly unlikely scenario. Likewise, the fermion ball hypothesis becomes less tenable (Viollier 2003). For the Galactic Center, it is still possible to make this scenario work by increasing the particle mass to 50 kev c^{-2}; however, this object is unlikely to survive for very long before becoming a black hole. More generally, these new observations impose a maximum mass for a fermion ball of $2 \times 10^8 M_\odot$, which is less massive than the most massive central dark concentrations found in other galaxies and therefore indicates that fermion balls are no longer viable alternatives to all cases of supermassive black holes. With the central dark mass now confined to a radius equivalent to $1,000 \times$ the Schwarzschild radius, the orbits present the strongest case yet for a supermassive black hole at the center of the Milky Way galaxy and, more broadly, at the center of any normal type galaxy.

Table 4.1. *Estimates of the central dark mass from fits to the stellar orbital motion. The reported values come from fits that assume a distance of 8 kpc. For the Keck results, only solutions with fractional uncertainties less than 30% are listed here.*

Star	Mass ($10^6 M_\odot$)	Reference
S0-2	3.7 ± 1.5	Schödel et al. (2002)
	4.1 ± 0.6	Ghez et al. (2003a)
S0-16	3.0 ± 0.7	Ghez et al. (2004)
S0-19	3.4 ± 0.9	Ghez et al. (2004)
Simultaneous	3.7 ± 0.4	Ghez et al. (2004)

4.4.2 Stellar Astrophysics

The detection of absorption lines by Ghez et al. (2003a) allows the ambiguities in spectral classification, which are present when only photometric information is available, to be sorted out. The average brightness at 2.2 μm for S0-2 is $K \approx 13.9$ mag and there is no evidence of brightening during or after periapse passage (Ghez et al.). With a distance of 8.0 kpc and K-band extinction of 3.3 mag (Rieke, Rieke, & Paul 1989), the 2.2 μm brightness of S0-2 implies that, if it is an ordinary star unaltered by its environment, it could either be an O9 main-sequence star or a K5 giant star; all supergiants are ruled out since they are too bright by at least 2 magnitudes in the K band. Kleinmann & Hall (1986) provide a 2.0–2.5 μm spectral atlas of late-type stars that demonstrates that if S0-2 is a K5 giant star, then it should have deep CO absorption lines, which definitively were not detected in either this experiment or in our earlier experiment reported by Gezari et al. (2002). In contrast, the spectral atlas of 180 O and B stars constructed by Hanson, Conti, & Rieke (1996) shows that an O9 main-sequence star both lacks the CO absorption and has Br γ and He I $\lambda2.1126$ μm consistent with the observed values. Furthermore, stars in this comparison sample earlier than O8 show N III $\lambda2.115$ μm in emission and He II $\lambda2.1885$ μm in absorption above our 3 σ thresholds; the lack of photospheric He I $\lambda2.058$ μm absorption does not provide any additional constraints. Similarly, dwarf B-type stars later than B0 have absorption-line equivalent widths that are too large. Together, photometry and absorption-line equivalent widths permit dwarf spectral types ranging from O8 to B0. Likewise, the rotational velocity of 224 km s^{-1} is reasonable for this range (Gathier, Lamers, & Snow 1981). S0-2, therefore, appears to have a spectral type, and hence temperature ($\sim$30,000 K), as well as luminosity ($\sim10^3 L_\odot$) that are consistent with a main-sequence star having a mass of $\sim15 M_\odot$ and an age <10 Myr.

It is challenging to explain the presence of such a young star in close proximity to a supermassive black hole. Assuming the black hole has not significantly affected S0-2's appearance or evolution, S0-2 must be younger than 10 Myr and thus formed relatively recently. If it has not experienced significant orbital evolution, its apoapse distances of $\sim$2,000 AU implies that star formation is possible in spite of the tremendous tidal forces presented by the black hole, which is highly unlikely. If this star formed at larger distances from the black hole and migrated inward, then the migration would have to be through a very efficient process. Current understanding of the distribution of stars, however, does not permit such efficient migration. This problem is similar to that raised by the He I emission-line stars (e.g., Sanders 1992, 1998; Morris 1993; Morris, Ghez, & Becklin 1999; Gerhard 2001; Kim & Morris 2004), which are also counter-revolving around the Galaxy (Genzel et al. 1997), but amplifies it with a distance from the black hole that is an order of magnitude smaller. Several other stars in the central $1'' \times 1''$, which collectively are known as the Sgr A* stellar cluster are likely to be similarly young, given their 2 μm luminosities and lack of CO absorption in spectra of individual stars (Genzel et al. 1997; Gezari et al. 2002) as well as in integrated spectra of the Sgr A* stellar cluster (Eckart, Ott, & Genzel 1999; Figer et al. 2000). Gould & Quillen (2003) recently suggested that these stars might be the captured companions from massive binary stars that fortuitously made a nearby encounter with the central black hole. An alternative explanation for these hot photospheres is that they may be significantly altered by their environment. While their periapse passage are too large for them to be tidally heated by the black hole, as explored by Alexander & Morris (2003), they may be affected by the high stellar densities found in this region. On the one hand, the high stellar densities might allow them to be older giant stars that have had their outer atmosphere stripped through collisions; however, to generate the necessary luminosity, significant external heating is required (Alexander 1999). On the other hand, high stellar densities might lead to a cascade of merger events (Lee 1996), which would allow these stars' formation process to have begun more than 10 Myr ago. However, a large number of collisions would have had to occur to provide the necessary lifetime to bring it in from sufficiently large radii. More exotically, it could be a "reborn" star, which occurs as the product of a merger of a stellar remnant with a normal star. None of these possibilities are altogether satisfactory, leaving the Sgr A* cluster stars as a paradox of apparent youth in the vicinity of a supermassive black hole.

4.5   Conclusions

After roughly a decade of high-spatial resolution imaging, the first three-dimensional orbital motion solutions have been obtained for several stars. For three of these stars, the orbital mass uncertainties are less than 30%. These measurements confine the central dark mass to within a radius of a mere 60 AU, or equivalently 1,000 R_{Sch}, dramatically strengthening the case for a supermassive black hole, whose location is now determined to within $\pm$ 1.5 milliarcsec.

The precision of the proper motion and radial velocity measurements opens up additional new realms for dynamical studies in the Galactic Center. First is the possibility of obtaining a direct measure of the distance to the Galactic Center, R_0, from orbital fits (Salim & Gould 1999; Ghez et al. 2003a). The existing radial velocity measurements were obtained just 30 days after the star's closest approach to the black hole when the radial velocity was changing very rapidly (see Fig. 4.4), which, along with the proper motion measurements, constrains M/R_0^3 very effectively ($\sim$15% uncertainty), but does not yet produce a meaningful measurement of R_0. Nonetheless, as Figure 4.4 shows, the radial velocities from the currently allowed orbits quickly diverge, producing a spread of a few hundred km s^{-1} in one year. With one more year of radial velocity measurements, the orbital fits based on both proper motions and additional radial velocity measurements should provide the most direct and precise estimate of the distance to the Galactic Center.

A second opportunity is the possibility of detecting deviations from a Keplerian orbit.

These might arise from precession of the periapse distance due to general relativistic effects (Jaroszyński 1998b; Fragile & Matthews 2000)* or an extended mass distribution (Rubilar & Eckart 2001), in the form of either an entourage of stellar remnants surrounding the central supermassive black hole, a spike of dark matter particles (Gondolo & Silk 1999; Ullio, Zhao, & Kamionkowski 2001), or a binary black hole.

While the Milky Way has become one of the most convincing cases of a supermassive black hole, it is one of the least massive. It therefore potentially has an important role to play in estimates of the $M_\bullet - \sigma$ relation (Ferrarese & Merrit 2000; Gebhardt et al. 2000)*. The higher mass value from the orbits brings our Galaxy into better agreement with the $M_\bullet - \sigma$ relationship derived from a large sample of galaxies.

Acknowledgements. Helpful comments on this review were provided by S. Hornstein, M. Morris, S. Salim, A. Tanner, S. Wright, and an anonymous referee. A. M. G.'s research on the Galactic Center is supported by the National Science Foundation through the individual grant AST99-88397 and the Science and Technology Center for Adaptive Optics, managed by the University of California at Santa Cruz under Cooperative Agreement No. AST-9876783.

References

Alexander, T. 1999, ApJ, 527, 835

Alexander, T., & Loeb, A. 2001, ApJ, 551, 223

Alexander, T., & Morris, M. 2003, ApJ, 590, L25

Binney, J., & Tremaine, S. 1987, Galactic Dynamics (Princeton: Princeton Univ. Press)

Chakrabarty, D., & Saha, P. 2001, AJ, 122, 232

Eckart, A., & Genzel, R. 1997, MNRAS, 284, 576

Eckart, A., Genzel, R., Ott, T., & Schödel, R. 2002, MNRAS, 331, 917

Eckart, A., Ott, T., & Genzel, R. 1999, A&A, 352, L22

Ferrarese, L., & Merritt, D. 2000, ApJ, 539, L9

Figer, D. F., Becklin, E. E., McLean, I. S., Gilbert, A. M., Graham, J. R., Larkin, J. E., Levenson, N. A., Teplitz, H. I., Wilcox, M., K., Morris, M. 1999, ApJ, 533, L49

Fragile, P. C., & Mathews, G. J. 2000, ApJ, 542, 328

Gathier, R., Lamers, H. J. G. L. M., & Snow, T. 1981, ApJ, 247, 173

Gebhardt, K., et al. 2000, ApJ, 539, L13

Genzel, R., Eckart, A., Ott, T., & Eisenhauer, F. 1997, MNRAS, 291, 219

Genzel, R., Pichon, C., Eckart, A., Gerhard, O. E., & Ott, T. 2000, MNRAS, 317, 348

Gerhard, O. E. 2001, ApJ, 546, L39

Gezari, S., Ghez, A. M., Becklin, E. E., Larkin, J., McLean, I. S., & Morris, M. 2002, ApJ, 576, 790

Ghez, A. M., et al. 2003a, ApJ, 586, L127

Ghez, A. M., Becklin, E. E., Duchêne, G., Hornstein, S., Morris, M., Salim, S., & Tanner, A. 2003b, Astron. Nachr., Vol. 324, No. S1, Special Supplement "The Central 300 Parsecs of the Milky Way", ed. A. Cotera et al.

Ghez, A. M., Hornstein, S., Salim, S., Tanner, A., Morris, M., & Becklin, E. E. 2004, in preparation

Ghez, A. M., Klein, B. C., Morris, M., & Becklin, E. E. 1998, ApJ, 509, 678

Ghez, A. M., Morris, M., Becklin, E. E., Tanner, A., & Kremenek, T. 2000, Nature, 407, 349

Gondolo, P., & Silk, J. 1999, Phys. Rev. Lett., 83, 1719

Gould, A., & Quillen, A. C. 2003, ApJ, 592, 935

Hanson, M. M., Conti, P. S., & Rieke, M. J. 1996, ApJS, 107, 281

Haller, J. W., Rieke, M. J., Rieke, G. H., Tamblyn, P., Close, L., & Melia, F. 1996, ApJ, 456, 194

Hornstein, S. D., Ghez, A. M., Tanner, A., Morris, M., Becklin, E. E., & Wizinowich, P. 2002, ApJ, 577, L9

Jaroszyński, M. 1998a, Acta Astron., 48, 413.

---

* This would require the discovery of a star with a significantly smaller periapse passage.
* The current impact of the Milky Way on the $M_\bullet - \sigma$ relation, however, is limited by uncertainties in the determination of the integrated bulge velocity dispersion.

——. 1998b, Acta Astron., 48, 653.

Kim, S. S., & Morris, M. 2004, ApJ, submitted

Kleinmann, S. G., & Hall, D. N. B 1986, ApJS, 62, 501

Lacy, J. H., Townes, C. H., Geballe, T. R., & Hollenbach, D. J. 1980, ApJ, 241, 132

Lee, H. M. 1996, in IAU Symp. 169, Unsolved Problems of the Milky Way, ed. L. Blitz & P. Teuben (Dordrecht: Kluwer), 215

Lo, K. Y., Backer, D. C., Ekers, R. D., Kellermann, K. I., Reid, M., & Moran, J. M. 1985, Nature, 315, 124

Maoz, E. 1998, ApJ, 494, L181

Matthews, K., et al. 2004, PASP, in preparation

McGinn, M. T., Sellgren, K., Becklin, E. E., & Hall, D. N. B., 1989, ApJ, 338, 82

Menten, K. M., Reid, M. J., Eckart, A., & Genzel, R. 1997, ApJ, 475, L111

Morris, M. 1993, ApJ, 408, 496

Morris, M., Ghez, A. M., & Becklin, E. E. 1999, Adv. Spa. Res., 23, 959

Munyaneza, F., & Viollier, R. D. 2002, ApJ, 564, 274

Narayan, R. 2002, in Lighthouses of the Universe: The Most Luminous Celestial Objects and Their Use for Cosmology, ed. M. Gilfanov et al. (Berlin: Springer), 405

Reid, M. J. 1993, ARA&A, 31, 345

Reid, M. J., Menten, K. M., Genzel, R., Ott, T., Schödel, R., & Brunthaler, A. 2003, Astron. Nachr., Vol. 324, No. S1, Special Supplement "The Central 300 Parsecs of the Milky Way", ed. A. Cotera et al.

Reid, M. J., Menten, K. M., Genzel, R., Ott, T., Schödel, R., & Eckart, A. 2003, ApJ, 587, 112

Reid, M. J., Readhead, A. C. S., Vermeulen, R. C., & Treuhaft, R. N. 1999, ApJ, 524, 816

Rieke, G. H., Rieke, M. J., & Paul, A. E. 1989, ApJ, 336, 752

Rubilar, G. F., & Eckart, A. 2001, A&A, 2001, 372, 95

Salim, S., & Gould, A. 1999, ApJ, 523, 633

Sanders, R. H. 1992, Nature, 359, 131

——. 1998, MNRAS, 294, 35

Schödel, R., et al. 2002, Nature, 419, 694

Tsiklauri, D., & Viollier, R. D. 1998, ApJ, 500, 591

Ullio, P., Zhao, H. S., & Kamionkowski, M. 2001, Phys. Rev. D, 64, 1302

Viollier, R. 2003, Astron. Nachr., Vol. 324, No. S1, Special Supplement "The Central 300 Parsecs of the Milky Way," ed. A. Cotera et al.

Wardle, M., & Yusef-Zadeh, F. 1992, ApJ, 387, L65

5

The first nonlinear structures and the reionization history of the Universe

ZOLTÁN HAIMAN
Department of Astronomy, Columbia University

Abstract

In cosmological models favored by current observations, the first astrophysical objects formed in dark matter halos at redshifts starting at $z \gtrsim 20$, and their properties were determined by primordial H_2 molecular chemistry. These protogalaxies were very abundant, but substantially less massive than typical galaxies in the local Universe. Extreme metal-poor stars, and massive black holes in their nuclei reionized the bulk of the hydrogen in the intergalactic medium. Reionization may have taken place over an extended redshift interval, ending around $z \approx 7$. Observational probes of the process of reionization may soon be afforded by studying the polarization of the cosmic microwave background anisotropies, as well as by studying the spectra and abundance of distant Lyα-emitting galaxies. Here we review theoretical expectations on how and when the first galaxies formed, and summarize future observational prospects of probing hydrogen reionization.

5.1   Introduction

Recent measurements of the cosmic microwave background (CMB) temperature anisotropies, determinations of the luminosity distance to distant Type Ia supernovae, and other observations have led to the emergence of a robust "best-fit" cosmological model with energy densities in cold dark matter (CDM) and "dark energy" of $(\Omega_m, \Omega_\Lambda) \approx (0.3, 0.7)$ (see Bahcall et al. 1999 for a review, and references therein). The growth of density fluctuations and their evolution into nonlinear dark matter structures can be followed in this cosmological model in detail from first principles by semi-analytic methods (Press & Schechter 1974; Sheth, Mo, & Tormen 2001). More recently, it has become possible to derive accurate dark matter halo mass functions directly in large cosmological N-body simulations (Jenkins et al. 2001). Structure formation in a CDM-dominated Universe is "bottom-up," with low-mass halos condensing first. Dark matter halos with the masses of globular clusters, $10^{5-6}M_\odot$, are predicted to have condensed from $\sim 3\sigma$ peaks of the initial primordial density field as early as $\sim 1\%$ of the current age of the Universe, or redshift $z \approx 25$.

It is natural to identify these condensations as the sites where the first astrophysical objects, such as stars, or quasars, were born. The nature of the objects that form in these early dark matter halos is currently one of the most rapidly evolving research topics in cosmology. Progress is being driven by recent observational and theoretical advances, and also by the promise of next generation instruments in several wavelength bands, such as the *James Webb Space Telescope (JWST)* in the infrared, the *Low Frequency Array (LoFAr)* in

© The Observatories of the Carnegie Institution of Washington 2004.

the radio, *XMM-Newton* in the X-rays, and *Laser Interferometer Space Antenna (LISA)* in gravity waves.

A comprehensive review of the status of the field two years ago was provided by Barkana & Loeb (2001); a more focused review on the role of H_2 molecules at high redshift was given by Abel & Haiman (2001). In the present paper, we briefly summarize the main theoretical issues in high-redshift structure formation, and then focus on progress in the last two years, both in theory and in observation. It is appropriate to single out[*] the recent discovery of a Gunn-Peterson (1965, hereafter GP) trough in the spectra of a few high-redshift quasars (Becker et al. 2001; Fan et al. 2003). The absence of any detectable flux shortward of $\sim (1+z)1216$ Å in the spectra of these $z > 6$ sources has raised the tantalizing possibility that at these redshifts we are directly probing into the epoch reionization. It has also brought into sharp focus the question of how to distinguish observationally various reionization histories. We will critically discuss these issues below.

5.2   Theoretical Expectations

5.2.1   *The First Galaxy: When and How Massive?*

Baryonic gas that falls into the earliest nonlinear dark matter halos is shock heated to the characteristic virial temperatures of a few hundred K. It has long been pointed out (Rees & Ostriker 1977; White & Rees 1978) that such gas needs to lose its thermal energy in order to continue contracting, or in order to fragment—in the absence of any dissipation, it would simply reach hydrostatic equilibrium, and would eventually be incorporated into a more massive halo further down the halo-merger hierarchy. While the formation of nonlinear dark matter halos can be followed from first principles, the cooling and contraction of the baryons, and the ultimate formation of stars or black holes (BHs) in these halos, is much more difficult to model *ab initio*. Nevertheless, it is useful to identify four important mass scales, which collapse at successively smaller redshifts: (1) gas contracts together with the dark matter only in dark halos above the cosmological Jeans mass, $M_J \approx 10^4[(1+z)/11]^{3/2} M_\odot$, in which the gravity of dark matter can overwhelm thermal gas pressure; (2) gas that condensed into Jeans-unstable halos can cool and contract further in halos with masses above $M_{H_2} \gtrsim 10^5[(1+z)/11]^{-3/2} M_\odot$ (virial temperatures of $T_{vir} \gtrsim 10^2$ K), provided there is a sufficient abundance of H_2 molecules, with a relative number fraction at least $n_{H_2}/n_H \approx 10^{-3}$); (3) in halos with masses above $M_H \gtrsim 10^8[(1+z)/11]^{-3/2} M_\odot$ (virial temperatures of $T_{vir} \gtrsim 10^4$ K), gas can cool and contract via excitation of atomic Lyα, even in the absence of any H_2 molecules; and (4) in halos with masses above $M_H \gtrsim 10^{10}[(1+z)/11]^{-3/2} M_\odot$ (virial temperatures of $T_{vir} \gtrsim 2 \times 10^5$ K), gas can cool and contract, even in the face of an existing photoionizing background.

The first of these scales is obtained simply by balancing gravitational and pressure forces. The second scale is obtained by requiring efficient cooling via roto-vibrational levels of H_2 molecules, on a time scale shorter than the age of the Universe at the appropriate redshift. The calculations of the appropriate cooling functions for molecular hydrogen seem to be converging (Galli & Palla 1998 has done the most recent computations; see Flower et al. 2001 for a review and other references). The third scale is obtained by requiring efficient

---

[*] As this article went to press, the first results by the *Wilkinson Microwave Anisotropy Probe* (WMAP; Bennett et al. 2003) experiment were announced. See the Appendix for a summary of the implications.

(Lyα line) cooling via atomic H. The fourth scale is obtained in detailed spherical collapse calculations (Thoul & Weinberg 1996).

In the earliest, chemically pristine clouds, radiative cooling is dominated by H_2 molecules. As a result, gas-phase H_2 "astro-chemistry" is likely to determine the epoch when the first astrophysical objects appear—a conclusion reached already in the pioneering works by Saslaw & Zipoy (1967) and Peebles & Dicke (1968). Several papers constructed complete gas-phase reaction networks and identified the two possible ways of gas-phase formation of H_2 via the H_2^+ or H^- channels. These were applied to derive the H_2 abundance under densities and temperatures expected in collapsing high-redshift objects (Hirasawa 1969; Matsuda, Sato, & Takeda 1969; Palla, Salpeter, & Stahler 1983; Lepp & Shull 1984; Shapiro & Kang 1987; Kang et al. 1990; Kang & Shapiro 1992; Shapiro, Giroux, & Babul 1994). Studies that incorporate H_2 chemistry into cosmological models and address issues such as nonequilibrium chemistry, dynamics, or radiative transfer, have appeared only relatively more recently. Haiman, Thoul, & Loeb (1996) and Tegmark et al. (1997) studied the masses and redshifts of the earliest objects that can collapse and cool via H_2. The first three-dimensional (3D) cosmological simulations that incorporate H_2 cooling date back to Gnedin & Ostriker (1996) and Abel et al. (1997).

The basic picture that emerged from these papers is as follows. The H_2 fraction after recombination in the smooth "protogalactic" gas is small ($x_{H_2} = n_{H_2}/n_H \approx 10^{-6}$). At high redshifts ($z \gtrsim 100$), H_2 formation is inhibited even in overdense regions because the required intermediaries H_2^+ and H^- are dissociated by CMB photons. However, at lower redshifts, when the CMB energy density drops, a sufficiently large H_2 abundance builds up inside collapsed clouds ($x_{H_2} \approx 10^{-3}$) at redshifts $z \lesssim 100$ to cause cooling on a time scale shorter than the dynamical time. Sufficient H_2 formation and cooling is possible only if the gas reaches temperatures in excess of ~ 200 K; or masses of few $\sim 10^5[(1+z)/11]^{-3/2} M_\odot$. The efficient gas cooling in these halos suggests that the first nonlinear object in the Universe was born inside a $\sim 10^5 M_\odot$ dark matter halo at redshift $z \approx 20$ (corresponding to a $\sim 3\sigma$ peak of the primordial density peak).

The nature of the first object is considerably more difficult to elucidate. Nevertheless, the two most natural possibilities are for stars or BHs, or perhaps both, to form. The behavior of gas in a cosmological "minihalo" is a well-defined problem that has recently been addressed in 3D numerical simulations (Abel, Bryan, & Norman 2000, 2002; Bromm, Coppi, & Larson 1999, 2002). These works have been able to follow the contraction of gas to much higher densities than previous studies. They have shown convergence toward a temperature/density regime of $T \approx 200$ K and $n \approx 10^4 \, \mathrm{cm}^{-3}$, dictated by the critical density at which the excited states of H_2 reach equilibrium and cooling becomes less efficient (Galli & Palla 1998). The 3D simulations suggest that the mass of the gas fragments exceeds $10^2 - 10^3 M_\odot$, and, therefore that the first stars in the Universe may have been unusually massive (but see Nakamura & Umemura 2002, who argue using 1D and 2D simulations that the initial mass function may have been bimodal, with a second peak around $1 - 2 M_\odot$). An important consequence of this conclusion is that the earliest stars had an unusually hard spectrum—because they were metal free (Tumlinson & Shull 2000) and also because they were massive (Bromm, Kudritzki, & Loeb 2001), possibly capable of ionizing helium in addition to hydrogen.

5.2.2   Radiative Feedback: Negative or Positive?

The first objects will inevitably exert prompt and significant feedback on subsequent structure formation. This is because any soft UV radiation produced below 13.6 eV and/or X-rays above $\gtrsim 1$ keV from the first sources can propagate across the smooth intergalactic hydrogen gas, influencing the chemistry of distant regions (Dekel & Rees 1987). Soft UV radiation is expected either from a star or an accreting BH, with a BH possibly contributing X-rays as well. Although recent studies find that metal-free stars have unusually hard spectra, these do not extend to $\gtrsim 1$ keV (e.g., Tumlinson & Shull 2000, but see also Glover & Brandt 2003, who find stellar X-rays to have a more significant effect). The first stars formed via H_2 cooling are also expected to explode as supernovae, also producing internal feedback within or near their own parent cloud (Ferrara 1998; Omukai & Nishi 1999).

External feedback from an early soft UV background were considered by Haiman, Rees, & Loeb (1997), Haiman, Abel, & Rees (2000), Ciardi, Ferrara, & Abel (2000), and Machacek, Bryan, & Abel (2001). It was found that H_2 molecules are fragile and are universally photodissociated even by a feeble background flux (although Ricotti, Gnedin, & Shull 2002 find a positive effect: relatively near the ionizing sources, H_2 formation can be enhanced behind the H II ionization front in dense regions). H_2 dissociation by the $E < 13.6$ eV photons occurs when the UV background is several orders of magnitude lower than the value needed for cosmological reionization at $z > 5$ (and also than the level $\sim 10^{-21}$ erg cm^{-2} s^{-1} Hz^{-1} sr^{-1} inferred from the proximity effect to exist at $z \approx 3$; Bajtlik, Duncan, & Ostriker 1988). The implication is a pause in the cosmic star formation history: the buildup of the UV background and the epoch of reionization are delayed until larger halos ($T_{\mathrm{vir}} \gtrsim 10^4$ K) collapse. This is somewhat similar to the pause caused later on at the hydrogen-reionization epoch, when the Jeans mass is abruptly raised from $\sim 10^4\ M_\odot$ to $\sim 10^{8-9}\ M_\odot$. An early background extending to the X-ray regime would change this conclusion, because it catalyzes the formation of H_2 molecules in dense regions (Haiman, Rees, & Loeb 1996; Haiman et al. 2000; Glover & Brandt 2003; but see also Machacek, Bryan, & Abel 2003 who find X-rays to have a less significant effect). If quasars with hard spectra ($\nu F_\nu \approx$ constant) contributed significantly to the early cosmic background radiation then the feedback might even be positive, and reionization can be caused early on by minihalos with $T_{\mathrm{vir}} < 10^4$ K.

5.2.3   The 2nd Generation: Atomic Cooling?

Whether the first sources of light were massive stars or accreting BHs (the latter termed "miniquasars" in Haiman, Madau, & Loeb 1999) is still an open question. Nevertheless, there is some tentative evidence that reionization was caused by stars, rather than quasars: high-redshift quasars appear to be rare, even at the impressive depths reached by the optical Hubble Deep Fields (Haiman et al. 1999) and the >1 Ms Chandra Deep Fields (Barger et al. 2003; see also Mushotzky et al. 2000; Alexander et al. 2001; Hasinger 2003 for earlier results on faint X-ray sources in the CDFs). There also seems to be a significant delay between the epochs of hydrogen and helium reionizations, with helium ionized only at $z \approx 3$ (Songaila 1998; Heap et al. 2000), but hydrogen already at $z > 6$ (see discussion below), implying a relatively soft ionizing background spectrum.

If indeed the first light sources were stars, without emitting a significant X-ray component above ~ 1 keV, then efficient and widespread star (and/or BH) formation, capable of reionizing the Universe, had to await the collapse of halos with $T_{\mathrm{vir}} > 10^4$ K, or $M_{\mathrm{halo}} > 10^8 [(1+z)/11]^{-3/2} M_\odot$. The evolution of such halos differs qualitatively from their less massive counterparts (Oh & Haiman 2002). Efficient atomic line radiation allows rapid cooling to ~ 8000 K; subsequently the gas can contract to high densities nearly isothermally at this temperature. In the absence of H_2 molecules, the gas would likely settle into a locally stable disk, and only disks with unusually low spin would be unstable. However, the initial atomic line cooling leaves a large, out-of-equilibrium residual free-electron fraction (Shapiro & Kang 1987; Oh & Haiman 2002). This allows the molecular fraction to build up to a universal value of $x_{H_2} \approx 10^{-3}$, almost independently of initial density and temperature (this is a nonequilibrium freeze-out value that can be understood in terms of time scale arguments; see Susa et al. 1998 and Oh & Haiman 2002).

Unlike in less massive halos, H_2 formation and cooling is much less susceptible to feedback from external UV fields. This is because the high densities that can be reached via atomic cooling. The H_2 abundance that can build up in the presence of a UV radiation field J_{21}, and hence the temperature to which the gas will cool, is controlled by the ratio J_{21}/n. For example, in order for a parcel of gas to cool down to a temperature of 500 K, this ratio has to be less than $\sim 10^{-3}$ (where J_{21} has units of $10^{-21} \mathrm{erg\,s^{-1}\,cm^{-2}Hz^{-1}\,sr^{-1}}$, and n has units of $\mathrm{cm^{-3}}$). Flux levels well below that required to fully reionize the Universe strongly suppresses the cold gas fraction in $T_{\mathrm{vir}} < 10^4$ K halos. In comparison, Oh & Haiman (2002) showed that UV radiation with the same intensity has virtually no impact on H_2 formation and cooling in $T_{\mathrm{vir}} > 10^4$ K halos, where all of the gas is able to cool to $T = 500$ K. Indeed, under realistic assumptions, the newly formed molecules in the dense disk can cool the gas to ~ 100 K, and allow the gas to fragment on scales of a few $\times 100\,M_\odot$.

Various feedback effects, such as H_2 photodissociation from internal UV fields and radiation pressure due to Lyα photon trapping, are then likely to regulate the eventual efficiency of star formation in these systems. These important questions can only be addressed with some degree of confidence by high-resolution numerical simulations that are able to track the detailed gas hydrodynamics, chemistry and cooling, paralleling the pioneering work already done for $T_{\mathrm{vir}} < 10^4$ K halos (Bromm et al. 1999; Abel et al. 2000).

Finally, it is worth noting that during the initial contraction of the gas in halos with $T_{\mathrm{vir}} > 10^4$ K a significant fraction of the cooling radiation may be emitted in the Lyα line, especially toward high redshifts, where the contracting gas has a low metallicity. The cooling radiation may be significant, and can be detectable for halos with circular velocities above ~ 100 km s^{-1}, as an extended, diffuse, low-surface brightness "fuzz," with an angular diameter of a few arcseconds (Haiman, Spaans, & Quataert 2000; Fardal et al. 2001). A quasar turning on during the early stages of the contraction of the gas can boost the surface brightness by a factor of ~ 100 (Haiman & Rees 2001).

5.3   Observational Prospects

5.3.1   *Implications of Known High-redshift Sources*

Observations over the last two years have uncovered a handful of objects at redshifts around, and exceeding, $z = 6$. Quasars discovered at $z \approx 6$ range from the very bright sources found in the Sloan Digital Sky Survey (SDSS; $M_B \approx -27.7$ mag at $5.8 \lesssim z \lesssim 6.4$; Fan et al. 2000, 2001, 2003) to the much fainter quasars found using the Keck telescope ($M_B = -22.7$ mag at $z = 5.5$; Stern et al. 2000). Galaxies around the same redshifts are being discovered via their Lyα emission lines* (Dey et al. 1998; Spinrad et al. 1998; Weymann et al. 1998; Hu, McMahon, & Cowie 1999), with some recent extreme examples ranging from star formation rates of $> 10-20 M_\odot\mathrm{yr}^{-1}$ at $z = 5.7-6.6$ (Rhoads & Malhotra 2001; Hu et al. 2002; Kodaira et al. 2003; Rhoads et al. 2003), to an exceptionally faint (with an inferred star formation rate of $0.5 M_\odot\mathrm{yr}^{-1}$) galaxy at $z = 5.56$ detected in a targeted search for gravitationally lensed and highly magnified sources behind an Abell cluster (Ellis et al. 2001).

The spatial volume that was searched to discover the various sources listed above spans ~ 9 orders of magnitude: the SDSS survey probed ~ 20 Gpc^3 to discover the bright $z \approx 6$ quasars, while the faint, strongly lensed Lyα galaxy was found in searching a mere ~ 100 Mpc^3. It is possible to associate these high-redshift sources with dark matter halos based simply on their inferred abundance: this suggests that the rare SDSS quasars reside in massive, $\sim 10^{13} M_\odot$ halos, corresponding to $4-5\sigma$ peaks in the primordial density field on these scales, while the faintest Lyα galaxies may correspond to nearly "M_*" halos with masses of $\sim 10^{10} M_\odot$. To first approximation, the existence of these objects can be naturally accommodated in the CDM structure formation models. We will next list some simple conclusions that can be drawn from the existence of the above sources.

5.3.1.1   Early Growth of Massive Black Holes

It is interesting to consider the sheer size of supermassive BHs required to power the bright SDSS quasars near $z \approx 6$. Assuming that these quasars are shining at their Eddington limit, and are not beamed or lensed*, their BH masses are inferred to be $M_\bullet \approx 4 \times 10^9 M_\odot$. The Eddington-limited growth of these supermassive BHs by gas accretion onto stellar-mass seed holes, with a radiative efficiency of $\epsilon \equiv L/\dot{m}c^2 \approx 10\%$, requires ~ 20 e-foldings on a time scale of $t_\mathrm{E} \approx 4 \times 10^7 (\epsilon/0.1)$ years. While the age of the Universe leaves just enough time ($\lesssim 10^9$ years) to accomplish this growth by redshift $z = 6$, it does mean that accretion has to start early, and the seeds for the accretion have to be present at ultra-high redshifts: $z \gtrsim 15$ or 20 for an initial seed mass of 100 or 10 $M_\odot$, respectively. This conclusion holds even when one considers that BH seeds may grow in parallel in many different early halos, which undergo subsequent mergers (Haiman & Loeb 2001). Furthermore, the radiative efficiency cannot be much higher than $\epsilon \approx 10\%$ (Haiman & Loeb 2001; Barkana, Haiman, & Ostriker 2001). Since an individual quasar BH could have accreted exceptionally fast (exceeding the Eddington limit), it will be important to apply this argument to a larger sample of high-redshift quasars. Nevertheless, we note that a comparison of the light output of quasars at the peak of their activity ($z \approx 2.5$) and the total masses of their remnant BHs at $z = 0$ (see the review by Richstone, this volume) shows that during the growth of most of the BH mass the radiative efficiency cannot be much smaller than 10%, and hence any "super-Eddington" phase must be typically restricted to building only a small fraction of the final BH mass (Yu & Tremaine 2002).

5.3.1.2   Amplification by Gravitational Lensing?

A caveat to any conclusion based on the observed fluxes of bright, distant quasars is that they may be gravitationally lensed, and strongly amplified. While the *a priori*

---

* This type of search was proposed over 30 years ago by Partridge & Peebles (1967).
* Strong lensing or beaming would contradict the large proximity effect around these quasars; see Haiman & Cen (2002) and discussion below.

probability of strong lensing, causing amplification by a factor of > 10, along a random line of sight is known to be small ($\tau \approx 10^{-3}$) even to high redshifts (e.g., Kochanek 1998; Barkana & Loeb 2000), the *a posteriori* probabilities for observed sources can be much higher due to magnification bias (see, e.g., Schneider 1992). Magnification bias depends strongly on the parameters of the intrinsic quasar luminosity function, which are poorly constrained at $z \approx 6$. As a result, the theoretically expected probability that the SDSS quasars are strongly amplified by lensing can be significant, even approaching unity if the quasar luminosity function has an intrinsic slope steeper than -dlogΦ/dlog$L \gtrsim 4$ and/or has a break at relatively faint characteristic luminosities (Comerford, Haiman, & Schaye 2002; Wyithe & Loeb 2002a,b). As a result, observed lensed fractions can be used to provide interesting constraints on the high-redshift quasar luminosity function (Comerford et al. 2002; Fan et al. 2003).

Haiman & Cen (2002) analyzed the flux distribution of the Lyα emission of the quasar with one of the highest known redshifts, SDSS 1030+0524 at $z = 6.28$, and argued that this object could not have been magnified by lensing by more than a factor of ~ 5. The constraint arises from the large observed size, ~ 30 (comoving) Mpc, of the ionized region around this quasar, and relies crucially only on the assumption that the quasar is embedded in a largely neutral intergalactic medium (IGM). Based on the line/continuum ratio of SDSS 1030+0524, this quasar is also unlikely to be beamed by a significant factor. The conclusion is that the minimum mass for its resident BH is $4 \times 10^8 \, M_\odot$ (for magnification by a factor of 5); if the mass is this low, then the quasar had to switch on prior to redshift $z_f \gtrsim 9$. From the large size of the ionized region, an absolute lower bound on the age of this quasar also follows at $t > 2 \times 10^7$ yrs (see also the review by Martini on quasar ages in this volume).

5.3.1.3   Lyα Emitters and Cold Dark Matter

The existence of the faintest Lyα emitters may have another interesting implication. Three faint sources were found by probing a volume of only about ~ 10 Mpc^3 (in a source plane area of $\Delta\Omega \approx 100$ arcsec^2 behind the cluster Abell 2218 and redshift range $\Delta z \approx 1$; Ellis et al. 2001). Associating the implied spatial abundance of a few $\times$ 0.1 Mpc^{-3} with those of CDM halos (Jenkins et al. 2001), these sources correspond to very low-mass ($M \approx 10^{10} M_\odot$) halos. This appears consistent with the very low star formation rates ($\sim 0.5 M_\odot \mathrm{yr}^{-1}$), inferred from the Ly$\alpha$ luminosity.

The existence of such low-mass halos is interesting from the perspective of other recent observations, which suggest that standard CDM models predict too much power for the primordial density fluctuations on small scales (see, e.g., Haiman, Barkana, & Ostriker 2001 for a brief review). Several modifications of the CDM models, exemplified by warm dark matter (WDM) models (e.g., Bode, Ostriker, & Turok 2001), have been proposed recently that reduce the small-scale power. Such modifications generally reduce the number of low-mass halos at high redshift, and if the WDM particle had a mass of $m_X \lesssim 1$ keV (or $z = 0$ velocity dispersion of $v_{\mathrm{rms}} \gtrsim 0.04$ km s^{-1}), then there may have been too few high-z sources to reionize the Universe by $z = 6$ (Barkana et al. 2001).

The faint Lyα-emitting galaxies are so far down on the mass function of halos that one can turn this into a similar constraint on the mass of the WDM particle. Indeed, for $m_X \lesssim 1$ keV, such low-mass halos would not exist at $z = 6$ (see Fig. 5 in Barkana et al. 2001). This constraint is of interest, since it is around the value of other current astrophysical limits (e.g., from the Lyα forest; Narayanan et al. 2000).

5.3.2   The Reionization History of the IGM

How and when the intergalactic plasma was reionized is one of the long-outstanding questions in astrophysical cosmology, likely holding many clues about the nature of the first generation of light sources and the end of the cosmological "Dark Age." The lack of any strong H I absorption (a GP trough) in the spectra of high-redshift quasars has revealed that the IGM is highly ionized at all redshifts $z \lesssim 6$ (Fan et al. 2000). On the other hand, the lack of a strong damping by electron scattering of the first acoustic peak in the temperature anisotropy of the CMB radiation has shown that the IGM was neutral between the redshifts $25 \lesssim z \lesssim 10^3$ (Kaplinghat et al. 2003). Together these two sets of data imply that most hydrogen atoms in the Universe were reionized during the redshift interval $6 \lesssim z \lesssim 25$.

It would be overly ambitious to provide a comprehensive review of the subject of reionization in this article. Instead, we will focus below on a few basic theoretical issues and discuss the implications of the most recent observations.

5.3.2.1   Models of Reionization

In the simplest models, an early population of stars or quasars drive expanding, discrete ionized regions. Once these regions overlap, the Universe has been reionized (e.g., Arons & Wingert 1972). In the context of CDM models, the ionizing sources form inside high-redshift dark matter halos. Since the formation and evolution of the halos is dictated by gravity alone, it is relatively well understood, and the main uncertainty in models of reionization is the "efficiency" (ionizing luminosity of stars and/or quasars) of each halo. Semi-analytical models (e.g., Shapiro et al. 1994; Tegmark, Silk, & Blanchard 1994; Haiman & Loeb 1997, 1998; Valageas & Silk 1999) and numerical simulations (Gnedin & Ostriker 1997; Nakamoto, Umemura, & Susa 2001; Gnedin 2000, 2003) adopt various, reasonably motivated, efficiencies and follow the evolution of the total volume-filling fraction of ionized regions. These models predict reionization to occur between $z = 7$ and 15, depending on the adopted efficiencies.

These studies have left significant uncertainties on the details of how reionization proceeds in an inhomogeneous medium. Since the ionizing sources are likely embedded in dense regions, one might expect that these dense regions are ionized first, before the radiation escapes to ionize the low-density IGM (Madau, Haardt, & Rees 1999; Gnedin 2000, 2003). Alternatively, most of the radiation might escape from the local, dense regions along low-column density lines of sight. In this case, the underdense "voids" are ionized first, with the ionization of the denser filaments and halos lagging behind (Miralda-Escudé, Haehnelt, & Rees 2000).

Apart from the topology of reionization, the current suite of models also leaves uncertainties about the redshift evolution of the mean ionized fraction. Most models predict a sharp increase whenever the discrete ionized regions percolate. However, if the formation rate of the ionizing sources does not parallel the collapse of high-σ peaks, reionization can be more gradual, and can have a complex history. For instance, the radiative H_2 feedback discussed above may result in two distinct episodes of reionization: (1) UV sources formed via H_2 cooling in minihalos partially ionize the Universe at $z \approx 20$, (2) the IGM recombines as these sources turn off, and (3) the Universe is reionized at $z \approx 7$ by UV sources in more massive halos. The first episode of reionization may be more pronounced, since the metal-free stars in minihalos are expected to have an unusually high ionizing photon production efficiency (see also Wyithe & Loeb 2003a; Cen 2003a).

Depending on the choice of the efficiency parameters, it is also possible that the IGM only partially recombines during stage (2), resulting in an extended episode of partial ionization (Cen 2003a). Finally, the decrease of the mean neutral fraction would be more gradual if the ionizing sources had a hard spectrum. Reionization by X-rays was considered recently by Oh (2001) and Venkatesan, Giroux, & Shull (2001). In contrast to a picture in which discrete H II regions eventually overlap, in this case the IGM is ionized uniformly and gradually throughout space. All of these uncertainties highlight the need for new and sensitive observational probes of the reionization history, which we will discuss in the next section.

A significant source of theoretical uncertainty in the above models is the average global recombination rate, or "clumping factor," which limits the growth of H II regions at high redshifts, when the Universe was dense. Although hydrodynamical simulations can compute gas clumping *ab initio* (Gnedin & Ostriker 1997), to date they have not been able to resolve the relevant small scales (the minihalos have typical masses below $10^7 M_\odot$). Gas clumping has been estimated semi-analytically (Chiu & Ostriker 2000; Benson et al. 2001; Haiman, Abel, & Madau 2001). In particular, Haiman et al. (2001) pointed out that the earliest ionizing sources are likely surrounded by numerous "minihalos" that had collapsed earlier, but had failed to cool and form any stars or quasars. The mean-free path of ionizing photons, before they are absorbed by a minihalo*, is about ~ 1 (comoving) Mpc. Simple models, summing over the expected population of minihalos, reveal that on average an H atom in the Universe recombines $\gtrsim 10$ times before redshift $z = 6$; as a result, the IGM had to be "reionized $\gtrsim 10$ times."

A naive extrapolation of the luminosity density of bright quasars toward $z = 6$ reveals that these sources fall short of this requirement (Haiman et al. 2001; see also Shapiro et al. 1994). Extrapolating the known population of Lyman-break galaxies (e.g., Steidel, Pettini, & Adelberger 2001) toward $z = 6$ comes closer: assuming that 15% of the ionizing radiation from Lyman-break galaxies escapes into the IGM (on average, relative to the escape fraction at 1500 Å), a naive extrapolation shows that Lyman-break galaxies emitted approximately one ionizing photon per hydrogen atom prior to $z = 6$. The implication is that the ionizing emissivity at $z > 6$ was ~ 10 times higher than provided by a straightforward extrapolation back in time of known quasar and galaxy populations. The Universe was likely reionized by a population of UV sources that is yet to be discovered!

5.3.2.2 Current Observations

The recent discovery of the bright quasar SDSS 1030+0524 in the SDSS at redshift $z = 6.28$ has, for the first time, revealed a full GP trough, i.e., a spectrum consistent with no flux at high S/N over a substantial stretch of wavelength shortward of $(1+z)\lambda_\alpha = 8850$ Å (Becker et al. 2001). At the time of this writing, a full GP trough has been discovered at high S/N in a second SDSS quasar at $z = 6.43$, and at a lower S/N in two other SDSS quasars (at $z = 6.23$ and $z = 6.05$; Fan et al. 2003; an "incomplete" trough was also reported at high S/N in a $z = 5.7$ source by Djorgovski et al. 2001).

These discoveries have raised the tantalizing possibility that we are detecting reionization occurring near redshift $z \approx 6.3$. The lack of any detectable flux indeed implies a strong lower limit, $x_H \gtrsim 0.01$, on the mean mass-weighted neutral fraction of the IGM at $z \approx 6$

---

* Photoionization unbinds the gas in these shallow potential wells (Shapiro, Raga, & Mellema 1998; Barkana & Loeb 1999). The gas acts as a sink of ionizing photons only before it is photoevaporated.

(Cen & McDonald 2002; Fan et al. 2002; Lidz et al. 2002; Pentericci et al. 2002). Still, the evolution of the IGM opacity inferred from quasar spectra does not directly reveal whether we have probed the neutral era. Nevertheless, comparisons with numerical simulations of cosmological reionization (Cen & McDonald 2002; Fan et al. 2002; Gnedin 2003; Lidz et al. 2002; Razoumov et al. 2002), together with the rapid rise toward high redshifts of the neutral fractions inferred from a sample of high-redshift quasars from $5.5 \lesssim z \lesssim 6$ (Songaila & Cowie 2002), suggest that the IGM is likely neutral at $z \gtrsim 6.5$.

While there may be a theoretical bias for reionization occurring close to $z \approx 6.3$, it is possible that the reionization history was nonmonotonic, and/or lasted over a considerably longer redshift interval, as was discussed above. An observational probe of the redshift history of reionization would be invaluable in constraining such scenarios, and to securely establish when the cosmic Dark Age ended. Below we consider prospects to probe the reionization history in future observations. Another interesting issue (not discussed further below) raised by the recent GP trough detections is: What is the best way to interpret quasar spectra? Using simply the wavelength extent of the "dark" region (without any detectable flux) in the spectrum is not, by itself, generally sufficient to give a strong constraint on the global topology of neutral versus ionized regions, because of stochastic variations. Other statistical measures need to be developed (Barkana 2002; Lidz et al. 2002; Nusser et al. 2002).

5.3.2.3   Future Probes of Reionization

CMB Polarization. An alternative way of probing deeper into the dark ages is the study of CMB anisotropies: the free electrons produced by reionization scatter a few percent of the CMB photons. Interesting results may be imminent from the ongoing CMB satellite experiment *WMAP*. The discussion below is based largely on the results of Kaplinghat et al. (2003); for a detailed review, see Haiman & Knox (1999).

Without reionization, the "primordial" polarization signal at large angles would be negligible. However, CMB photons scattering in a reionized medium boost the polarization signal—likely making it measurable in the future. CMB polarization anisotropy at large angles is very sensitive to the optical depth to electron scattering, and the future experiment *Planck* (and if the optical depth is large, then *WMAP* as well), will have the power to discriminate between different reionization histories even when they lead to the same optical depth.

One of the advantages of studying the CMB is that it probes the presence of free electrons, and can therefore detect a neutral hydrogen fraction of $x_H = 0.1$ and $x_H = 10^{-3}$ with nearly equal sensitivity. Physically, the CMB and the GP trough therefore probe two different stages of reionization. The CMB is sensitive to the initial phase when x_H first decreases below unity and free electrons appear, say, at redshift z_e. On the other hand, the (hydrogen) GP trough is sensitive to the end phase, when neutral hydrogen atoms finally disappear, say, at z_H. In most models, these two phases coincide to $\lesssim 10\%$ of the Hubble time, such that $z_e \approx z_H$. However, as argued above, one can conceive alternative theories in which the two phases are separated by a large redshift interval, and $z_e \gg z_H$.

One of the difficulties with CMB is that the effect of electron scattering on the temperature anisotropies is essentially an overall suppression, nearly degenerate with the intrinsic amplitude of the fluctuation power spectrum. However, Kaplinghat et al. (2003) showed that for most models constrained by current CMB data and by the discovery of a GP trough (i.e., requiring that reionization occurred at $z > 6.3$), *WMAP* can break this degeneracy, and detect the reionization signature in the polarization power spectrum.[*] The expected 1σ error on the measurement of the electron optical depth is around $\delta\tau \approx 0.03$, with only a weak dependence on the actual value of τ. This will also allow *WMAP* to achieve a 1σ error on the amplitude of the primordial power spectrum of 6%. As an example, *WMAP* with two years (*Planck* with one year) of observation can distinguish a model with 50% (6%) partial ionization between redshifts of 6.3 and 20 from a model in which hydrogen was completely neutral at redshifts greater than 6.3. *Planck* will be able to distinguish between different reionization histories even when they imply the same optical depth to electron scattering for the CMB photons (Holder et al. 2003; Kaplinghat et al. 2003).

Lyα Emitters. An alternative method to probe the reionization history is to utilize the systematic changes in the profiles of Lyα emission lines toward higher redshift. The increased hydrogen IGM opacity beyond the reionization redshift makes the emission lines appear systematically more asymmetric, and the apparent line center systematically shifts toward longer wavelengths, as absorption in the IGM becomes increasingly more important and eliminates the blue side of the line (Haiman 2002; Madau 2003). Because of the intrinsically noisy Lyα line shapes, this method will require a survey that delivers a large sample of Lyα emitters (Rhoads et al. 2002).

Lyα photons injected into a neutral IGM are strongly scattered, and the red damping wing of the GP trough can strongly suppress, or even completely eliminate, the Lyα emission line (Miralda-Escudé 1998; Miralda-Escudé & Rees 1998; Loeb & Rybicki 1999). Resonant absorption by the IGM may itself extend to the red side of the line, if there is still significant cosmological gas infall toward the source (Barkana & Loeb 2003). The reionization of the IGM may therefore be accompanied by a rapid decline in the observed space density of Lyα emitters beyond the reionization redshift z_r (Haiman & Spaans 1999). Indeed, such a decline could by itself provide a useful observational probe of the reionization epoch in a large enough sample of Lyα emitters (Haiman & Spaans 1999; Rhoads & Malhotra 2001), complementary to methods utilizing the GP trough.

As shown by Cen & Haiman (2000) and Madau & Rees (2000), a source with a bright ionizing continuum can create a large ($\gtrsim 30$ comoving Mpc) cosmological H II region. For a sufficiently luminous source, and/or for a sufficiently wide intrinsic Lyα line width, the size of the H II region corresponds to a wavelength range $\Delta\lambda$ that exceeds the width of the emission line, allowing most of the intrinsic Lyα line to be transmitted without significant scattering. Furthermore, even for faint sources with little ionizing continuum, a significant fraction of the emission line can remain observable if the intrinsic line width is $\Delta v \gtrsim 300$ km s^{-1} (Haiman 2002).

The recent discovery of Lyα-emitting galaxies with the Keck and Subaru telescopes at redshifts as high as $z = 6.56$ (Hu et al. 2002) and $z = 6.58$ (Kodaira et al. 2003) illustrates the fact the Lyα lines can indeed be detected even at these high redshifts. These redshifts exceed those at which the GP troughs were discovered in the SDSS quasar spectra, and hence these galaxies could be located *beyond* the reionization redshift of the IGM. The Hu et al. spectrum is consistent with being embedded in a neutral IGM that only partially obscures the line, and

---

[*] A detailed morphological study of the effects of reionization on maps of the temperature anisotropy may also be helpful to break this degeneracy (Gnedin & Shandarin 2002).

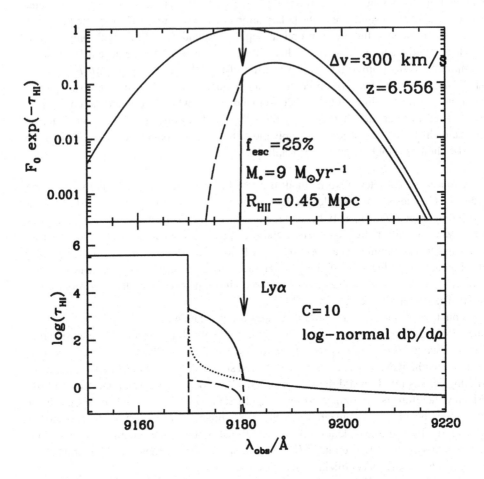

Fig. 5.1. *Upper panel:* Toy model profile of the Lyα line from a $z = 6.56$ galaxy. The top solid curve shows the adopted intrinsic profile, and the bottom solid curve shows the profile including absorption in the IGM and by the neutral atoms inside the 0.45 (proper) Mpc H II region surrounding the source. *Lower panel:* The optical depth as a function of wavelength from within the H II region (short-dashed curve), from the neutral IGM outside the H II region (dotted curve), as well as from the sum of the two (solid curve). In both panels, the long-dashed curves describe an alternative, more realistic treatment of the H I opacity within the H II region (see Haiman 2002 for a discussion), and the arrows indicate the central wavelength of the unobscured Lyα line.

allows ∼ 20% of the total line flux to be transmitted (Haiman 2002; see Fig. 5.1). This would imply that the source is somewhat surprisingly bright, given the inferred abundance and estimates of the number density of $z \approx 5.7$ Lyα emitters from a different survey (Rhoads et al. 2003).

Although the lines are detectable, the IGM still has a significant effect on the Lyα line profile. A statistical sample of Lyα emitters that spans the reionization redshift should be a useful probe of reionization, through the study of the correlations between the luminosity of the sources and the properties of the emission lines, such as their total line/continuum ratio (if a continuum is measured), the asymmetry of the line profile, and the offset of the peak of the line from the central Lyα wavelength (for sources that have redshift measurements from other emission lines).

Redshifted 21 cm Features. Future radio telescopes could observe 21 cm emission or absorption from neutral hydrogen at the time of reionization (see a recent review by Madau 2004). This would provide a direct measure of the physical state of the neutral hydrogen and its evolution through the time of reionization. Recently, Carilli, Gnedin, & Owen (2002) considered the radio equivalent of the GP trough, using numerical simulations. Unlike the Lyα case, the mean absorption by the neutral medium is about 1% at the redshifted 21 cm. Furlanetto & Loeb (2002) and Iliev et al. (2002) have used semi-analytic methods to look at the observable features (in 21 cm) of minihalos and protogalactic disks. These studies suggest that the 21 cm observations would yield robust information about the thermal history of collapsed structures and the ionizing background, provided that a sufficiently bright radio-loud quasar can be found at $z_e > 6.3$. In addition, characteristic angular fluctuations that trace early density fluctuations of the 21-cm emitting gas (e.g., Madau, Meiksin, & Rees 1997; Tozzi et al. 2000) may be detectable in the future.

Gunn-Peterson Trough in Metal Lines. Although detections of the hydrogen GP trough suffer from the "saturation problem" discussed above, an alternative possibility may be to use corresponding absorption troughs caused by heavy elements in the high-redshift IGM. Recently, Oh (2002) showed that if the IGM is uniformly enriched by metals to a level of $Z = 10^{-2} - 10^{-3} Z_\odot$, then absorption by resonant lines of O I or Si II may be detectable. The success of this method depends on the presence of oxygen and silicon at these levels, and on these species being neutral or once-ionized in regions where hydrogen has not yet been ionized. It may be more natural for hydrogen reionization to precede metal enrichment, rather than vice versa, however; because of the short recombination time at high redshift, the gas can remain neutral even in the metal-enriched regions, and may provide the absorption features necessary for its detection.

Acknowledgements. I thank Carnegie Observatories for their kind invitation, and my recent collaborators Gil Holder, Manoj Kaplinghat, Lloyd Knox, and Peng Oh for many fruitful discussions, and for their permission to draw on joint work.

Appendix: Implications of First WMAP Results

As this article went to press, the first results by the *WMAP* experiment were announced. These new results have significant implications for reionization, which we only briefly summarize here. *WMAP* has measured the optical depth to electron scattering from the cross-correlation between the the temperature and E-mode polarization angular power spectra (TE; Spergel et al. 2003), yielding the high value of $\tau = 0.17 \pm 0.06$. The value reflects marginalization over all other relevant cosmological parameters. A reassuringly similar value, $\tau = 0.17 \pm 0.04$, is obtained by predicting the TE power spectrum directly from the temperature power spectrum (Kogut et al. 2003).

The immediate conclusion that can be drawn from the high value of τ is that the IGM was significantly ionized at redshifts as early as $z \approx 15$. This discovery has important implications for the sources of reionization, and allows, for the first time, constraints to be placed on physical reionization scenarios out to redshift $z \approx 20$. The *WMAP* results have been interpreted by a flurry of papers presenting models of the reionization history in the few weeks following the data release (Cen 2003b; Ciardi, Ferrara, & White 2003; Fukugita & Kawasaki 2003; Haiman & Holder 2003; Sokasian, Abel, & Hernquist 2003; Somerville & Livio 2003; Wyithe & Loeb 2003b). The reader is referred to these papers for many interesting details; no doubt the list of papers devoted to reionization will continue to grow by the time this article appears in print.

The main implications of the *WMAP* result can be summarized as follows (based on Haiman & Holder 2003, but broadly consistent with the conclusions reached in all the interpretive papers listed above).

• Previous evidence has shown that the IGM is highly ionized at least out to redshift $z \approx 6$. As argued in the body of this article, there is also evidence that a "percolation epoch" is taking place near $z \approx 6-7$.* Abrupt reionization at $z = 6$ would yield $\tau \approx 0.04$, significantly lower than the *WMAP* value.

• Reionization models predict a "percolation" redshift that depends on the combination of efficiency parameters, essentially on $\epsilon \equiv N_\gamma f_* f_{esc}/C$, where $f_* \equiv M_*/(\Omega_b M_{halo}/\Omega_m)$ is the fraction of baryons in the halo that turns into stars; N_γ is the mean number of ionizing photons produced by an atom cycled through stars, averaged over the initial mass function of the stars; f_{esc} is the fraction of these ionizing photons that escapes into the IGM; and C is the mean clumping factor of ionized gas. A value of $\epsilon \approx 10$ is required for percolation to occur at $z \approx 6$. This value is quite reasonable, and it produces a natural "tail" of ionization at redshifts exceeding $z = 6$ in these models (due to the gradual turn-on of ionizing sources associated with dark halos; see Figure 5.2). This "tail" increases the optical depth to $\tau \approx 0.08$, which is still discrepant with the *WMAP* value at the $\sim 3\sigma$ level.

• As a result, no simple reionization model can be consistent with the combination of the central *WMAP* value of $\tau = 0.17$ *and* a percolation occurring at $z \approx 6$. Satisfying both constraints requires either of the following: (1) H_2 molecules form efficiently at $z \approx 20$, survive feedback processes, and allow UV sources in halos with virial temperatures $T_{vir} < 10^4$ K to contribute substantially to reionization, or (2) the efficiency ϵ in halos with $T_{vir} > 10^4$ K decreased by a factor of $\gtrsim 30$ between $z \approx 20$ and $z \approx 6$. The latter may be a natural result of a switch-over from a metal-free to a normal stellar population (Wyithe & Loeb 2003a; Cen 2003a). These options are illustrated by the upper three curves in Figure 5.2.

• As apparent from above, there are interesting implications for the formation history of ionizing sources, but there is no "crisis" for cosmology: ΛCDM cosmogonies can still accommodate the high value of τ measured by *WMAP*. However, interesting limits can be drawn on cosmological models with reduced small-scale power. As an example, the combination of *WMAP* and other large-scale structure data has provided tentative evidence

---

* Hui & Haiman (2003) have argued that there is additional evidence for percolation at $z < 10$ from the *thermal* history of the IGM.

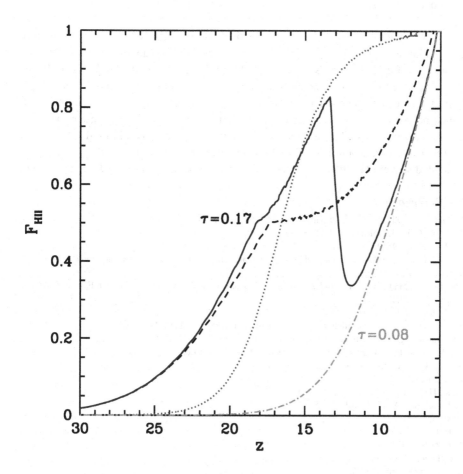

Fig. 5.2. Model reionization histories (evolution of ionized fraction with redshift). The dot–dashed curve is a simple model with "universal" efficiencies calibrated to produce reionization at $z = 6$. It produces a low optical depth of $\tau = 0.08$. The other three models satisfy the dual requirement of (1) a low–redshift ($z \sim 6$) percolation epoch, and a (2) high value of electron scattering optical depth ($\tau \sim 0.17$) by (1) increased efficiency in "mini-halos" (solid curve), (2) excluding minihalos, but increasing the efficiencies in larger halos (dotted curve), and (3) a sudden drop in efficiencies, (e.g. due to a transition from metal-free to normal stellar population). The three models produce large–angle polarization signatures in the CMB that will be distinguishable by Planck. (Adapted from Haiman & Holder 2003).

for a running spectral index in the power spectrum $P(k)$. Haiman & Holder (2003) and Somerville et al. (2003) have shown that in the favored running-index model achieving $\tau = 0.17$ is impossible without extreme efficiencies of ionizing photon production in halos with virial temperatures $T_{vir} < 10^4$ K; a significantly stronger curvatures could be ruled out. Similar conclusions can be drawn (Somerville, Bullock, & Livio 2003) about models with a strong tilt in the scalar power-law index n, or about WDM models (see also Barkana et al. 2001 and Yoshida et al. 2003 for limits even on "lukewarm" dark matter models from reionization at $z \approx 20$.)

It is also worth emphasizing that the reionization history is likely to have been complex enough so that it will have distinctive features that *Planck* can distinguish at $> 3\sigma$ significance (see Fig. 5.2; and also Holder et al. 2003). At the high *WMAP* value for τ, *Planck* will be able to provide tight statistical constraints on reionization model parameters and help elucidate the nature of the sources ending the Dark Ages. In addition to the large-angle polarization signatures from reionization, small-scale fluctuations in the IGM temperature may be observable (a "Sunyaev-Zel'dovich" effect from high-redshift ionized regions; Oh, Cooray, & Kamionkowski 2003). Finally, the sources responsible for the high optical depth discovered by *WMAP* should be directly detectable out to $z \approx 15$ by the *JWST*.

References

Abel, T., Anninos, P., Zhang, Y., & Norman, M. L. 1997, NewA, 2, 181
Abel, T., Bryan, G. L., & Norman, M. L. 2000, ApJ, 540, 39
——. 2002, Science, 295, 93
Abel, T., & Haiman, Z. 2001, in Molecular Hydrogen in Space, ed. F. Combes & G. Pineau des Forêts (Cambridge: Cambridge Univ. Press), 237
Alexander, D. M., Brandt, W. N., Hornschemeier, A. E., Garmire, G. P., Schneider, D. P., Bauer, F. E., & Griffiths, R. E. 2001, AJ, 122, 2156
Arons, J., & Wingert, D. W. 1972, ApJ, 177, 1
Bahcall, N. A., Ostriker, J. P., Perlmutter, S., & Steinhardt, P. J. 1999, Science, 284, 1481
Bajtlik, S., Duncan, R. C., & Ostriker, J. P. 1988, ApJ, 327, 570
Barger, A. J., Cowie, L. L., Capak, P., Alexander, D. M., Bauer, F. E., Brandt, W. N., Garmire, G. P., & Hornschemeier, A. E. 2003, ApJ, 584, L61
Barkana, R. 2002, NewA, 7, 85
Barkana, R., Haiman, Z., & Ostriker, J. P. 2001, ApJ, 558, 482
Barkana, R., & Loeb, A. 1999, ApJ, 523, 54
——. 2000, ApJ, 531, 613
——. 2001, Physics Reports, 349, 125
——. 2003, Nature, 421, 341
Becker, R. H., et al. 2001, AJ, 122, 2850
Bennett, C. L., et al. 2003, ApJ, 583, 1
Benson, A. J., Nusser, A., Sugiyama, N., & Lacey, C. G. 2001, MNRAS, 320, 153
Bode, P., Ostriker, J. P., & Turok, N. 2001, ApJ, 556, 93
Bromm, V., Coppi, P. S., & Larson, R. B. 1999, ApJ, 527, 5
——. 2002, ApJ, 564, 23
Bromm, V., Kudritzki, R. P., & Loeb, A. 2001, ApJ, 552, 464
Carilli, C. L., Gnedin, N. Y., & Owen, F. 2002, ApJ, 577, 22
Cen, R. 2003a, ApJ, 591, 12
——. 2003b, ApJ, 591, L5
Cen, R., & Haiman, Z. 2000, ApJ, 542, L75
Cen, R., & McDonald, P. 2002, ApJ, 570, 457
Chiu, W. A., & Ostriker, J. P. 2000, ApJ, 534, 507
Ciardi, B., Ferrara, A., & Abel, T. 2000, ApJ, 533, 594
Ciardi, B., Ferrara, A., & White, S. D. M. 2003, MNRAS, 344, L7
Comerford, J. M., Haiman, Z., & Schaye, J. 2002, ApJ, 580, 63
Dekel, A., & Rees, M. J. 1987, Nature, 326, 455
Dey, A., Spinrad, H., Stern, D., Graham, J. R., & Chaffee, F. 1998, ApJ, 498, L93
Djorgovski, S. G., Castro, S., Stern, D., & Mahabal, A. A. 2001, ApJ, 560, 5
Ellis, R. S., Santos, M., Kneib, J.-P., & Kuijken, K. 2001, ApJ, 560, L119
Fan, X., et al. 2000, AJ, 120, 1167
——. 2001, AJ, 122, 2833

——. 2002, AJ, 123, 1247

Fan, X., Narayanan, V. K., Strauss, M. A., White, R. L., Becker, R. H., Pentericci, L., & Rix, H.-W. 2002, AJ, 123, 1247

Fardal, M. A., Katz, N., Gardner, J. P., Hernquist, L., Weinberg, D. H., & Davé, R. 2002, ApJ, 562, 605

Ferrara, A. 1998, ApJ, 499, L17

Flower, D., Le Bourlot, J., Pineau Des Forêts, G., & Roueff, E. 2001, in Molecular Hydrogen in Space, ed. F. Combes & G. Pineau des Forêts (Cambridge: Cambridge Univ. Press), 23

Fukugita, M., & Kawasaki, 2003, MNRAS, 343, L25

Furlanetto, S. R., & Loeb, A. 2002, ApJ, 579, 1

Galli, D., & Palla, F. 1998, A&A, 335, 403

Glover, S. C. O., & Brandt, P. W. J. L. 2003, MNRAS, 340, 210

Gnedin, N. Y. 2000, ApJ, 535, 530

——. 2003, MNRAS, submitted (astro-ph/0110290)

Gnedin, N. Y., & Ostriker, J. P. 1996, ApJ, 472, 63

——. 1997, ApJ, 486, 581

Gnedin, N. Y., & Shandarin, S. F. 2002, MNRAS, 337, 1435

Gunn, J. E., & Peterson, B. A. 1965, ApJ, 142, 1633

Haiman, Z. 2002, ApJ, 576, L1

Haiman, Z., Abel, T., & Madau, P. 2001, ApJ, 551, 599

Haiman, Z., Abel, T., & Rees, M. J. 2000, ApJ, 534, 11

Haiman, Z., Barkana, R., & Ostriker, J. P. 2001, in The 20th Texas Symposium on Relativistic Astrophysics, ed. H. Martel & J. C. Wheeler (Melville: AIP), 136

Haiman, Z., & Cen, R. 2002, ApJ, 578, 702

Haiman, Z., & Holder, G. P. 2003, ApJ, 595, 1

Haiman, Z., & Knox, L. 1999, in Microwave Foregrounds, ed. A. de Oliveira-Costa & M. Tegmark (San Francisco: ASP), 227

Haiman, Z., & Loeb, A. 1997, ApJ, 483, 21

——. 1998, ApJ, 503, 505

——. 2001, ApJ, 552, 459

Haiman, Z., Madau, P., & Loeb, A. 1999, ApJ, 514, 535

Haiman, Z., & Rees, M. J. 2001, ApJ, 556, 87

Haiman, Z., Rees, M. J., & Loeb, A. 1996, ApJ, 467, 522

——. 1997, ApJ, 476, 458 (erratum: 1997, ApJ, 484, 985)

Haiman, Z., & Spaans, M. 1999, ApJ, 518, 138

Haiman, Z., Spaans, M., & Quataert, E. 2000, ApJ, 537, L5

Haiman, Z., Thoul, A. A., & Loeb, A. 1996, ApJ, 464, 523

Hasinger, G. 2003, in New Visions of the X-ray Universe in the XMM-Newton and Chandra Era, ed. F. Jansen (Nordwijk: ESA), ESA SP-488 (astro-ph/0202430)

Heap, S. R., Williger, G. M., Smette, A., Hubeny, I., Sahu, M., Jenkins, E. B., Tripp, T. M., & Winkler, J. N. 2000, ApJ, 534, 69

Hirasawa, T. 1969 Prog. Theor. Phys., 42(3), 523

Holder, G. P., Haiman, Z., Kaplinghat, M., & Knox, L. 2003, ApJ, 595, 13

Hu, E. M., Cowie, L. L., McMahon, R. G., Capak, P., Iwamuro, F., Kneib, J.-P., Maihara, T., & Motohara, K. 2002, ApJ, 568, L75 (erratum: 2002 ApJ, 576, L99)

Hu, E. M., McMahon, R. G., & Cowie, L. L. 1999, ApJ, 522, L9

Hui, L., & Haiman, Z. 2003, ApJ, submitted (astro-ph/0302439)

Iliev, I. T., Shapiro, P. R., Ferrara, A., & Martel, H. 2002, 572, 123

Jenkins, A., Frenk, C. S., White, S. D. M., Colberg, J. M., Cole, S., Evrard, A. E., Couchman, H. M. P., & Yoshida, N. 2001, MNRAS, 321, 372

Kang, H., & Shapiro, P. R. 1992, ApJ, 386, 432

Kang, H., Shapiro, P. R., Fall, S. M., & Rees, M. J. 1990, ApJ, 363, 488

Kaplinghat, M., Chu, M., Haiman, Z., Holder, G. P., Knox, L., & Skordis, C. 2003, ApJ, 583, 24

Kochanek, C. S. 1998, in Science With The NGST, ed. E. P. Smith & A. Koratkar (San Francisco: ASP), 96

Kodaira, K., et al. 2003, PASJ, 55, L17

Kogut, A., et al. 2003, ApJ, 148, 161

Lepp, S., & Shull, J. M. 1984, ApJ, 280, 465

Lidz, A., Hui, L., Zaldarriaga, M., & Scoccimarro, R. 2002, ApJ, 579, 491

Loeb, A., & Rybicki, G. B. 1999, ApJ, 524, 527

Machacek, M. E., Bryan, G. L., & Abel, T. 2001, ApJ, 548, 509

——. 2003, MNRAS, 338, 273

Madau, P. 2003, in ESO-CERN-ESA Symposium on Astronomy, Cosmology, and Fundamental Physics, 39

——. 2004, in Galaxy Evolution: Theory and Observations, ed. V. Avila-Reese et al., in press (astro-ph/0212555)

Madau, P., Haardt, F., & Rees, M. J. 1999, ApJ, 514, 648

Madau, P., Meiksin, A., & Rees, M. J. 1997, ApJ, 475, 429

Madau, P., & Rees, M. J. 2000, ApJ, 542, L69

Matsuda, T., Sato, H., & Takeda, H. 1969, Prog. Theor. Phys., 42(2), 219

Miralda-Escudé, J. 1998, ApJ, 501, 15

Miralda-Escudé, J., Haehnelt, M., & Rees, M. J. 2000 ApJ, 530, 1

Miralda-Escudé, J, & Rees, M. J. 1998, ApJ, 497, 21

Mushotzky, R. F., Cowie, L. L., Barger, A. J., & Arnaud, K. A. 2000, Nature, 404, 459

Nakamoto, T., Umemura, M., & Susa, H. 2001, MNRAS, 321, 593

Nakamura, F., & Umemura M. 2002, ApJ, 569, 549

Narayanan, V. K., Spergel, D. N., Davé, R., & Ma, C.-P. 2000, ApJ, 543, L103

Nusser, A., Benson, A. J., Sugiyama, N., & Lacey, C. 2002, ApJ, 580, 93

Oh, S. P. 2001, ApJ, 553, 499

——. 2002, MNRAS, 336, 1021

Oh, S. P., Cooray, A., & Kamionkowski. M. 2003, MNRAS, 342, L20

Oh, S. P., & Haiman, Z. 2002, ApJ, 569, 558

Omukai, K., & Nishi, R. 1999, ApJ, 518, 64

Palla, F., Salpeter, E. E., & Stahler, S. W. 1983, ApJ, 271, 632

Partridge, R. B., & Peebles, P. J. E. 1967, ApJ, 147, 868

Peebles, P. J. E., & Dicke, R. H. 1968, ApJ, 154, 891

Pentericci, L., et al. 2002, AJ, 123, 2151

Press, W. H., & Schechter, P. L. 1974, ApJ, 181, 425

Razoumov, A. O., Norman, M. L., Abel, T., & Scott, D. 2002, ApJ, 572, 695

Rees, M. J., & Ostriker, J. P. 1977, ApJ, 179, 541

Rhoads, J. E., et al. 2002, BAAS, 201, 5221

——. 2003, AJ, 125, 1006

Rhoads, J. E., & Malhotra, S. 2001, ApJ, 563, L5

Ricotti, M., Gnedin, N. Y., & Shull, J. M. 2002, ApJ, 575, 49

Saslaw, W. C., & Zipoy, D. 1967, Nature, 216, 976

Schneider, P. 1992, A&A, 254, 14

Shapiro, P. R., Giroux, M. L., & Babul, A. 1994, ApJ, 427, 25

Shapiro, P. R., & Kang, H. 1987, ApJ, 318, 32

Shapiro, P. R., Raga, A. C., & Mellema, G. 1998, in Molecular Hydrogen in the Early Universe, ed. E. Corbelli, D. Galli, & F. Palla (Florence: Soc. Ast. Italiana), 463

Sheth, R. K., Mo, H. J., & Tormen, G. 2001, MNRAS, 323, 1

Sokasian, A., Abel, T., & Hernquist, L. 2003, MNRAS, 344, 607

Somerville, R. S., & Livio, M. 2003, ApJ, 593, 611

Somerville, R. S., Bullock, J. S., & Livio, M. 2003, ApJ, 593, 616

Songaila, A. 1998, AJ, 115, 2184

Songaila, A., & Cowie, L. L. 2002, AJ, 123, 2183

Spergel, D. N., et al. 2003, ApJ, 149, 175

Spinrad, H., Stern, D., Bunker, A. J., et al. 1998, AJ, 117, 2617

Stern, D., Spinrad, H., Eisenhardt, P., Bunker, A. J., Dawson, S., Stanford, S. A., & Elston, R. 2000, ApJ, 533, L75

Steidel, C. C., Pettini, M., & Adelberger, K. L. 2001, 546, 665

Susa, H., Uehara, H., Nishi, R., & Yamada, M. 1998, Prog. Theor. Phys., 100, 63

Tegmark, M., Silk, J., & Blanchard, A. 1994, ApJ, 420, 484

Tegmark, M., Silk, J., Rees, M. J., Blanchard, A., Abel, T., & Palla, F. 1997, ApJ, 474, 1

Thoul, A., & Weinberg, D. H. 1996, ApJ, 465, 608

Tozzi, P., Madau, P., Meiksin, A., & Rees, M. J. 2000, ApJ, 528, 597

Tumlinson, J., & Shull, J. M. 2000, ApJ, 528, L65

Valageas, P., & Silk, J. 1999, A&A, 347, 1

Venkatesan, A., Giroux, M. L., & Shull, J. M. 2001, ApJ, 563, 1

Weymann, R. J., Stern, D., Bunker, A., et al. 1998, ApJ, 505, L95
White, S. D. M., & Rees, M. J. 1978, MNRAS, 183, 341
Wyithe, J. S. B., & Loeb, A. 2002a, Nature, 417, 923
——. 2002b, ApJ, 577, 57
——. 2003a, ApJ, 586, 693
——. 2003b, ApJ, 588, L69
Yoshida, N., Sokasian, A., Hernquist, L., & Springel, V. 2003, ApJ, 591, L1
Yu, Q., & Tremaine, S. 2002, MNRAS, 335, 965

6

Adiabatic growth of massive black holes

STEINN SIGURDSSON
Department of Astronomy & Astrophysics, Pennsylvania State University

Abstract
We discuss the process of adiabatic growth of central black holes in the presence of a stationary, pre-existing distribution of collisionless stars. Within the limitations of the assumptions, the resulting models make robust physical predictions for the presence of a central cusp in the stellar and dark matter density, a Keplerian rise in the velocity dispersion, and a significant tangential polarization of the velocity tensor. New generations of numerical models confirm and extend previous results, permit the study of axisymmetric and triaxial systems, and promise new insights into the dynamics of the central regions of galaxies. These studies enable detailed comparisons with observations, further our understanding on the fueling processes for AGNs and quiescent black holes, and help elucidate the secular evolution of the inner regions and spheroids of galaxies.

6.1   Introduction
Given the premise that the massive central dark objects in normal galaxies in the local Universe are in fact supermassive black holes (Lynden-Bell 1969; Rees 1990; Kormendy & Richstone 1995; Richstone et al. 1998), we can entertain a number of conjectures about the interaction of the central black hole with its environment. Obvious questions to consider include: formation scenarios for the black hole (e.g., Rees 1984; Shapiro, this volume); the demographics of the present population of black holes (Richstone and Ho, this volume); the fueling of active nuclei (Blandford, this volume); the interaction of the active nucleus with its environment (Begelman, this volume); and, the effect of the central object upon the surrounding stellar population and the larger-scale structure of the host galaxy (Burkert, Gebhardt, Haehnelt, and Merritt, this volume). Hence, one can also test whether the inferred effects of the central object are consistent with observations, and whether additional observational constraints can be placed on either the presence or the evolutionary history of the central black hole.

A particular assumption can be made (with the caveat, that, as with all assumptions, it may be false) that a substantial increase in the mass of the central black hole takes place after initial formation, and that the mass is in some rigorous sense (to be established) added slowly to a pre-existing seed black hole. This is the assumption of *adiabatic growth*, which will be reviewed here. It leads to some nontrivial, testable predictions for the effects of black holes on their environments.

A central supermassive black hole dynamically dominates the surrounding stellar population inside some characteristic radius, $r_h = GM_\bullet/\sigma^2$, where $M_\bullet$ is the mass of the black hole and σ is the velocity dispersion of the stars outside the radius of influence. A natural "shortest" time scale for growth of a black hole is the "Salpeter" time scale, $t_S = M_\bullet/\dot{M}_{Edd} \approx 5 \times 10^7$ yr, where $\dot{M}_{Edd}$ is the usual Eddington accretion rate. The dynamical time scale inside r_h is just $t_{dyn} = r_h/\sigma$. For the Milky Way, $t_{dyn}(r_h) \sim 10^4$ yr, for $M_\bullet \sim 2 \times 10^6 M_\odot$ and $\sigma \approx 66\,\mathrm{km\,s}^{-1}$. If the observed correlation between dispersion and black hole mass holds, then $M_\bullet \propto \sigma^4$ (Gebhardt et al. 2000a; Ferrarese & Merritt 2000; Tremaine et al. 2002), and hence $t_{dyn} \propto \sigma$. Hence, we conclude that for reasonable black hole masses ($M_\bullet \lesssim 10^{10} M_\odot$), the dynamical time scale inside r_h is always much shorter than the Salpeter time scale, and therefore the likely time scale for black hole growth through accretion of baryonic matter is much longer than the dynamical time scale inside the radius at which the black hole dominates the dynamics. We thus conclude that there may be a broad range of situations under which black hole growth is "adiabatic" and the assumptions of these studies hold. The stellar population will generally form a density cusp, $\rho \propto r^{-A}$, inside r_h, with the stellar velocity dispersion showing a Keplerian rise $\sigma(r) \propto r^{-1/2}$ inside the cusp (Peebles 1972; Bahcall & Wolf 1976; Young 1980; Quinlan, Hernquist, & Sigurdsson 1995).

6.1.1   Assumptions

We consider the response of a stellar distribution function (DF) to the slow growth of a massive central black hole. The initial conditions assume that there is no central black hole to begin with (or more realistically a seed black hole with a negligible initial mass), and there are usually implicit assumptions that the mass is "magically" added to the black hole— that is to say, the mass of the central object is increased without necessarily withdrawing the mass from some explicit reservoir. The underlying assumption here is that the mass is accreting from some diffuse medium, such as cool gas, that is distributed like the stellar population but with a density much lower than that of the stellar mass density, and is replaced by some inflow that is an implicit outer boundary condition (e.g., Young 1980; Quinlan et al. 1995). This is *not* a necessary assumption; it is just a simplifying assumption. It is trivial to extend it to scenarios where the mass is explicitly withdrawn from some reservoir, with the added complication of having to specify the physical nature of the mass reservoir. In most situations modeled so far, where the mass comes from is not important; the response of the system is robust, independent of the source of the mass. The exception is if all the mass comes only from the black hole swallowing the most tightly bound stars, in which case the conclusions are somewhat different and the process effectively violates our assumptions of adiabatic growth.

An additional implicit assumption is that the black hole is *central*; that is, it is at the center of mass of the stellar system. In practice, the surrounding stellar system is discrete and the black hole mass is finite, so we expect the black hole to undergo quasi-Brownian motion away from the center (see, e.g., Chatterjee, Hernquist, & Loeb 2001). For masses of astrophysical interest, the displacement is typically much larger than the black hole Schwarzschild radius, r_S, but much smaller than r_h, and the time scale for wandering is short enough that the outer cusp is not carried with the black hole as it moves; this can lead to modification of the cusp profile at small radii ($r \lesssim 10^{-3} r_h$ for typical $M_\bullet$) as the black hole wandering produces rapid fluctuations in the central potential seen by stars in the inner cusp (see Sigurdsson 2003).

A final fundamental consideration is whether the dynamics of the stellar population are "collisionless"—that is, whether the relaxation time scale for a population of N stars, $t_R \sim N t_{dyn}/8 \ln \Lambda$, is shorter or longer than the evolutionary time scale of the stellar system, usually taken to be the Hubble time, $t_H \sim 10^{10}$ yr (e.g., Spitzer 1971; Hills 1975). The response of a relaxed stellar system to the presence of a central massive black hole has been extensively considered, primarily in the context of globular clusters, or in the context of initial black hole formation and rapid growth in protogalaxies (Bahcall & Wolf 1976, 1977; Lightman & Shapiro 1977; Cohn & Kulsrud 1978; Shapiro & Marchant 1978; Shapiro 1985; Amaro-Seoane & Spurzem 2001 and Freitag & Benz 2001 and citations therein). For supermassive black holes in normal, evolved, galaxies the relaxation time scales in the inner spheroid, but outside the black hole cusp, are generally longer than the Hubble time; inside the cusp the relaxation time may be constant, increase, or decrease with decreasing radius. For those cases where the relaxation time decreases with decreasing radius, the dynamics of the stellar population surrounding the black hole *may* undergo a transition to the fully collisional regime in the inner cusp, and the discussion in the papers cited above then becomes appropriate but is beyond the scope of this review. The mean central relaxation time can be approximated as $t_R \sim 2 \times 10^9 (\sigma/200 \mathrm{km\,s^{-1}})^3 / (\rho/10^6 M_\odot \mathrm{pc^{-3}})$ (Young 1980). It is not sufficient that $t_R < t_H$ for non-adiabatic growth. For such relaxed cusps the relaxation time at small radii may become shorter, and, if $t_R \lesssim t_S$ at some small radius, which may well occur for a significant fraction of galactic nuclei or proto-nuclei at some point in their evolution, then any central black hole may grow by tidal disruption of stars or by swallowing stars whole, more rapidly then the cusp can dynamically readjust its structure; in such a situation, the growth is definitely non-adiabatic.

The underlying physical assumption of the "adiabatic growth" model is that as the integrals of motion change smoothly in response to the increase in central mass, the action variables for the surrounding stellar population remain invariant (Binney & Tremaine 1987). This is to be contrasted with the opposite extreme assumption of "violent relaxation," in which the potential is assumed to fluctuate rapidly compared to the dynamical time, and the DF evolves to some final statistical equilibrium state (Lynden-Bell 1967; Stiavelli 1998). The resulting "final distribution" may then be compared with observations. It should be noted that real galaxies may not have "initial" DFs that are well represented by any of the analytic or numerical distributions assumed in these models, nor is it necessarily the case that significant increase in black hole mass ever takes place under conditions in which the adiabatic approximation holds. In particular, an implicit assumption is that a relaxed stellar population is in place as an initial condition, and that significant increase in black hole mass takes place *after* (the inner region of) the galaxy is assembled. The adiabatic models are physically distinct from *ab initio* models, where a DF including a central black hole, by design, is required to satisfy the Boltzmann equation (e.g., Huntley & Saslaw 1975; Tremaine et al. 1994). The adiabatic models are also distinct from the "orbit assembly" models used to construct kinematic models of observed galaxies (Schwarzschild 1979; Richstone & Tremaine 1984, 1988; Magorrian et al. 1998).

An interesting question is whether any of these models in some sense rigorously represents real stellar systems. Nature need not settle on the analytically or numerically derived solutions of the Boltzmann equation, out of the infinite number that exist. As found by Quinlan et al. (1995), apparently small differences in some phase-space values can lead to large changes in the averaged properties of the evolved system. We may also worry whether the different techniques for constructing stationary solutions of the

Boltzmann equation representing collisionless stellar objects surrounding a central black hole are actually equivalent, or whether the different techniques produce wholly distinct families of solutions, as opposed to solutions with an overlap in properties or formally identical for some range of parameters.

6.2 Spherical Growth

The response of a spherical distribution of stars to the adiabatic growth of a central black hole in the collisionless limit was first considered by Peebles (1972) for an isothermal sphere. Young (1980) confirmed the primary result that a density cusp $\rho \propto r^{-3/2}$ would form, with an associated velocity dispersion cusp, $\sigma(r) \propto r^{-1/2}$; he also pointed out that the velocity anisotropy, $\beta(r) = 1 - \langle v_t^2 \rangle / \langle 2v_r^2 \rangle$, becomes negative (tangentially biased) at small radii, where v_t and v_r is the tangential and radial velocity, respectively. Goodman & Binney (1984) showed that when $\beta(0) = 0$ the distribution is isotropic at the center for an initial isothermal distribution (see also Binney & Petit 1989), and Lee & Goodman (1989) generalized the approximate solution of the problem to axisymmetric rotating distributions. The basic physics of the problem for a spherical system are discussed in Shapiro & Teukolsky (1983 and reference therein), as a simple application of Liouville's theorem. Their Equation 14.2.9 shows the response of a spherical system, with some initial DF $f(E)$, to a central black hole. The final density $n(R) = 4\pi \int f(E)\sqrt{[2(E-\Phi)]}dE \propto r^{-1/2} \times r^{-1}$ for $f(E) \rightarrow$ constant, appropriate for the $n = 0$ case discussed by Quinlan et al. (1995).

Quinlan et al. (1995) generalized the result to a broad range of initially spherical DFs and found that for different DFs the final cusp slope may be very different, even for near-identical initial spatial density profiles. They also found that, in contrast with the result for initially isothermal distributions, for some initial DFs the polarization of the velocity distribution is generic and always tangentially biased, and that the tangential bias may persist to zero radius. The velocity distribution is in general non-Gaussian, and initially non-Gaussian distributions may evolve to be either closer to or farther from Gaussian in response to the black hole growth (Sigurdsson, Hernquist, & Quinlan 1995). The net results are distinct, but unfortunately do not provide a simple or unique prediction for the final spherical distribution of a stellar population responding adiabatically to the growth of a central black hole. The semi-analytic results of Quinlan et al. were confirmed numerically in a companion paper by Sigurdsson et al. (1995), who extended the numerical methodology to a family of non-spherical models.

A major purpose for producing a broad range of adiabatic growth models is for comparison with observations, for example to establish robust estimators for central black hole masses from the observed surface density profiles or spectroscopically determined projected velocity dispersion profiles. In addition to the intrinsic degeneracies between the DFs and the density and dispersion profiles, we are mostly restricted to observing projected quantities, the line integrals of the light density and velocity distribution, which lead to degeneracies in the inversion to the full volume distribution (e.g., Romanowsky & Kochanek 1997 and references therein). We are further restricted to observing the dominant light-emitting population (mainly giant, sub-giant and post-AGB stars), and the mass may be distributed differently, with different stellar populations (or dark matter) having different density profiles. Still, with the use of higher moments of the velocity distribution (van der Marel & Franx 1993; Dehnen & Gerhard 1994; van der Marel 1994a, b; van der Marel et al. 1994) strong constraints can be put on the true stellar DF; by making some "natural" assumptions (e.g., the unobserved dark matter distribution is consistent with the light distribution), strong constraints can be put on the total mass of any inferred central dark object.

6.2.1   Action

In a spherical potential, we consider some initial DF f specified by the energy E and angular momentum L. The quantity f is then also a function of the actions L and $J_r = \oint v_r dr$. As the integrals change under the adiabatic growth of the black hole, the action, by assumption, remains invariant, and f evolves to remain a fixed function of the actions (see Young 1980 and Quinlan et al. 1995 for discussion).

We want to consider some initial DF with an explicitly specified form and a corresponding density profile (see Binney & Mamon 1982; Binney & Tremaine 1987). Of particular interest is the asymptotic behavior of the density profile at small radii $[\rho(r) \propto r^{-\gamma}; \text{ as } r \to 0]$ and the corresponding asymptotic behavior of $f(E)$ in the limit $E \to \Phi(0)$, which in general is some power law $f(E) \sim [E - \Phi(0)]^{-n}$. [But note that real galaxies need not be nicely monotonic power laws, even asymptotically, but may, for example, have density inversions at small radii (e.g., Peebles 1972; Lauer et al. 2002).] For an isothermal density profile the central density approaches a constant at small radii, as does the DF at the lowest energies. In general, $0 \leq \gamma \leq 3$, and it is useful to distinguish between models with "analytic" cores (in the nomenclature of Quinlan et al. 1995), which have a density profile $\rho(r) \approx \rho_0 + \frac{1}{2}\rho'' r^2 + \dots$, as $r \to 0$, and non-analytic models, which do not approximate a harmonic potential at the origin. Particular examples of analytic models include a non-singular isothermal sphere, a King model, a Plummer model, or an isochrone. Non-analytic models include (1) a singular isothermal sphere (Cipollina & Bertin 1994); (2) the $\gamma = 2$ Jaffe (1983) or $\gamma = 1$ Hernquist (1990) model; (3) the generalized "gamma" models ($0 \leq \gamma \leq 3$; Dehnen 1993; Tremaine et al. 1994); (4) and further spherical generalizations of these models, such as those of Navarro, Frenk, & White (1997) and Zhao (1997).

It turns out that the final density cusp slope A generally depends both on the initial density slope γ and the asymptotic divergence of the DF n; the relationship among the variables is given analytically by*

$$A = \frac{3}{2} + n\left(\frac{2-\gamma}{4-\gamma}\right). \tag{6.1}$$

As illustrated in Figure 6.1, which compares the adiabatic growth of a black hole in a $\gamma = 0$ model with an isochrone model (Hénon 1960), analytic and non-analytic models with the same, or very nearly the same, initial density profiles produce qualitatively different final density profiles. More generally, the response to the adiabatic growth of a black hole produces a cusp with a slope as low as $A = 3/2$, as originally found, up to values as steep as $A = 3$, although $A = 5/2$ is probably the steepest physically sustainable slope before collisional effects in the inner cusp necessarily dominate the dynamics. Table 6.1 lists the values of some initial and final slopes (Quinlan et al. 1995); C is the final cusp slope in the limit of an initially completely tangentially biased DF. Note that the presence of a density cusp by itself is *not* a robust indicator of a central supermassive black hole; this is clearly so, since, for example, the Jaffe (1983) model or a singular isothermal sphere, with no central

---

* For the full derivation, see Appendix A of Quinlan et al. (1995), Gondolo & Silk (1999), and Ullio, Zhao, & Kamionkowski (2001).

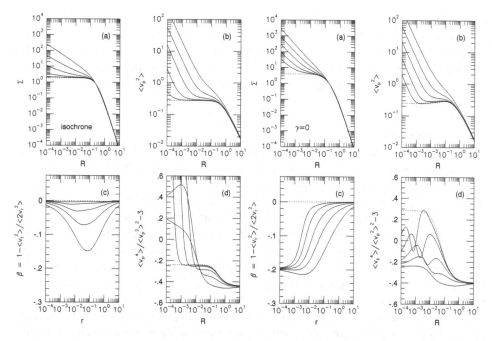

Fig. 6.1. Results of the growth of black holes of different masses, $M_\bullet = 10^{-3} - 10^{-1}$ of the total (spheroid) galaxy mass, in an isochrone (*left*) and $\gamma = 0$ (*right*) model, respectively. The dotted lines show the initial (near-identical) models. The panels show (*a*) surface density, (*b*) projected velocity dispersion, (*c*) projected anisotropy $\beta(r)$, and (*d*) kurtosis $\kappa - 3$. Note the very different final profiles despite near-identical observable initial conditions. (From Quinlan et al. 1995.)

Table 6.1. *Adiabatic density cusps*

Model	γ	n	A	C
isochrone	0	0	3/2	9/4
$\gamma = 0$	0	1	2	9/4
$\gamma = 1$	1	5/2	7/3	7/3
$\gamma = 3/2$	3/2	9/2	12/5	12/5
$\gamma = 2$	2	–	5/2	5/2

black holes, have $\gamma = 2$ cusps, steeper than the $A = 3/2$ cusps predicted for the response of a non-singular isothermal sphere to a central black hole. On the other hand, the presence of a density cusp, a Keplerian rise in the velocity dispersion, and the kinematic signature of tangential anisotropy at small radii *are* robust indicators of a central supermassive black hole.

This does not preclude the possibility that in the absence of a central black hole actual stellar systems tend toward flat, constant density cores, whether through formation or relaxation, and that in practice cusps are in fact signatures of central black holes. We know that for a broad range of formation scenarios, stellar cusps form around central black holes (with cusps as shallow as $A = 1/2$ or as steep as $A = 5/2$); however, it is possible that in some situations binary black holes completely destroy cusps, leaving density inversions (e.g., Peebles 1972), and we know it is possible for cuspy stellar systems to exist in the absence of central black holes. Assuming that cusps are tracers of black holes, van der Marel (1999) has explored the use of the adiabatic growth models in matching observed density profiles.

6.2.2   Anisotropy

Quinlan et al. (1995) also experimented with spherical, radially anisotropic distributions (Osipkov 1979; Tonry 1983; Merritt 1985; Dejonghe 1987; Cudderford 1991; Gerhard 1993), but found it impossible to generate physical distributions with significant radial anisotropy persisting to zero radius. The general conclusion is therefore that adiabatic growth induces tangential bias at small radii, and that an initial tangential bias can, but does not necessarily, lead to steeper final cusps, compared to the equivalent isotropic model. The Keplerian velocity cusp is a robust prediction of spherical adiabatic growth models. As noted by Duncan & Wheeler (1980), however, a strong radial velocity anisotropy can mimic a Keplerian rise in velocity in projection, although there are severe concerns about the stability of any such model (Merritt 1987; Palmer & Papaloizou 1988). A robust prediction of a tangential bias induced by a central black hole is therefore potentially important, although the anisotropy is not a directly observable quantity but must be inferred from the projected moments of the velocity distribution.

As shown in Figure 6.1, the final anisotropy, $\beta(r)$, may be either zero at the black hole or remain negative at small radii. The deviation from Gaussianity is conveniently measured by the kurtosis, κ (by construction, the skew is zero for these models), or equivalently, the fourth Gauss-Hermite moment, $h_4 \approx (\kappa - 3)/8\sqrt{6}$, for $h_4 \lesssim 0.03$ (van der Marel & Franx 1993; Dehnen & Gerhard 1994; Quinlan et al. 1995).

With $A = 3/2 + p > 3/2$, the relaxation time scale at small radii decreases as $t_R \propto r^p$. The cusps induced in isotropic, analytic models have $p = 0$ and constant t_R; more generally, $p > 0$ and t_R can be small close to the black hole. Very close to the black hole, relaxation and collision time scales get short for strong cusps, and strong collisional effects may lead to rapid growth of the black hole, with corresponding associated depletion of the stellar population. This is certainly the case for cusps as steep as $A = 3$, and may even be a problem for shallower cusps (Frank & Rees 1976; Quinlan & Shapiro 1990; Quinlan et al. 1995; Sigurdsson & Rees 1997; Freitag & Benz 2001).

6.2.3   Non-adiabatic Growth

Formally, the adiabatic growth model implies an infinitely long time scale for accretion. In practice, of course, any growth in mass occurs on a finite time scale. We can investigate the nature of non-adiabatic growth without losing the predictive power of the adiabatic models.

Sigurdsson et al. (1995; see also Hernquist & Ostriker 1992; Hernquist, Sigurdsson, & Bryan 1995; Sigurdsson et al. 1997a) explored the time scale for adding mass, and concluded that, for time scales $t \gtrsim 10t_{dyn}(r_h)$, the adiabatic approximation was satisfied for the resolution of the models. The use of N-body modeling also showed the final distributions after adiabatic growth were stable; stability is not guaranteed by the adiabatic growth process, nor is there a general analytic criterion for stability of arbitrary DF.

Adiabatic growth formally also implies reversibility. Sigurdsson & Hernquist (unpublished) experimented with numerical models in which a central black hole grown adiabatically in a spherical stellar distribution was *removed* adiabatically. The original distribution was in fact recovered to within the resolution of the models.

In general, violent formation can lead to either galaxies with constant-density cores (e.g., Lynden-Bell 1967; van Albada 1982; Norman, May, & van Albada 1985; Burkert, this volume) or singular profiles (e.g., Aarseth 1966; Fillmore & Goldreich 1984; Bertschinger 1985; Navarro et al. 1997).

Stiavelli (1998) and Ullio et al. (2001) explored non-adiabatic growth with a pre-existing black hole and found results that did not deviate strongly from the case of adiabatic growth. More recently, MacMillan & Henriksen (2002) suggested that non-adiabatic accretion of dark matter might account for the $M_\bullet - \sigma$ relation, which is not explained by a simple adiabatic compression of the dark matter halo (Dubinski & Carlberg 1991). Adiabatic compression of the dark matter in the inner regions by the formation of a central black hole is potentially interesting, as it can lead to increased rates of dark matter accretion onto the black hole, and to higher rates of dark matter self-interaction, for models in which such interactions may occur (Gondolo & Silk 1999; Ostriker 2000).

Sigurdsson et al. (1995) also found that steep initial density cusps were vulnerable to violent disruption by the "wandering" of the central black hole. Black hole mergers will also efficiently destroy steep stellar cusps around a black hole (Makino & Ebisuzaki 1996; Quinlan & Hernquist 1997; Faber et al. 1997; Milosavljević & Merritt 2001; Zier & Biermann 2001; Hemsendorf, Sigurdsson, & Spurzem 2002; Ravindranath, Ho, & Filippenko 2002).

6.3   Non-spherical Systems

The obvious next approximation beyond spherical (isotropic and anisotropic) models is to consider axisymmetric ones. A number of families of two- and three-integral axisymmetric models exist in the literature (e.g., Evans 1993; Hunter & Qian 1993; Kuijken & Dubinski 1994; Qian et al. 1995; Gebhardt et al. 2000b; Lynden-Bell 2002).

Van der Marel et al. (1997a) constructed a detailed model for the central black hole in M32, and van der Marel, Sigurdsson, & Hernquist (1997b) ran a numerical model, using techniques developed for adiabatic growth simulations, to demonstrate its stability. A concern remains that such models may be unstable to $m = 1$ modes, which are typically suppressed in numerical simulations (if not, they can arise spontaneously through numerical artifacts, which make it difficult in general to identify physical instabilities). Lee & Goodman (1989) modeled adiabatic growth of black holes in approximate rotating, isothermal axisymmetric models. Rather interestingly, they found that the rotation curve rises more rapidly than the dispersion curve, but not enough to account for the observed high rotation in the inner regions of some systems. A more general exploration of adiabatic growth in non-rotating axisymmetric models was done by Sigurdsson & Hernquist (unpublished, see Fig. 6.2). They found that axisymmetric models are quite similar to spherical models, particularly in that the tangential anisotropy is induced in the cusp and the density profile becomes rounder at small radii (Fig. 6.2). Leeuwin & Athanassoula (2000) simulated adiabatic growth in Lynden-Bell (1962) models and obtained results consistent with those of Lee & Goodman (1989), including rounding of the inner density profile and

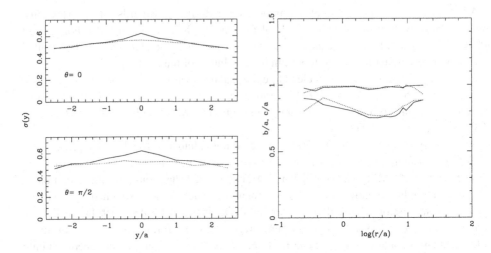

Fig. 6.2. Results of the growth of a black hole in an "Evans model." The panels show the projected density and axis ratios for edge-on and face-on inclinations before (*dotted line*) and after (*solid line*) black hole growth. The inner density profile becomes rounder in response to the black hole, and the characteristic Keplerian dispersion profile is seen in projection. (From Sigurdsson & Hernquist, unpublished.)

a significant rise in the rotation velocity inside the cusp, consistent with the tangential polarization of the central black hole.

6.3.1   Triaxial Systems

Real galaxies are generally not spherical or axisymmetric, but triaxial (Binney 1976; Franx, Illingworth, & de Zeeuw 1991; Ryden 1992, 1996; Tremblay & Merritt 1995; Bak & Statler 2000). We expect triaxial galaxies to form from general cosmological initial conditions (e.g., Norman et al. 1985; Dubinski & Carlberg 1991) and from galaxy mergers (Barnes 1988, 1992; Hernquist 1992, 1993). Exact analytic models exist for triaxial galaxies with cores (Schwarzschild 1979; de Zeeuw 1985; Statler 1987; van de Ven et al. 2003). Observationally, we also see that the density profiles of the spheroidal component of galaxies generally continue to rise toward the center, with $0.5 \lesssim \gamma \lesssim 2.3$ (e.g., Lauer et al. 1995; Gebhardt et al. 1996; Faber et al. 1997; Ravindranath et al. 2001). There are also observed correlations between the cusp slope γ and the global properties of the galaxy, including shape in the form of boxy or disky isophotes (e.g., Faber et al. 1997).

Historically, dynamical arguments suggest that the presence of a strong central cusp ($\gamma > 1$) induces chaos in the orbit families that populate the galaxy, driving the system away from strong triaxiality (e.g., Gerhard & Binney 1985; Norman et al. 1985; Merritt & Valluri 1996; Merritt & Quinlan 1998; Merritt 1997, 1999). The argument is that central cusps or central point masses scatter the box orbits that support triaxiality in galaxies, inducing a population of chaotic orbits that drive figure evolution toward axisymmetry (Miralda-Escudé & Schwarzschild 1989; Lees & Schwarzschild 1992; Fridman & Merritt 1997; Valluri & Merritt 1998). Hence, central supermassive black holes should preclude the presence of triaxiality at small radii, and might drive global figure evolution of the system (Norman et al. 1985; Merritt & Quinlan 1998). This is potentially very important because triaxial potentials support fueling of the central black hole through material falling into it on box orbits (e.g., Norman & Silk 1983; Valluri & Merritt 1998) or by gas traveling on intersecting orbits that drive dissipation and inflow, thus providing a direct link between the dynamics in the center of the galaxy and its global properties. In the extreme case of disk systems, analogous instabilities exist (e.g., Hasan & Norman 1990; Sellwood & Valluri 1997).

6.3.2   Adiabatic Growth and Triaxiality

Holley-Bockelmann et al. (2001; see also Sigurdsson et al. 1997b, 1998) showed that applying numerical adiabatic growth techniques to "squeeze" an initially spherically symmetric cuspy DF could produce a stable, stationary, cuspy triaxial configuration with well-characterized phase-space properties. A key aspect of the models is that they contain a central cusp of near-constant slope and near-constant axis ratios, with significant triaxiality at all radii resolved by the models (Holley-Bockelmann et al. 2001, 2002). Galaxies with density cusps support different stellar orbits than, for instance, $\gamma = 0$ core models (Gerhard & Binney 1985; Gerhard 1986; Pfenniger & de Zeeuw 1989; Schwarzschild 1993; de Zeeuw 1995; Merritt 1999; Holley-Bockelmann et al. 2001). The set of models thus produced provide a starting point for investigation of the adiabatic growth of central black holes in cuspy, triaxial potentials. A black hole is then grown using the previously developed numerical N-body techniques (Sigurdsson et al. 1995; Holley-Bockelmann et al. 2002).

Following Holley-Bockelmann et al. (2001, 2002), consider a black hole grown in a triaxial Hernquist model with initial cusp slope $\gamma = 1$. As the black hole grows, both the cusp slope γ and central velocity dispersion σ_p increase, as in spherical and axisymmetric models. The cusp settles to an equilibrium value $\gamma \simeq 2.05$, measured at projected ellipsoidal radius $Q = 10^{-1.3}$, with projected central dispersion $\sigma_p \simeq 0.7$, measured at projected ellipsoidal radius $Q = 10^{-2.3}$. These results are characteristic of adiabatic black hole growth in cuspy galaxies and can be compared both to analytic estimates for adiabatic black hole growth in a spherical $\gamma = 1.0$ model, which predict $\gamma = 7/3$ and $\sigma_p = 0.75$ (Quinlan et al. 1995), and to the results from N-body simulations where $\gamma \approx 2.2$ and $\sigma_p \approx 0.65$ (Sigurdsson et al. 1995). The fact that the measured cusp slope is less than the analytic value is to be expected, since the cusp slope is measured over a finite radial range near the center, and it is not the asymptotic $q = 0$ value.

As the black hole grows, the inner regions become rounder (Fig. 6.3b); the central 10% of the mass, corresponding to an ellipsoidal radius $q = \sqrt{x^2 + (y/b)^2 + (z/c)^2} < 0.1$, is close to spherical with axis ratios $a : b : c = 1.0 : 0.95 : 0.92$. The shape evolution in the outer regions is much less dramatic. Following the growth of the black hole, the model exhibits a marked shape gradient, becoming more strongly triaxial with increasing radius. Despite the nearly axisymmetric shape at the center, the inner region is still triaxial enough to influence the stellar-orbital dynamics (Statler 1987; Hunter & de Zeeuw 1992; Arnold, de Zeeuw, & Hunter 1994).

The final state of this model features several hallmarks of a black hole-embedded triaxial figure. Figure 6.3 shows the properties of this object as a function of ellipsoidal radius q at $T = 40$ (12.8 t_{dyn} at $q = 1$), well after the model black hole has stopped growing. Figure 6.3a shows the $\gamma \approx 2$ density cusp induced by the black hole inside $\log q = -1$. At a larger radii $\log q > -1$, however, this plot demonstrates that the system retains the original Hernquist density profile. Figure 6.3b shows explicitly the strong shape gradient in the model. Inside r_h, both the projected and intrinsic velocity dispersions exhibit a strong

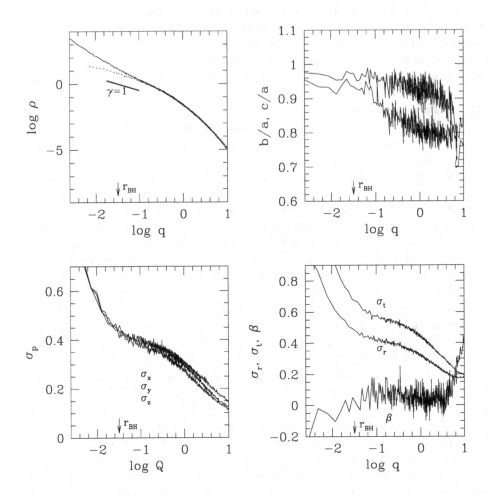

Fig. 6.3. The structural and kinematic properties of a cuspy triaxial galaxy model with a massive central black hole. *Upper left*: density profile; *upper right*: intermediate and minor axis lengths as a function of ellipsoidal radius; *lower left*: projected velocity dispersion along the fundamental axes, as a function of projected ellipsoidal radius; *lower right*: true radial and tangential velocity dispersion, and velocity anisotropy parameter, as a function of ellipsoidal radius. (From Holley-Bockelmann et al. 2002.)

central cusp (panels *c* and *d*). In the outskirts, where the model maintains its triaxiality, the projected velocity distributions follow $\sigma_x > \sigma_y > \sigma_z$, in accord with a triaxial model where $a > b > c$. However, inside the cusp the projected velocity dispersions are commensurate. Interestingly, the anisotropy parameter, $\beta = 1 - \langle v_t^2 \rangle / \langle 2v_r^2 \rangle$, becomes negative near the black hole. This is consistent with models of stellar orbits around a black hole that is adiabatically grown, where $\beta = -0.3$ (Goodman & Binney 1983; Quinlan et al. 1995). Exterior to the black hole's radius of influence, the system is radially anisotropic ($\beta > 0$), as expected for a triaxial galaxy.

Poon & Merritt (2001, 2002), using Schwazschild's orbit-assembly technique, have now also found models with triaxiality at small radii in the presence of a central black hole and a surrounding cusp.

Clearly, it is possible for some significant triaxiality to persist both in the presence of a central cusp, and, more importantly, in the presence of a central black hole.

6.3.3 Chaos

If a significant fraction of the orbits in triaxial models containing central black holes become chaotic, then by the ergodic theorem the shape of the distribution must evolve toward sphericity (possibly halting when axisymmetry is reached). It is clear that the onset of chaos in the most tightly bound orbit families leads to a rapid change in the inner structure of the model galaxies. In the outer regions, orbits stay regular even after repeated passages near the potential center. It is possible that many of these orbits are actually chaotic orbits that are "sticky" (Siopis & Kandrup 2000), with a very long diffusion time scale. The course grainedness of the numerical model potential seems to argue against this explanation; a course-grained potential effectively creates holes in the Arnold web (Arnold 1964) through which an otherwise confined orbit may escape. Figure 6.4 illustrates what is probably happening in these models; strong scattering is taking place, inducing some chaos, but triaxiality is sustained by the persistence of resonant boxlet orbits that scatter between each other, rather than into true chaotic orbits.

There are two important issues to be explored here.

- When a spherical (or axisymmetric) model is squeezed adiabatically into a triaxial configuration, one of the implicit assumptions of adiabatic growth is violated. The evolution of the potential has discretely broken a symmetry underlying the second and third integrals of motion; incidentally, the reversibility of the process is also destroyed. However, the action is still conserved, at least approximately, so the DF must bifurcate, leaving an excluded region of phase space. In 2-D this region would be forbidden; in 3-D other bifurcating branches can cross-over into the newly created vacant region of phase space, but do not in general fill it. By construction, this technique leaves vacant islands in phase space, and consequently we reach stable and stationary triaxial configurations despite the presence of the black hole. Numerically these are robust solutions, but they are not guaranteed to be robust physically. It is possible that small amounts of relaxation or potential fluctuation could rapidly refill these vacated phase-space regions, leading to boxlet–chaos transitions, breaking the dynamical equilibrium constructed for boxlet–boxlet transitions. Some of this is seen through chaos induced by numerical scattering in the models.

  We cannot yet be sure that our triaxial adiabatic solutions are physically robust ones achieved by natural systems, as distinct from mathematical curiosities; nevertheless, they are potentially very interesting solutions for triaxial systems.

- We also do not understand well how a transition to chaos occurs in these systems. The scattering by the central singular potential, in and of itself, need not induce chaos. After all, perturbed orbits in Keplerian potentials are regular. Some insight can be gained by considering a toy 2-D model (to be compared with the dynamics in the principal plane of a triaxial system).

  Following Devaney (1982), consider a homogeneous, *anisotropic* potential of degree k, $\Phi(r, \psi) \propto r^{-k}$. There are three special values of $k = \{0, 1, 2\}$, the first two corresponding to the isothermal and Keplerian potential, respectively. We consider the characteristic

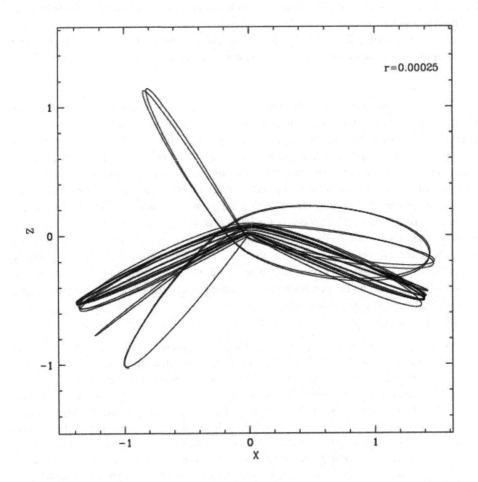

Fig. 6.4. The evolution of a banana boxlet orbit due to scattering after close approach to a central black hole in a triaxial potential. The banana orbit flips to a fish orbit, another resonant boxlet, and does not become chaotic due to strong scattering. (From Holley-Bockelmann et al. , in preparation.)

exponents in the linearized theory for 2-D orbits and explore the effects of varying k and the degree of anisotropy. Solving for the characteristic exponent λ, we find

$$\lambda(k,r) = \frac{1}{2}[(\frac{1}{2}k-1)v_r - [(\frac{1}{2}k-1)^2 v_r^2 - 4\Phi''(\psi)]^{1/2}] \tag{6.2}$$

where Φ'' is the second angular derivative of the potential, which is by hypothesis self-similar (i.e., the shape of the isopotential contours is independent of radius). Clearly, realistic galaxy models are not homogeneous, but at large and small radii they well approximate a homogeneous potential, and the dynamics, in particular the orbit divergences, are due to *local* potential gradients.

Instability formally occurs when λ is imaginary, so the critical points occur when

$[(\frac{1}{2}k-1)^2 v_r^2 - 4\Phi''(\psi)] = 0$. Using Poisson's equation, $\Phi'' = 4\pi r^2 \rho - 2v_c^2 - k(k-1)\Phi$, where, by hypothesis, $\Phi''(\psi)$ is independent of radius. We therefore conclude, that for Keplerian and isothermal potentials orbits in this model go chaotic at large r, not crossing through the center. For $k = 2$, appropriate for the outer regions of galaxy models, orbits that are chaotic anywhere are chaotic everywhere.

If the dynamics of the toy 2-D homogeneous model are a good indicator, then the onset of chaos in triaxial models occurs not at small radii, but in the transition region between the inner cusp and the region outside r_h.

6.3.4 Uses of Triaxial Models

Much work remains to be done. The following are some implications worth exploring.

- The gas dynamics on intermediate scales and possibilities for AGN fueling may depend strongly on even mild triaxialities on small scales.
- The dynamical evolution of merging dwarf galaxies and the fate of low-mass black holes merging with massive galaxies may be sensitive to triaxiality persisting as the central black hole becomes more massive. Dynamical friction processes may be expedited through centrophilic orbits in triaxial potentials.
- On small scales, triaxiality inside r_h may promote rapid loss-cone refilling and be critical for sustaining high tidal-disruption rates and the influx of low-mass compact objects coalescing with the central massive black hole (Sigurdsson 2003).

6.4 Conclusions

Adiabatic growth models provide valuable physical insights into the dynamical processes and interactions of central massive black holes with their surroundings. Over the last 30 years, a broad range of physically robust results on the dynamical influence of the black hole on the surrounding stellar population have contributed to our confidence in the reality of supermassive black holes and helped provide strong quantitative constraints on black hole masses. Additional physical insight has been gained in understanding the ever-fascinating subtleties of Newtonian dynamics of many-body systems. New generations of N-body models will allow a broader exploration of more realistic, less symmetric systems, more tests of stability and secular evolution, and possibly a deeper understanding of dynamical processes on small and large scales.

Acknowledgements. The author gratefully acknowledges the support of the NSF under grant PHY-0203046, the Center for Gravitational Wave Physics at Penn State, an NSF-supported Physics Frontier Center, and the hospitality of Carnegie Observatories. I would like to thank my collaborators Gerry Quinlan, Lars Hernquist, Roeland van der Marel, Chris Mihos, Colin Norman, and especially Kelly Holley-Bockelmann for her hard work and for letting me use unpublished results.

References

Aarseth, S. J. 1966, MNRAS, 132, 35
Amaro-Seoane, P., & Spurzem, R. 2001, MNRAS, 327, 995
Arnold, V. I. 1964, Russ. Math. Surveys, 18, 8
Arnold, R., de Zeeuw, P. T., & Hunter, C. 1994, MNRAS, 271, 924
Bahcall, J. N., & Wolf, R. A. 1976, ApJ, 209, 214

———. 1977, ApJ, 216, 883

Bak, J., & Statler, T. 2000, AJ, 120, 110

Barnes, J. E. 1988, ApJ, 331, 699

———. 1992, ApJ, 393, 484

Bertschinger, E. 1985, ApJS, 58, 39

Binney, J. 1976, MNRAS, 177, 19

———. 1985, MNRAS, 212, 767

Binney, J., & Mamon, G. A. 1982, MNRAS, 200, 361

Binney, J., & Petit, J.-M. 1989, in Dynamics of Dense Stellar Systems, ed. D. Merritt (Cambridge: Cambridge Univ. Press), 43

Binney, J., & Tremaine, S. 1987, Galactic Dynamics (Princeton: Princeton Univ. Press)

Chatterjee, P., Hernquist, L., & Loeb, A. 2001, ApJ, 572, 371

Cipollina, M., & Bertin, G. 1994, A&A, 288, 43

Cohn, H., & Kulsrud, R. M. 1978, ApJ, 226, 1087

Cudderford, P. 1991, MNRAS, 253, 414

Dehnen, W. 1993, MNRAS, 265, 250

Dehnen, W., & Gerhard, O. E. 1994, MNRAS, 268, 1019

Dejonghe, H. 1987, MNRAS, 224, 13

Devaney, R. L. 1982, Am. Math. Month, Oct. '82, 535

de Zeeuw, P. T. 1985, MNRAS, 216, 273

———. 1995, in Gravitational Dynamics, ed. O. Lahav, E. Terlevich, & R. J. Terlevich (Cambridge: Cambridge Univ. Press), 1

Dubinski, J., & Carlberg, R. G. 1991, ApJ, 378, 496

Duncan, M. J., & Wheeler, J. C. 1980, ApJ, 237, L27

Evans, N. W. 1993, MNRAS, 260, 191

Faber, S. M., et al. 1997, AJ, 114, 1771

Ferrarese, L., & Merritt, D. 2000, ApJ, 539, L9

Fillmore, J. A., & Goldreich, P. 1984. ApJ, 281, 1

Frank, J., & Rees, M. J. 1976, MNRAS, 176, 633

Franx, M., Illingworth, G., & de Zeeuw, P. T. 1991, ApJ, 383, 112

Freitag, M., & Benz, W. 2001, A&A, 394, 345

Fridman, T., & Merritt, D. 1997, ApJ, 114, 1479

Gebhardt, K., et al. 2000a, ApJ, 539, L13

———. 2000b, AJ, 119, 1157

Gebhardt, K., Richstone, D., Ajhar, E. A., Kormendy, J., Dressler, A., Faber, S. M., Grillmair, C., & Tremaine, S. 1996, AJ, 112, 105

Gerhard, O. E. 1986, MNRAS, 219, 373

———. 1993, MNRAS, 265, 213

Gerhard, O. E., & Binney, J. 1985, MNRAS, 216, 467

Gondolo, P., & Silk, J. 1999, Phys. Rev. Lett., 83, 1719

Goodman, J., & Binney, J. 1984, MNRAS, 207, 511

Hasan, H., & Norman, C. A. 1990, ApJ, 361, 69

Hemsendorf, M., Sigurdsson, S., & Spurzem, R. 2002, ApJ, 581, 1256

Hénon, M. 1960, Ann. d'Astrophys., 23, 474

Hernquist, L. 1990, ApJ, 356, 359

———. 1992, ApJ, 400, 460

———. 1993, ApJ, 409, 548

Hernquist, L., & Ostriker, J. P. 1992, ApJ, 386, 375

Hernquist, L., Sigurdsson, S., & Bryan, G. L. 1995, ApJ, 446, 717

Hills, J. G. 1975, Nature, 154, 295

Holley-Bockelmann, K., Mihos, J. C., Sigurdsson, S., & Hernquist, L. 2001, ApJ, 549, 862

Holley-Bockelmann, K., Mihos, J. C., Sigurdsson, S., Hernquist, L., & Norman, C. 2002, ApJ, 567, 817

Hunter, C., & de Zeeuw, P. T. 1992, ApJ, 389, 79

Hunter, C., & Qian, E. 1993, MNRAS, 262, 401

Huntley, J. M., & Saslaw, W. C. 1975, ApJ, 199, 328

Jaffe, W. 1983, MNRAS, 202, 995

Kormendy, J., & Richstone, D. 1995, ARA&A, 33, 581

Kuijken, K., & Dubinski, J. 1994, MNRAS, 269, 13

Lauer, T. R., et al. 1995, AJ, 110, 2622

——. 2002, AJ, 124, 1975

Lee, M. H., & Goodman, J. 1989, ApJ, 343, 594

Lees, J., & Schwarzschild, M. 1992, ApJ, 384, 491

Leeuwin, F., & Athanassoula, E. 2000, MNRAS, 317, 79

Lightman, A. P., & Shapiro, S. L. 1977, ApJ, 211, 244

Lynden-Bell, D. 1967, MNRAS, 136, L101

——. 1969, Nature, 223, 690

——. 2002, MNRAS, 338, L208

MacMillan, J. D., & Henriksen, R. N. 2002, ApJ, 569, 83

Magorrian, J., et al. 1998, AJ, 115, 2285

Makino, J., & Aarseth, S. J. 1992, PASJ, 44, 141

Makino, J., & Ebisuzaki, T. 1996, ApJ, 465, 527

Merritt, D. 1985, MNRAS214, 25

——. 1987, ApJ, 319, 55

——. 1997, ApJ, 486, 102

——. 1999, PASP, 111, 247

Merritt, D., & Quinlan, G. 1998, ApJ, 498, 625

Merritt, D., & Valluri, M. 1996, ApJ, 471, 82

Milosavljević, M. & Merritt, D. 2001, ApJ, 563, 34

Miralda-Escudé, J., & Schwarzschild, M. 1989, ApJ, 339, 752

Navarro, J. F., Frenk, C. S., & White, S. D. M. 1997, ApJ, 490, 493

Norman, C. A., May, A., & van Albada, T. S. 1985, ApJ, 296, 20

Norman, C. A., & Silk, J. 1983, ApJ, 266, 502

Osipkov, L. P. 1979, Sov. Astr. Let., 5, 42

Ostriker, J. P. 2000, Phys. Rev. Lett., 84, 5258

Palmer, P. L., & Papaloizou, J. 1988, MNRAS, 231, 935

Peebles, P. J. E. 1972, General Relativity and Gravitation, 3, 63

Pfenniger, D., & de Zeeuw, P. T. 1989, in Dynamics of Dense Stellar Systems, ed. D. Merritt (Cambridge:
    Cambridge Univ. Press), 81

Poon, M. Y., & Merritt, D. 2001, ApJ, 549, 192

——. 2002, ApJ, 568, L89

Qian E., de Zeeuw P. T., van der Marel, R. P., & Hunter C. 1995, MNRAS, 274, 602

Quinlan, G. D., & Hernquist, L. 1997, NewA, 2, 533

Quinlan, G. D., Hernquist, L., & Sigurdsson, S. 1995, ApJ, 440, 554

Quinlan, G. D., & Shapiro, S. L. 1990, ApJ, 356, 483

Ravindranath, S., Ho, L. C., & Filippenko, A. V., 2002, ApJ, 566, 801

Ravindranath, S., Ho, L. C., Peng, C. Y., Filippenko, A. V., & Sargent, W. L. W. 2001, AJ, 122, 653

Rees, M. J. 1984, ARA&A, 22, 471

——. 1990, Science, 247, 817

Richstone, D. O., et al. 1998, Nature, 395, A14

Richstone, D. O., & Tremaine, S. 1984, ApJ, 286, 27

——. 1988, ApJ, 327, 82

Romanowsky, A. R., & Kochanek, C. S. 1997, MNRAS, 287, 35

Ryden, B. S. 1992, ApJ, 396, 445

——. 1996, ApJ, 461, 146

Schwarzschild, M. 1979, ApJ, 232, 236

——. 1993, ApJ, 409, 563

Sellwood J. A., & Valluri M. 1997, MNRAS, 287, 124

Shapiro, S. L. 1985, in IAU Symp. 113, Dynamics of Star Clusters, ed. J. Goodman & P. Hut (Dordrecht: Reidel),
    373

Shapiro, S. L., & Marchant, A. B. 1978, ApJ, 255, 603

Shapiro, S. L., & Teukolsky, S. A. 1983, Black Holes, White Dwarfs, and Neutron Stars: The Physics of Compact
    Objects (New York: John Wiley)

Sigurdsson, S. 2003, Classical and Quantum Gravity, 20, S45

Sigurdsson, S., He, B., Melhem, R., & Hernquist, L. 1997a, Comp. in Phys., 11.4, 378

Sigurdsson, S., Hernquist, L., & Quinlan, G. D. 1995, ApJ, 446, 75

Sigurdsson, S., Mihos, J. C., Hernquist, L., & Norman, C. 1997b, BAAS, 191, 491

——. 1998, in Galactic Halos, ed. D. Zaritsky (San Francisco: ASP), 388

Sigurdsson, S., & Rees, M. J. 1997, MNRAS, 284, 318

Spitzer, L., Jr. 1971, in Galactic Nuclei, ed. D. J. O'Connell (New York: North-Holland), 443

Statler, T. 1987, ApJ, 321, 113

Stiavelli, M. 1998, ApJ, 495, L91

Tonry, J. L. 1983, ApJ, 266, 58

Tremaine, S., et al. 2002, ApJ, 574, 740

Tremaine, S., Richstone, D. O., Byun, Y.-I., Dressler, A., Faber, S. M., Grillmair, C. J., Kormendy, J., & Lauer, T. R. 1994, AJ, 107, 634

Tremblay, M., & Merritt, D. 1995, AJ, 110, 1039

Ullio, P., Zhao, H. S., & Kamionkowski, M. 2001, Phys. Rev. D, 64, 3504

Valluri, M., & Merritt, D. 1998, ApJ, 506, 686

van Albada, T. S. 1982, MNRAS, 201, 939

van der Marel, R. P. 1994a, ApJ, 432, L91

——. 1994b, MNRAS, 270, 271

——. 1999, AJ, 117, 744

van der Marel, R. P., de Zeeuw, P. T., Rix, H.-W., & Quinlan, G. D. 1997a, Nature, 385, 610

van der Marel, R. P., & Franx, M. 1993, ApJ, 407, 525

van der Marel, R. P., Rix, H.-W., Carter, D., Franx, M., White, S. D. M., & de Zeeuw, P. T. 1994, MNRAS, 268, 521

van der Marel, R. P., Sigurdsson, S., & Hernquist, L. 1997b, ApJ, 487, 153

van de Ven, G., Hunter, C., Verolme, E. K., & de Zeeuw, P. T. 2003, MNRAS, 342, 1056

Young, P. J. 1980, ApJ, 242, 1232

Zhao, H. S. 1997, MNRAS, 287, 525

Zier, C., & Biermann, P. L. 2001, A&A, 377, 23

7

Formation of supermassive black holes: simulations in general relativity

STUART L. SHAPIRO
The University of Illinois at Urbana–Champaign

Abstract

There is compelling evidence that supermassive black holes exist. Yet the origin of these objects, or their seeds, is still unknown. We discuss several plausible scenarios for forming the seeds of supermassive black holes. These include the catastrophic collapse of supermassive stars, the collapse of relativistic clusters of collisionless particles or stars, the gravothermal evolution of dense clusters of ordinary stars or stellar-mass compact objects, and the gravothermal evolution of self-interacting dark matter halos. Einstein's equations of general relativity are required to describe key facets of these scenarios, and large-scale numerical simulations are performed to solve them.

7.1   Introduction

There is substantial evidence that supermassive black holes (SMBHs) of mass $\sim 10^6 - 10^{10} \, M_\odot$ exist and are the engines that power active galactic nuclei (AGNs) and quasars (Rees 1998, 2001; Macchetto 1999). There is also ample evidence that SMBHs reside at the centers of many, and perhaps most, galaxies (Richstone et al. 1998; Ho 1999), including the Milky Way (Genzel et al. 1997; Ghez et al. 2000; Schödel et al. 2002).

Since quasars have been discovered out to redshift $z \gtrsim 6$ (Fan et al. 2000, 2001), the first SMBHs must have formed by $z_\bullet \gtrsim 6$, or within $t_\bullet \lesssim 10^9$ yrs after the Big Bang. However, the cosmological origin of SMBHs is not known. This issue remains one of the crucial, unresolved components of structure formation in the early universe. Gravitationally, black holes are strong-field objects whose properties are governed by Einstein's theory of relativistic gravitation—general relativity. General relativistic simulations of gravitational collapse to black holes therefore may help reveal how, when and where SMBHs, or their seeds, form in the universe. Simulating plausible paths by which the first seed black holes may have arisen is the underlying motivation of our investigation (see Fig. 7.1).

7.2   The Boltzmann Equation

Various routes have been proposed over the years by which SMBHs or their seeds might arise by conventional physical processes (see, e.g., Fig. 1 in Rees 1984). Some routes are hydrodynamical in nature, such as the formation and collapse of supermassive stars (SMSs), while others are stellar dynamical, like the evolution and collapse of collisionless clusters. The Boltzmann equation provides a common mathematical framework for comparing competing scenarios:

© The Observatories of the Carnegie Institution of Washington 2004.

Fig. 7.1. The formation of a black hole is a strong-field gravitational phenomenon in curved spacetime that requires Einstein's equations of general relativity for a description and, in nontrivial cases, numerical simulations for a solution.

$$\frac{Df}{Dt} = \left(\frac{\partial f}{\partial t}\right)_{collisions}. \qquad (7.1)$$

In equation (7.1) f is the phase-space distribution function for the matter, which might be in the form of a gaseous fluid, collisionless particles, and/or stars. The left-hand side of the equation represents the total time derivative of f following a matter element along its trajectory in phase space. The right-hand side describes the role of collisions in modifying the phase-space distribution along the trajectory. To treat different scenarios the Boltzmann equation must be solved in different physical regimes, all of which share gravitation as the the dominant long-range interaction.

Table 7.1 summarizes some of the SMBH formation simulations that we have performed in recent years. Typically, every scenario falls into one of three distinct regimes. In the Vlasov (collisionless Boltzmann) regime the dynamical timescale of the system, t_d, which is the time for matter to cross from one side of the system to the other, as well as the time to achieve virial equilibrium by violent relaxation, is much shorter than the relaxation timescale t_r, the time for the system to reach thermal equilibrium via collisions. In pure Vlasov simulations the integration time t may exceed t_d but always remains much shorter than t_r. In

Table 7.1. *Boltzmann Simulations of SMBH Formation Scenarios*

REGIME	Vlasov (Collisionless)	Fokker-Planck (Secularly Collisional)	Fluid (Collision Dominated)
TIME SCALE ORDERING	$t_r \gg t \gg t_d$	$t \gg t_r \gg t_d$	$t \gg t_d \gg t_r$
SCENARIOS	dynamical collapse of a relativistic cluster of (1) compact stars (NSs or stellar-mass BHs) or (2) collisionless particles	"gravothermal catastrophe" drives core contraction of (1) a dense cluster of compact stars; or (2) a dense cluster of ordinary stars; or (3) an SIDM halo	hydrodynamical collapse of an SMS
GRAVITATION	GR	Newtonian	GR, PN*
SPATIAL SYMMETRY	Spherical; Axisymmetrical	Spherical	Spherical; Axisymmetrical; Arbitrary*
COMPUTATIONAL DIMENSIONS	1 + 1; 2 + 1	1 + 1	1 + 1; 2 + 1; 3 + 1*
COMPUTATIONAL TECHNIQUE	particle simulation (matter) + finite-differencing (field)	finite-differencing	finite-differencing

such cases collisions can be ignored. The system can be evolved to dynamical (virial) equilibrium, but not thermal equilibrium. In the secularly collisional regime, t_d again is much shorter than t_r but the integration time is now much longer than t_r. Here collisions are crucial in driving the quasi-stationary evolution of the virialized system. Since the timescale for collisions remains much longer than the dynamical timescale, collisions can often be handled perturbatively by tracking the secular drift of the system from one, nearly collisionless, virialized state to the next. This is the approach adopted in the Fokker-Planck approximation to the Boltzmann equation. In the collision-dominated regime, t_r is much shorter than t_d and the system behaves as a fluid. This regime embraces all of hydrodynamics.

Scenarios considered to date for forming SMBHs, or their seeds, in the Vlasov regime focus on the dynamical collapse of a radially unstable, relativistic cluster of compact stars (neutrons stars or stellar-mass black holes) or collisionless particles. Scenarios in the secularly collisional regime typically involve the gravothermal contraction of a dense cluster of ordinary stars or compact stars which undergo collisions and mergers, leading to a build-up of massive black holes. The gravothermal contraction of a self-interacting dark matter halo (SIDM) in the early universe may also produce a SMBH. Hydrodynamical scenarios typically focus on the collapse of a SMS or gas cloud. Not surprisingly, simulations performed in the different regimes require very different computational approaches. Table 7.1 indicates that the different scenarios have been tackled by adopting various degrees of spatial symmetry to simplify the calculations. A summary of the results of these simulations is given in the sections below.

7.3   Numerical Relativity

Numerical relativity—the art and science of developing computer algorithms to solve Einstein's field equations of general relativity—is the principal tool needed to simulate plausible black hole formation processes. The underlying equations are multidimensional, highly nonlinear, coupled partial differential equations in space and time. They have in common with other areas of computational physics, like fluid dynamics and MHD, all of the usual problems associated with solving such nontrivial equations. However, solving Einstein's equations poses some additional complications that are unique to general relativity. The first complication concerns the choice of coordinates. In general relativity, coordinates are merely labels that distinguish points in spacetime; by themselves coordinate intervals have no physical significance. To use coordinate intervals to determine physically measurable (proper) distances and times requires the spacetime metric, but the metric is determined only after Einstein's equations have been solved. Moreover, as the integrations proceed, it often turns out that the original (arbitrary) choice of coordinates turns out to be bad, because, for example, singularities eventually are encountered in the equations. The gauge freedom inherent in general relativity—the ability to choose coordinates in an arbitrary way—is not always easy to exploit successfully in a numerical routine.

The appearance of black holes always poses a complication in a numerical relativity simulation. Black holes inevitably contain spacetime singularities—regions where the gravitational tidal field, the matter density, and the spacetime curvature all become infinite. Encountering such singularities results in some of the terms in Einstein's equations becoming infinite, causing overflows in the computer output and premature termination of the numerical integration. Thus, when dealing with black holes, it is crucial to choose a technique which avoids the spacetime singularities inside. Some of the techniques involve choosing appropriate coordinate gauges that avoid or postpone the appearance of singularities inside black holes. Others involve excising the black hole interiors altogether from the numerical grid.

One of the main goals of a numerical relativity simulation is to determine the gravitational radiation generated from a dynamical scenario. However, the gravitational wave components usually constitute small fractions of the background spacetime metric. Moreover, to extract the waves from the background requires that one probe the spacetime in the far-field or radiation zone, which is typically at large distance from the strong-field central source. Yet it is the strong-field, near-zone region that usually consumes most the computational resources (e.g., spatial grid) to guarantee accuracy. Furthermore, waiting for the wave to propagate to the far-field region usually takes nonnegligible integration time. Overcoming these difficulties to reliably measure the wave content thus requires that a code successfully cope with the problem of dynamic range inherent in such a simulation.

For a recent review of the status of numerical relativity, and a summary of the key equations, see Baumgarte & Shapiro (2003) and references therein.

7.4   Collapse of a Rotating SMS to a SMBH

SMBHs must be present by $z_\bullet \gtrsim 6$ to power quasars. It has been suggested (Gnedin 2001) that even if they grew by accretion from smaller seeds, SMBH seeds $\gtrsim 10^5 \, M_\odot$ must have formed at $z \approx 9$ to have had sufficient time to build up to a typical mass of $\sim 10^9 \, M_\odot$. A likely progenitor is a very massive object (e.g., an SMS) supported by radiation pressure.

SMSs ($10^3 \lesssim M/M_\odot \lesssim 10^{13}$) may form when contracting or colliding primordial gas builds up sufficient radiation pressure to inhibit fragmentation and prevent star formation (see, e.g., Bromm & Loeb 2003). SMSs supported by radiation pressure will evolve in a quasi-stationary manner to the point of onset of dynamical collapse due to general relativity (Chandrasekhar 1964a,b; Feynmann, unpublished, as quoted in Fowler 1964). Unstable SMSs with $M \gtrsim 10^5 \, M_\odot$ and metallicity $Z \lesssim 0.005$ do not disrupt due to thermonuclear explosions during collapse (Fuller, Woosley, & Weaver 1986). In fact, recent Newtonian simulations suggest that evolved zero-metallicity (Pop III) stars $\gtrsim 300 \, M_\odot$ do not disrupt but collapse with negligible mass loss (Fryer, Woosley, & Heger 2001). This finding could be important since the first generation of stars may form in the range $10^2 - 10^3 \, M_\odot$ (Bromm, Coppi, & Larson 1999; Abel, Bryan, & Norman 2000). A combination of turbulent viscosity and magnetic fields likely will keep a spinning SMS in uniform rotation (Bisnovatyi-Kogan, Zel'dovich, & Novikov 1967; Wagoner 1969; Zel'dovich & Novikov 1971; Shapiro 2000; but see New & Shapiro 2001 for an alternative). As they cool and contract, uniformly rotating SMSs reach the maximally rotating *mass-shedding limit* and subsequently evolve in a quasi-stationary manner along a mass-shedding sequence until reaching the instability point. At mass-shedding, the matter at the equator moves in a circular geodesic with a velocity equal to the local Kepler velocity (Baumgarte & Shapiro 1999).

It is straightforward to understand the radial instability induced by general relativity in a SMS by using an energy variational principle (Zel'dovich & Novikov 1971; Shapiro & Teukolsky 1983). Let $E = E(\rho_c)$ be the total energy of a momentarily static, spherical fluid configuration characterized by central mass density ρ_c. The condition that $E(\rho_c)$ be an extremum for variations that keep the total rest mass and specific entropy distribution fixed is equivalent to the condition of hydrostatic equilibrium and establishes the relation between the equilibrium mass and central density:

$$\frac{\partial E}{\partial \rho_c} = 0 \implies M_{eq} = M_{eq}(\rho_c) \qquad \text{(equilibrium).} \qquad (7.2)$$

The condition that the second variation of $E(\rho_c)$ be zero is the criterion for the onset of dynamical instability. This criterion shows that the turning point on a curve of equilibrium mass vs. central density marks the transition from stability to instability:

$$\frac{\partial^2 E}{\partial \rho_c^2} = 0 \iff \frac{\partial M_{eq}}{\partial \rho_c} = 0 \qquad \text{(onset of instability).} \qquad (7.3)$$

Consider the simplest case of a spherical Newtonian SMS supported solely by radiation pressure and endowed with zero rotation. This is an $n = 3$, ($\Gamma = 1 + 1/n = 4/3$) polytrope, with pressure

$$P = P_{rad} = \frac{1}{3} a T^4 = K \rho^{\frac{4}{3}}, \qquad (7.4)$$

where $K = K(s_{rad})$ is a constant determined by the value of the (constant) specific entropy $s_{rad} = \frac{4}{3} a T^3 / n$ in the star. Here T is the temperature, n is the baryon number density, and a is the radiation constant. Consider a sequence of configurations with the same specific entropy but different values of central density. The total energy of each configuration is

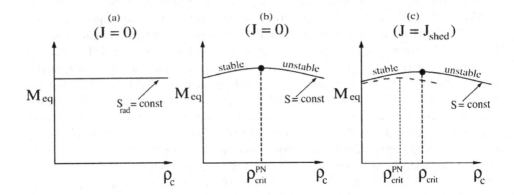

Fig. 7.2. A sketch of mass versus central density along an equilibrium sequence of SMSs of fixed entropy. Panel (*a*) shows nonrotating, spherical Newtonian models supported by pure radiation pressure; (*b*) shows nonrotating, spherical PN models supported by radiation pressure plus thermal gas pressure; (*c*) shows rotating PPN models spinning at the mass-shedding limit.

$$E(\rho_c) = U_{\text{rad}} + W, \tag{7.5}$$

where U_{rad} is the total internal radiation energy and W is the gravitational potential energy. Applying the equilibrium condition (7.2) to this functional yields $M_{\text{eq}} = M_{\text{eq}}(s_{\text{rad}})$, i.e. the equilibrium mass depends only on the specific entropy and is independent of central density (see Fig. 7.2*a*). Applying the stability condition (7.3) then shows that all equilibrium models along this sequence are marginally stable to collapse.

Now let us account for the effects of general relativity. If we include the small (de-stabilizing) Post-Newtonian (PN) correction to the gravitational field, we must also include a comparable (stabilizing) correction to the equation of state arising from thermal gas pressure:

$$P = P_{\text{rad}} + P_{\text{gas}} = \frac{1}{3}aT^4 + 2nkT, \tag{7.6}$$

where we have taken the gas to be pure ionized hydrogen. Note that $P_{\text{gas}}/P_{\text{rad}} = 8/(s_{\text{rad}}/k) \ll 1$. The energy functional of a star now becomes

$$E(\rho_c) = U_{\text{rad}} + W + \Delta U_{\text{gas}} + \Delta W_{\text{PN}}, \tag{7.7}$$

where ΔU_{gas} is the internal energy perturbation due to thermal gas energy and ΔW_{PN} is the PN perturbation to the gravitational potential energy. Applying the equilibrium condition (7.2) now yields $M_{\text{eq}} \approx M_{\text{eq}}^{\text{Newt}}$ times a slowly varying function of ρ_c (see Fig. 7.2*b*). The turning point on the equilibrium curve marks the onset of radial instability; the marginally stable critical configuration is characterized by

$$\begin{aligned} \rho_{c,\text{crit}} &= 2 \times 10^{-3} M_6^{-7/2} \text{gm cm}^{-3}, \\ T_{c,\text{crit}} &= (3 \times 10^7) M_6^{-1} \text{ K}, \end{aligned} \tag{7.8}$$

$$(R/M)_{\rm crit} = 1.6 \times 10^3 M_6^{1/2},$$

where M_6 denotes the mass in units of $10^6\ M_\odot$. (Here and throughout we adopt gravitational units and set $G = 1 = c$.)

Finally, let us consider a uniformly rotating SMS spinning at the mass-shedding limit (Baumgarte & Shapiro 1999). A centrally condensed object like an $n = 3$ polytrope can only support a small amount of rotation before matter flys off at the equator. At the mass-shedding limit, the ratio of rotational kinetic to gravitational potential energy is only $T/|W| = 0.899 \times 10^{-2} \ll 1$. Most of the mass resides in a nearly spherical interior core, while the low-mass (Roche) envelope bulges out in the equator: $R_{\rm eq}/R_{\rm pole} = 3/2$. When we include the contribution of rotational kinetic energy to the energy functional, we must now also include the effects of relativistic gravity to Post-Post-Newtonian (i.e. PPN) order, since both T and $\Delta W_{\rm PN}$ scale with ρ_c to the same power. The energy functional becomes

$$E(\rho_c) = U_{\rm rad} + W + \Delta U_{\rm gas} + \Delta W_{\rm PN} + \Delta W_{\rm PPN} + T \tag{7.9}$$

Applying the equilibrium condition (7.2), holding M, angular momentum J and s fixed, now yields $M_{\rm eq} \approx M_{\rm eq}^{\rm Newt}$ times a slowly varying function of ρ_c (see Fig. 7.2c). If we restrict our attention to rapidly rotating stars with $M > 10^5\ M_\odot$ the influence of thermal gas pressure is unimportant in determining the critical point of instability. The turning point on the equilibrium curve then shifts to higher density and compaction than the critical values for nonrotating stars, reflecting the stabilizing role of rotation:

$$\begin{aligned}
\rho_{c,\rm crit} &= 0.9 \times 10^{-1} M_6^{-2} {\rm gm\ cm}^{-3}, \\
T_{c,\rm crit} &= (9 \times 10^7) M_6^{-1/2}\ {\rm K}, \\
(R_{\rm pole}/M)_{\rm crit} &= 427, \\
(J/M^2)_{\rm crit} &= 0.97.
\end{aligned} \tag{7.10}$$

The actual values quoted above for the critical configuration were determined by a careful numerical integration of the general relativistic equilibrium equations for rotating stars (Baumgarte & Shapiro 1999); they are in close agreement with those determined analytically by the variational treatment. The numbers found for the nondimensional critical compaction and angular momentum are quite interesting. First, they are universal ratios that are independent of the mass of the SMS. This means that a single relativistic simulation will suffice to track the collapse of a marginally unstable, maximally rotating SMS of arbitrary mass. Second, the large value of the critical radius shows that a marginally unstable configuration is nearly Newtonian at the onset of collapse. Third, the fact that the angular momentum parameter of the critical configuration J/M^2 is below unity suggests that, in principle, the entire mass and angular momentum of the configuration could collapse to a rotating black hole without violating the Kerr limit for black hole spin (but see below!).

There are several plausible outcomes that one might envision *a priori* for the dynamical collapse of a uniformly rotating SMS once it reaches the marginally unstable critical point identified above. It could collapse to a clumpy, nearly axisymmetric disk, similar to the one arising in the Newtonian SPH simulation of Loeb & Rasio (1994) for the isothermal ($\Gamma = 1$) implosion of an initially homogeneous, uniformly rotating, low-entropy cloud. Alternatively, the disk might develop a large-scale, nonaxisymmetric bar. After all, the onset of a dynamically unstable bar mode in a spinning equilibrium star occurs when the ratio $T/|W| \approx 0.27$ (see, e.g., Chandrasekhar 1969 and Lai, Rasio, & Shapiro 1993 for Newtonian treatments and Saijo et al. 2001 and Shibata, Baumgarte, & Shapiro 2000a for simulations in general relativity). Since $T/|W|$ is 0.899×10^{-2} at the onset of collapse and scales roughly as R^{-1} during collapse due to conservation of mass and angular momentum, this ratio climbs above the dynamical bar instability threshold when the SMS collapses to $R/M \approx 20$, well before the horizon is reached. The growth of a bar might begin at this point. Indeed, a weak bar forms in simulations of rotating supernova core collapse (Rampp, Müller, & Ruffert 1998; Brown 2001), but here the equation of state stiffens ($\Gamma > 4/3$) at the end of the collapse, triggering a bounce and thereby allowing more time for the bar to develop. A rapidly rotating unstable SMS might not form a disk at all, but instead collapse entirely to a Kerr black hole; not surprisingly, a nonrotating spherical SMS has been shown to collapse to a Schwarzschild black hole (Shapiro & Teukolsky 1979). Alternatively, the unstable rotating SMS might collapse to a rotating black hole *and* an ambient disk.

Two recent simulations have resolved the fate of a marginally unstable, maximally rotating SMS of arbitrary mass M. Saijo et al. (2002) followed the collapse in full $3D$, assuming PN theory. They tracked the implosion up to the point at which the central spacetime metric begins to deviate appreciably from flat space at the stellar center. They found that the massive core collapses homologously during the Newtonian epoch of collapse, and that axisymmetry is preserved up to the termination of the integrations. This calculation motivated Shibata & Shapiro (2002) to follow the collapse in full general relativity by assuming axisymmetry from the beginning (see Fig. 7.3). They found that the final object is a Kerr-like black hole surrounded by a disk of orbiting gaseous debris. The final black hole mass and spin were determined to be $M_h/M \approx 0.9$ and $J_h/M_h^2 \approx 0.75$. The remaining mass goes into the disk of mass $M_{\text{disk}}/M \approx 0.1$. A disk forms even though the total spin of the progenitor star is safely below the Kerr limit. This outcome results from the fact that the dense inner core collapses homologously to form a central black hole, while the diffuse outer envelope avoids capture because of its high angular momentum. Specifically, in the outermost shells, the angular momentum per unit mass j, which is strictly conserved on cylinders, exceeds j_{ISCO}, the specific angular momentum at the innermost stable circular orbit about the final hole. This fact suggests how the final black hole and disk parameters can be calculated *analytically* from the initial SMS density and angular momentum distribution (Shapiro & Shibata 2002). The result applies to the collapse of *any* marginally unstable $n = 3$ polytrope at mass-shedding. Maximally rotating stars which are characterized by stiffer equations of state and smaller n (higher Γ) do not form disks, typically, since they are more compact and less centrally condensed at the onset of collapse (Shibata, Baumgarte, & Shapiro 2000b).

The above calculations show that a SMBH formed from the collapse of a maximally rotating SMS is always born with a "ready-made" accretion disk. This disk might provide a convenient source of fuel to power the central engine. The calculations also show that the SMBH will be born rapidly rotating. This fact is intriguing in light of suggestions that observed SMBHs rotate rapidly (e.g., Wilms et al. 2001; Elvis, Risaliti, & Zamorani 2002).

7.5   Collapse of Collisionless Matter to a SMBH

Zel'dovich & Podurets (1965) speculated that sufficiently compact, relativistic clusters of collisionless particles (e.g., relativistic star clusters) would be unstable to gravitational collapse. It has taken considerable theoretical effort to prove that this

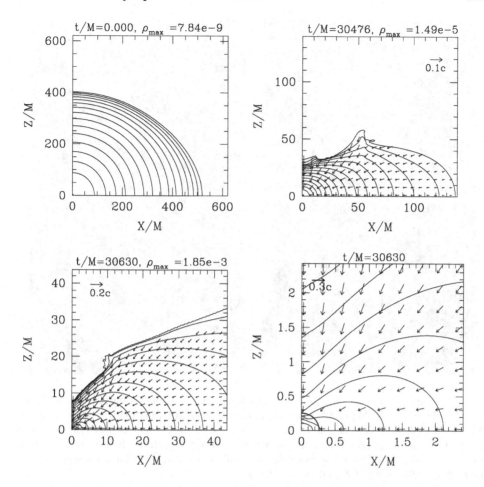

Fig. 7.3. Snapshots of density and velocity profiles during the implosion of a marginally unstable SMS of arbitrary mass M rotating uniformly at break-up speed at $t = 0$. The contours are drawn for $\rho/\rho_{max} = 10^{-0.4j}$ ($j = 0 - 15$), where ρ_{max} denotes the maximum density at each time. The fourth figure is the magnification of the third one in the central region: the thick solid curve at $r \approx 0.3M$ denotes the location of the apparent horizon of the emerging SMBH. (From Shibata & Shapiro 2002.)

speculation is correct. For a given distribution function $f = f(E)$, one can construct a sequence of spherical equilibrium clusters parametrized by the gravitational redshift at the cluster center, z_c. One can then plot a curve of fractional binding energy $E_b/M_0 = 1 - M/M_0$. vs. z_c along the sequence. Linear perturbation theory, implemented via a variational principle and trial functions, shows that the onset of radial instability occurs near the first turning point on such a binding energy curve (Ipser & Thorne 1968; Ipser 1969; Fackerell 1970). A rigorous theorem has been proven that spherical equilibrium configurations are stable, at least up to the first turning point on the binding energy curve (Ipser 1980). Numerical simulations have shown that all spherical configurations beyond the first turning point are dynamically unstable (Shapiro & Teukolsky 1985a,b,c; 1986). Most

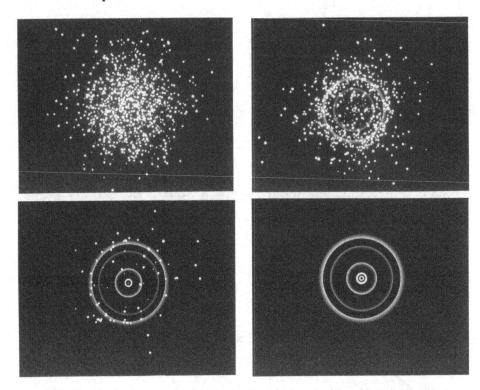

Fig. 7.4. The collapse of a marginally unstable gas of collisionless particles of arbitrary mass M which at $t = 0$ obeys a Maxwell-Boltzmann distribution function with an areal radius $R/M = 9.0$. Spherical flashes of light are used to probe the spacetime geometry; at late times the light rays are trapped by the gravitational field. Their trajectories help locate the black hole event horizon, which in this example eventually reaches $r_s/M = 2$ and encompasses all the matter. (After Shapiro & Teukolsky 1988.)

significantly, these simulations have tracked the nonlinear evolution of unstable spherical and axisymmetric clusters, including those with rotation (Abrahams et al. 1994; Shapiro, Teukolsky, & Winicour 1996), and have determined their final fate (see Fig. 7.4). This computational enterprise ("relativistic stellar dynamics on a computer;" see Shapiro & Teukolsky 1992 for a review and references) has demonstrated the following:

(1) the dynamical collapse of a collisionless cluster leads to the formation of a Kerr black hole whenever $J/M^2 \lesssim 1$;

(2) collapse leads instead to a bounce followed by virialization of the cluster by relativistic violent relaxation whenever $J/M^2 \gtrsim 1$;

(3) in extreme core-halo systems, collapse leads to a new stationary, equilibrium system consisting of a central black hole surrounded by an extensive, nearly Newtonian halo containing most of the mass.

Conclusion (3) is very tantalizing as a plausible route for forming SMBHs (see Fig. 7.5 and 7.6). But a key question remains: under what circumstances can a cluster be driven to a dynamically unstable, relativistic state to trigger such a collapse?

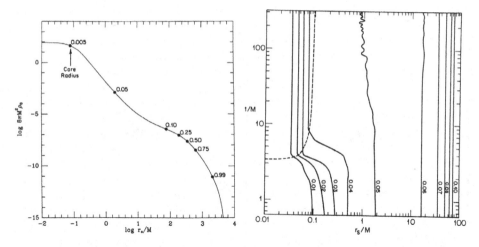

Fig. 7.5. An extreme core-halo configuration of arbitrary mass M constructed from an $n = 4$ relativistic polytropic distribution function. The *left* panel shows the initial equilibrium rest-mass density profile. The points label the interior rest-mass fraction. This cluster has a highly relativistic core and an extensive Newtonian halo and is marginally unstable to collapse. The *right* panel is a spacetime diagram showing the worldlines of imaginary Lagrangian matter tracers. The dashed line shows the event horizon of the black hole that forms at the center. Note that the central core and its surroundings undergo collapse but that 95% of the cluster mass settles into stable dynamical equilibrium about the central hole. (From Shapiro & Teukolsky 1986.)

One possibility may involve the "gravothermal catastrophe," the runaway core contraction on a relaxation timescale of a stable, virialized cluster due to the perturbative influence of collisions (Chandrasekhar 1942; Lynden-Bell & Wood 1968; see Spitzer 1975, 1987 and Lightman & Shapiro 1978 for reviews and references). Collisional scattering is a source of kinetic energy transport (heat conduction). Self-gravitating, nearly collisionless clusters with $t_d/t_r \ll 1$ have negative heat capacity: as their high-temperature cores lose energy to their low-temperature halos by heat conduction, the cores contract and, in accord with the virial theorem, become hotter still, leading to a thermal runaway. The result is that the clusters undergo homologous core contraction (Lynden-Bell & Eggleton 1980), as depicted in Figure 7.7 for Newtonian clusters composed of identical particles. Contraction to a singular state is complete in a time $\approx 300 t_r(0)$, where the central relaxation timescale $t_r(0)$ is measured from an arbitrary initial time $t = 0$ (Cohn 1979). Specifically, as $t \to 300 t_r(0)$, the central density as well as the the the central redshift, potential and velocity dispersion (temperature) all blow up: $\rho_c \to \infty$ and $z_c \sim \Phi_c \sim v_c^2 \to \infty$. At the same time the core radius and mass both shrink: $r_c \to 0$ and $M_c \to 0$. The required number of relaxation timescales to reach this singular state, as well as the $r^{-2.2}$ fall-off in the halo density profile, are nearly independent of the velocity dependence of the collision cross section (Balberg & Shapiro, unpublished). This gravothermal catastrophe can, in principle, drive a core to a highly relativistic state, at which point it could become dynamically unstable to catastrophic collapse on a dynamical timescale.

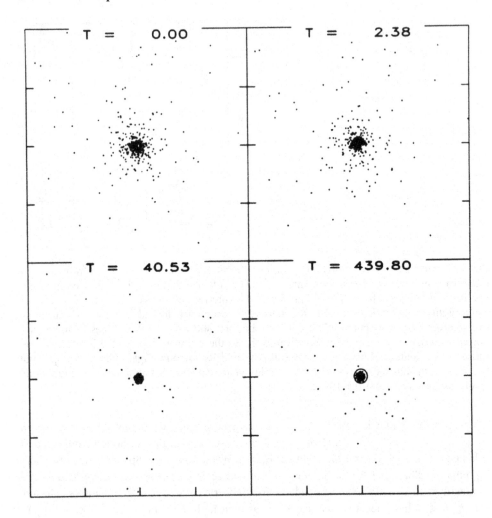

Fig. 7.6. Snapshots of the central particle distribution inside $r_s/M = 2$ at selected times during the collapse described in Figure 7.5. The cluster does not evolve appreciably after $t/M = 40$. The circle in the last frame shows the black hole event horizon at $r_s/M = 0.1$. (From Shapiro & Teukolsky 1986.)

Detailed numerical calculations have been performed for several scenarios to test whether the gravothermal catastrophe can trigger the formation of SMBHs. We will summarize three of them below.

7.5.1   *Gravothermal Evolution of a Cluster of Compact Stars*

Here we describe simulations involving the evolution of clusters of compact stars (neutron stars or stellar-mass black holes) and the build-up of massive black holes. This route has been discussed by several authors (e.g., Zel'dovich & Podurets 1965; Rees 1984; Shapiro & Teukolsky 1985c; Quinlan & Shapiro 1987, 1989). We briefly summarize below the key results of the multi-mass Fokker-Planck calculations of Quinlan & Shapiro (1989).

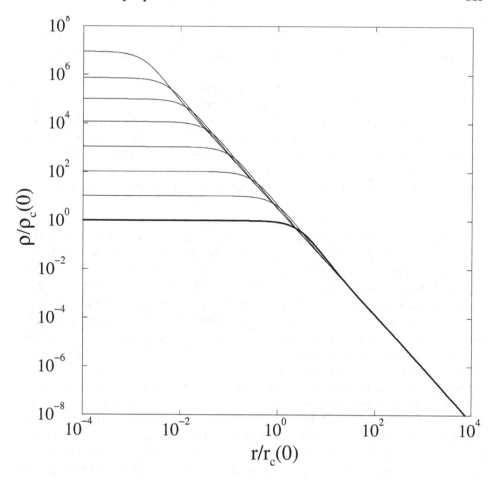

Fig. 7.7. Snapshots of the self-similar density profile at selected times during *secular* core collapse of a nearly collisionless Newtonian cluster (the gravothermal catastrophe). The thick line shows the profile at $t = 0$ and successive profiles with higher central densities correspond to later times. (From Balberg, Shapiro, & Inagaki 2002.)

Consider a dense cluster of compact stars composed initially of identical neutron stars or black holes of mass $m_* = 1.4 \, M_\odot$ in virial equilibrium. Take the central mass density to be $\rho_c \approx 10^8 \, M_\odot \, \mathrm{pc}^{-3}$ and the central velocity dispersion to be $v_c \approx 500 \, \mathrm{km \, s^{-1}}$. Here gravothermal evolution is driven by the cumulative effect of repeated, distant, small-angle, gravitational (Coulomb) scatterings between the stars. Binary formation is significant and is dominated by dissipative, 2-body capture by gravitational radiation.

The simulations reveal that mass segregation causes significant departures from single-component homological evolution models. For example, v_c does not increase at the center and the cluster is not driven to a relativistic state. However, there is an inevitable build-up of massive BHs via successive binary mergers. The evolution is followed up to the formation of BHs of mass $M_\bullet \gtrsim 100 \, M_\odot$, at which point the number of stars in the core become sufficiently small that the Fokker-Planck treatment breaks down. The intermediate mass

BH binaries produced in this scenario would be prime sources of gravitational waves for the ground-based network of laser interferometers now under construction (LIGO, VIRGO, GEO, and TAMA) and for the space-based interferometer currently being designed (LISA).

7.5.2   *Gravothermal Evolution of a Dense Cluster of Ordinary Stars*

Here we discuss simulations that start from more typical initial conditions. These calculations treat the gravothermal evolution of a dense cluster of ordinary, main-sequence stars (Spitzer & Saslaw 1966; Colgate 1967; Sanders 1970; Begelman & Rees 1978; Lee 1987; Quinlan & Shapiro 1990; Gao et al. 1991; Portegies Zwart & McMillan 2002). Here we briefly summarize the multi-mass Fokker-Planck calculations of Quinlan & Shapiro (1990), which are among the most detailed for galactic nuclei that do not assume the presence of a SMBH *a priori*.

Consider a dense galactic nucleus initially consisting of main-sequence stars with mass $m_* = 1.0\,M_\odot$ in virial equilibrium. Take the central mass density to be in the range $\rho_c \approx 10^6 - 10^8\,M_\odot\,\mathrm{pc}^{-3}$ and the central velocity dispersion to be in the range $v_c \approx 100 - 400$ km s^{-1}. This velocity is below the escape velocity from the surface of the stars, so that collisions will lead to mergers and not disruptions. Collisions and mergers, stellar evolution and the formation of new stars from gas liberated by supernovae are included in the calculation, as are the formation of binaries by 3-body encounters and the interaction between hard binaries and single stars. All of this activity takes place in the context of the gravothermal evolution of the cluster, which again is driven by the cumulative effect of repeated, distant, small-angle, gravitational scattering of the stars.

The outcome of the evolution is that stars with $m_* \gtrsim 100\,M_\odot$ form easily, then merge and collapse to form seed BHs with masses $M_\bullet \approx 100 - 1000\,M_\odot$ in a time $t \lesssim 10^{10}$ yrs. The end result is the formation of a dense cluster of compact remnants comprised of intermediate mass black holes. The cluster is characterized by frequent binary mergers ($\gg 1$ per year when integrated throughout the visible universe of $\sim 10^{10}$ galaxies). These intermediate mass black hole binaries are again promising sources of gravitational waves for the ground-based laser interferometers like LIGO and for the proposed space-based interferometer LISA.

7.5.3   *Gravothermal Contraction of an SIDM Halo to a Relativistic State*

Dark matter comprises about 90% of the matter in the universe. The simplest description of dark matter which accounts for many features of the large-scale structure of the universe is the "cold dark matter" (CDM) model, in which the dark matter particles are essentially collisionless. However, the possibility that dark matter particles interact with each other strongly and have a substantial scattering cross section has been revived recently (Spergel & Steinhardt 2000) to explain several observations of dark matter structures on the order of $\lesssim 1$ Mpc. Dynamical studies confirm that halos formed from such "self-interacting" dark matter (SIDM) have flatter density cores in better agreement with the observations than the more cuspy profiles predicted by standard CDM.

Balberg & Shapiro (2002) have recently demonstrated that SMBH formation may be an inevitable consequence of dynamical core collapse following the gravothermal catastrophe in SIDM halos. This conclusion follows from their earlier dynamical study (Balberg et al. 2002) which tracked the gravothermal evolution of a virialized, spherical SIDM halo by employing the fluid formalism of Lynden-Bell & Eggleton (1980). In the early universe,

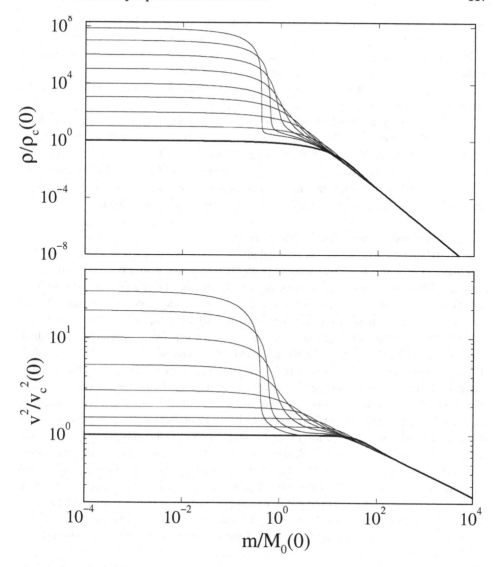

Fig. 7.8. Snapshots of the (a) density and (b) velocity dispersion profiles of an SIDM halo at selected times during gravothermal evolution. The thick line shows the profile at $t = 0$ and successive profiles with higher central densities correspond to later times. Bifurcation into a fluid inner core and a collisionless outer core is evident at late times. (From Balberg et al. 2002.)

halos form with a Press-Schechter (1974) distribution. Typical halos are characterized by a central mass density $\rho_c \gtrsim 10^{-2} \, M_\odot \, \text{pc}^{-3}$, a central velocity dispersion $v_c \gtrsim 100 \, \text{km s}^{-1}$ and an elastic scattering cross section $\sigma \gtrsim 0.1 \, \text{cm}^2 \, \text{gm}^{-1}$. The ratio of the scattering mean free path λ to the gravitational scale height H everywhere satisfies the long mean free path inequality $\lambda / H \gg 1$ initially. This results in homologous gravothermal contraction at first (see Fig. 7.7). However, in SIDM halos the interactions are large-angle scatterings between close neighbors, not cumulative, small-angle Coulomb scatterings by distant particles, as in star clusters. Consequently, in contrast to star clusters, the inner core of an SIDM halo evolves into the short mean free path (fluid) regime where $\lambda/H \ll 1$. There is a bifurcation of the halo into a fluid inner core surrounded by a collisionless outer core and halo (see Fig. 7.8). Continued heat conduction out of the inner core drives it to a relativistic state, which eventually becomes unstable to collapse to a black hole.

The initial mass of the black hole will be $10^{-8} - 10^{-6}$ of the total mass of the halo. Very massive SIDM halos form SMBHs with masses $M_\bullet \gtrsim 10^6\ M_\odot$ directly. Smaller halos believed to form by redshift $z \approx 5$ produce seed black holes of mass $M_\bullet \approx 10^2 - 10^3\ M_\odot$ which can merge and/or accrete to reach the observed SMBH range. Significantly, this scenario for SMBH formation requires no baryons, no prior star formation and no other black hole seed mechanisms (cf. Ostriker 2000; Hennawi & Ostriker 2002).

7.6   Conclusions and Final Thoughts

Relativistic gravitation—general relativity—induces a dynamical, radial instability to collapse in *all* forms of self-gravitating matter whenever the matter becomes sufficiently compact. Plausible scenarios have been proposed that can trigger this instability to form SMBHs or their seeds. SMBHs are believed to reside at the centers of quasars, AGNs, and many, if not all, normal galaxies with bulges, including the Milky Way. According to recent observations, even globular clusters may contain black holes of intermediate mass $\sim 10^3 - 10^4\ M_\odot$ (Gebhardt, Rich, & Ho 2002; Gerssen et al. 2002), although alternative interpretations have been proposed (Baumgardt et al. 2003). Explaining the origin of SMBHs is thus a fundamental, unresolved issue of modern cosmology and structure formation. Some of the scenarios proposed for forming SMBHs are "hydrodynamical" in nature, as in the collapse of fluid SMSs to SMBHs. Some of them are "stellar dynamical," as in the collapse of a relativistic cluster of collisionless particles or compact stars. Still others are "hybrids," as in the case of the collapse of massive stars, built-up by collisions and binary mergers in dense clusters undergoing gravothermal contraction; or the case of the collapse of the fluid inner core of an SIDM halo, following its gravothermal contraction to a relativistic state. The challenge in exploring the competing scenarios is that the different physical regimes characterizing them are described by very different sets of equations, requiring very different numerical techniques for solution. Yet numerical simulations have begun to explore many of the proposed routes.

At the present time we do not know for certain what is the dominant route by which observed SMBHs are formed: Are they born supermassive, or do they grow supermassive from small seeds? Do the seed black holes grow by merger, by gas accretion or both? Do the first black holes arise from the collapse of ordinary baryonic matter, collisionless dark matter, or from some more exotic form of mass-energy (e.g., scalar fields? gravitational waves?).

The growth of black hole seeds by gas accretion is supported by the consistency between the total energy density in QSO light and the BH mass density in local galaxies, adopting a reasonable accretion rest-mass–to–energy conversion efficiency (Soltan 1982; Yu & Tremaine 2002). But quasars have been discovered out to redshift $z \approx 6$, so it follows that the first SMBHs must have formed by $z_\bullet \gtrsim 6$ or within $t_\bullet \lesssim 10^9$ yr after the Big Bang. This timescale provides a tight constraint on SMBH formation. For example, if SMBHs indeed grew by accretion, black hole seeds of mass $\gtrsim 10^5 \, M_\odot$ must have formed by $z \approx 9$ to have had sufficient time to reach a mass of $\sim 10^9 \, M_\odot$ by $z \approx 6$ (Gnedin 2001).

The correlations $M_\bullet \propto L_{\text{bulge}}$ (Kormendy & Richstone 1995) and $M_\bullet \propto v_c^4$ (Gebhardt et al. 2000; Ferrarese & Merritt 2000) inferred for galaxies provide important additional constraints. For example, SMBH formation by mergers of smaller seed holes during the hierarchical build-up of galaxies can account for these scaling laws (Haehnelt & Kauffmann 2000). But some observations suggest that SMBHs spin rapidly. This conclusion might restrict the significance of merger scenarios, since black holes are typically spun down by repeated mergers (Hughes & Blandford 2003). On the other hand, a single final merger with a binary companion of comparable mass could drive the spin of a black hole back up to a large value.

Further observations, including the detection of gravitational waves from distant coalescing black hole binaries, might establish the evolutionary tracts and merging histories of SMBHs and help identify their principle formation mechanism.

Acknowledgements. This work was supported in part by NSF Grants PHY-0090310 and PHY-0205155 and NASA Grants NAG5-8418 and NAG5-10781 at the University of Illinois at Urbana-Champaign.

References

Abel, T., Bryan, G. L., & Norman, M. L. 2000, ApJ, 540, 39

Abrahams, A. M., Cook, G. B., Shapiro, S. L., & Teukolsky, S. A. 1994, Phys. Rev. D, 49, 5153

Balberg, S., & Shapiro, S. L. 2002, Phys. Rev. Lett., 88, 101301

Balberg, S., Shapiro, S. L., & Inagaki, S. 2002, ApJ, 568, 475

Baumgardt, H., Hut, P., Makino, J., McMillan, S., & Portegies Zwart, S. 2003, ApJ, 582, L21

Baumgarte, T. W., & Shapiro, S. L. 1999, ApJ, 526, 941

——. 2003, Phys. Reports, 376/2, 41

Begelman, M. C., & Rees, M. J. 1978, MNRAS, 185, 847

Bisnovatyi-Kogan, G. S., Zel'dovich, Ya. B., & Novikov, I. D. 1967, Soviet Astron., 11, 419

Bromm, V., Coppi, P. S., & Larson, R. B. 1999, ApJ, 527, L5

Bromm, V., & Loeb, A. 2003, ApJ, in press (astro-ph/0212400)

Brown, J. D. 2001, in Astrophysical Sources for Ground-based Gravitational Wave Detectors, ed. J. M. Centrella (New York: AIP), 234

Chandrasekhar, S. 1942, Principles of Stellar Dynamics (Chicago: Univ. of Chicago Press)

——. 1964a, Phys. Rev. Lett., 12, 114, 437E

——. 1964b, ApJ, 140, 417

——. 1969, Ellipsoidal Figures of Equilibrium (New Haven: Yale Univ. Press)

Cohn, H. 1979, ApJ, 234, 1036

Colgate, S. A. 1967, ApJ, 150, 163

Elvis, M., Risaliti, G., & Zamorani, C. 2002, ApJ, 565, L75

Fackerell, E. D. 1970, ApJ, 160, 859

Fan, X., et al. 2000, AJ, 120, 1167

——. 2001, AJ, 122, 2833

Ferrarese, L., & Merritt, D. 2000, ApJ, 539, L9

Fowler, W. A. 1964, Rev. Mod. Phys., 36, 545, 1104E

Fryer, C. L., Woosley, S. E., & Heger, A. 2001, ApJ, 550, 372

Fuller, G. M., Woosley, S. E., & Weaver, T. A. 1986, ApJ, 307

Gao, B., Goodman, J., Cohn, H., & Murphy, B. 1991, ApJ, 370

Gebhardt, K., et al. 2000, ApJ, 539, L13

Gebhardt, K., Rich, R. M., & Ho, L. C. 2002, ApJ, 578, L41

Genzel, R., Eckart, A., Ott, T., & Eisenhauer, F. 1997, MNRAS, 291, 219

Gerssen, J., van der Marel, R. P., Gebhardt, K., Guhathakurta, P., Peterson, R. C., & Pryor, C. 2002, AJ, 124, 3270

Ghez, A. M., Morris, M., Becklin, E. E., Tanner, A., & Kremenek, T. 2000, Nature, 407, 349

Gnedin, O. Y. 2001, Class. & Quant. Grav., 18, 3983

Haehnelt, M. G., & Kauffmann, G. 2000, MNRAS, 318, L35

Hennawi, J. F., & Ostriker, J. P. 2002, ApJ, 572, 41

Ho, L. C. 1999, in Observational Evidence for Black Holes in the Universe, ed. S. K. Chakrabarti (Dordrecht: Kluwer), 157

Hughes, S. A., & Blandford, R. D. 2003, ApJ, 585, L101

Ipser, J. R. 1969, ApJ, 158, 17

———. 1980, ApJ, 238, 1101

Ipser, J. R., & Thorne, K. S. 1968, ApJ, 154, 251

Kormendy, J., & Richstone, D. 1995, ARA&A, 33, 581

Lai, D., Rasio, F. A, & Shapiro, S. L. 1993, ApJS, 88, 205

Lee, H. M. 1987 ApJ, 319, 801

Lightman, A. P., & Shapiro, S. L. 1978, Rev. Mod. Phys., 50, 437

Loeb, A., & Rasio, F. A. 1994, ApJ, 432, 52

Lynden-Bell, D., & Eggleton, P. P. 1980, MNRAS, 483, 191

Lynden-Bell, D., & Wood, R. 1968, MNRAS, 138, 495

Macchetto, F. D. 1999, Ap&SS 269, 269

New, K. C. B., & Shapiro, S. L. 2001, ApJ, 548, 439

Ostriker, J. P. 2000, Phys. Rev. Lett., 84, 5258

Portegies Zwart, S. F., & McMillan, S. L. W. 2002, ApJ, 576, 899

Press, W. H., & Schechter, P. L. 1974, ApJ, 190, 253

Quinlan, G. D., & Shapiro, S. L. 1987, ApJ, 321, 199

———. 1989, ApJ, 343, 725

———. 1990, ApJ, 356, 483

Rampp, M., Müller, E., & Ruffert, M. 1998, A&A, 332, 969

Rees, M. J. 1984, ARA&A, 22, 471

———. 1998, in Black Holes and Relativistic Stars, ed. R. M. Wald (Chicago: Chicago Univ. Press), 79

———. 2001, in Black Holes in Binaries and Galactic Nuclei, ed. L. Kaper, E. P. J. van den Heurel, & P. A. Woudt (New York: Springer-Verlag), 351

Richstone, D., et al. 1998, Science, 395, A14

Saijo, M., Shibata, M., Baumgarte, T. W., & Shapiro, S. L. 2001, ApJ, 548, 919

———. 2002, ApJ, 569, 349

Sanders, R. H. 1970, ApJ, 162, 791

Schödel, R., et al. 2002, Nature, 419, 694

Shapiro, S. L. 2000, ApJ, 544, 397

Shapiro, S. L., & Shibata, M. 2002, ApJ, 577, 904

Shapiro, S. L., & Teukolsky, S. A. 1979, ApJ, 234, L177

———. 1983, Black Holes, White Dwarfs, and Neutron Stars: The Physics of Compact Objects (New York: Wiley Interscience)

———. 1985a, ApJ, 298, 34

———. 1985b, ApJ, 298, 58

———. 1985c, ApJ, 292, L41

———. 1986, ApJ, 307, 575

———. 1988, Science, 241, 421

———. 1992, Phil. Trans. Roy. Soc. Ser. A., A340, 365

Shapiro, S. L., Teukolsky, S. A., & Winicour J. 1996, Phys. Rev. D, 52, 6982

Shibata, M., Baumgarte, T. W., & Shapiro, S. L. 2000a, ApJ, 542, 453

———. 2000b, Phys. Rev. D, 61, 44012

Shibata, M., & Shapiro, S. L. 2002, ApJ, 572, L39

Soltan, A. 1982, MNRAS, 200, 115

Spergel, D. N., & Steinhardt, P. J. 2000, Phys. Rev. Lett., 84, 3760

Spitzer, L. 1975, in IAU Symp. 69, Dynamics of Stellar Systems, ed. A. Hayli (Dordrecht: Reidel), 3

———. 1987, Dynamical Evolution of Globular Clusters (Princeton: Princeton Univ. Press)

Spitzer, L., & Saslaw, W. C. 1966, ApJ, 143, 400

Wagoner, R. V. 1969, ARA&A, 7, 553

Wilms, J., Reynolds, C. S., Begelman, M. C., Reeves, J., Molendi, S., Staubert, R., & Kendziorra, E. 2001, MNRAS, 328, L27

Yu, Q., & Tremaine, S., 2002, MNRAS, 335, 965

Zel'dovich, Ya. B., & Novikov, I. D. 1971, Relativistic Astrophysics, Vol. 1 (Chicago: Univ. of Chicago Press)

Zel'dovich, Ya. B., & Podurets, M. A. 1965, Astron. Zh., 42, 963 (English translation in Soviet Astr.-A. J., 9, 742)

8

Gas-dynamical processes in dense nuclei

CATHIE J. CLARKE
Institute of Astronomy, Cambridge

Abstract
We review recent progress in simulating star formation on a small scale and ask if the insights gained can be applied to the formation of supermassive black holes in dense nuclei. Topics discussed include the origin of the Salpeter initial mass function (IMF), the role of feedback, and a proposal for how the characteristic stellar mass might vary in a cosmological context. We also present surprising evidence that the black hole-bulge mass relation is entirely consistent with such objects representing the tail of a Salpeter IMF in these systems, but remain agnostic about the interpretation of this result.

8.1   Introduction

This review describes some recent work that explores a number of hydrodynamical processes in the context of star formation. Many of the simulations involved are sufficiently recent that their implications have scarcely been digested in their original context, so one should be properly sceptical about the attempts made here to map out some tentative applications to the formation of supermassive black holes (SMBHs).

It is currently unclear how much star formation and SMBH formation have in common. In a general sense, one can see that both processes involve a generic question—given a mass of self-gravitating gas, how does this distribute itself among Jeans unstable objects? Does the bulk of the mass collapse into a single mass concentration or are objects produced over a large dynamic range in mass scales? If the latter occurs, are there processes subsequent to the initial fragmentation process (such as coalescence and competitive accretion) which might drive one toward the former scenario? Recently, and more controversially, Larson (2002, 2003) has hinted that SMBH formation may be "no more than" the high-mass end of the star formation process, involving "no more than" a scaled-up version of the processes operating at the stellar-mass scale. Here we remain agnostic on this issue, although we briefly examine if there is any empirical evidence that this is *not* the case.

Most of the simulations described here have modeled rather small-scale systems with masses in the range of hundreds to thousands of $M_\odot$, as typify star-forming complexes such as Taurus and the Orion Nebula cluster. There are two reasons why such environments have been explored rather than the massive/low-metallicity regions that might provide a more suitable environment for producing SMBHs. First, these local regions produce stars with a reasonably well-calibrated mass-luminosity relation and thus the mass spectrum of the objects produced is quite well constrained observationally (see, e.g., Hillenbrand & Carpenter 2000; Luhman et al. 2000). The second reason is a technical one. In nearby

© The Observatories of the Carnegie Institution of Washington 2004.

star-forming regions, the typical stellar mass is around a solar mass, whereas the minimum ("opacity limit") mass for fragmentation is about a factor 100 lower than this. In the case of star formation with primordial abundances, however, the typical stellar mass scale is around $1000 M_\odot$, whereas the opacity limit is 6 orders of magnitude lower than this (Schneider et al. 2002). Thus, a much larger dynamic range of mass scales has to be captured in order to model the primordial problem. Therefore, notwithstanding the seminal simulations of Population III star formation by Bromm, Coppi, & Larson (1999, 2002) and Abel, Bryan, & Norman (2000), the first fully resolved calculations of *ensembles* of stars have been in the context of Population I systems (Bate, Bonnell, & Bromm 2002a,b, hereafter BBB).

We start by briefly describing these first fully resolved simulations of Population I clustered star formation. Since these simulations are at the borderline of what is computationally feasible, it is impossible to explore a wide range of initial conditions, or to treat large ensembles of stars. Consequently, many of the further conclusions about the nature of cluster formation, and, in particular, an analytical understanding of the process, have been derived from more idealized, computationally cheaper, calculations. Before indulging in extrapolations to star formation in different environments, therefore, we should be reasonably assured that the calculations are roughly compatible with observations in the Population I case.

The computational expense in these simulations derives from the necessity of resolving the minimum Jeans mass that can develop in the calculations. These calculations do not treat radiative transfer in the gas, but instead model its thermal evolution by a piecewise polytropic equation of state derived from detailed radiative transfer calculations of spherically symmetric collapse (Larson 1969; Masunaga & Inutsuka 2000). This equation of state (involving a transition from isothermal to adiabatic behavior at the point that dust opacity renders the collapse optically thick) imparts the gas with a minimum Jeans mass corresponding to conditions at this transition. This mass (known as the opacity limit; Low & Lynden-Bell 1976; Rees 1976) is a few Jupiter masses. The BBB calculations (Fig. 8.1) employ smoothed particle hydrodynamics (SPH: see Monaghan & Gingold 1983; Benz et al. 1990), in which all fluid quantities are evaluated by weighted averaging over ensembles of 50 particles. Therefore, in order to resolve all structures that can conceivably collapse in the calculation, one must assign at least ~ 100 particles to the mass scale of the opacity limit (see Bate & Burkert 1997; Whitworth 1998). Evidently, the number of particles required for the simulation of an astronomically interesting cluster of stars is enormous. The BBB simulations employ 3.5 million particles to model $50 M_\odot$ of gas and require 95000 CPU hours on a SGI Origin 3800 supercomputer.

The BBB calculations are currently unique in that they resolve down to the opacity limit, but are in other respects the natural extension of a number of pioneering simulations that have modeled the evolution of turbulent molecular gas (Vázquez-Semadeni, Ballesteros-Paredes, & Rodriguez 1997; Mac Low et al. 1998; Stone, Ostriker, & Gammie 1998; Padoan & Nordlund 1999; Klessen, Heitsch, & Mac Low 2000). Following these authors, the BBB calculations start with initially homogeneous gas that is endowed with a divergence-free random Gaussian velocity field with power spectrum $P(k) \propto k^{-4}$, this choice being motivated by the desire to reproduce the observed size-linewidth relation in molecular clouds (Larson 1981). (Note that, contrary to some of the calculations cited above, the turbulence is not driven, but instead decays on roughly a cloud-crossing time scale.) The initial supersonic turbulent velocity field produces a complex web of sheets and filaments of shock-

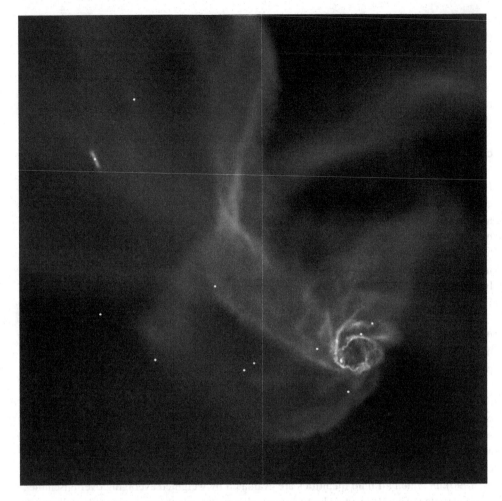

Fig. 8.1. Snapshot from BBB simulations depicting the small clusters that form at the intersection of gas filaments. Note the slingshot ejection of brown dwarf (and edge-on disk) in upper left. The box is 5100 AU on a side. (Courtesy of Matthew Bate.)

compressed gas: stars form, initially at the opacity limit, through fragmentation of both these filaments and of circumstellar disks, and thereafter grow in mass by accretion. This accretion is *competitive* in the sense that protostars accrete from a common gas reservoir, and their orbital history determines their relative success in acquiring mass. Thus, whereas some stars remain close to the center of mass of the ensemble, and accrete rapidly, others suffer slingshot encounters and are ejected, being thereafter deprived of accretable material. Thus, the chaotic orbital dynamics and the consequently diverse accretion histories of the various stars result in a large dynamic range of final masses.

The results of these simulations (i.e., the mass spectrum of stars and brown dwarfs, the kinematics of young stars, the presence of circumstellar disks, and the statistics of binary and multiple systems) can be compared in detail with observations. At a crude level, the results appear to be generally consistent with observations (see Bate, Bonnell, & Bromm

2003), although there are a number of detailed areas (particularly regarding the separations and stability of multiple systems) where further work is required.

Having assured ourselves that these calculations are not grossly incompatible with observations, we shall use this "success" to justify many of the subsequent discussions. This "success" does not mean that these calculations incorporate all the relevant physics. For example, they omit magnetic fields because of the well-known difficulty of incorporating magnetic fields in SPH (see Phillips & Monaghan 1985). Magnetic fields are clearly present in molecular clouds, and a number of authors have constructed models in which these fields play a central role in regulating star formation (e.g., Shu, Adams, & Lizano 1987; McKee 1989). It is now known, however, following Mac Low et al. (1998) and Stone et al. (1998), that magnetic fields are ineffective in one of the roles originally assigned to them, namely the "support" of molecular clouds against global collapse due to the magnetic cushioning of shocks. These simulations (see also Ostriker, Stone, & Gammie 2001) imply that magnetic fields can support molecular clouds only in regions where the mass to flux ratio is subcritical (i.e., roughly at the point where the magnetic energy density exceeds the gravitational potential energy density of the cloud). Regions of the cloud that are not threaded by such strong fields instead collapse on a dynamical time scale (as in the BBB simulations), although fragmentation may be modified by the local magnetic field structure. Clearly, the incorporation of magnetic fields in cluster formation simulations remains an important goal. For now, noting also that magnetic fields may be less important in the early Universe than in Population I systems, we do not consider them further.

8.2   The IMF and the Maximum Object Mass

The IMF in the local Universe is well constrained observationally (see, e.g., Kroupa 2002 and references therein) and is characterized by a power law at high masses such that the number of stars in the mass range m to $m+dm$ is $\propto m^{-\alpha}dm$ with $\alpha = 2.35$ (Salpeter 1955). At masses below around a solar mass, the slope flattens somewhat to give α in the range 1 to 1.5 (Kroupa, Tout, & Gilmore 1990). The mean mass of a stellar population is chiefly set by the upper (lower) limit of the power law if α is $< (>)2$. It thus follows that the transition in α from > 2 to < 2 at around a solar mass imprints the local IMF with a characteristic mass scale of around $1 M_\odot$.

Over the years, fashions concerning the universality of the IMF have come and gone, and the types of departure from universality discussed have likewise varied, with bimodal IMFs and IMFs with variable upper slope having once been popular parameterizations (see, e.g., Güsten & Mezger 1982; Larson 1986, 1998a, b). As the evidence has gathered that the IMF in the local Universe is surprisingly uniform (at least with regard to its upper power law, which may be the only part that is observationally accessible in more distant systems), a popular view is that any variation is likely to be accommodated in a variable value of the characteristic mass (Larson 1998a, b, 2003; Clarke & Bromm 2003). The justification for this view is that a power law is by its nature scale free and may reflect the operation of processes that are themselves blind to scale. A characteristic stellar mass, by contrast, is a dimensional quantity that is likely to depend on some property of the star-forming system.

In an IMF in which the upper power law has a slope ~ 2.35, massive stars are rare and do not dominate the overall mass budget of the system. If there are no *physical constraints* on the upper end of this power law, then the maximum mass in a given stellar system is simply set by the statistics of randomly drawing a finite number of stars from such an IMF.

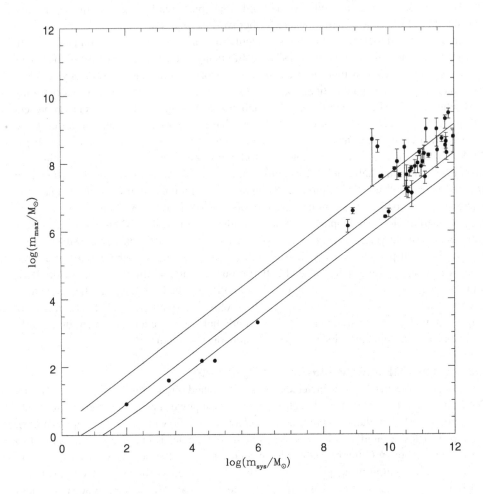

Fig. 8.2. The 95th, 50th, and 5th centiles of the distribution of maximum object mass as a function of system mass, assuming objects are selected from a Salpeter IMF for $M > 0.1 M_{\odot}$. See text for the source of observational data.

Evidently, in a populous system, one is more likely to draw objects from the upper reaches of the IMF than in a small-N system, and, given an IMF, one may readily construct the probability density function of the most massive star as a function of N. The IMF, however, also defines a mean stellar mass $\bar{m}$, such that the system mass is $M_{sys} = N\bar{m}$. In Figure 8.2 we depict the 5th, 50th and 95th centile of the probability density function of the most massive object as a function of M_{sys}, assuming a Salpeter IMF ($\alpha = 2.35$) for object mass $> 0.1 M_{\odot}$. The data points correspond to massive stars in local star-forming regions (Ophiuchus, Orion Nebula cluster, and 30 Doradus) taken from the compilation in Larson (2003) for objects less than $\sim 200 M_{\odot}$, the inferred black hole in M15 (Gerssen et al. 2002) with mass $\sim 2000 M_{\odot}$ and, for the objects more massive than $\sim 10^6 M_{\odot}$, the masses of SMBHs as a function of bulge mass derived from Kormendy & Gebhardt (2001).

The surprising aspect of Figure 8.2 is that the data are so consistent with the "random drawing" hypothesis over about 10 orders of magnitude in system mass! This consistency is unsurprising in the low-mass (stellar) regime, since the IMF has been derived from star counts in such regions. [The fact that the data points in this region suggest a rather shallower slope than predicted by the Salpeter model, may provide some evidence that the IMF slope is somewhat steeper at high masses than the Salpeter value, as suggested by Larson (1982, 2003), although the small number of data points do not make the case compelling.] What is surprising—and quite possibly fortuitous—is the excellent fit to the SMBH data, with the slope, normalization and scatter in the data all being reproduced by this most simple model.

What significance, if any, should one attach to this result? "Random drawing" is of course not a physical description of either the formation of massive stars or SMBHs, but provides a way of quantifying the statistical consequences of a given spectrum of object mass. We have seen that the existence of such a spectrum (combined with a lack of any physical limitations on the masses produced) implies a relation between maximum object mass and the total mass of the reservoir involved. Therefore any physical model that produces such a spectrum must have a way of communicating the total reservoir mass to the region forming the most massive object (even though this object comprises a small fraction of the total reservoir mass). This is reasonable in observed star-forming complexes where the dynamical time is less than the lifetime of the stars concerned and where there is thus time for the reservoir to communicate with the central region where the most massive object is generally formed (see discussion in § 8.4 and § 8.5 of how accretion into the central regions controls the history of competitive accretion and coalescence in these regions and dictates the maximum mass of stars produced). In the case of SMBHs forming in the bulges of galaxies, the dynamical time scales are considerably longer, but on the other hand, the accreting lifetime of the SMBH is indefinite. It is not clear that we can reject the analogy between the two processes on these grounds.

If such a correlation tells us something about analogies between SMBH formation and massive star formation, would this be incompatible with current wisdom (Haehnelt & Kauffmann 2000; Kauffmann & Haehnelt 2000) about the hierarchical assembly of galactic bulges and the build up of black holes through accretion and mergers? The answer here is "probably not." A process that produced a Salpeter-like IMF extending up to supermassive objects could either function in the monolithic collapse of a bulge or, equally, in its constituent progenitors. In the latter case, it would not make much difference to the $M_\bullet - M_{bulge}$ relation if the final black hole was the most massive object in the progenitor halos or the merged sum of such objects (because, for a Salpeter IMF, the mass contained in equal decades of object mass is roughly constant, and therefore one could merge a large dynamic range of hole masses and only increase the final hole mass by a factor of a few). One might even ascribe the fact that the data points are somewhat above the predicted centiles in Figure 8.2 to this effect, although this is probably over-interpretation of the data.

In summary, then, the data on SMBH mass versus bulge mass are, somewhat surprisingly, not incompatible with SMBHs being formed as a massive tail of a mass function extending down to stellar masses. We have not found any obvious physical reasons why a communication between the total reservoir mass and the mass of the maximum object should not occur in both the stellar and the supermassive regime, although the arguments about the extent of the analogy are not at all well developed. Are there other observational grounds for ruling out this conjecture? One implication is that in massive (galactic-scale) systems, there should be a number of objects intermediate in mass between the most massive (central) black hole and objects of stellar mass. Although objects with $M > 10^6 M_\odot$ are likely to have spiralled in and merged with the central black hole, this still leaves a large dynamic range of objects with masses in the range $10^2 - 10^6 M_\odot$ that would be free floating in the galaxy. A number of authors have considered the detectability, or otherwise, of a population of black holes in the halo of the Galaxy (e.g., Carr & Sakellariadou 1999), but were attempting to explain the bulk of the dark matter by such objects. In the current case, the massive black holes would be a minority contributor to the mass budget of the galaxy—considerably less than that contained in stars—and so it is not clear that such a population could be ruled out.

8.3   The Origin of the Characteristic Stellar Mass

We noted above that the IMF in the local Universe is imprinted with a characteristic mass of around a solar mass. An obvious question regards the physics that controls this value, and how one might expect this to vary as a function of environment. Are there, for example, regions of the Universe where very massive stars form not as part of the Salpeter upper-mass tail of a distribution but instead represent the characteristic mass of that population?

In local star-forming regions, this mass scale may readily be identified with the Jeans mass corresponding to dense ($n \approx 10^5$ cm^{-3}), cool ($T \approx 10$ K) molecular gas in star-forming cores (Benson & Myers 1993). This temperature is set by thermal balance between heating (by cosmic rays or UV dust heating) and cooling (by molecular line radiation and infrared dust cooling; see Whitworth, Boffin, & Francis 1998). The density is apparently controlled by a state of rough balance between the thermal pressure in the cores and the mean internal pressure of bulk motions ("turbulence") within the parent giant molecular cloud (GMC; Larson 1996, 1998a). This conjecture is supported by hydrodynamic simulations of turbulent clouds. For example, Padoan, Nordlund, & Jones 1997 (see also Padoan & Nordlund 2002) have shown that driven supersonic turbulence in an isothermal medium produces a characteristic density at which the thermal pressure roughly balances the turbulent pressure in the cloud and have argued that the characteristic stellar mass is the corresponding isothermal Jeans mass. Similar conclusions may be derived from the BBB simulations described in § 8.1, where the turbulence is not driven and where collapse and fragmentation are fully resolved. Here ,too, the mean stellar mass appears to be set by the gas temperature and the turbulent pressure of the initial conditions.

What, then, sets the "turbulent pressure" in star-forming clouds? Two factors may be relevant. First, the attainment of the low temperatures required for star formation requires that molecular coolants are self-shielded against photodissociation by the ambient UV field (e.g., van Dishoeck & Black 1988). This imposes a minimum column for molecular clouds of around 3×10^{21} cm^{-2}. In a self-gravitating system in which the flow velocities are comparable to their free-fall values, the pressure depends on the column density according to $P \sim G\Sigma^2$ and thus the self-shielding requirement implies a minimum viable pressure of around 2×10^4 K cm^{-3}. On the other hand, the ambient pressure of the ISM sets a lower limit to the mean internal pressure (local GMCs are somewhat self-gravitating and thus the internal pressure of molecular clouds exceeds the ambient pressure by around an order of magnitude). The pressure within GMCs (or, equivalently, their column densities) does, however, appear to scale with the ambient interstellar pressure, in that the pressures within

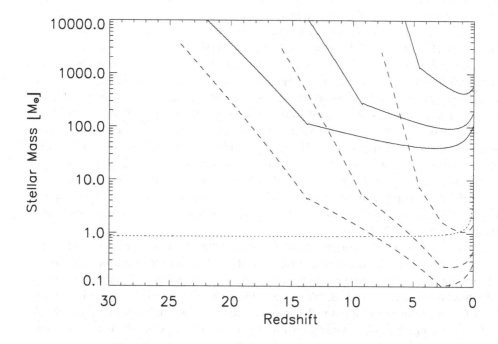

Fig. 8.3. The dependence of characteristic stellar mass on redshift of virialization. (From Clarke & Bromm 2003; see text for details.)

GMCs in the Galactic Center are higher than those in local clouds (Sanders, Scoville, & Solomon 1985).

Recently, Clarke & Bromm (2003) have extended these arguments so as to estimate the characteristic mass scales associated with the bursts of star formation accompanying the assembly of proto-galaxies. Again, the characteristic mass scale is identified with the Jeans mass at given temperature and pressure. The temperature is jointly set by the appropriate cooling agent (~ 10 K for CO line cooling, ~ 200 K for H_2 cooling in primordial gas) and the floor set by the temperature of the cosmic microwave background. The pressure, in this model, is set by the weight of overlying baryons in collapsing halos of given mass and virialization redshift. This latter choice is motivated by the results of hydrodynamical simulations of star formation in low-mass halos, which show that gas that has been heated to the virial temperature collapses nearly isobarically (Bromm et al. 2001). The problem thus reduces to determining the mean pressure of the baryons at the point of virialization. The results of this exercise are illustrated in Figure 8.3, where the solid and dashed lines represent lines of constant overdensity (1, 2, and 3σ from top to bottom) in a ΛCDM cosmology, with cooling provided by H_2 and CO, respectively. The characteristic stellar mass rises toward high z, due both to the increasing temperature of the cosmic microwave background and the decreasing pressure associated with the lower mass halos virializing at high redshift.

This simple prescription roughly reproduces both the near-solar values of the characteristic mass, M_{char}, inferred in the local Universe and the high masses found in numerical simulations of Population III star formation. To see if it is consistent with all observational constraints, it has to be included in hierarchical merger models, and the variation of the functional form of the IMF with redshift must be specified. If we assume (see § 8.2), that the characteristic stellar mass may vary with cosmic conditions but that the functional form of the IMF should remain constant, then the sorts of observations that can test the model are those that link the relative numbers of stars above and below the characteristic mass scale. For example, Hernandez & Ferrara (2002) have argued that the relative paucity of very metal poor stars in the halo of the Milky Way requires a rather high characteristic mass scale (around $10 M_\odot$) in low-mass halos ($10^8 - 10^9 M_\odot$) at redshifts $5 - 10$, in good agreement with the model. On the other hand, the element *ratios* in low metallicity objects are expected to tend to the Type II plateau value (Gibson 1998; Wyse et al. 2002) in all cases that M_{char} is below the minimum progenitor mass of a Type II supernova, since in that case all the objects providing the enrichment would lie on the Salpeter tail of the IMF. Likewise, population synthesis models for Lyman-break galaxies (which imply a Salpeter IMF for stars down to early B spectral type; Pettini et al. 2002) only require that M_{char} is less than $15 - 20 M_\odot$.

An important implication of this model for the production of very massive black holes (and possibly SMBHs) is that even in the case of H_2 cooling, the characteristic stellar mass falls with decreasing redshift due purely to the higher pressures involved in the higher mass halos that collapse at later epochs. We see from Figure 8.3 that for z in the range $10 - 15$ the characteristic stellar mass for gas in $2 - 3\sigma$ peaks falls to a few hundred solar masses. Stars in this mass range are able to undergo pair-production supernovae, implying that very low metallicity conditions are unlikely to prevail at more recent epochs. We therefore conclude that it is very unlikely that an IMF biased toward very high masses could persist at redshifts less than $10 - 15$. According to Schneider et al. (2002), this would be sufficient to account for the density of SMBHs in nearby galaxies (assuming that very massive black holes formed in this mode eventually merged into SMBHs that settled in galactic nuclei) but would not produce enough remnant black holes to account for the bulk of baryonic dark matter.

8.4   The Origin of the Upper End of the Power-law IMF

A wide variety of scale-free processes have been invoked to explain the power-law nature of the upper IMF, including coalescence (Silk & Takahashi 1979; Murray & Lin 1996), fragmentation (Silk 1978), and the result of feedback (Silk 1995; Adams & Fatuzzo 1996). In the present section we focus on a process that produces such an IMF in the BBB simulations described in § 8.1, namely *competitive accretion*. In this case, final stellar masses are not set predominantly by their mass scale when they first become Jeans unstable but instead by their subsequent accretion history from a distributed gaseous reservoir which they share with other accreting protostars. (Note that the success of competitive accretion in providing such a power law does not preclude the importance of other processes for the production of the observed IMF. It does imply, however, that even if these other processes are important, one cannot use the IMF argument to justify their necessity.)

The term "competitive accretion" was first coined by Zinnecker (1992) when he showed that a power-law IMF is readily generated from Bondi-Hoyle (1944) accretion on to seed masses with a small spread in initial masses. The argument may be simply reproduced: for an object of instantaneous mass M, the Bondi-Hoyle accretion rate (assuming the object is moving at constant speed with respect to an infinite, homogeneous background) scales as $\dot{M} \propto M^2$. Thus, at time t

$$\frac{1}{M} = \frac{1}{M_{\text{init}}} - kt,$$

for some constant k. This implies that an initial spread in masses, dM_{init}, maps on to a range of final masses (at some time t) according to

$$dM = \left(\frac{M}{M_{\text{init}}}\right)^2 dM_{\text{init}}.$$

Now, if the number of seeds remains constant as they grow, we thus deduce that the number of seeds per unit mass range must decline as M^{-2}. Thus, in its simplest form, competitive accretion generates an IMF of slope -2, tantalizingly close to the Salpeter slope of -2.35.

Evidently, a real cluster environment is rather different from the idealized model considered by Zinnecker (for example, the gas is likely to be inhomogeneous or self-gravitating and the stars' velocity with respect to the gas will not remain constant). The role of competitive accretion in clusters has been explored by "plum pudding" models (Bonnell et al. 1997, 2001a,b; Delgado-Donate, Clarke, & Bate 2003), which represent an intermediate state of realism. In these numerical (SPH) models, a large number of gravitational "sinks" are distributed within an initially homogeneous self-gravitating gas cloud, with most of the mass being initially in the distributed gas. The evolution of the gas/sink system is then followed hydrodynamically, with accretion of gas onto the sinks (stars) being handled by the sink-particle formulation of Bate, Bonnell, & Price (1995).

Analysis of such simulations reveals two regimes in which stars may accrete the bulk of their mass. Initially, the whole system collapses (because it contains a large number of Jeans masses) and the stars collapse with the gas. As the stars are in this case essentially passengers in the gas flow, the relative velocity, v_{rel}, between the stars and gas is low, and hence the Bondi-Hoyle radius (GM/v_{rel}^2, which sets the cross section for Bondi-Hoyle accretion) is correspondingly very high. In such circumstances, gas approaching at this large impact parameter would be prevented from accreting by the tidal field of the parent cluster, and it is found that the accretion cross section is set by the tidal (Roche) radius of each sink. At late times, and in the central regions of the cluster, the stars start to dominate the potential locally and can undergo violent relaxation, resulting in a virialized stellar core. Gas continues to flow into this central region, but now the relative velocity between the inflowing gas and the virialized stars is high ($\sim$ the free-fall velocity). In this case, the Bondi-Hoyle radius is less than the tidal radius and now sets the accretion cross section.

These two accretion regimes give rise to two different regimes in the resulting IMF. The Bondi-Hoyle accretion that occurs in the virialized core gives rise to an IMF of slope around -2 (see Bonnell et al. 2001a,b) for a discussion of why the slopes in the simulations are rather steeper than this, in even better agreement with the Salpeter value). In the other case, where the accretion cross section is set by the tidal radius one may show analytically that the resulting IMF has slope -1.5 (Bonnell et al. 2001b). This picture is confirmed when one calculates for each star whether it acquired the bulk of its mass in an environment where the local mass density is mainly contributed by the gas or in a virialized core where the stars are locally dominant. It is found (Fig. 8.4) that stars that gain their mass mainly in the gas-dominated regime are lower mass stars, which follow an IMF slope of -1.5, whereas higher mass stars gain most of their mass in the star-dominated, virialized core of

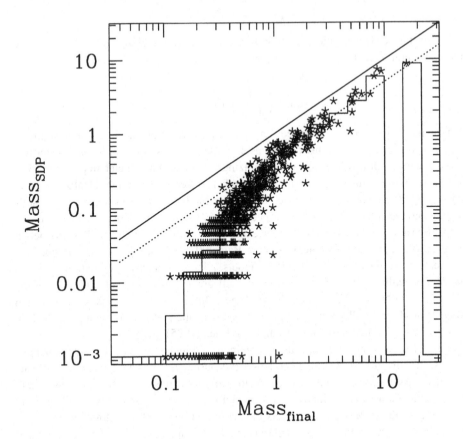

Fig. 8.4. The mass each star acquired in the star-dominated regime (see text) is plotted against its final mass (both in arbitrary units). The histogram plots binned averages of the same quantities and the solid and dashed lines depict the case that 100% and 50% of the final mass is acquired in this regime. The IMF in these simulations shows a transition to a steeper (Salpeter-like) IMF at a mass ~ 1 in these units. (From Bonnell et al. 2001b.)

the cluster, and follow an IMF with slope ~ -2. [Note: the position of the "knee" in the composite IMF is set purely by the average mass per star, since (see § 8.2), the mean mass is always close to the location where the IMF makes a transition between a slope greater or less than -2. In these "plum pudding" models, the mean final stellar mass is put into the initial conditions. In a more realistic model, where the seeds result from Jeans instability in an inhomogeneous and dynamic medium, one might expect the mean stellar mass to be set instead by the considerations set out in § 8.3—i.e., by the minimum gas temperature and dynamical pressure in the parent cloud. The BBB simulations (see § 8.1) appear to corroborate this conjecture.]

8.5   The Highest Mass Stars

In the simulations of populous cluster formation by Bonnell et al. (see § 8.4), the most massive stars form in the center of the cluster. This outcome is consistent with the widely observed tendency of massive stars to be located in cluster cores (Clarke, Bonnell, & Hillenbrand 2000). Since, however, two-body relaxation also leads to the same outcome, one can only *prove* that primordial mass segregation is required in a few well-observed and very young clusters (see the *N*-body simulations of Bonnell & Davies 1998 for such a demonstration in the Orion Nebula cluster).

In cluster formation simulations, stars near the cluster center accrete their local gas reservoir and hence undergo violent relaxation ahead of the rest of the cluster, thereby forming a kinematically distinct core in a state of rough virial equilibrium. Gas, however, continues to flow into this central region. If the core's dynamical time scale becomes less than the gas inflow time scale, then the core subsequently evolves adiabatically and its radius shrinks strongly with accreted mass ($R \propto M^{-3}$). Thus, the accretion of relatively modest quantities of gas produces a dramatic shrinkage of the central core and a very steep ($\propto M^9$) increase of the number density of stars in the core.

Bonnell, Bate, & Zinnecker (1998) first suggested that this shrinkage could ultimately lead to stellar collisions and that the possible merger of collision products could provide a promising mechanism for the formation of the most massive stars. The conditions required for mergers are extreme (stellar number densities of at least 10^8 pc^{-3}) and exceed the observed stellar densities in the densest young clusters by at least 3 orders of magnitude. It is therefore necessary to posit that the high-density phase in which collisions occur is very brief, and that the cluster core rapidly reexpands. Bonnell et al. invoked the destructive effects of feedback (e.g., from ionizing radiation) to expel gas from the core and reexpand it to more reasonable stellar densities.

There seems little doubt that simulations show an inexorable runaway toward high-mass objects in the cores of populous clusters, due to the combined effects of accretion and mergers. The runaway is favored in clusters where the gas flow has a single focus (usually the cluster's center of mass) and where the virialized core is relatively populous. In the BBB simulations, by contrast, the flow (at least on the time scale followed in the calculation) is focused on multiple centers (set up by the pattern of shocks following the dissipation of the initial turbulence), and each center contains only a small number of stars. Under these conditions, the small stellar systems tend to break up (by dynamical encounters and slingshot ejections) before they are driven by accretion to the very high densities required for mergers. From this we conclude (rather unsurprisingly) that the formation of very massive objects is likely to be favored by the monolithic collapse of massive systems.

The simulations, however, neglect feedback and assume that stars stick when they collide. This latter assumption is probably reasonable at the rather low velocities (tens of km s^{-1}) of these cluster cores (M. Davies, private communication). Note that it is found (Bonnell & Bate 2002) that mergers rarely result from a hyperbolic flyby, but instead follow the hardening of a close massive binary by continued accretion and by perturbations from loosely bound companion stars. An anticipated feature of this scenario is that massive stars should frequently be found in close binary pairs with other massive stars, in good agreement with observations (Mermilliod & García 2001).

Because effective feedback mechanisms are associated with more massive stars, feedback breaks the scale-free nature of the processes described above and allows one to consider

Fig. 8.5. The effect of ionizing radiation from a very massive ($220 M_\odot$) star on a proto-cluster. Gas column density is plotted in the central $1.4\ \text{pc} \times 1.4\ \text{pc}$ of a cluster containing $\sim 500 M_\odot$ of gas. The central gas density prior to the switch on of the ionizing source exceeded $10^8\ \text{cm}^{-3}$. (Courtesy of Jim Dale.)

physical (as opposed to statistical) limits on the maximum stellar mass. Although the effects of supernovae on young clusters can be dramatic, the time that must elapse before supernovae explode (2 Myr even in the case of the most massive stars) is comfortably longer than the dynamical time scales of observed clusters (and indeed *much* longer than the dynamical time scales in the extremely dense cores posited in the merger models). The remaining relevant feedback mechanisms are stellar winds, ionization, and the effect of radiation pressure on dust, and each of these will both oppose accretion onto individual stars and expel gas from the cluster core, thus reducing the incidence of stellar collisions.

Simulations are currently being developed that model the effect of radiation pressure on dust during Bondi-Hoyle accretion onto massive stars (Edgar et al., in prep.), while others are treating the effect of ionizing radiation on the gas dynamics in the cluster core (Dale et al., in prep.; see also Kessel-Deynet & Burkert 2000). It appears that feedback limits the range of initial cluster parameters (cloud mass and radius and number of stars forming) that can undergo this runaway to very high densities and consequent stellar collisions. If the local gas density and escape velocity when massive stars form is insufficiently high, ionizing radiation can effectively heat and expel gas from the core (Fig. 8.5) and the runaway is reversed. At higher densities, however, ionizing radiation may be confined to very compact regions around the stars, and the effect on the cluster dynamics may be small.

8.6   Conclusions

In this paper we have reviewed recent progress in modeling the hydrodynamics of star cluster formation and have asked if there is any relevance of the processes discussed to the formation of SMBHs. If there is a link, it is clearly in the area of IMF predictions. Can one scale up the parameters considered here so as to predict the incidence of very massive objects, say stars of hundreds of solar masses, that may collapse to form very massive black holes? To be more provocative, can one even apply these arguments to the formation of objects that have never been stars, such as SMBHs? And, if not, why not? Several themes have run through this paper, and although they may be applied in combination, they are not inextricably linked. Thus, one might accept the arguments about the statistics of drawing objects from a scale-free distribution, but favor a different physical origin for this scale-free IMF than the one advocated here. Likewise, the speculations about how the characteristic mass for star formation may vary with cosmic conditions are independent of a specific model for the upper IMF. Below, we briefly summarize the main themes.

(1) In systems where the distribution of resulting masses follows a power law at high masses, the *maximum object mass* is a simple function of this power-law slope, the lower limit of this power law, and the total mass of the system. This just represents the obvious statistical truth that one is likely to pick a higher mass from the distribution if one starts with a larger mass reservoir (and hence a larger number of random drawings). We have examined if there is any empirical evidence that one cannot apply this argument to systems with a wide range of masses. To our surprise, we have found that if one simply assumes that all systems follow a Salpeter IMF (to arbitrarily high masses), one predicts a distribution of maximum object mass as a function of system mass that is in remarkable agreement with observations over 10 orders of magnitude in system mass. In particular, the masses of SMBHs as a function of host bulge mass and the masses of the putative black holes in globular clusters are in excellent agreement with the hypothesis that the SMBHs represent the high-mass tail of the Salpeter IMF in these systems. We remain agnostic about the significance of this result, but note that observations provide no empirical disproof of this simple hypothesis.

(2) We furthermore speculate that whereas the slope of the upper IMF may be a universal constant, produced, for example, by the scale-free processes described in (3) below, the *characteristic stellar mass* (i.e., the point at which the IMF flattens to a slope less than −2), *may* be a function of environment and epoch. We present a simple model that predicts the characteristic mass for star formation during the assembly of galaxies as a function of halo mass and redshift. This model successfully reproduces both the stellar mass scale associated with simulations of Population III star formation, as well as star formation in the

local Universe. We discuss the implications of this model for the production of very massive black holes at high redshift.

(3) We discuss a model for the IMF suggested by simulations of cluster formation. Competitive accretion plays a central role in these simulations: the final mass of an object depends on its history of competing with neighboring protostars from a common gas reservoir. The upper (Salpeter) tail to the IMF is produced by Bondi-Hoyle accretion in the virialized core of the cluster. We furthermore discuss how such a core may enter a regime of adiabatically invariant shrinkage in response to gas inflow, and how this may lead to the very high densities required for stellar collisions. If these collisions lead to coalescence, the uppermost end of the IMF may be populated by the products of stellar mergers in short-lived, ultra-dense cluster cores.

(4) We briefly discuss how feedback may modify the scale-free nature of the processes described in (3), and may thus undermine the assumptions behind the exercise described in (1). Ongoing simulations of the role of ionizing radiation and radiation pressure on dust on the dynamics of cluster cores suggest that the avoidance of strong feedback (and hence the possibility of growing to very high masses) may put particular constraints on the parameters of the parent cluster.

References

Abel, T., Bryan, G. L., & Norman, M. L. 2000, ApJ, 540, 39

Adams, F. C., & Fatuzzo, M. 1996, ApJ, 464, 256

Bate, M. R., Bonnell, I. A., & Bromm, V. 2002a, MNRAS, 32, L65

——. 2002b, MNRAS, 336, 705

——. 2003, MNRAS, 339, 557

Bate, M. R., Bonnell, I. A., & Price, N. M., 1995, MNRAS, 277, 362

Bate, M. R, & Burkert, A., 1997, MNRAS, 288, 1060

Benson, P. J., & Myers, P. C. 1989, ApJS71, 89

Benz, W., Bowers, R. L., Cameron, A. G. W., & Press, W. 1990, ApJ, 348, 647

Bondi, H., & Hoyle, F. 1944, MNRAS, 104, 273

Bonnell, I. A., & Bate M. R. 2002, MNRAS, 336, 659

Bonnell, I. A., Bate, M. R., Clarke, C. J., & Pringle, J. E. 1997, MNRAS, 285, 201

——. 2001a, MNRAS, 323, 785

Bonnell, I. A., Bate, M. R., & Zinnecker, H. 1998, MNRAS, 298, 93

Bonnell, I. A., Clarke, C. J., Bate, M. R., & Pringle, J. E. 2001b, MNRAS, 324, 573

Bonnell, I. A., & Davies, M. B. 1998, MNRAS, 295, 691

Bromm, V., Coppi, P. S., & Larson, R. B. 1999, ApJ, 527, L5

——. 2002, ApJ, 564, 23

Bromm, V., Ferrara, A., Coppi, P. S., & Larson, R. B. 2001, MNRAS, 328, 969

Carr, B. J., & Sakellariadou, M. 1999, MNRAS, 516, 195

Clarke, C. J., Bonnell, I. A., & Hillenbrand, L. A. 2000, in Protostars and Planets IV, ed. Mannings, V., Boss, A. P., & Russell, S. S. (Tucson: Univ. Arizona Press), 151

Clarke, C. J., & Bromm, V. 2003, MNRAS, 343, 1224

Delgado-Donate, E. J., Clarke, C. J., & Bate , M. R. 2003, MNRAS, 342, 926

Gebhardt, K., Pryor, C., O'Connell, R. D, Williams, T. B., & Hesser, J. E. 2000, AJ, 119, 268

Gerssen, J., van der Marel, R. P., Gebhardt, K., Guhathakurta, P., Peterson, R. C., & Pryor, C. 2002, AJ, 124,327 (addendum: 2003, AJ, 125, 376)

Gibson, B. K. 1998, ApJ, 501, 675

Güsten, R., & Mezger, P. G. 1982, Vistas Astron., 26, 159

Haehnelt, M., & Kauffmann, G. 2000, MNRAS, 318, L35

Hernandez, X., & Ferrara, A. 2001, MNRAS, 324, 484

Hillenbrand, L. A., & Carpenter, J. M. 2000 ApJ, 540, 236

Kauffmann, G., & Haehnelt, M. 2000, MNRAS, 576, 588

Kessel-Deynet, O., & Burkert, A. 2000, MNRAS, 315, 713

Klessen, R. S., Heitsch, F., & Mac Low, M.-M. 2000, ApJ, 535, 887

Kormendy, J., & Gebhardt, K. 2001, in The 20th Texas Symposium on Relativistic Astrophysics, ed. H. Martel & J. C. Wheeler (New York: AIP), 363

Kroupa, P. 2002, Science 295, 82

Kroupa, P., Tout, C. A., & Gilmore, G. G. 1990, MNRAS, 244, 76

Larson, R. B. 1969, MNRAS, 145, 271

——. 1981, MNRAS, 194, 809

——. 1982, MNRAS, 200, 159

——. 1986, MNRAS, 218, 409

——. 1996, in The Interplay between Massive Star Formation, the ISM and Galaxy Evolution, ed. D. Kunth et al. (Gif-sur-Yvette: Editions Frontières), 3

——. 1998a, in The Orion Complex Revisited, ed. M. J. McCaughrean & A. Burkert (San Francisco: ASP), in press

——. 1998b, MNRAS, 301, 569

——. 2002, MNRAS, 332, 155

——. 2003, in Galactic Star Formation Across the Stellar Mass Spectrum, ed. J. M. De Buizer & N. S. van der Bliek (San Francisco: ASP), 65

Low, C., & Lynden-Bell, D. 1976, MNRAS, 176, 367

Luhman, K. L., Rieke, G. H., Young, E. T., Cotera, A. S., Chen, H., Rieke, M. J., Schneider, G., & Thompson, R. I. 2000, ApJ, 540, 1016

Mac Low, M.-M., Klessen, R., Burkert, A., Smith, M. D., & Kessel, O. 1998, Phys. Rev. Lett, 80, 2754

Masunaga, H., & Inutsuka, S. 2000, ApJ, 531, 350

McKee, C. F. 1989, ApJ, 345, 782

Mermilliod, J. C., & García, B. 2001, in IAU Symp. 200, The Formation of Binary Stars, ed. H. Zinnecker & R. D. Mathieu (San Francisco: ASP), 191

Monaghan, J. J., & Gingold, R. A. 1983, J. Comput. Phys., 52, 378

Murray, S. D., & Lin, D. N. C. 1996, ApJ, 467, 728

Ostriker, E. C., Stone, J. M., & Gammie, C. F. 2001, ApJ, 546, 980

Padoan, P., & Nordlund, A. P. 1999, ApJ, 526, 279

——. 2002, ApJ, 576, 870

Padoan, P., Nordlund, A. P., & Jones, B. J. T. 1997, MNRAS, 288, 145

Pettini, M., Rix, S. A., Steidel, C. C., Shapley, A. E., & Adelberger, K. L. 2002, in IAU Symp. 212, From Main Sequence to Supernova, ed. K. A. Van der Hucht, A. Herrero, & C. Esteban (San Francisco: ASP), 617

Phillips, G. J., & Monaghan, J. J. 1985, MNRAS, 216, 883

Rees, M. J. 1976, MNRAS, 176, 483

Salpeter, E. E. 1955, ApJ, 121, 161

Sanders, D. B., Scoville, N. Z., & Solomon, P. M. 1985, ApJ, 289, 373

Schneider, R., Ferrara, A., Natarajan, P., & Omukai, K. 2002, ApJ, 571, 30

Shu, F. H., Adams, F. C., & Lizano, S. 1987, ARA&A, 25, 23

Silk, J. 1978, in Protostars and Planets (Tucson: Univ. Arizona Press), 172

——. 1995, ApJ, 438, 41

Silk, J., & Takahashi, T. 1979, ApJ, 229, 242

Stone, J. M., Ostriker, E. C., & Gammie, C. F. 1998, ApJ, 508, L99

van Dishoeck, E. F., & Black, J. H. 1988, ApJ, 334, 171

Vázquez-Semadeni, E., Ballesteros-Paredes, J., & Rodriguez, L. F. 1997, ApJ, 474, 292

Whitworth, A. P. 1998, MNRAS, 296, 442

Whitworth, A. P., Boffin, H. M. J., & Francis, N. 1998, MNRAS, 299, 554

Wyse, R. F. G., et al. 2002, NewA, 7, 395

Zinnecker, H. 1992, in Symposium on the Orion Nebula to Honour Henry Draper, ed. A. E. Glassgold et al. (New York: New York Academy of Sciences), 226

9

Formation of massive black holes in dense star clusters

FREDERIC A. RASIO[1], MARC FREITAG[1,2], and M. ATAKAN GÜRKAN[1]

(1) Department of Physics and Astronomy, Northwestern University, Evanston, IL 60208, USA

(2) Astronomisches Rechen-Institut, Mönchhofstrasse 12-14, D-69120 Heidelberg, Germany

Abstract

We review possible dynamical formation processes for central massive black holes in dense star clusters. We focus on the early dynamical evolution of young clusters containing a few thousand to a few million stars. One natural formation path for a central seed black hole in these systems involves the development of the Spitzer instability, through which the most massive stars can drive the cluster to core collapse in a very short time. The sudden increase in the core density then leads to a runaway collision process and the formation of a very massive merger remnant, which must then collapse to a black hole. Alternatively, if the most massive stars end their lives before core collapse, a central cluster of stellar-mass black holes is formed. This cluster will likely evaporate before reaching the highly relativistic state necessary to drive a runaway merger process through gravitational radiation, thereby avoiding the formation of a central massive black hole. We summarize the conditions under which these different paths will be followed, and present the results of recent numerical simulations demonstrating the process of rapid core collapse and runaway collisions between massive stars.

9.1 Introduction

The main focus of this chapter is on the formation of massive black holes (BHs) through stellar-dynamical processes in young star clusters. Here by "massive" we mean BHs with masses in the range $\sim 10^2 - 10^4 M_\odot$, which could be "intermediate mass" BHs in small systems such as globular clusters (van der Marel, this volume) or "seed" BHs in larger systems such as proto-galactic nuclei. The later growth of seed BHs by gas accretion or stellar captures to form supermassive BHs in galactic nuclei is discussed by Blandford (this volume). The early dynamical evolution of dense star clusters on a time scale $t \lesssim 10^7$ yr is completely dominated by the most massive stars that were formed in the cluster. This early phase of the evolution is therefore very sensitive to the stellar initial mass function (IMF), particularly at the high-mass end (see Clarke, this volume). One possibility, which we will discuss in some detail here, is that the massive stars could drive the cluster to core collapse before evolving and undergoing supernova explosions. Successive collisions and mergers of these massive stars during core collapse can then lead to a runaway process and the rapid formation of a very massive object containing the entire mass of the collapsing cluster core. Although the fate of such a massive merger remnant is rather uncertain, direct "monolithic" collapse to a BH with little or no mass loss is a likely outcome, at least for sufficiently low metallicities (Heger et al. 2003).

© The Observatories of the Carnegie Institution of Washington 2004.

Alternatively, if the most massive stars in the system evolve and produce supernovae before the onset of this runaway collision process, the collapse of the cluster core will be reversed by the sudden mass loss, and a cluster of stellar-mass BHs will be formed. The final fate of this cluster is also rather uncertain. For small systems such as globular clusters, complete evaporation is likely (with all the stellar-mass BHs ejected from the cluster through 3-body and 4-body interactions in the dense core). This is expected theoretically on the basis of simple qualitative arguments (Kulkarni, Hut, & McMillan 1993; Sigurdsson & Hernquist 1993) and has been demonstrated recently by direct *N*-body simulations (Portegies Zwart & McMillan 2000). However, for larger systems such as proto-galactic nuclei, contraction of the cluster to a highly relativistic state could again lead to successive mergers (driven by gravitational radiation) and the formation of a single massive BH (Quinlan & Shapiro 1989).

Because of the great complexity and variety of dynamical processes involved, questions related to BH formation in dense star clusters are best studied using numerical simulations. On today's computers, this can be done with direct *N*-body simulations for $N \approx 10^3 - 10^5$ stars (see, e.g., Aarseth 1999) or with Monte Carlo (MC) techniques (see Sec. 1.2) for up to $N \approx 10^7$ stars. Observationally, this large range covers a variety of well-studied young star clusters, from "young populous clusters" (e.g., Arches, R136) to "super star clusters" (Whitmore 2003). However, realistic, star-by-star simulations of larger systems, on the scales of entire galactic nuclei, are not yet possible. For these systems, one must rely on more qualitative analyses based on extrapolations from numerical results for smaller *N*.

The role of stellar collisions in dense galactic nuclei was first consider to explain the quasar/active galactic nucleus phenomenon (e.g., Spitzer & Saslaw 1966). Colgate (1967) pointed out that, when collisions first set in a stellar cluster evolving from "reasonable" initial conditions, they are unlikely to be disruptive so that runaway mergers should lead to the formation of massive stars. He estimated that growth would saturate at $\sim 50 M_\odot$ because small stars could fly across a massive star without being stopped. In "particle-in-a-box"-type simulations using a semi-analytical model for the outcome of individual collisions, Sanders (1970) found clear runaway growth up to a few hundred $M_\odot$. No proper account of the stellar dynamics was included, however; the cluster was treated as a homogeneous sphere of constant mass (the gas ejected in collisions being recycled into stars) contracting through collisional energy loss.

Stellar dynamics must be playing an important role, however, especially for massive stars, which can be affected significantly by mass segregation. Indeed, massive stars of mass *m* undergo mass segregation and concentrate into the cluster core on a time scale $t_{\text{segr}} \simeq (\langle m \rangle / m) t_{\text{rh}}$, where $\langle m \rangle$ is the average stellar mass and the overall relaxation time (at the half-mass radius r_{h}) for a cluster of total mass *M* is given by

$$t_{\text{rh}} \simeq 10^8 \, \text{yr} \left(\frac{r_{\text{h}}}{1 \, \text{pc}} \right)^{3/2} \left(\frac{M}{10^6 M_\odot} \right)^{1/2} \left(\frac{\langle m \rangle}{1 M_\odot} \right)^{-1}.$$

It is clear that the most massive stars in a dense cluster can undergo mass segregation on a time scale much shorter then their stellar evolution time. When they eventually dominate the density in the cluster core, these massive stars will then *decouple dynamically* from the rest of the cluster (go out of thermal equilibrium, evolving *away* from energy equipartition) and evolve very quickly to core collapse. This process is often referred to as Spitzer's "mass-segregation instability."

The possibility of a "mass-segregation instability" in simple two-component systems (clusters containing only two kinds of stars, one much more massive than the other) was first predicted by Spitzer (1969), and the first dynamical simulations revealing mass segregation at work were performed in the 70's (Spitzer & Hart 1971; Spitzer & Shull 1975). The physics of this instability is now very well understood theoretically (Watters, Joshi, & Rasio 2000). In a remarkably prescient paper, Vishniac (1978) showed correctly for the first time that this instability must affect the early dynamical evolution of any star cluster born with a reasonable "Salpeter-like" IMF. He concluded that, as a result, "most globular clusters may undergo core collapse at an early time in the evolution of the universe. This is clearly significant as a possible mechanism for creating collapsed bodies in the center of globular clusters."[*] In their classic paper the same year, Begelman & Rees (1978) also mentioned the combination of mass segregation and runaway merging as "one of the quickest routes to the formation of a massive object in a dense stellar system."

However, the first detailed dynamical simulations of dense star clusters including stellar collisions and mergers considered clusters where all stars are initially identical (Lee 1987; Quinlan & Shapiro 1990, hereafter QS90). These Fokker-Planck (FP) simulations also included the dynamical formation of binaries through 3-body interactions and their subsequent hardening (and ejection) as a central source of energy capable of reversing core collapse and turning off collisions in clusters with a relatively low number of stars. Furthermore, in these simulations, collisions themselves can stop collapse when collisionally produced massive stars lose mass in a supernova explosion at the end of their life. The results of these early simulations suggested that runaway collisions would occur provided that $t_{rh} < 10^8$ yr (to beat stellar evolution) and $N \geq 3 \times 10^6$ (to avoid binary heating). QS90 stressed that, as a result of mass segregation, the rise in the central velocity during collapse is only moderate and collisions do not become disruptive. Although not very realistic, these early studies made plausible the idea that successive collisions and mergers of main-sequence stars could lead to the formation of a $\sim 10^2 - 10^3 M_\odot$ object.

More recently, through direct N-body simulations of clusters containing 2000 to 65,000 stars, Portegies Zwart & McMillan (2002) showed that, in such low-N systems, dynamically formed binaries, far from preventing collisions (by heating the cluster and reversing collapse), actually *encourage* them by increasing the effective cross section. In these small systems, once the few massive stars have segregated to the center, one of them will repeatedly form a binary with another star and later collide with its companion when an

---

[*] Historical note contributed by G. Burbidge: "When we first were trying to understand what was responsible for violent events, originally in radio sources, in the early 1960s we (Hoyle, Fowler and the Burbidges') discussed the problem of gravitational collapse of massive objects (superstars) after Hoyle and Fowler had argued that the mechanism I had proposed in 1961, chain reactions of supernovae, would not work. In this early period Hoyle and Fowler wrote several papers on the collapse, and in 1964 HFB^2 published a paper in ApJ discussing all of the ramifications. We did not know how to form the beast — the work of Colgate and Spitzer showed how it would work if the star density was high enough, but the densities required were far greater than any that could reasonably exist in the centers of galaxies. But we were convinced that such phenomena must exist in the centers, and Hoyle was quite angry when much later in 1969 Lynden-Bell was given credit for the idea after a press conference. By then Hoyle and I had become convinced that the usual mechanism of accretion would not give enough energy because the efficiency of the process in terms of what *we see* must be very low, and much below 10%. Thus we turned to new and fundamental ideas concerning creation in galaxy centers which can only take place in regions of very strong gravitational fields, i.e., next to classical BHs. This is my position today. Fred and I worked on it until his death. When Vishniac's paper appeared in 1978, Fred and I discussed it and realized that mass segregation might allow a BH to form rapidly. But this was in the context of total masses of $10^5 - 10^6 M_\odot$ — globular clusters — much less than is required for massive BHs in nuclei. Still we talked about doing more work on it. It was not pursued, probably because we were neither in a position to do it ourselves; I was just moving to Kitt Peak as director, and Fred had left Cambridge and was living up in Cumberland. But we both thought it was an important step forward."

interaction with a third star increases the binary's eccentricity. The growth of this star is ultimately stopped by stellar evolution, or by the dissolution of the cluster in the tidal field of the parent galaxy. Given the small number of stars in these simulations, the maximum mass of the collision product is only $\sim 200 M_\odot$ when mass loss from stellar winds is negligible. The "pistol star" in the Galactic Center Quintuplet cluster (Figer et al. 1998) may provide an example of a directly observable runaway collision product of this type in a small, young star cluster.

Work in progress by the authors is now attempting for the first time to study numerically these processes for much larger star clusters, containing up to $N \approx 10^7$ stars, and resolving in detail the core collapse and runaway collisions. Section 9.2 provides a summary of the numerical methods that we are using, based on a MC technique for collisional stellar dynamics. Section 9.3 presents a few of our initial results. Our main conclusions can be summarized as follows.

Our numerical simulations show that, in the absence of stellar evolution, the core collapse time in a dense star cluster is always given by

$$t_{cc} \simeq 0.1 \, t_{rh},$$

for any "reasonable" IMF (i.e., not too different from a Salpeter IMF) and initial cluster structure (basically, any initial density profile that is "not too centrally concentrated"). If we assume that core collapse corresponds to the onset of runaway collisions, then the condition for a runaway to occur can be written very simply

$$t_{cc} \simeq 0.1 \, t_{rh} < \tau_*(m_{max}),$$

where $\tau_*(m_{max})$ is the lifetime of the most massive stars, of mass m_{max}. From current stellar evolution calculations (e.g., Schaller et al. 1992) we know that

$$\tau_*(m_{max}) \simeq 3 \, \text{Myr} \qquad \text{for } m_{max} > 30 M_\odot,$$

i.e., nearly *constant* (this is simply because these massive stars are nearly Eddington-limited*), and also nearly independent of metallicity and rotation (within $\sim 10\%$). Therefore, under very general conditions, the simple criterion for a runaway process can be written

$$t_{rh} \lesssim 3 \times 10^7 \, \text{yr}.$$

This is in perfect agreement with the results of the direct N-body simulations by Portegies Zwart & McMillan (2002), although they cannot resolve the core collapse in their simulations, and instead *define* core collapse to be the onset of collisions†.

Perhaps the most interesting result from our new simulations is that the central (BH) mass produced by this runaway process may well be determined largely by the Spitzer instability: the total mass in massive stars going into core collapse will be the final BH mass, at least in the absence of significant mass loss from stellar evolution. Remarkably, we find that this mass is always (within the same "reasonable" assumptions about the IMF and initial cluster

---

* If the mass segregation time scale $t_s \propto 1/m$ is compared to the "usual" $\tau_* \propto 1/m^3$, one could conclude erroneously that massive stars never play a role in core collapse (Applegate 1986). However, the approximate $1/m^3$ scaling of the stellar lifetime applies only for $m \lesssim 10 M_\odot$.

† For any runaway collision process to occur, the IMF must extend to $m_{max} \gg 10 M_\odot$ and the total number of stars must be sufficiently large, $N \gg 10^3$. Otherwise there will simply not be enough (or any) massive stars to collide.

structure as above) around 10^{-3} of the total cluster mass, as suggested by observations (see Kormendy, Richstone, and van der Marel, this volume).

9.2   Monte Carlo Simulations of Dense Star Cluster Dynamics

Mass segregation and collisional runaways are driven by the most massive stars in a cluster, which, for a "normal" IMF, represent a very small fraction of the stellar population. For instance, in a Salpeter IMF from 0.2 to $120\,M_\odot$, only a fraction $\simeq 3 \times 10^{-4}$ of stars are more massive than $60\,M_\odot$. Resolving these processes thus requires numerical simulations with very large numbers of particles (at least $\sim 10^5 - 10^6$). In the last few years, new MC codes have been developed that make possible simulations of stellar clusters with such high resolutions. The results reported here have been obtained with two independent MC codes that we call, for convenience, MCglob and MCnucl, as they were devised to follow the evolution of globular clusters and galactic nuclei, respectively. Both are based on the scheme first proposed by Hénon (1973), and they rely on very similar principles that we summarize below. In Sections 9.2.1 and 9.2.2, a few important aspects specific to each code are described.

The MC technique assumes that the cluster is spherically symmetric and can be modeled by a set of discrete particles. Each particle in the simulation could represent an individual star (as in MCglob), or an entire spherical shell of stars sharing the same orbital and stellar properties (as in MCnucl and most earlier MC codes). In the latter case, the number of particles may be lower than the number of stars in the simulated cluster, but the number of stars per particle has to be the same for all particles. Another important assumption in these codes is that the system is always in dynamical equilibrium, so that orbital time scales need not be resolved and the natural time step is a fraction of the relaxation (or collision) time. Instead of being determined by integration of its orbit, the position R of a particle (star, or the radius R of the shell) is picked at random, with a probability density that reflects the time spent at R: $dP/dR \propto 1/V_r(R)$ where V_r is the radial velocity along the orbit.

The relaxation is treated as a diffusive process in the usual FP approximation (Chandrasekhar 1960; Binney & Tremaine 1987). The long-term effects on orbits of departures of the gravitational forces from a smooth quasi-stationary potential, ϕ_s, are assumed to be those of a large number of uncorrelated small-angle scatterings. If a particle of mass M_1 travels with relative velocity v_{rel} through a homogeneous field of particles of mass M_2 with number density n during δt, in the center-of-mass reference frame, its trajectory will be deflected by an angle $\theta_{\delta t}$ with

$$\langle \theta_{\delta t} \rangle = 0 \quad \text{and} \quad \langle \theta_{\delta t}^2 \rangle = 8\pi \ln \Lambda\, G^2 n\, (M_1 + M_2)^2\, v_{rel}^{-3}\, \delta t,$$

where G is the gravitational constant and $\ln \Lambda \simeq 10 - 15$ is the familiar Coulomb logarithm (Binney & Tremaine 1987). In the MC codes, at each time step, a pair of neighboring particles are selected and their velocities are modified through an effective hyperbolic encounter with deflection angle $\theta_{\mathrm{eff}} = \sqrt{\langle \theta_{\delta t}^2 \rangle}$. As any given particle will be selected many times, at various positions on its orbit, the MC scheme will actually integrate the effect of relaxation over the particle's orbit and over all possible field particles. Proper averaging is ensured if the time steps are sufficiently short for the orbit to be significantly modified only after a large number of effective encounters. The gravitational potential ϕ_s is approximated as the potential generated by all the particles. This potential is not completely smooth because the particles are razor-thin spherical shells whose radii change discontinuously.

Through test computations, it can be shown that the corresponding spurious relaxation is negligible if the number of particles is $\gtrsim 10^4$ (Freitag & Benz 2001).

In contrast to methods based on the direct integration of the FP equation in phase space, the particle-based MC approach allows for the natural inclusion of many additional stellar and dynamical processes, such as stellar evolution, collisions and mergers, tidal disruptions, captures, large-angle scatterings, and strong interactions with binaries. The dynamical effects of binaries (the dominant 3- and 4-body processes), which may be crucial in the evolution of globular clusters, have been included in several MC codes through the use of approximate analytic cross sections and simple recipes (Stodołkiewicz 1986; Giersz & Spurzem 2000; Fregeau et al. 2003). Codes are currently in development by Giersz, Rasio and collaborators that will treat binaries with much higher realism by explicitly integrating 3- or 4-body interactions on the fly, a "brute-force" approach required to tackle the full diversity of unequal-mass binary interactions.

Among methods that can be used to follow the dynamical evolution of collisional stellar systems, only direct N-body integrations do not require assumptions on the geometry of the system or dynamical equilibrium and, being also particle based, rival the ability of MC codes to incorporate realistic physics. Unfortunately, as they are based on explicit integration of the orbits of N particles and require the computation of all 2-body forces, they are extremely computationally demanding, with a CPU time scaling like N^{2-3}. In practice, even with the use of special-purpose computers (such as the current GRAPE-6) and in spite of continuous progress in the development of N-body algorithms, the simulation of a cluster containing $\sim 10^5$ stars still requires months of computer time (Makino 2001). In contrast, MC codes, with CPU times scaling like $N \ln(N)$, routinely use up to a few million particles. Such high numbers of particles imply that globular clusters can actually be modeled on a star-by-star basis (Giersz 1998, 2001; Joshi, Rasio, & Portegies Zwart 2000; Watters, Joshi, & Rasio 2000; Joshi, Nave, & Rasio 2001). However, galactic nuclei typically contain $N_* \approx 10^7 - 10^8$ stars, and, for such systems, one has to take advantage of the fact that each physical process is included in the simulation with its explicit N_* scaling so that a single particle can represent many stars (in `MCnucl`).

9.2.1 *A Monte Carlo Code for Globular Cluster Simulations*

One of our MC codes, `MCglob`, based directly on the ideas of Hénon (1973) described in the previous section, was developed to study the dynamical evolution of globular clusters using a realistic number of stars (Joshi et al. 2000, 2001; Fregeau et al. 2003). One important characteristic of this code is that each particle represents a single star (or binary star) in the cluster, which allows us to incorporate stellar evolution processes in a completely realistic manner. The code uses a single time step for the evolution of all the stars (typically a small fraction of the core relaxation time), which allows for effective parallelization of the algorithm. Typical simulations for $\sim 10^5$ stars over $\sim 10^{10}$ yr can be performed in just a few hours of computing time.

This code has been tested extensively using comparisons to previous FP and direct N-body results, and it has been used to study a variety of fundamental dynamical processes such as the Spitzer instability (Watters et al. 2000) and mass segregation (Fregeau et al. 2002) in simple two-component clusters. It was also used to re-examine the question of globular cluster lifetimes in the tidal field of a galaxy (Joshi et al. 2001). Work is in progress to incorporate a full treatment of primordial binaries, including all dynamical interactions (binary–binary and binary–single) as well as binary stellar evolution (Rasio, Fregeau, & Joshi 2001; Fregeau et al. 2003). Another major advantage of these star-by-star MC simulations (e.g., compared to direct FP schemes) is that it is straightforward to include a realistic, continuous stellar-mass spectrum. With a broad IMF, this requires adjusting carefully the Coulomb logarithm, which can be done by using the results of large N-body simulations for calibration (Giersz & Heggie 1996, 1997). Another difficulty introduced by a broad mass spectrum is the necessity of adjusting carefully the time step to treat correctly encounters between stars of very different masses. When pairs of stars are selected to undergo an effective hyperbolic encounter as described above, one has to make sure that the deflection angle remains small for *both* stars. In situations where the mass ratio of the pair can be extreme, one has to decrease the time step accordingly (Stodołkiewicz 1982). In practice, for the simulations described here, we find that the time step has to be reduced by a factor of up to $\sim 10^3$ compared to what would be appropriate for a cluster of equal-mass stars.

9.2.2   A Monte Carlo Code for Galactic Nuclei Simulations

To the best of our knowledge, MCnucl is the only Hénon-like MC code specifically aimed at the study of galactic nuclei (Freitag 2001; Freitag & Benz 2001, 2002b). In addition to relaxation and stellar evolution, it also incorporates a detailed treatment of collisions between single main-sequence stars. A central massive BH can also be included (together with a full treatment of tidal disruptions and stellar captures) but this is not necessary for the work presented here.

Individual time steps are used, with particles updated more frequently where the evolution is faster. Specifically, the time steps are set to some small fraction f of the *local* relaxation (or collision) time: $\delta t(R) \simeq f\left(t_{rel}^{-1} + t_{coll}^{-1}\right)^{-1}$. At each step, a pair of neighboring particles is selected randomly with probability $P_{selec} \propto 1/\delta t(R)$, ensuring that the *average* residence time at R is $\delta t(R)$. After a particle is modified, the potential, stored in a binary-tree structure, is updated.

Unlike relaxation, collisions cannot be treated as a continuous process. They are discrete events that can affect very significantly the orbits and masses, or even the existence of particles. When a pair is selected, the collision probability between stars from each particle (shell),

$$P_{coll} = S_{coll} v_{rel} n \, \delta t \quad \text{with} \quad S_{coll} = \pi b_{max}^2 = \pi (R_1 + R_2)^2 \left(1 + \frac{2G(M_1 + M_2)}{(R_1 + R_2)v_{rel}^2}\right),$$

is compared with a uniform-deviate random number to decide whether a collision has occurred. If so, the impact parameter b is determined by picking another random number X, with $b = b_{max}\sqrt{X}$. The other parameters of the collision, i.e., M_1, M_2 and v_{rel}, are known from the particle properties. The final outcome of the collision (new velocities and masses) is determined very accurately by interpolation from a table containing the results of more than 14,000 3-D SPH (smoothed particle hydrodynamics) calculations of encounters between two main-sequence stars (Freitag 2000; Freitag & Benz 2002c, 2004).

The structure and stellar evolution of collision products is an intricate problem that has only been studied in some detail for low-velocity collisions, relevant to globular clusters (Sills et al. 1997, 2001; Lombardi et al. 2002). The main issues are the importance of entropy stratification versus collisional mixing, and the rapid rotation of merger remnants.

In view of these difficulties, we use one of two simple prescriptions to treat the stellar evolution of collision products. *(1) Maximal rejuvenation,* where the remnant is assumed to be completely mixed and is brought back on the zero-age main sequence. As hydrodynamic simulations show only very little mixing, this assumption leads obviously to an overestimate of the stellar lifetime of collision products. *(2) Minimal rejuvenation.* Here we assume that, during a merger, the helium cores of both parent stars merge together, while the hydrogen envelopes combine to form the new envelope; no hydrogen is brought to the core. An effective age on the main sequence is given by adopting a linear dependence of the helium core mass on age.

9.3   Core Collapse in Young Star Clusters

We have used our code `MCglob` to study the onset of core collapse in young star clusters with a wide variety of initial structures and stellar IMFs (Gürkan, Freitag, & Rasio 2004). The results of a typical simulation are illustrated in Figs. 9.1–9.3. This simulation was performed for a cluster containing 2.5×10^6 stars with a Salpeter IMF between $m_{\min} = 0.2 M_\odot$ and $m_{\max} = 120 M_\odot$. The rapid mass segregation of the most massive stars toward the cluster center is evident, as is the abrupt onset of core collapse once the central region becomes dominated by massive stars. In Fig. 9.2 we show the evolution of the mean radius of stars in various mass bins. We see that the mass segregation starts right at the beginning of the evolution and proceeds at a steady rate, even though the overall mass distribution (total density profile) in the cluster is hardly changing. Significant changes in the Lagrange radii become apparent only when the heaviest stars reach the center of the cluster. This is because these heaviest stars account for only a very small fraction of the total cluster mass. In all our simulations we find that the ratio of the core collapse time to initial half-mass relaxation time, $t_{cc}/t_{rh}(0)$, is always within the range 0.05–0.20 so long as the heaviest stars have masses $\gtrsim 20$ times the average stellar mass in the cluster. We used this result $[t_{cc}/t_{rh}(0) \simeq 0.1]$ in Section 9.1 to derive our simple criterion for the onset of a runaway. In Fig. 9.3, we show the core collapse in more detail. We see that massive stars representing about 0.1% of the total cluster mass are driving the final core collapse. This ratio of the collapsing core mass to total cluster mass, M_{cc}/M_{tot}, also appears to be confined to a narrow range, of about 0.1%–0.3%, in all our calculations to date. These were performed for both Plummer models and King models with varying initial concentration, and for a variety of IMFs including simple power-law IMFs with exponents in the range 2–3 (the standard Salpeter value being 2.35). We have also checked the robustness of this result by varying the number of stars and the time step in our simulations. As pointed out in Section 9.1, the apparent agreement between this ratio and the ratio of BH mass to total cluster mass in many observed systems is very encouraging. Work is in progress to study more systematically the dependence on initial cluster structure, including the possibility of initial mass segregation.

9.4   The Runaway Collision Process

The formation of a massive central object by runaway collisions and mergers has been demonstrated in idealized FP models of proto-galactic nuclei by QS90. Unfortunately, because of the limitations of FP codes, these authors had to rely on a highly simplified treatment of collisions (see below). Furthermore, they assumed an initial single-mass stellar population, which significantly reduces the impact of mass segregation and stellar evolution.

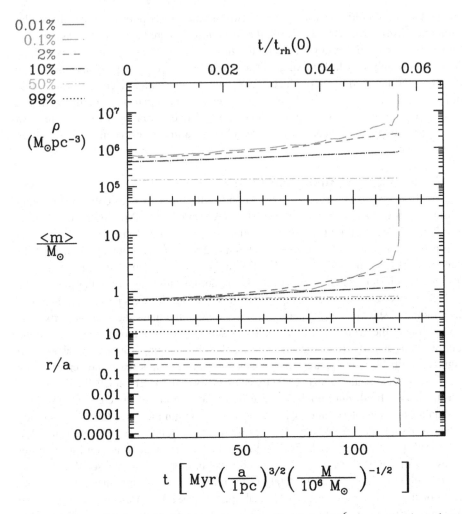

Fig. 9.1. The evolution of a cluster containing $N = 2.5 \times 10^6$ stars, terminated at core collapse. Time is given both in years (bottom) and in units of the initial half-mass relaxation time (top). The initial configuration is a Plummer sphere with a Salpeter IMF spanning the range from $m_{min} = 0.2\,M_\odot$ to $m_{max} = 120\,M_\odot$. The bottom panel shows the evolution of Lagrange radii (enclosing a constant fraction of the total cluster mass, indicated in the top left), in units of the Plummer length (core radius) of the initial model. The middle panel shows the average stellar mass within each Lagrange radius, and the top panel shows the average density within each Lagrange radius. Core collapse takes place at $t = 0.056\,t_{rh}(0)$.

Indeed, these processes come into play only when more massive stars are formed through collisions.

Using MCnucl, we have started re-examining this problem with more realistic simulations. In the high-velocity environment of galactic nuclei, the formation and/or survival of binaries is unlikely (i.e., most binaries are "soft"), so it is reasonable to neglect them in the computations. As the stellar density rises abruptly to very high values during core collapse, collisions are bound to occur even in the absence of binaries.

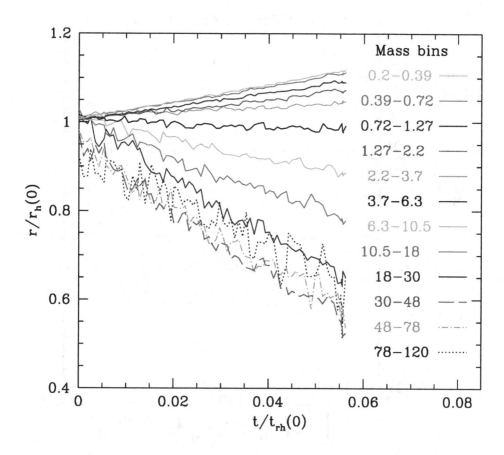

Fig. 9.2. The evolution of the mean radius (in units of the initial half-mass radius) for stars in various mass bins (indicated on the right; values are in units of $M_\odot$). The simulation is the same as in Fig. 9.1. Mass segregation is very clear, concentrating more massive stars near the center of the cluster.

For definiteness, we consider first QS90's model E4A, starting from a Plummer sphere with initial central density and 3D velocity dispersion of $3 \times 10^8 M_\odot \text{pc}^{-3}$ and 400km s^{-1}. QS90 started their FP simulations with all stars having $1 M_\odot$ and assumed that all collisions lead to mergers with no mass loss and maximal rejuvenation. Not surprisingly, if we use the same, highly simplified treatment of collisions as QS90, we get clear runaway growth of one or a few particles. When we switch to our realistic prescription for the collisions and minimal rejuvenation, the runaway still occurs, although later. However, if we adopt a more realistic, Kroupa-type IMF (Kroupa 2001), significant mass loss from the massive stars occurs before core collapse has proceeded to high stellar densities. As we assume that the gas is not retained in the cluster, this mass loss drives a re-expansion of the whole system. A second, deeper core collapse occurs later, when the remnant (stellar-mass) BHs segregate

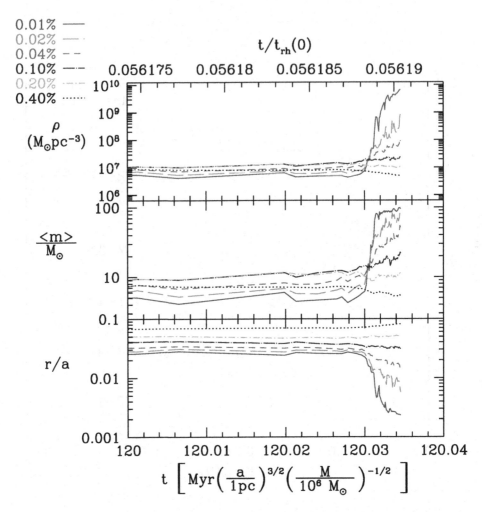

Fig. 9.3. Same as figure 9.1, but concentrating on the evolution of the cluster core near collapse.

to the center. The evolution of this dense cluster of stellar BHs cannot be treated with the present version of MCnucl because dynamically formed binaries will play a central role.

In addition to models with the same densities and velocity dispersions as the class "A" considered by QS90, we also simulated clusters with densities 3 times (models "Z") and 9 times (models "Y") larger (but the same velocity dispersion) with correspondingly shorter relaxation times (see Fig. 9.4). As shown in Fig. 9.5, models of class "Z" have a core collapse time slightly larger than 3 Myr and do not enter a runaway phase. One the other hand, runaway growth occurs in all simulations for clusters of class "Y," which collapse in about 1 Myr. Figure 9.6 shows such a case. The growth of the runaway particle(s) is limited to a few hundred $M_\odot$ (650$M_\odot$ in the "best" case), probably by some numerical artifact. Note that 500,000 particles were used for all these computations, independent of the true number of stars in the cluster. Hence, every particle represents many stars (12

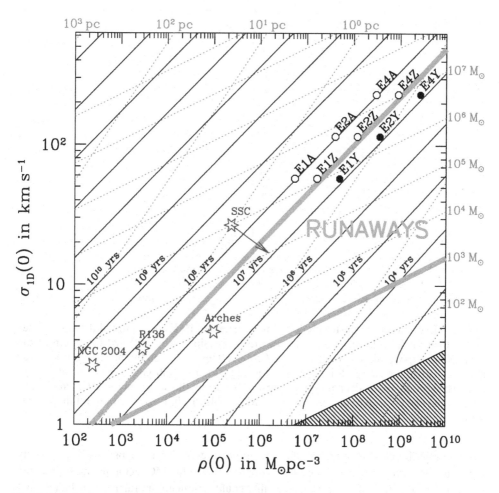

Fig. 9.4. Initial conditions for the clusters considered in Sec. 9.4. We use a notation similar to that of QS90, but our models have a broad IMF. Here $\rho(0)$ and $\sigma_{1D}(0)$ are the initial central density and velocity dispersion, respectively. The thin dotted lines show the values of the total mass (right labels) and the initial Plummer scale (top labels). We also show lines of constant half-mass relaxation time (solid lines labeled from 10^4 to 10^{10} yrs). Clusters born in the region between the thick diagonal lines have a core collapse time shorter than the lifetime of their most massive stars and are expected to undergo a runaway collision process. The solid and open round dots in the upper right region show the results of our MC simulations: an open dot indicates that a runaway was avoided, while a solid dot indicates that the onset of runaway collisions was detected. The open star symbols indicate the positions of a few observed young star clusters. The one labeled "SSC" indicates the position of a typical "super star cluster." Most of these systems have sizes at the resolution limit of the observations, with an upper limit on their radius of $\sim 1\,pc$.

to 36), a numerical treatment whose validity becomes obviously questionable as soon as a single particle detaches from the overall mass spectrum. In addition, we note that, before

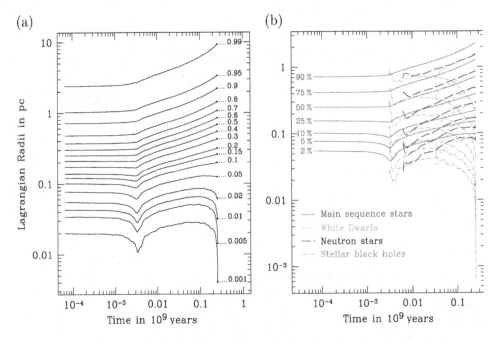

Fig. 9.5. Evolution of a cluster model with an extended IMF and an initial density 3 times higher than model E4A but the same velocity dispersion (see text). *(a)* Evolution of the overall Lagrange radii. The first, very mild core collapse is driven by the segregation of the massive stars to the center. It is quickly reversed by their evolution and the associated mass loss. The second, much deeper core collapse is driven by the stellar-mass BHs, which have become the most massive species after a few Myrs ($7\,M_{\odot}$). *(b)* Lagrange radii for the various stellar species.

the runaway abruptly saturates in our simulations, the growth rate observed is extremely rapid and the basic orbit-averaging assumption implied in the MC technique must probably break down. Possible physical processes that could terminate the runaway include: rapid mass loss from some of the massive stars, increased inefficiency of collisional merging for very massive objects*, depletion of the "loss-cone" orbits that bring stars to the center, or some combination of these factors. In any case, a robust conclusion can be drawn from these simulations, namely that runaway merging can produce stars at least as massive as $\sim 500\,M_{\odot}$ in the centers of clusters with $t_{cc} \lesssim 3\,\mathrm{Myr}$, even when central velocity dispersions are as high as $\sim 400\,\mathrm{km\,s^{-1}}$.

Acknowledgements.   We are very grateful to Douglas Heggie, Steve McMillan, and Simon Portegies Zwart for many helpful discussions. This work was supported by NASA ATP Grant NAG5-12044, NSF Grant AST-0206276, and NCSA Grant AST980014N at Northwestern University. The work of MF was supported in part by Sonderforschungsbereich (SFB) 439 'Galaxies in the Young Universe' of the German Science Foundation (DFG) at the University of Heidelberg, performed under subproject A5.

---

* We have not computed collisions for stars more massive than $75\,M_{\odot}$, so considerable extrapolation of our results is required in these simulations.

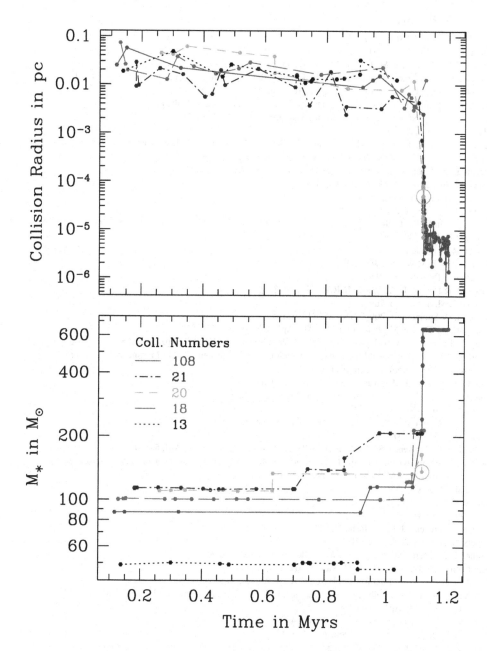

Fig. 9.6. Histories of a few particles that suffered from a large number of collisions. The initial model, E4Y, has an extended IMF and is 9 times denser than E4A but with the same velocity dispersion (see text). The top panel shows the distance from the center at which each collision occurred. In the bottom panel we plot the mass of the star after each collision. One particle (solid line) undergoes runaway growth up to about $650 M_\odot$. The reason why the growth saturates abruptly at this value is still unclear to us. The circle indicates that the particle has merged with a more massive one, probably the runaway star.

References

Aarseth, S. J. 1999, PASP, 111, 1333

Applegate, J. H. 1986, ApJ, 301, 132

Begelman, M. C., & Rees, M. J. 1978, MNRAS, 185, 847

Benz, W. 1990, in Numerical Modeling of Nonlinear Stellar Pulsations Problems and Prospects, ed. J. R. Buchler
    (NATO ASI Ser. C, 302; Dordrecht: Kluwer Academic Publishers), 269

Binney, J., & Tremaine, S. 1987, Galactic Dynamics (Princeton: Princeton Univ. Press)

Chandrasekhar, S. 1960, Principles of Stellar Dynamics (New York: Dover)

Colgate, S. A. 1967, ApJ, 150, 163

Figer, D. F., Najarro, F., Morris, M., McLean, I. S., Geballe, T. R., Ghez, A. M., & Langer, N. 1998, ApJ, 506, 384

Fregeau, J. M., Gürkan, M. A., Joshi, K. J., & Rasio, F. A. 2003, ApJ, 593, 772

Fregeau, J. M., Joshi, K. J., Portegies Zwart, S. F., & Rasio, F. A. 2002, ApJ, 570, 171

Freitag, M. 2000, Ph.D. Thesis, Université de Genève

——. 2001, Classical and Quantum Gravity, 18, 4033

Freitag, M. & Benz, W. 2001, A&A, 375, 711

——. 2002a, in Stellar Collisions, Mergers and their Consequences, ed. M. M. Shara (San Francisco: ASP), 261

——. 2002b, A&A, 394, 345

——. 2004, in preparation

Giersz, M. 1998, MNRAS, 298, 1239

——. 2001, MNRAS, 324, 218

Giersz, M. & Heggie, D. C. 1996, MNRAS, 279, 1037

——. 1997, MNRAS, 286, 709

Giersz, M., & Spurzem, R. 2000, MNRAS, 317, 581

Gürkan, M. A., Freitag, M., & Rasio, F. A. 2004, ApJ, in press

Heger, A., Fryer, C. L., Woosley, S. E., Langer, N., & Hartmann, D. H. 2003, ApJ, 591, 288

Hénon, M. 1973, in Dynamical Structure and Evolution of Stellar Systems, ed. L. Martinet & M. Mayor
    (Sauverny: Observatoire de Genéve, 183

Joshi, K. J., Nave, C. P., & Rasio, F. A. 2001, ApJ, 550, 691

Joshi, K. J., Rasio, F. A., & Portegies Zwart, S. 2000, ApJ, 540, 969

Kroupa, P. 2001, MNRAS, 322, 231

Kulkarni, S. R., Hut, P., & McMillan, S. 1993, Nature, 364, 421

Lee, H. M. 1987, ApJ, 319, 801

Lombardi, J. C., Warren, J. S., Rasio, F. A., Sills, A., & Warren, A. R. 2002, ApJ, 568, 939

Makino, J. 2001, in Dynamics of Star Clusters and the Milky Way, ed. S. Deiters et al. (San Francisco: ASP), 87

Portegies Zwart, S. F. & McMillan, S. L. W. 2000, ApJ, 528, L17

——. 2002, ApJ, 576, 899

Quinlan, G. D. & Shapiro, S. L. 1989, ApJ, 343, 725

——. 1990, ApJ, 356, 483 (QS90)

Rasio, F. A., Fregeau, J. M., & Joshi, K. J. 2001, Ap&SS, 264, 387

Sanders, R. H. 1970, ApJ, 162, 791

Schaller, G., Schaerer, D., Meynet, G., & Maeder, A. 1992, A&AS, 96, 269

Sigurdsson, S., & Hernquist, L. 1993, Nature, 364, 423

Sills, A., Faber, J. A., Lombardi, J. C., Rasio, F. A., & Warren, A. R. 2001, ApJ, 548, 323

Sills, A., Lombardi, J. C., Bailyn, C. D., Demarque, P., Rasio, F. A., & Shapiro, S. L. 1997, ApJ, 487, 290

Spitzer, L. J., Jr. 1969, ApJ, 158, L139

Spitzer, L. J., Jr., & Hart, M. H. 1971, ApJ, 166, 483

Spitzer, L. J., Jr., & Saslaw, W. C. 1966, ApJ, 143, 400

Spitzer, L. J., Jr., & Shull, J. M. 1975, ApJ, 201, 773

Stodołkiewicz, J. S. 1982, Acta Astron., 32, 63

——. 1986, Acta Astron., 36, 19

Vishniac, E. T. 1978, ApJ, 223, 986

Watters, W. A., Joshi, K. J., & Rasio, F. A. 2000, ApJ, 539, 331

Whitmore, B. C. 2003, in The Formation of Star Clusters, ed. M. Livio (Baltimore: STScI), in press
    (astro-ph/0012546)

10

Accretion onto black holes

ROGER D. BLANDFORD
California Institute of Technology

Abstract

Recent improvements in the observational study of quasar luminosity functions and the relationship between the masses of massive black holes and the properties of their host galaxies motivate the development of a more specific and quantitative model of accretion onto black holes. Some recent ideas in the theory of black hole accretion are summarized. Three cases are distinguished. A: The mass supply rate is well below the Eddington rate, the flow is adiabatic, and most mass is driven away by a combination of thermal and hydromagnetic stress. B: The mass supply rate is comparable with the Eddington rate, and the gas is mostly accreted with high radiative efficiency. C: The mass supply rate is high, the flow is again adiabatic, and most of the mass supplied is blown away by radiation pressure. A fluid model of adiabatic accretion is described, and it is conjectured that it captures the essential dynamical features of magnetic and radiative accretion. A procedure for incorporating the results from existing and future numerical simulations of accretion disks into global models is outlined. In particular, it is suggested that the boundary conditions at large and small radii determine the flow of mass, angular momentum, and energy at intermediate radii.

This (over)simple model of accretion implies a quantitative relationship between the time integral of the quasar luminosity function and the local mass function of dormant holes that ought to be testable.

10.1   The Cosmological Context

Although the notion that active galactic nuclei (AGNs, including quasars within this class) are powered by accretion onto black holes has been around since the discovery of quasars (Salpeter 1964; Zel'dovich & Novikov 1964), we have only recently developed a quantitative understanding of their relationship to their host galaxies and their role in galaxy evolution. In fact, we see quasars as far as the most distant galaxies, and four examples have been found with redshifts between $z = 6$ and 6.4 (Fan 2004). These high-redshift quasars are surprisingly luminous and indicate that holes more massive than a billion solar masses can accumulate in less than a Gyr (or the Eddington limit can be broken at will). Holes of this size cannot fail to exert a major influence on the dynamics and thermal properties of the nascent galaxy, and may even dictate its morphology (Blandford 1999).

AGNs, or more specifically quasars, are collectively and individually powerful. They can account for several percent of the luminosity density of the Universe (Fabian, this volume). A bright quasar radiates a bolometric power of $\sim 10^{13-14} \, L_\odot$ and can release enough energy

($\sim 10^{62}$ erg) to unbind $\sim 10^{12}$ $M_\odot$ of interstellar gas a hundred times over. Equivalently, the $\sim 10^{73}$ ultraviolet photons that are released can reionize intergalactic gas out to a radius of ~ 30 Mpc, and so quasars contribute to the reionization of the intergalactic medium, particularly at later epochs, and therefore influence subsequent generations of galaxies. For these and other reasons, AGN are now seen as being in symbiosis with galaxies (Haehnelt, Somerville, this volume).

The study of accretion onto black holes has become much more quantitative in recent years, as it has become possible to relate the mass of the central black hole to the properties of the surrounding galaxy. The most recent version of this relationship has been the strong correlation reported between the hole mass and the bulge velocity dispersion (Richstone, this volume). A careful application of this relation to compare the local energy density of radiant energy from the totality of accreting black holes (corrected for redshift of the photon energies) to the mass density in local black holes leads to the conclusion that the overall radiative efficiency of accretion onto black holes is $\sim 0.2 - 0.3$ (Yu & Tremaine 2002; Yu, this volume). In other words, it looks very much as though $(2-3) \times 10^{20}$ erg of energy is produced for every gram by which a black hole increases its mass in a luminous AGN. This is a very powerful, though not unexpected, constraint on the nature of the accretion process.

10.2   Modes of Disk Accretion

I will follow Blandford & Begelman (1999, 2004) and take a simple-minded approach to describing the accretion of gas onto a black hole of mass M. We suppose that there is always sufficient angular momentum for the mass supplied to form a viscous disk. We distinguish three modes of accretion. These can be identified semi-quantitatively by the ratio of the mass supply rate to the Eddington rate. Accordingly, we define

$$\dot{m} = \frac{\dot{M}\sigma_T c}{4\pi G M m_p} = \frac{\dot{M} t_E}{M},$$
(10.1)

where the Eddington time $t_E \approx 400$ Myr. Note that we distinguish the mass supply rate from the accretion rate onto the black hole, from which it may differ by many orders of magnitude.

10.2.1   Case A: Low Mass Supply Rate

When $\dot{m} < m_1$, with $m_1 \approx 0.3$, it appears that the energy dissipation associated with the viscous torque does not heat the electrons efficiently close to the black hole. In practice, it is supposed that the energy either is transported way from the disk by large-scale magnetic fields or is absorbed by the ions, which are coupled to the electrons by collisionless processes that are not more effective than Coulomb scattering. There is now quite a lot of phenomenological support for this conclusion from analyses of supernova remnant shock fronts, accreting systems, and laboratory plasmas, as well as theory (e.g., Quataert & Gruzinov 2000), although the details remain to be filled in. When $\dot{m} \ll \dot{m}_1$, this results in a very low radiative efficiency. For example, it appears that the rate of mass supply to the black hole at the center of our Galaxy is at least $\sim 10^{21}$ g s^{-1}, equivalent to $m_1 \sim 0.003$ for a three million solar mass hole (Ghez, this volume). Yet the bolometric luminosity is only $\sim 10^{36}$ erg s^{-1}, giving an apparent radiative efficiency of $\gtrsim 10^{15}$ erg g^{-1} $\approx 10^{-6}c^2$, hardly a good advertisement for "gravity power"! I will shortly argue that this efficiency is misleading, but the qualitative point remains true. Under these circumstances one can model the flow quite accurately by ignoring the cooling all together. We call this adiabatic accretion, in parallel with the usage in describing supernova remnants. The pressure is dominated by the ions, and the relevant specific heat ratio is $\gamma = 5/3$.

10.2.2   Case B: Intermediate Mass Supply Rate

When $\dot{m}_1 < \dot{m} < \dot{m}_2$, with m_2 estimated as ~ 30, a conventional, radiative accretion disk is supposed to form. Here the viscous torque G in the disk dissipates energy at a rate $-Gd\Omega/dr$, where Ω is the angular velocity, and transports energy outward through the disk at a rate $G\Omega$. The dissipated energy is converted into radiation locally within the disk, and this radiation escapes. The innermost parts of the disk, where most of the energy is liberated, are dominated by radiation pressure and electron scattering opacity (e.g., Frank, King, & Raine 1992). The overall radiative efficiency is $\sim 0.1 - 0.3$, as discussed above. The flow is radiative and most bright AGNs are probably in this state.

There has been quite a lot of progress in describing radiative accretion disk models in recent years. As most of the energy is released close to the hole, it is important to decide if the hole is spinning rapidly or not. This is because the disk can be stable closer to the horizon when the specific angular momentum of the hole approaches its maximal value $a = m$ in geometrical units. This increases the limiting binding energy of the disk and the overall radiative efficiency. The best direct evidence that we have that holes do spin rapidly is provided by *Chandra* and *XMM-Newton* observations of broad iron lines in the X-ray spectra of nearby Seyfert galaxies (e.g., Fabian 2002). It is generally envisaged that the disk itself creates a relatively soft, ultraviolet corona, which is Comptonized to form a power-law spectrum by hot electrons in an active disk corona.

The situation has become somewhat more complicated since the first *ASCA* observations, but there are still several *bona fide* observations of very broad lines, most notably by Wilms et al. (2001) in MCG –6-30-15. In the case of this last investigation, not only was it concluded that the hole had to be spinning rapidly but that energy had to be extracted from the spinning Kerr spacetime to power the line emission. However, these lines do not seem to be varying in response to the variation of the continuum that excites them through resonance fluorescence. This is unfortunate, as it had been hoped to perform reverberation analyses that would have improved the description of the disk geometry.

Another area where there has been good progress is in the creation of disk atmosphere models (e.g., Blaes et al. 2001; Hubeny et al. 2001). These are now able to reproduce the observed spectra of bright quasars like 3C273 quite well and provide good evidence radiative disks really do form.

10.2.3   Case C: High Mass Supply Rate

When $\dot{m}_2 < \dot{m}$, the mass supply rate is so high that, although there is no difficulty in emitting radiation, the photons are trapped in the gas as it flows inward onto the black hole and cannot escape. This trapping is effective out to a radius $\sim \dot{m}m$, where m is the gravitational radius and $\dot{m}$ is to be interpreted as the scaled local mass accretion rate. Broad absorption-line quasars, and perhaps most radio-quiet quasars, are examples of Case C accretion. (Galactic objects like SS433 and GRS1915+115 may also provide smaller scale examples of accretion flows in this regime.) For the purposes of understanding the gas dynamics, we can treat the flow as approximately adiabatic, with an important caveat, discussed below; as the gas is radiation dominated, $\gamma = 4/3$.

10.3   Transport Mechanisms in Disks

10.3.1   Fluid Disks

Before discussing these three accretion modes in more detail, I would like to describe the manner by which mass, angular momentum, and energy are transported within accretion disks.   The fundamental principle, which has transformed our understanding of disk accretion, is that orbiting, conducting gas is generically unstable to the magnetorotational instability (MRI; Balbus & Hawley 1998).   We are all magnetohydrodynamicists now! Although it is probably true that most disk torque derives from some variant of the MRI, I want to spend a lot of this review describing a model of disk accretion that is almost surely invalid, but which, I argue, allows one to discuss some general principles in a quite transparent fashion. I will start by ignoring the MRI altogether and just consider adiabatic accretion. Numerical simulations (e.g., Stone, Pringle, & Begelman 1999) show that adiabatic disks quickly evolve to become convectively unstable.

The most familiar example of convection is the Schwarzschild case, which is appropriate when the entropy decreases upward in a gravitational field. When two fluid elements are interchanged slowly on a dynamical time scale, but rapidly with respect to the time it takes heat to be exchanged, they will quickly come into pressure equilibrium, and the one moving up will find itself lighter than its surroundings and keep on going under the force of buoyancy. This is the physical content of a small-amplitude instability calculation. Stars are unstable when the entropy decreases outward, as happens in the outer envelopes of solar-type stars.

The next most familiar type of convection is the Rayleigh case. Here we imagine a thin, differentially rotating disk orbiting under gravity and ignore the entropy and the pressure. Imagine a thin ring displaced radially and axisymmetrically outward. If we suppose that this happens slowly on a dynamical time scale, though too fast for viscous torques to operate, the specific angular momentum L will be constant. If the centrifugal force L^2/R^3 is less than that of its surroundings, the ring will return to its initial radius. Disks are unstable when the specific angular momentum decreases outward. Consequently, Keplerian disks are quite stable with respect to the Rayleigh criterion.

Adiabatic, fluid disks are thick, as the pressure is large. They combine features of thin disks and stars, and their convective stability must be handled using considerations that combine the Schwarzschild and Rayleigh criteria. The entropy naturally increases inward, as this is where most of the energy is released, which is destabilizing However, the angular momentum increases outward, and this is stabilizing. To understand the stability, consider displacing a slender ring of gas in the poloidal plane. (Note that the displacement is now two dimensional.) The displacement $\delta \vec{r}$ must be made fast enough to be adiabatic and inviscid, though slow enough for the displaced gas to remain in pressure equilibrium with its surroundings and for inertial terms to be ignored. Next, evaluate the poloidal acceleration linear in the displacement and the virtual work per unit mass. This gives a quadratic form in the two-dimensional displacement, $\delta W = U_{ij}\delta r_i \delta r_j/2$, where

$$U = \nabla \left(\frac{P^{1-1/\gamma}}{1-1/\gamma} \right) \otimes \nabla S + \nabla \left(\frac{1}{2R^2} \right) \otimes \nabla L^2, \tag{10.2}$$

ρ is the gas density, and $S = P^{1/\gamma}/\rho$ is an entropy function that turns out to be more convenient to use than the true entropy and to which it is monotonically related. The flow will be stable if δW is negative definite (see Blandford & Begelman 2004 and references therein for details). There are two "Høiland" criteria for instability (e.g., Tassoul 1978).

The first of these is

$$\nabla \left(\frac{P^{1-1/\gamma}}{1-1/\gamma} \right) \cdot \nabla S + \nabla \left(\frac{1}{2R^2} \right) \cdot \nabla L^2 > 0. \tag{10.3}$$

This is really telling us if the flow is unstable to radial perturbations, and it turns out that most reasonable disk models are quite stable to this criterion. Rayleigh stabilization wins. The second instability criterion is

$$(\nabla P \times \nabla R) \cdot (\nabla S \times \nabla L) > 0. \tag{10.4}$$

This describes vertical perturbations in the disk, and it seems that fluid disks can easily evolve to a condition of marginal stability where the isentropes, the surfaces of constant entropy, and the isogyres, the surfaces of constant specific angular momentum, almost coincide. It is easy to show that this implies that the surfaces of constant Bernoulli function,

$$B = L^2/2R^2 + \Phi + H, \tag{10.5}$$

where Φ is the gravitational potential and H is the enthalpy, also coincide with the gyrentropes ($S = L = $ constant), as

$$\nabla B = H \nabla \ln S + \Omega \nabla L. \tag{10.6}$$

It is helpful to consider other surfaces inside the disk—isochores ($\rho = $ constant), isobars ($P = $ constant) and isorotational surfaces ($\Omega = $ constant)—which are nested within the gyrentropes (Fig. 10.1). We therefore argue that the principle governing the structure of adiabatic fluid disks is that they be gyrentropic, so that $S = S(L)$. This is analogous to modeling the solar convection zone as isentropic. An examination of the character of the unstable modes shows that the nature of the motion has slender ribbons of gas sliding past each other along the gyrentropes. Some axisymmetric, numerical simulations of adiabatic disks have shown this type of transport (Stone et al. 1999), with mass, angular momentum, and energy being carried toward the disk surface where they can drive an outflow. On this basis, it is possible to calculate two-dimensional, analytic, self-similar models of fluid accretion disks, in hydrostatic equilibrium (Fig. 10.2.). We develop this disk model further below.

10.3.2 *Magnetorotational Instability*

As emphasized above, real disks almost certainly evolve under the action of hydromagnetic stresses. Essentially a differentially rotating flow with a small vertical magnetic field will be generically unstable to radial motions, which cause the magnetic field to grow exponentially on a dynamical time scale. The nonlinear evolution of this instability and the asymptotic state of the gas are still not very well understood, although considerable progress has been made in recent years, thanks to numerical simulations of ever increasing sophistication (e.g., Hawley & Balbus 2002). It is very important to perform three-dimensional, as opposed to two-dimensional, simulations, as the latter may be quite misleading. In addition, it is necessary to include a serious energy equation when dealing with adiabatic disks, not just solve the momentum equations assuming a polytropic relation between the density and the pressure, as was done in early calculations.

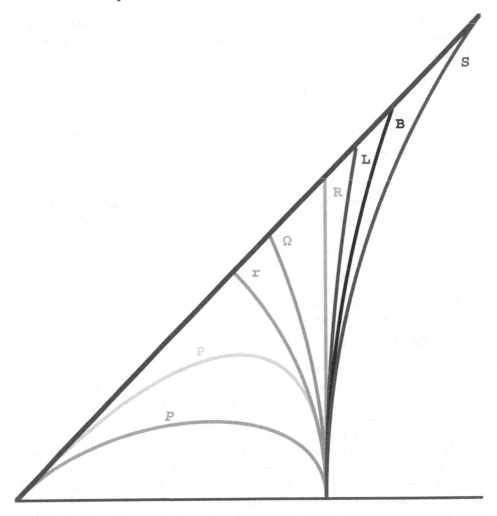

Fig. 10.1. Structure of a convectively stable fluid accretion disk. Level surfaces of density ($\mathcal{P}$), pressure (P), spherical radius (r), angular frequency (Ω), cylindrical radius (R), angular momentum (L), Bernoulli function (B), and entropy (S) passing through an equatorial ring of radius r. The configuration shown is dynamically stable according to both Høiland criteria. When the order of the isentropes and isogyres is reversed, the flow becomes convectively unstable.

We still lack a good prescription for characterizing the MRI torque G in terms of local physics. In general in these simulations, the magnetic stresses appear to saturate at a level corresponding to roughly several percent of the gas pressure. The overall momentum transport is probably not well described by a simple α prescription, in which the ratio of the shear stress to the total pressure P is supposed to be fixed number, α. Also, to repeat the point made above, we do not have a good understanding of how the associated dissipation proceeds. In one approach, it has been argued that an inertial range hydromagnetic turbulence spectrum is established with gas heating occurring at some inner scale (e.g., Quataert & Gruzinov 2000). Reconnection may also be important under some

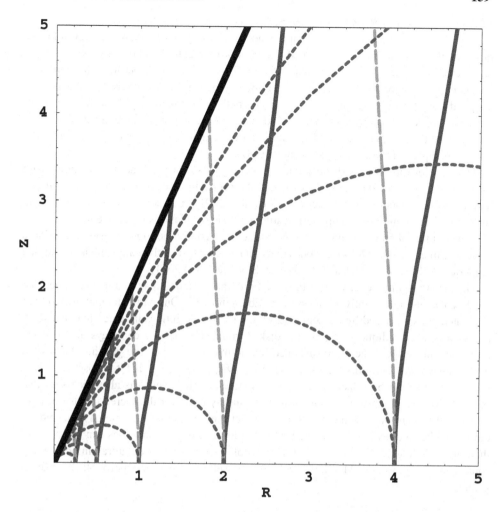

Fig. 10.2. Hydrostatic structure of an adiabatic, self-similar, ion-dominated, gyrentropic accretion disk. The gyrentropic surfaces, on which the entropy, specific angular momentum, and Bernoulli function are constant, are shown with solid lines. The isorotational and isobaric surfaces are denoted by long-dashed and short-dashed lines, respectively. Magnetic and radiation-dominated accretion disks are conjectured to have an analogous structure.

circumstances. In an alternative proposal, there is an inverse cascade and little direct heating of the disk (e.g., Tout & Pringle 1996).

One expected feature of these disks, which is exhibited by some simulations, is the presence of outflow (e.g., Hawley & Balbus 2002). There are really two mechanisms for bringing this about. Firstly, there may be a large dissipation rate and entropy production in an active disk corona, and the outflow can be essentially thermally driven, although perhaps eventually guided by magnetic stress. Alternatively, the external magnetic field may be so strong that the MRI is suppressed on the disk surface, such that a cold outflow may be initiated as a consequence of centrifugal force flinging out plasma along open field lines.

10.3.3   Transport in Radiation-dominated Disks

There have been heroic attempts made in recent years to address the even harder problem of understanding the combined transport of momentum and radiation in radiation-dominated disks near the central black hole. This is important because this emission accounts for a major fraction of the the luminosity density of the Universe, rivaling that supplied by nuclear physics inside stars! The strength of the magnetic field within radiation-dominated disks has long been a matter of some controversy. For example, it has sometimes been argued that the magnetic field saturates at a value that is a fraction of the gas pressure, not the total pressure (Sakimoto & Coroniti 1981).

The way forward is to perform numerical simulations that combine gas dynamics with radiative transfer. To date these have had to be pretty limited, involving shearing boxes and ignoring vertical gravity (e.g., Agol et al. 2001), but still they have been quite instructive. The single most important feature of these simulations is that the disk naturally becomes quite inhomogeneous. This is not so surprising when one considers the large variety of linearly unstable wave modes that can be described analytically and demonstrated numerically (Blaes & Socrates 2001; Turner, Stone, & Sano 2001).

This strong inhomogeneity raises the possibility that radiation may be able to escape from the disk at a rate significantly in excess of the Eddington rate. One particular manifestation of this outcome is proposed by Begelman (2002), who finds solutions in which magnetosonic modes propagate obliquely through the disk, forming low-density chimneys along which radiation can escape. These are sustained by the magnetic field. The gas that defines the high-density chimney walls is held in place by gravity. It is argued that it is possible to beat the Eddington limit by a factor ~ 100 in this manner. However, it is also necessary that the radiation escape anisotropically so that the supply route for the accreting gas does not intersect the beam of escaping radiation, and that the power that is scattered isotropically out of this beam does not exceed the Eddington limit. If these conditions can be satisfied, then super-Eddington accretion provides a possible explanation for the ultraluminous X-ray sources (see, e.g., van der Marel, this volume), as well as perhaps the narrow-line Seyfert 1 galaxies.

10.4   Adiabatic Accretion Models

I would now like to return to adiabatic fluid accretion under Case A, and, more controversially, Case C. However, in order to make a central point, I should first consider regular, radiative accretion. In the simplest model of a stationary, Keplerian disk, the mass flow *inward* $\dot{M}$ and the net angular momentum flow $F_L = \dot{M}L - G$ *inward* are conserved. As $L \propto r^{1/2}$, conservation of angular momentum over several octaves of radius implies that $F_L \approx 0$, except very close to the hole. Combining these two equations, we can obtain an energy equation

$$\frac{dF_E}{dr} = G\frac{d\Omega}{dr}, \tag{10.7}$$

where $F_E = G\Omega - \dot{M}E$ is the net flow of energy *outward*, and E is the sum of the kinetic and gravitational potential energy. The term on the right-hand side corresponds to the rate of loss of energy from the flow through radiative loss and is easily shown to be 3 times the rate of release of binding energy in a thin, Keplerian disk. The origin of this extra energy release is the work done by the torque G, and it derives from the binding energy release at small radii, where this equation has to be modified.

Now, consider what happens when the accretion is stationary, adiabatic, and conservative. The mass conservation and angular momentum conservation are unchanged. However, the gas is no longer cold and E must be augmented with the thermal energy so that $F_E \to G\Omega - \dot{M}B$. In addition, as the gas cannot radiate, there is no sink of energy, and so F_E must be conserved and must also be very close to zero, as $G\Omega \propto r^{-3/2}$. This then implies that

$$B = \frac{G\Omega}{\dot{M}} = \Omega L. \tag{10.8}$$

The Bernoulli function must be positive and the gas has enough energy to escape to infinity with positive kinetic energy. Three resolutions of this paradox have been proposed.

In the ADAF prescription (Narayan & Yi 1994), it is proposed that the flow is essentially conservative and closes off along the rotation axis so as to prevent the escape of the unbound gas. Gas then crosses the horizon with positive energy. However, this requires the velocity field to become singular. In another variation on the basic model, it is also proposed that the flow is quasi-spherical so that little angular momentum transport is required. However, this seems very difficult to arrange in a galactic nucleus.

In the CDAF prescription (e.g., Quataert & Narayan 1999), it is recognized that the flow becomes convective, and it is proposed that convection transports angular momentum inward, while transporting energy outward. Independent of the likelihood of this happening in practice, this prescription leads to a mass accretion rate that increases with radius. If the flow is conservative, then it must be non-stationary. The mass supply can back up but eventually it must find some other mode of accretion.

Finally, in the ADIOS prescription (Blandford & Begelman 1999, 2004), it is proposed that the flow is non-conservative. The most important requirement is that the energy be removed in an outflow of some form. In principle this could be a magnetohydrodynamic wind with very little mass content. However, in practice, we strongly suspect that the outflow involves a large mass loss, and the rate at which gas accretes onto the black hole is orders of magnitude smaller than the rate at which it is supplied. This is not a small effect, and if this model is essentially correct, then many of the papers describing the radiative properties of conservative flows will need to be modified. As discussed above, observations of Sgr A* and other nearby (inactive) galactic nuclei do suggest that the flow is non-conservative.

10.5 ADiabatic Inflow-Outflow Solutions

It is quite instructive to consider how to make a model of disk accretion onto a black hole. As discussed above, I will restrict attention, initially, to stationary, adiabatic, hydrodynamic, convective models as these appear to be simpler (Blandford & Begelman 2004).

10.5.1 Hydrostatics

The first step is to make a model of a hydrostatic disk. Following the discussion above, we suppose that the disk is convective and consequently gyrentropic so that $L = L(S)$, or equivalently $L = L(B)$. We can make things simple by assuming that the structure of the bulk of the Newtonian disk is self-similar:

$$P = r^{n-5/2} p(\theta), \quad \mathcal{P} = r^{n-3/2} \rho(\theta), \quad L = r^{1/2} \ell(\theta), \tag{10.9}$$

where θ is the polar angle and the fixed exponent n has a value in $[0, 1]$. If we use the self-similar scaling of the radial velocity $v_r \propto r^{-1/2}$, then we can identify n as the accretion exponent $\dot{M} \propto r^n$. It turns out that the parameter n can be replaced by a parameter b_0, which measures the Bernoulli function of the disk in the equatorial plane. b_0 is negative for a bound disk (although it can be positive in a funnel). As the disk becomes thicker and loses more mass, $b_0 \to 0$. The disk models are further parametrized by the equatorial angular momentum in units of the Keplerian value, ℓ_0. These parameters suffice to specify the disk structure (Fig. 10.3). The disk surface, where $P \to 0$ and $\mathcal{P} \to 0$, is a cone with an opening angle that decreases with ℓ_0.

10.5.2   Convection and Torque

Gyrentropic convection also provides a rule for transporting mass, angular momentum and energy from the body of the disk to the disk surface—the direction of transport is parallel to the gyrentropic surfaces. However, in order to make a more detailed model we must make additional assumptions. A convenient and natural choice in this context is that the effective Prandtl number is large, which means that the moving ribbons are in much stronger thermal than mechanical contact with their environment. In other words, the convection is only efficient at transporting heat directly to the surface of the disk.

Now, a steady disk flow also requires that angular momentum be transported across the gyrentropes, and so it is necessary to supplement the convection with a phenomenological torque. Self-similarity dictates a constant value for the α parameter, or, equivalently, a kinematic viscosity that scales with the product of the disk thickness and the sound speed. The torque per unit length in the meridional plane then satisfies

$$\vec{G} = -2\pi\alpha(r\sin\theta)^3 P\nabla \ln \Omega. \tag{10.10}$$

10.5.3   Circulation

Given these rules, or others like them, it is possible to use conservation of mass, angular momentum, and energy to determine the mass flow in the disk. This is accomplished by solving the equations

$$\nabla \cdot \vec{J} = 0 \tag{10.11}$$

$$\nabla \cdot (L\vec{J} + \vec{G}) = 0 \tag{10.12}$$

$$\nabla \cdot (B\vec{J} + \Omega\vec{G} + \vec{Q}) = 0, \tag{10.13}$$

where $\vec{J}$ is the mass flux per unit length in the poloidal direction and $\vec{Q}$ is the convective energy flux along the gyrentropes. Solving these equations leads to the general conclusion that there is a general inflow of gas through the disk and that superimposed upon this is a meridional circulation. Combining the circulation with the convection, we find that there is a steady flow of mass, angular momentum, and energy to the disk surface. The most common pattern is to have inflow near the equatorial plane and a quadrupolar circulation that can lead to outflow near the disk surface. However this sense of circulation is not required, and it is possible to reverse it by changing assumptions like the latitude dependence of α. The net inflow of gas in the disk matches the net loss of mass at the disk surface. Likewise, the difference between the angular momentum and energy carried inward by the flow and

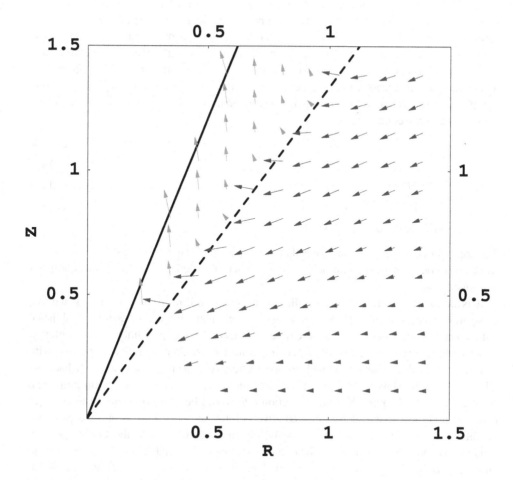

Fig. 10.3. Flow vectors for an axisymmetric, self-similar, fluid model of an ion-dominated, adiabatic accretion disk. The accretion disk is separated from the outflow by a thermal front, which is designated by a dashed line. The disk flow exhibits a poloidal circulation pattern with a net radial inflow. In addition, there is a net flow incident upon the disk surface, where entropy is generated, which feeds the outflow. Centrifugal forces acting in the outflow define an excluded cone which could be filled by a jet. As discussed in the text, radiation-dominated and hydromagnetic disk models should exhibit similar features.

transported outward across gyrentropes by the torque is equal to the energy and angular momentum flux crossing the disk surface.

10.5.4 *Thermal Front*

Now let us consider the consequences of having a flow of mass, angular momentum, and energy incident upon the disk surface. As the density vanishes at the surface, the flow velocity must diverge, signifying the ultimate failure of our model. What we actually believe would happen under these circumstances is that there is a rapid production of entropy where the flow speeds become supersonic. This allows gas that is bound in the disk to become unbound at the disk surface and supply an outflow. Something similar, but much more gentle, happens in the Sun, although the actual details of what is going on remain obscure. However, independent of the details, mass, momentum, and energy must be conserved and so we can introduce a "thermal front," analogous to a deflagration wave, across which we can compute the jump in these quantities, assuming that the convective energy flux Q vanishes downstream from the thermal front. Specifically, we solve the jump conditions

$$[J_\theta] = 0 \tag{10.14}$$

$$[J_\theta V_r] = 0 \tag{10.15}$$

$$[2\pi RP + J_\theta V_\theta] = 0 \tag{10.16}$$

$$[G_\theta + J_\theta L] = 0 \tag{10.17}$$

$$[G_\theta \Omega + J_\theta B + Q_\theta] = 0. \tag{10.18}$$

The approximation that is made here is that entropy is created quickly as gas flows across the front, analogous to what is commonly assumed in deriving the Rankine-Hugoniot conditions at a shock front.

It is important to emphasize that there can be no outflow from a fluid disk without dissipation near its surface. If there is no dissipation, then a circulation will be established with no net inflow and the flow resembles the CDAF solution, although we prefer to distinguish the organized poloidal circulation from the oscillatory motion associated with convection. A flow like this would be non-stationary—the mass supplied will back up all the way to the Bondi radius. Behavior like this can be seen in some recent numerical simulations. At this point the mass accretion rate would be shut down almost to zero. In practice, flows are likely to become radiative before they back up too far, except under conditions when the mass accretion rate is tiny in comparison with the Eddington rate. This radius, beyond which disks are radiative as opposed to adiabatic, is known as the "transition radius." It marks the outer boundary of an adiabatic flow. It is our contention that supersonic convection near the disk surface is replaced by strong dissipation and that, under the assumptions we have made in this model, there will be a strong outflow.

10.5.5   Outflow

The gas downstream from the thermal front is hot and flows away from the disk with a subsonic speed. Unlike the gas in the disk, it has positive Bernoulli function and can therefore escape to infinity with positive energy. We model the outflow as non-dissipative and ignore the effects of the torque as well as any convective motion. Clearly this is an approximation. Note that the quantities L, S, and B are conserved along the flow lines, and so the outflow is gyrentropic, just like the disk, though for a quite different reason. The disk speed quickly becomes supersonic, although there is no critical point in the solution due to the imposition of self-similarity, which only allows the velocity component resolved in the θ direction to vary freely, and this remains subsonic (Fig. 10.3).

This prescription for a simple, thermally driven wind does enable us to augment the disk model with an analytic, self-similar outflow that is asymptotic to the surface of a cone with a smaller opening angle than the disk. The excluded region owes its existence to the centrifugal barrier. If we allow a torque to develop in the outflow, the cone would be narrower, though should not disappear.

We have therefore described a large family of self-similar fluid disk models. At this point an interesting matter of principle arises. If we suppose that mass is supplied at a certain rate at an outer, transition radius and accretes adiabatically onto a massive black hole, and that microphysical parameters like γ and α are known, then our fluid disk models are described by two independent parameters, which we have chosen to be b_0 and ℓ_0. However, the disk structure ought to be unique. (Actually, in general there will be three independent parameters describing the central densities of mass, energy, and angular momentum. However, in a fluid disk the outflowing gas carries off only its own specific angular momentum and does not exert any significant reaction torque on the remaining gas. This implies that the transport of mass and angular momentum through the disk are not independent, and the number of independent parameters is reduced to two.)

So what fixes the remaining two degrees of freedom? The answer appears to be that the conditions in the innermost, relativistic portion of the accretion disk (especially the spin of the hole), where self-similarity breaks down, essentially act as a boundary condition and determine the variation of the Bernoulli parameter b_0 throughout the disk. To be more specific, a certain fraction of the energy released by the gas as it flows into the hole will be transported out into the disk by viscous torque and large-scale circulation. If the energy so transported is large, the next annular ring of gas will be relatively thick. It will then transport a large amount of energy to the next ring, and so on. Self-similarity propagates this information from small radius to large radius. Likewise, we suppose that the conditions at the outer radius of the adiabatic disk determine the angular momentum (and mass) flow inward.

Is this prescription an artificiality derived from self-similarity? I suspect that this is not the case. If the rule for deciding what fraction of the energy flow propagating outward through a disk actually leaves the disk over the next octave of radius really is independent of radius, then the structure of the disk will be dictated by the conditions at small radius. This is somewhat similar to what happens in a boundary layer on a solid surface in a flow, where the rule for determining the rate of thickening is local, but the actual thickness of the boundary layer will respond to what happens upstream where the surface first encounters the flow.

10.5.6 *Radiation-dominated Disks*

The modification of this model for the radiation-dominated outflow that exists under Case C is fairly straightforward. Firstly, it is necessary to change the specific heat ratio, γ, from $5/3$ to $4/3$. Secondly, the relevant rule for determining the torque may be quite different from that for an ion-dominated gas, and the influence of inhomogeneity may be large, as we described above. However, if we ignore this possibility, then the radiation will be trapped and the disk flow will be roughly adiabatic as long as $\dot{m} > r/m$. The trapping radius, r_{tr}, where this inequality becomes an equality, is the effective transition radius. The power emitted from $r > r_{tr}$ is roughly the Eddington limit. If the mass accretion exponent n is not much less than unity, the flow will be radiatively quite efficient with respect to the accreted mass, with most of the mass supplied leaving in a wind from within the trapping radius. Only a fraction $\sim 1/\epsilon \dot{m}$ of the mass supplied will actually accrete. The outflow will be driven by electron scattering, and the fluid model will certainly become invalid when the optical depth of the wind falls below c/V, where V is the outflow speed, so that the photons can escape on an outflow time scale. However, resonance lines are likely to be important in accelerating the outflow beyond this radius.

10.5.7   Magnetic Accretion

Now let us return to the more complex question of the accretion of magnetized gas. We have argued that the MRI implies that hydromagnetic stresses dominate the torque within the disk. However, despite considerable progress in performing numerical simulations, we still do not have a good, global model of the transport of mass, energy, and angular momentum. (There are some hints that the disk adjusts so as to become barytropic, rather than gyrentropic. If so, the implications for the disk structure are relatively minor.) However, the primary "convective" (i.e., oscillatory, as opposed to circulatory) transport is no longer necessarily vertical, as we have argued is the case in a fluid model. Another important change is that there can be a magnetic reaction torque connecting the disk to the outflow. In other words, magnetic outflows can be centrifugally as well as thermally driven. This means that there are now three eigenvalues that must be to be fixed through the imposition of an inner boundary condition on the energy and outer boundary conditions on the mass and the angular momentum.

Adopting the methodology advocated in this review, all of this means that we must first determine the rules for describing the *local*, averaged behavior of accretion disks. This will almost surely require numerical, hydromagnetic simulations with sufficient dynamic range to encompass many scale heights of the disk and the corona. In particular, it may be necessary to design the difference scheme so as to accommodate a structure like a thermal front in much the same way that schemes that have to handle supersonic flows are able to locate shock fronts. This scheme would have to create the necessary entropy at the disk surface. We note that the jump conditions at the thermal front naturally generalize to the hydromagnetic case.

The next task will be to use these prescriptions in global models of accretion disk evolution. In general, the flows will be time dependent. Let us denote the mass per unit spherical radius in the disk by $\mathcal{M}$. The angular momentum and energy per unit radius can likewise be defined by

$$S = \langle \mathcal{M}L \rangle \quad \text{and} \quad \mathcal{E} = \langle \mathcal{M}(B - P/\rho) \rangle, \tag{10.19}$$

respectively, where the averages are performed on spheres. If there are no interior radial scales in the adiabatic flow, then we expect there to be simple scaling relations such as $F_L \propto \dot{M}R^{1/2}$, etc.

We can also define the flows of mass inward, angular momentum inward, and energy outward across spheres in the disk by $\dot{M}$, $F_L = \langle \dot{M}L - G \rangle$, and $F_E = \langle G\Omega - \dot{M}B \rangle$, respectively. Naturally we want to know the functional forms of $\dot{M}(\mathcal{M}, L, B)$, etc. Finally, we need the mass, angular momentum, and energy carried off by the outflow per unit spherical radius, which we denote by K_M, K_L, K_E. An adiabatic disk will then evolve approximately according to the three coupled "kinematic" one-dimensional equations

$$\frac{\partial \mathcal{M}}{\partial t} = \frac{\partial \dot{M}}{\partial r} - K_M \tag{10.20}$$

$$\frac{\partial S}{\partial t} = \frac{\partial F_L}{\partial r} - K_L \tag{10.21}$$

$$\frac{\partial \mathcal{E}}{\partial t} = -\frac{\partial F_E}{\partial r} - K_E. \tag{10.22}$$

These equations must be supplemented with boundary conditions on $\mathcal{M}$ and $\mathcal{S}$ at the outer radius and $\mathcal{E}$ at the inner radius, as discussed above. They make the implicit assumption that the local adjustment time scales are short compared with the flow times, which are not very accurate for thick disks.

It will be interesting to see if reasonable choices for these unknown functions lead to stationary or episodic inflows and outflows.

10.6 Implications for Black Hole Growth

Let us return to our simple scheme for describing accretion onto black holes. We suspect that, early in the life of an AGN, the hole mass will be low and the mass supply rate will be high, $\dot{m} \gg \dot{m}_2$, so that we are solidly in Case C. If our conjecture about these flows being adiabatic is correct, and it is not, in practice, possible to exceed the Eddington limit by a large factor, then almost all of the mass supplied must be blown off in an outflow and the hole will grow (exponentially) on a time scale $\sim \epsilon t_E \approx 40$ Myr. The accretion is "demand limited" instead of "supply driven," as in the "feast or famine" model of Small & Blandford (1992). There is time for the black hole to grow by a factor of at least a thousand before the highest redshift quasars are observed. (There is a possible exception to this rule, which may operate when the hole mass is small. If the trapping radius exceeds the accretion radius, $\sim GM/s^2$, with s the effective sound speed in the gas, and typically this requires that $\dot{M} > (m/100 M_\odot) \, M_\odot \, \mathrm{yr}^{-1}$, then the outflow may be trapped by the infalling gas and the energy radiated away at a rate much greater than the Eddington limit for the hole. This may allow a stellar mass hole to grow quickly to an intermediate mass, very early in the life of the nascent galaxy.)

In general, however, we expect that the mass supply will decline while the hole mass grows. There will then come a time when the trapping radius shrinks so that it is not much larger than the event horizon, and Case B accretion will commence. Let us suppose that this only happens once in the lifetime of a galaxy. (If holes are frequently resurrected, then what follows must be modified.) Most of the black hole mass will have been built up around the time of this transition. Let us further assume that, at the time when holes of final mass M were mostly assembled, the luminosity function for the corresponding Eddington luminosity is dominated by holes transitioning from Case C to Case B. (Note that this does not have to be true now.) This implies that the local, hole mass function $N(M)$ should reflect the integral of the comoving quasar luminosity function $\Phi(L)$, evaluated at $L \approx L_{\mathrm{Edd}}(M)$ over time. Specifically,

$$N(M) \approx \mu \int dt \, \frac{\Phi(\lambda M c^2/t_E)}{t_E^2 c^2}, \tag{10.23}$$

where the coefficients $\lambda \approx 1$ and $\mu \approx \epsilon^{-1}$ can, in principle, be determined observationally and used to parameterize a more detailed model.

This somewhat impressionistic portrait of hole growth should eventually lead to a more specific prediction concerning the relationship between the contemporary hole mass function and the historical quasar luminosity function than the integral Sołtan (1982) relationship. This next challenge in the study of black demography—to determine both when *and* how the dormant holes we see around us acquired their masses. Success in this endeavor will allow us to understand in more detail both the physics of the accretion process and its environmental impact on the host galaxy.

Acknowledgements.   I am indebted to my collaborator Mitch Begelman for his contributions to the above research and for his patience. I thank Luis Ho for organizing a stimulating meeting and also for his patience. Support under NASA grant NAGW5-12032 is gratefully acknowledged.

References

Agol, E., Krolik, J. H., Turner, N., & Stone, J. 2001, ApJ, 558, 543

Balbus, S. A., & Hawley, J. F. 1998, Rev. Mod. Phys., 70, 1

Begelman, M. C. 2002, ApJ, 568, L97

Blaes, O., Hubeny, I., Agol, E., & Krolik, J. H. 2001, ApJ, 563, 560

Blaes, O., & Socrates, A. 2001, ApJ, 553, 987

Blandford, R. D. 1999, in Astrophysical Disks, ed. J. A. Sellwood & J. Goodman (San Francisco: ASP), 265

Blandford, R. D., & Begelman, M. C. 1999, MNRAS, 303, L1

——. 2004, MNRAS, in press

Fabian, A. C. 2002, Phil. Trans. R. Soc., 360, 21035

Fan, X. 2004, in Carnegie Observatories Astrophysics Series, Vol. 1: Coevolution of Black Holes and Galaxies, ed. L. C. Ho (Pasadena: Carnegie Observatories, http://www.ociw.edu/ociw/symposia/series/symposium1/proceedings.html)

Frank, J., King, A. R., & Raine, D. J. 1992, Accretion Power in Astrophysics (Cambridge: Cambridge Univ. Press)

Hawley, J. F., & Balbus, S. A. 2002, ApJ, 573, 738

Hubeny, I., Blaes, O., Krolik, J. H., & Agol, E. 2001, ApJ, 559, 680

Narayan, R., & Yi, I. 1994, 428, L13

Quataert, E., & Gruzinov, A. 2000, ApJ, 539, 809

Quataert, E., & Narayan, R. 1999, ApJ, 520, 298

Sakimoto, P. J., & Coroniti, F. V. 1981, ApJ, 247, 19

Salpeter, E. E. 1964, ApJ, 140, 796

Small, T. A., & Blandford, R. D. 1992, MNRAS, 259, 725

Sołtan, A. 1982, MNRAS, 200, 115

Stone, J. M., Pringle, J. E., & Begelman, M. C. 1999, MNRAS, 310, 1002

Tassoul, J.-L. 1978, Theory of Rotating Stars (Princeton: Princeton Univ. Press)

Tout, C. A., & Pringle, J. E. 1996, MNRAS, 281, 219

Turner, N. J., Stone, J. M., & Sano, T. 2002, ApJ, 566, 148

Wilms, J., Reynolds, C. S., Begelman, M. C., Reeves, J., Molendi, S., Staubert, R., & Kendziorra, E. 2001, MNRAS, 328, 27

Yu, Q., & Tremaine, S. 2002, MNRAS, 335, 965

Zel'dovich, Ya. B., & Novikov, I. D. 1964, Sov. Phys. Dokl., 158, 811

11

QSO lifetimes

PAUL MARTINI
The Observatories of the Carnegie Institution of Washington

Abstract

The QSO lifetime t_Q is one of the most fundamental quantities for understanding black hole and QSO evolution, yet it remains uncertain by several orders of magnitude. If t_Q is long, then only a small fraction of galaxies went through a luminous QSO phase. In contrast, a short lifetime would require most galaxies today to have undergone a QSO phase in their youth. The current best estimates or constraints on t_Q from black hole demographics and the radiative properties of QSOs vary from at least 10^6 to 10^8 years. This broad range still allows both possibilities: that QSOs were either a rare or a common stage of galaxy evolution. These constraints also do not rule out the possibility that QSO activity is episodic, with individual active periods much shorter than the total active lifetime.

In the next few years a variety of additional observational constraints on the lifetimes of QSOs will become available, including clustering measurements and the proximity effect. These new constraints can potentially determine t_Q to within a factor of 3 and therefore answer one of the most fundamental questions in black hole evolution: Do they shine as they grow? This precision will also test the viability of our current model for accretion physics, specifically the radiative efficiency and need for super-Eddington luminosities to explain the black hole population.

11.1 Introduction

Shortly after the discovery of luminous QSOs, several authors suggested that they could be powered by accretion onto supermassive black holes (e.g., Salpeter 1964; Zel'dovich & Novikov 1964; Lynden-Bell 1969). The strong evidence that dormant, supermassive black holes exist at the centers of all galaxies with a spheroid component (e.g., Richstone et al. 1998) supports this hypothesis and suggests these galaxies go through an optically luminous accretion phase. More recent observations suggest that the masses of supermassive black holes are correlated with the sizes of their host galaxy spheroids, which implies some connection between the growth of the subparsec-scale supermassive black holes and the kiloparsec-scale spheroids. The critical parameter for understanding how quickly these present-day black holes grew to their present size is the lifetime of the QSO phase t_Q. The QSO lifetime determines the net growth of supermassive black holes during an optically luminous phase. It may also have implications for the growth rate of galactic spheroids.

For the purpose of the present review, the lifetime of a QSO is defined to be the total

© The Observatories of the Carnegie Institution of Washington 2004.

amount of time that accretion onto a supermassive black hole is sufficiently luminous to be classified as a QSO—that is, the net time the luminosity in some wavelength range $L > L_{Q,min}$, where $L_{Q,min}$ is the minimum luminosity of a QSO. This is an observational definition, chosen because QSOs are categorized on the basis of luminosity in existing surveys. A more physically motivated definition of a QSO would define the lifetime as the time a black hole is accreting above some minimum mass accretion rate, or emitting radiation above some minimum fraction of the Eddington luminosity (and this definition would also require a minimum black hole mass), as these two parameters are more directly relevant to the details of the accretion physics. The measured QSO luminosity instead depends on the product of one of these quantities with the black hole mass.

A valuable fiducial time scale for supermassive black hole growth is the Salpeter (1964) or *e*-folding time scale:

$$t_S = M/\dot{M} = 4.5 \times 10^7 \left(\frac{\epsilon}{0.1}\right) \left(\frac{L}{L_{Edd}}\right) \text{yr}, \qquad (11.1)$$

where $\epsilon = L/\dot{M}c^2$ is the radiative efficiency for a QSO radiating at a fraction L/L_{Edd} of the Eddington luminosity. Commonly accepted values of these two key parameters for luminous QSOs are $\epsilon = 0.1$ and $L/L_{Edd} = 1$. The importance of an optically luminous or QSO mode of black hole growth relative to significantly less efficient, or less luminous modes is not well established, primarily due to uncertainties in these two parameters and the uncertainty in the QSO lifetime. If the QSO lifetime is long, then QSOs were exceptionally rare objects and only a small number of the present-day supermassive black holes went through a QSO phase. In this scenario, essentially all of the mass in these black holes can be accounted for by luminous accretion. The remaining supermassive black holes in other galaxies must instead have accreted their mass through less radiatively efficient or less luminous modes. If instead the QSO lifetime is short, most present-day black holes went through an optically luminous accretion phase, yet this phase made a relatively small contribution to the present masses of the black holes.

One long-standing upper limit on the lifetime of QSOs is the lifetime of the entire QSO population. The evolution of the QSO space density is observed to rise and fall on approximately a 10^9 year time scale (see, e.g., Osmer 2004). More refined demographic arguments, described in detail in the next section, find values on order $10^6 - 10^7$ years. Lower limits to the QSO lifetime are less direct, such as that provided by the proximity effect in the Lyα forest, and they argue for a lower limit of $t_Q > 10^4$ years. Unlike demographic arguments, which constrain the net time a black hole accretes matter as a luminous QSO, the lower limits are based on radiative measures that do not take into account the possibility that a given black hole may go through multiple luminous accretion phases. These radiative methods, described in §11.3, constrain the episodic lifetime of QSOs and place a lower bound on the net lifetime.

Observations of QSOs and the less-luminous Seyfert and LINER galaxies demonstrate that AGNs have a range of different accretion mechanisms and these mechanisms depend on AGN luminosity. In the present review I will only discuss the luminous QSOs, defined to be objects with absolute B magnitude $M_B > -23$ mag. These objects may form a relatively homogeneous subclass of AGNs that accrete at an approximately fixed fraction of the Eddington rate with constant radiative efficiency. This luminosity limit corresponds to $L \approx L_{Edd}$ for a $10^7 M_\odot$ black hole.

11.2   The Net Lifetime

The main methods used to estimate the net lifetime of QSOs are demographic estimates, which typically take one of two forms: integral or counting. Integral estimates are based on the integrated properties of QSOs over the age of the Universe, while counting arguments compare the numbers of objects at different epochs. The classic integral constraint on t_Q is that the amount of matter accreted onto QSOs during their lifetime t_Q, as represented by the luminosity density due to accretion, should be less than or equal to the space density of remnant black holes in the local Universe (Sołtan 1982). This type of estimate is most sensitive to the assumed value of the radiative efficiency ϵ. One example of a counting argument is to ask what value of t_Q is require if all bright galaxies go through a QSO phase in their youth (e.g., Rees 1984). Counting arguments are not sensitive to ϵ, but are affected by the assumed value of L/L_{Edd} and galaxy (black hole) mergers.

The predicted upper bound for ϵ is 0.3, attained in models of thin disk accretion onto a maximally rotating Kerr black hole (Thorne 1974). The value of ϵ is more commonly set to 0.1, based on expectations for thin-disk accretion onto a non-rotating black hole. Determination of L/L_{Edd} is more tractable observationally, at least once the mass of the central black hole is known. The most common assumption is that $L/L_{\text{Edd}} \sim 1$ for luminous QSOs. If they are substantially sub-Eddington, then this implies the presence of a population of massive ($> 10^{10} M_\odot$) black holes not seen in the Universe today. In contrast, substantial super-Eddington luminosities (greater than a factor of a few) have historically been considered unlikely due to the limitations of radiation pressure that define the Eddington limit. However, recent work has suggested that the Eddington luminosity could be exceeded by as much as a factor of 100 in some objects due to small-scale inhomogeneities in thin accretion disks (Begelman 2002).

The demographic constraint set by the evolution of the entire QSO population, $t_Q \leq 10^9$ years, is essentially the only demographic constraint on t_Q that does not depend on some assumptions about the parameters ϵ and L/L_{Edd} and is neither a counting nor an integral constraint. If the conventional values for these parameters are assumed and $t_Q = 10^9$ years, then the black holes in QSOs grow by ~ 25 *e*-folds during this QSO phase. This rapid, prolonged growth would produce a population of extremely massive black holes ($> 10^{10} M_\odot$) that have not been observed in the local Universe (Cavaliere & Padovani 1988, 1989), assuming the initial black hole seeds are stellar remnants with $M > 1 M_\odot$.

A simple counting argument was considered by Richstone et al. (1998), who compared the space density of QSOs at $z \approx 3$ to supermassive black holes in the local Universe. As the space density of QSOs at their peak is approximately 10^{-3} times that of present-day supermassive black holes, the implied lifetime is $t_Q \approx 10^6$ years. This lifetime is much shorter than the Salpeter time scale and therefore implies that only a small percentage of the present-day mass in black holes was accreted during the QSO epoch.

11.2.1   *QSO Evolution and the Local Black Hole Population*

Many more detailed demographic models have been put forth, particularly in the last decade (Small & Blandford 1992; Haehnelt & Rees 1993; Haehnelt, Natarajan, & Rees 1998; Salucci et al. 1999; Yu & Tremaine 2002). These models are mostly based on counting arguments that simultaneously address the present-day black hole mass function and the space density evolution of QSOs with a recipe for the luminosity evolution and

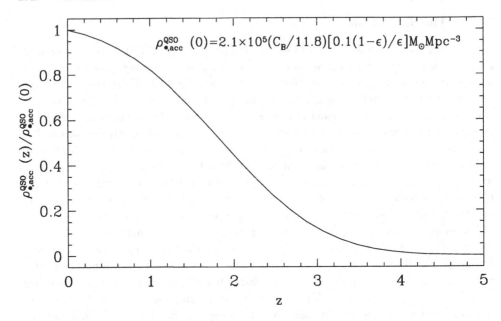

Fig. 11.1. History of accretion onto supermassive black holes by optically luminous QSOs (from Yu & Tremaine, their Fig. 1).

growth of supermassive black holes in the QSO phase. Haiman & Loeb (1998) and Haehnelt et al. (1998) find that they can match the present-day black hole mass function and the QSO luminosity function at $z = 3$ with t_Q between 10^6 and 10^8 years, depending on the relationship between the mass of the central black hole and the host halo. They describe how a lifetime as short as 10^6 years requires that much of the present-day black hole mass is not accreted as an optically luminous QSO.

Yu & Tremaine (2002; see also Yu 2004) develop a similar model that is a variation on Sołtan's (1982) integral approach using new data for the present-day black hole mass function, as well as the luminosity function and space density evolution of QSOs. They calculate the present-day black hole mass function with the $M_\bullet - \sigma$ relation (Ferrarese & Merrit 2000; Gebhardt et al. 2000) and velocity dispersion measurements for a large number of elliptical galaxies and massive spheroids from the SDSS (Bernardi et al. 2003) and take the QSO luminosity function from the 2dF (Boyle et al. 2000) and the Large Bright QSO Survey (Hewett, Foltz, & Chaffee 1995). Their estimate of the current mass density in black holes is $\rho_\bullet^{QSO}(z = 0) = (2.5 \pm 0.4) \times 10^5 h_{0.65}^2 M_\odot \text{Mpc}^3$. This can be compared to the integrated mass accretion onto black holes due to luminous QSOs, which they calculate to be

$$\rho_{\bullet,\text{acc}}^{QSO}(z) = \int_z^\infty \frac{dt}{dz} dz \int_0^\infty \frac{(1-\epsilon)L_{bol}}{\epsilon c^2} \Phi(L,z)dL, \qquad (11.2)$$

where $\Phi(L,z)$ is the QSO luminosity function. The total mass density accreted by optically luminous QSOs is then $\rho_{\bullet,\text{acc}}^{QSO}(z = 0) = 2.1 \times 10^5 (C_B/11.8)[0.1(1-\epsilon)/\epsilon] M_\odot \text{Mpc}^3$, where C_B is the bolometric correction in the B band scaled by the value of 11.8 calculated by Elvis et al. (1994). The cumulative accretion history they derive is shown in Figure 11.1. Note that half of the total mass density accreted by QSOs occurred by $z \approx 1.9$.

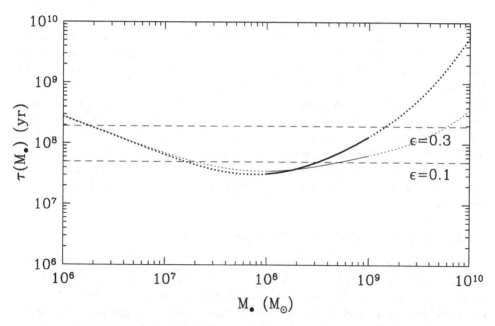

Fig. 11.2. Mean QSO lifetime from Yu & Tremaine (2002; their Fig. 5). The two lines represent the calculated lifetime for two values for the radiative efficiency ϵ. The solid parts of the lines show the range of local black hole masses estimated from observations of early-type galaxies. The dotted lines represent regions where the local black hole mass function and QSO luminosity function were extrapolated. The two horizontal, dashed lines are the Salpeter time scale for $\epsilon = 0.3$ and 0.1.

The mean lifetime of QSOs above some black hole mass is then approximately equal to the integrated luminosity of accreted matter by these black holes, divided by their space density:

$$\tau(> M_\bullet) \simeq \frac{\int_{L(M_\bullet)}^{\infty} dL \int_0^{t_0} \Phi(L,t)dt}{\int_{M_\bullet}^{\infty} n_{M_\bullet}^{\text{early}}(M_\bullet',t_0)dM_\bullet'}. \tag{11.3}$$

This method is therefore essentially an integral method, but by considering the lifetime above a range of black hole masses Yu & Tremaine (2002) capture some of the elements of a counting argument as well. Figure 11.2 shows the mean lifetime derived by Yu & Tremaine (2002) as a function of black hole mass for two values of the radiative efficiency. The solid lines mark the range of black hole mass constrained by the velocity dispersions of early-type galaxies. The dotted lines are based on extrapolations of the local black hole mass function and the luminosity function of QSOs. This calculation assumes that QSOs are radiating at the Eddington luminosity and that the spectral energy distributions of the entire QSO population are well represented by the Elvis et al. (1994) sample. For $\epsilon = 0.1-0.3$ and $10^8 < M_\bullet < 10^9 M_\odot$, Yu & Tremaine calculate that the mean QSO lifetime lies in the range $t_Q = 3-13 \times 10^7$ years.

For the standard radiative efficiency of $\epsilon = 0.1$ and a bolometric correction $C_B = 11.8$, optically selected QSOs can account for $\sim 75\%$ of the current matter density in black holes. The X-ray and infrared backgrounds offer a potentially better constraint on the total mass accretion onto black holes, as these wavelengths are less susceptible to obscured or highly beamed AGNs. Barger et al. (2001) combined these multiple backgrounds and estimated $\rho_{\bullet,acc} \approx 9 \times 10^5 (0.1/\epsilon) M_\odot \text{Mpc}^3$, higher than Yu & Tremaine's estimate due to optically selected QSOs alone. The discrepancy may be due to a genuine population of obscured AGN, but then comparison of this higher accreted mass density with that estimated for local black holes implies $\epsilon \geq 0.3$, in conflict with the maximum efficiency estimated for accretion onto a Kerr black hole. Other potential causes of this discrepancy include (1) uncertainties in the local black hole mass density, which has been extrapolated to $M_\bullet < 10^8 M_\odot$ and $M_\bullet > 10^9 M_\odot$, (2) a different population responsible for these backgrounds, (3) the possibility of a free-floating supermassive black hole population that has been ejected from galaxies, and (4) the possibility that a significant amount of black hole mass has been lost as gravitational radiation in mergers.

11.2.2   Coevolution of QSOs, Black Holes, and Galaxies

An alternative approach to modeling the coevolution of QSOs, black holes, and galaxies is to build detailed models with some adjustable parameters and ask what the lifetime is in that context. This is the approach adopted by Kauffmann & Haehnelt (2000), who model the formation of QSOs during major mergers of gas-rich galaxies. Any merger between galaxies with a mass ratio greater than 0.3 results in some fraction M_{acc} of gas accreted onto the black hole over a time scale t_Q. The galaxy mergers and their cold gas content are then obtained from their semi-analytic model of galaxy evolution, combined with the merging history of dark matter halos from extended Press-Schechter (1974) theory.

Kauffmann & Haehnelt let the peak luminosity of each QSO scale as the mass of gas accreted onto the central black hole, which in turn scales as the mass of available cold gas in the merger and the lifetime of the accretion phase. The radiative efficiency is taken to be constant and chosen so that the most luminous QSOs do not exceed the Eddington luminosity. The luminosities of QSOs after the merger do not remain constant, but decline exponentially from their peak on a time scale t_Q. While their high-redshift QSOs emit at close to the Eddington luminosity, the QSOs at low redshift have $L/L_{Edd} = 0.01 - 0.1$. In addition to a QSO lifetime that is constant with redshift, these authors also parametrize the lifetime to scale as the dynamical time: $t_Q(z) = t_Q(0)(1+z)^{-1.5}$, where $t_Q(0)$ is the lifetime at $z = 0$. The lifetime that best fits the evolution of the QSO population, the merger history of galaxies, and the evolution of their cold gas fraction is $t_Q(0) = 3 \times 10^7$ years (for a given merger event).

11.3   Limits on Episodic Activity

11.3.1   QSOs and Starbursts

The presence of many QSOs in merging systems and the detection of luminous hard X-ray emission from the cores of some merging galaxies strongly suggest that at least some QSOs are triggered by interactions of giant, gas-rich galaxies. The merging process removes angular momentum from a significant fraction of the two galaxies' ISM, leading it to flow inward toward the circumnuclear region and central black hole. This angular momentum loss is a natural mechanism to ignite QSO activity, as well as significant amounts of circumnuclear star formation.

In extreme circumstances Norman & Scoville (1988) proposed that mass loss from a sufficiently large star cluster formed in such a merger may provide a continuous fuel supply for a long t_Q. These authors specifically explore the implications of a star cluster of mass $4 \times 10^9 M_\odot$ within the central 10 pc of the merger remnant. In $\sim 10^8$ years, mass loss from this cluster can power accretion rates of up to $10 M_\odot$ yr^{-1}, corresponding to QSO luminosities $L > 10^{12} L_\odot$ due to the black hole alone, and much higher than the stellar luminosity of the central star cluster. This model therefore predicts an episodic QSO lifetime of $\sim 10^8$ years after a major merger between two massive, gas-rich galaxies.

A number of examples of so-called transition QSOs, which show evidence for recent and substantial star formation, were studied in detail by Canalizo & Stockton (2001). They derived ages for the post-starburst population as high as 300 Myr in some objects, while Brotherton et al. (1999) derived an age of 400 Myr for the post-starburst population in UN J1025-0040. If the the QSO and starburst were triggered at the same time, the implied lifetimes for the QSOs are greater than a few $\times 10^8$ years.

Kawakatu, Umemura, & Mori (2004) recently considered a similar coevolution model for black holes and galaxies. They model the formation and growth of the central black hole via radiation drag on the host galaxy ISM. This process gradually increases the mass of the central black hole for $\sim 10^8$ years, during which time the galaxy may appear as an ultraluminous infrared galaxy. Once the central region is no longer obscured, the galaxy appears as an optically identifiable QSO. This evolutionary scenario corresponds to a great deal of supermassive black hole growth, which these authors propose would lead to larger velocities in the broad-line region.

11.3.2 Size of Narrow-Line Regions

The size of the narrow-line region (NLR) around AGNs potentially provides a straightforward, geometric estimate of their lifetime. Bennert et al. (2002) recently studied a sample of Seyferts and QSOs and found that the size of the NLR increases approximately as the square root of the [O III] luminosity and the square root of an estimate of the Hβ luminosity. For a constant ionization parameter, electron density, and covering factor, the size of the ionized region should scale as the square root of the luminosity. The NLR is therefore ionization bounded and the size of the ionized region sets a lower limit to the lifetime of the ionizing source. As the most luminous QSOs in their sample have NLRs approximately 10 kpc in radius, the episodic lifetime of the most luminous QSOs must be greater than 3×10^4 yr.

11.3.3 Lengths of Jets

Jets are another macroscopic product of accretion onto a supermassive black hole. Their expansion time therefore sets a lower limit to the lifetime of the accretion. The main uncertainty in the use of jets to constrain the lifetime is the expansion speed, although the lengths of the jets themselves are also somewhat uncertain due to the unknown inclination angle.

The expansion speed of jets as a population can be estimated by the mean ratio of the length of the jet to the counterjet (Longair & Riley 1979), although this requires that we assume that all jets have the same expansion speed. For an expansion speed $v = \beta c$ and inclination to the line of sight θ, the observed length ratio of the jet to counterjet (following Scheuer 1995) is

$$Q = (1 + \beta \cos \theta)/(1 - \beta \cos \theta). \tag{11.4}$$

If we assume that the inclination angle of the population is uniformly distributed between zero and θ_{max} degrees, the mean of log Q in the limit of small β is

$$\langle \ln Q \rangle = \beta(1 + \cos \theta_{max})[1 + \frac{1}{6} \beta^2 (1 + \cos^2 \theta_{max})...]. \tag{11.5}$$

Combining studies of radio jets by a number of authors, Scheuer (1995) obtains $\beta = 0.03 \pm 0.02$ and sets strong bounds on β such that $0 < \beta < 0.15$.

Blundell, Rawlings, & Willott (1999) compiled data from three separate, flux-limited radio surveys and studied the evolution of the classical double population, using this large database to decouple redshift and luminosity degeneracies. With the expansion speed constraints of Scheuer (1995), they find ages for classical double sources as large as a few $\times 10^8$ years in the lowest redshift sample (such large sources are selected against at higher redshift due to the steepening of their spectra). The existence of these objects therefore appears to set a long lower limit to the lifetime of nuclear activity.

The main caveat in the use of maximum jet lengths to place lower limits on the episodic lifetime of QSOs is that jets are commonly found in very low-luminosity AGNs, and not only in systems of QSO luminosities. While nearly all of the sources used in Scheuer's study had QSO luminosities, this is not true of all of the sources studied by Blundell et al. (1999). The lifetime derived from the lengths of jets therefore do not exclude the possibility that a high-luminosity QSO phase is a relatively short-lived stage in the lifetime of an AGN.

11.3.4 Proximity Effect

QSOs can create a sphere of ionized matter in their vicinity whose radius is set by the episodic lifetime or recombination time scale, whichever is shorter. This ionized region is observed as the proximity effect in the Lyα forest (Bajtlik, Duncan, & Ostriker 1988; Scott et al. 2000), which quantifies the decrease in the number of neutral hydrogen clouds of a given column density per unit redshift in the vicinity of the QSO. In order to produce this zone of increased ionization and decreased number density of the high-column density Lyα forest clouds, the ionization field from the QSO must have been operating for at least the recombination time scale of these relatively high-column density clouds. Because the proximity effect is observed along the line of sight to the QSO, the physical extent of the region only depends on the observed luminosity of the QSO, and not the episodic lifetime.

The equilibration time scale for Lyα clouds is:

$$\tau \equiv n_{HI} \left(\frac{dn_{HI}}{dt} \right)^{-1}, \tag{11.6}$$

where n_{HI} is the number density of neutral hydrogen atoms. The relevant time scale is

$$\tau = 1 \times 10^4 (1 + \omega)^{-1} J_{21}^{-1} \quad \text{yr}, \tag{11.7}$$

where ω is the increase in ionizing flux due to the QSO relative to the background and J_{21} the intensity of the background ionizing radiation at the Lyman-limit in units of 10^{-21} erg cm^{-2} Hz^{-1} sr^{-1}. The equilibration time scale is $\sim 10^4$ years for typical values of ω and J_{21}.

11.4   Future Prospects

11.4.1   Luminous QSOs at $z > 6$

Observation of QSOs in the early Universe when the optical depth in neutral hydrogen is large ($z > 6$) offers a more powerful constraint on the lifetime than the proximity effect because these QSOs exist in a neutral, predominantly lower-density IGM. Above $z \approx 6$, the Gunn-Peterson trough in the IGM will erase all flux blueward of Lyα in the absence of ionization of the surrounding medium by the QSO, i.e. an H II region. The size of this region is set by the lifetime of the QSO as the sole ionizing source, unlike the case for the proximity effect, where the lifetime is only constrained to be the recombination time because only high density regions remain to be (re)ionized.

Haiman & Cen (2002; see also Haiman & Loeb 2001) applied this concept to the $z = 6.28$ QSO SDSS 1030+0524. They used the expected density distribution of the IGM with a hydrodynamic simulation (Cen & McDonald 2002) and found that a quasar lifetime of 2×10^7 years and an H II region ~ 4.5 Mpc (proper) radius produced a good match to the observed flux in SDSS 1030+524 blueward of Lyα .

The masses of the black holes powering QSOs such as this one and others with $z > 6$ provide some additional information on the early evolution of QSOs. For example, one of the key conclusions of Haiman & Cen (2002) was that the $z = 6.28$ QSO was not significantly lensed or beamed. Therefore if the QSO is shining at or below the Eddington rate, the mass of the central black hole is on order $2 \times 10^9 M_\odot$. The other $z > 6$ QSOs discovered by the SDSS have comparable luminosities, and therefore are presumably similarly massive. To form such a massive black hole at this early epoch requires that it has been accreting at the Eddington rate for nearly the lifetime of the Universe, less than 10^9 years at $z = 6$ assuming $\epsilon = 0.1, L \approx L_{\mathrm{Edd}}$, and a $100 M_\odot$ seed. Depending on the mass of the seed black hole population, either the QSO lifetime must be quite long or super-Eddington accretion must occur in the early Universe.

11.4.2   Transverse Proximity Effect

The proximity effect can be used to set a stronger constraint on the episodic lifetime with a direct measure of its extent in the plane of the sky. This measurement could be made through observations of multiple QSOs that lie in close proximity on the sky. The proximity effect due to a lower-redshift QSO in the spectrum of higher-redshift QSO would provide a lower limit on the lifetime of the lower-redshift QSO equal to the time required to ionize the intervening neutral medium, where the ionization front expands at a speed $v \simeq c$ in the primarily low-density IGM. The angular diameter distance between the two QSOs is then a lower-limit to the lifetime. Such a measurement was first discussed by Bajtlik et al. (1988), although in practice the statistical significance of the proximity effect in neutral hydrogen of any given QSO at $z < 6$ is sufficiently weak that this measurement would require combining data from large numbers of QSOs to obtain a mean lower limit to the episodic lifetime.

The proximity effect in neutral hydrogen is difficult to measure because its mean opacity of the intergalactic medium is so low (at least below $z \approx 6$) and the recombination times for the remaining neutral, high-density clouds are quite short. Helium offers a better chance of success as the mean opacity of neutral helium is quite high above $z \approx 2.9$. Individual helium ionizing sources, i.e. QSOs, above this redshift will therefore produce distinct opacity gaps.

Unfortunately, relatively few He II opacity gaps are known as the He II Lyα λ304 Å line is still in the space UV regime at $z \approx 3$. Current instrumentation therefore requires them to be observed in the spectra of very bright QSOs. These systems are also easily obscured by the much more common Lyman-limit systems.

There are currently five known helium absorption systems in the spectra of four QSOs. These systems offer potentially excellent lower limits on the episodic lifetime if the origins of these systems are identified with QSOs a significant distance from the He II system. Jakobsen et al. (2003) identified such a candidate QSO for the He II opacity gap at $z \approx 3.056$ in the spectrum of the $z = 3.256$ QSO Q0302-003. They identified a $z = 3.050 \pm 0.003$ QSO at the approximate redshift of the helium feature with an angular separation of $6\rlap{.}'5$ from the line of sight to Q0302-003. If the new QSO is responsible for the helium opacity gap, then the light travel time sets a lower limit of $t_Q > 10^7$ yr.

The widths of all of the known He II opacity gaps are $\Delta z \simeq 0.01 - 0.02$, which corresponds to a size of $2-5$ Mpc (Jakobsen et al. 2003). These sizes set a strong lower limit of $\sim 10^6$ years for the lifetime of the ionizing source. An alternative explanation of these opacity gaps is that they are due to low-density regions in the IGM. Heap et al. (2000) studied the opacity gap in Q0302-003 and argue against this interpretation, as their simulations cannot reproduce the distribution and amplitude of the gap with a low-density region.

11.4.3   *Clustering*

The clustering of QSOs can provide a strong constraint on their net lifetime, under the assumption that QSOs reside in the most massive dark matter halos. If QSOs are long-lived, then they need only reside in the most massive, and therefore most strongly clustered, halos in order to match the observed space density of luminous QSOs. In contrast, if QSOs are short-lived phenomena, then a much larger population of host halos is required to match the observed QSO space density at the epoch of observation. As this large population of hosts will be dominated by less massive halos, due to the shape of the halo mass function, the QSO population will not be as strongly clustered. These simple arguments therefore predict that the larger the observed correlation length, the longer the lifetime of QSOs (Haehnelt et al. 1998).

Martini & Weinberg (2001) calculated the expected relation between QSO clustering and lifetime in detail in anticipation of forthcoming measurements from the 2dF and SDSS collaborations (see also Haiman & Hui 2001). The main assumptions of this model are that the luminosity of a QSO is a monotonic function of the mass of its host dark matter halo and that all sufficiently massive halos go through a QSO phase. For some absolute magnitude-limited sample of QSOs at redshift z, the probability that a given halo currently hosts a QSO is simply t_Q/t_H, where t_H is the lifetime of the host halo. The QSO therefore does not necessarily turn on when the halo forms, which differs slightly from the model developed by Haehnelt et al. (1998).

Calculation of the clustering corresponding to some lifetime t_Q requires knowledge of the minimum halo mass, which can be found by setting the integral of the mass function of dark matter halos, multiplied by the fraction that host QSOs, equal to the observed space density $\Phi(z)$:

$$\Phi(z) = \int_{M_{min}}^{\infty} dM \frac{t_Q}{t_H(M,z)} n(M,z). \tag{11.8}$$

The halo lifetime t_H depends on both the halo mass and redshift and was calculated by solving for the median time a halo of mass M at redshift z will survive until it is incorporated into a new halo of mass $2M$. The mass function of dark matter halos $n(M,z)$ is calculated with the Press-Schechter formalism:

$$n(M,z)\,dM = -\sqrt{\frac{2}{\pi}}\frac{\rho_0}{M}\frac{\delta_c(z)}{\sigma^2(M)}\frac{d\sigma(M)}{dM}\exp\left[-\frac{\delta_c^2(z)}{2\sigma^2(M)}\right]dM. \tag{11.9}$$

The predicted clustering of QSOs depends directly on the minimum halo mass. The bias factor for halos of a given mass was calculated by Mo & White (1996) as

$$b(M,z) = 1 + \frac{1}{\delta_{c,0}}\left[\frac{\delta_c^2(z)}{\sigma^2(M)}-1\right]. \tag{11.10}$$

The effective bias factor for the host halo population is then the integral of the bias factor for all masses greater than M_{min} and weighted by the halo number density and lifetime:

$$b_{\mathrm{eff}}(M_{min},z) = \left(\int_{M_{min}}^{\infty}dM\,\frac{b(M,z)n(M,z)}{t_H(M,z)}\right)\left(\int_{M_{min}}^{\infty}dM\,\frac{n(M,z)}{t_H(M,z)}\right)^{-1}. \tag{11.11}$$

The clustering amplitude of QSOs can be parametrized as the radius of a top-hat sphere in which the rms fluctuations of QSO number counts σ_Q is unity:

$$\sigma_Q(r_1,z) = b_{\mathrm{eff}}(M_{min},z)\sigma(r_1)D(z) = 1, \tag{11.12}$$

where $\sigma(r_1)$ is the rms linear mass fluctuation in spheres of radius r_1 at $z = 0$ and $D(z)$ is the linear growth factor. The radius r_1 is similar to the correlation length r_0, at which the correlation function $\xi(r)$ is unity, but can be determined more robustly because it is an integrated quantity and does not require fitting $\xi(r)$.

For some specified cosmology and comoving QSO space density, r_1 depends directly on t_Q. Figure 11.3 shows the minimum mass, effective bias factor, and r_1 as a function of t_Q for a range of cosmologies at $z = 2, 3,$ and 4. The bottom panels, which show the relationship between r_1 and t_Q, demonstrate that the clustering length increases as a function of QSO lifetime and that the clustering length can be used to obtain a good estimate of the lifetime if the cosmological parameters are relatively well known. The difference in the relationship between r_1 and t_Q as a function of redshift mostly reflects the evolution in the QSO space density (taken from Warren, Hewett, & Osmer 1994). QSOs are rarer at $z = 4$ than at $z = 2$ or 3 and therefore the observed space density can be matched with a larger M_{min} at the same t_Q. At all of these redshifts the most massive halos correspond to the masses of individual galaxy halos. At lower redshifts, the most massive halos could contain multiple galaxies and the assumption of one QSO per halo may break down. The relation between clustering and lifetime is much shallower for the $\Omega_M = 1$ models because of their smaller mass fluctuations. For these models the values of M_{min} also lie out on the steep, high-mass tail of the halo mass function, where a smaller change in M_{min} is required to compensate for the same change in t_Q.

The relationship between lifetime and clustering is sensitive to several model details, which can be used to either determine the precision of the estimated lifetime or observationally test and further refine the relation. One uncertainty lies in the definition of the halo lifetime, which was taken to be the median time before the halo was incorporated into a new halo of twice the original mass. If the definition were increased to a factor of five

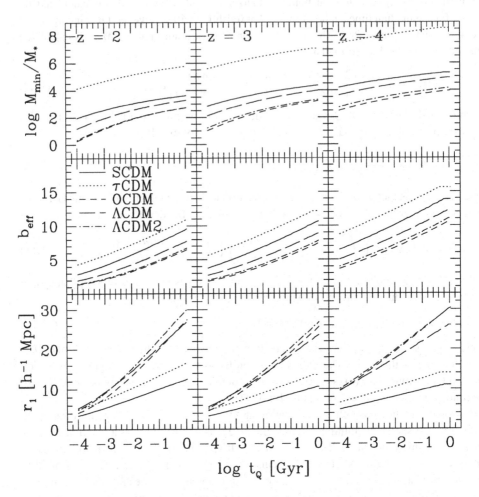

Fig. 11.3. Minimum halo mass, effective bias, and clustering length as a function of t_Q for a range of CDM models at $z = 2, 3,$ and 4 (from Martini & Weinberg 2001; their Fig. 7).

increase in mass (a strong upper limit), then the QSO lifetimes corresponding to the same observed r_1 would increase by a factor of 2–4. Another potential complication is that the correlation between halo mass and QSO luminosity is not perfect. Scatter in this relation will lead to the inclusion of a larger number of less massive halos in an absolute-magnitude limited sample of QSOs. Because this scatter always serves to increase the number of lower-mass halos, it decreases the clustering signal from the ideal case of a strict relationship and effectively makes the clustering predictions in Figure 11.3 into lower limits for t_Q for a measured r_1. The presence of scatter in this relation could be determined and measured observationally. In the absence of scatter, there should be a direct relationship between the clustering length and QSO luminosity at fixed redshift. The presence of scatter will flatten the predicted relation between clustering and luminosity and could be used to estimate and correct for the amount of scatter in the determination of t_Q.

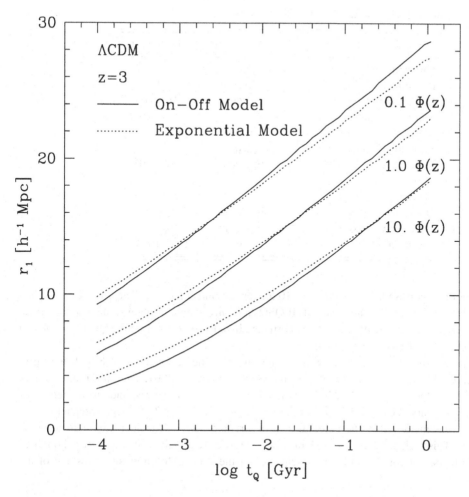

Fig. 11.4. Clustering length vs. t_Q for the ΛCDM model for two luminosity evolution models and three values of the space density (from Martini & Weinberg 2001; their Fig. 8).

This model also assumes that the luminosity evolution of QSOs is a step function, that is they are either on at some constant luminosity, or off. If QSOs are instead triggered by a mechanism that drives a great deal of fuel toward the center, such as a galaxy merger, this fuel supply will probably gradually diminishes over time and cause the fading of the QSO. An alternative parameterization of QSO luminosity evolution, employed by Haehnelt et al. (1998) is an exponential decay: $L(t) \propto \exp(-t/t_Q)$. Martini & Weinberg (2001) found that if this luminosity evolution is used instead of the "on-off" model, and t_Q redefined to represent the e-folding time of the QSO, there is very little change in the relation between t_Q and r_1 (see Fig. 11.4). This figure also demonstrates the scalable nature of the model to QSO surveys with different absolute magnitude limits. A decrease in the absolute magnitude limit is equivalent to increasing the space density of QSOs and therefore the number of lower-mass halos. To first approximation, increasing the space density of the sample by a

Table 11.1. *Constraints on the QSO Lifetime*

Method	Net/Episodic	Lifetime [yr]
Current:		
Evolution	Net	$< 10^9$
Local Black Holes	Net	$10^7 - 10^8$
Merger Scenario	Net	few $\times 10^7$
Mergers & Starbursts	Episodic	few $\times 10^8$
Proximity Effect	Episodic	$> 10^4$
Radio Jets	Episodic	few $\times 10^8$
Future:		
QSOs at $z > 6$	Episodic	few $\times 10^7$?
Transverse Proximity Effect	Episodic	10^7 ??
Clustering	Net	$10^6 - 10^7$??

factor of 10 corresponds to a factor of 10 longer lifetime at fixed r_1. These results scale in a similar manner if QSOs are beamed. If QSOs are only visible through some beaming angle f_B, then the space density needs to be corrected by a factor f_B^{-1} before the QSO lifetime can be determined from the clustering.

The 2dF and SDSS surveys will provide the best measurements of QSO clustering to date. Preliminary results from these surveys (Croom et al. 2001; Fan 2004) indicate that $r_0 \approx 6h^{-1}$Mpc, or $r_1 \approx 9h^{-1}$Mpc for $\xi(r) = (r/r_0)^{-1.8}$. This corresponds to a lifetime of $t_Q \approx 10^6$ years. A much larger correlation length $r_0 = 17.5 \pm 7.5h^{-1}$Mpc (Stephens et al. 1997) was measured in the Palomar Transit Grism Survey (Schneider, Schmidt, & Gunn 1994). This high correlation length may be a statistical anomaly as it is only based on three QSO pairs, but could also be a consequence of the very high luminosity threshold of the survey.

11.5   Summary

Current estimates of the QSO lifetime lie in the range $t_Q = 10^6 - 10^8$ years, still uncertain by several orders of magnitude. Estimates of the net lifetime are primarily demographic and rely on assumptions about the radiative efficiency of accretion ϵ and the luminosity of QSOs relative to the Eddington rate L/L_{Edd}. The new models developed in the last few years (Kauffmann & Haehnelt 2000; Yu & Tremaine 2002) have started to relax the common assumptions about the values of ϵ and L/L_{Edd}, but a complex relationship between ϵ, L/L_{Edd}, and $M_{\bullet}$ has by no means been ruled out.

While the physics of the accretion process and the potential for significant obscured black hole growth may appear daunting obstacles, there are reasons for optimism. Demographic estimates of the net lifetime will be substantially improved as more supermassive black hole masses are measured and the validity of the $M_{\bullet} - \sigma$ relation better established, including its application to active galaxies. Measurement of the black hole mass in more QSOs via reverberation mapping (e.g., Barth 2004), or estimates using the $M_{\bullet} - \sigma$ method will lead to direct estimates of L/L_{Edd} for many QSOs. These steps will help to eliminate some of the main uncertainties in the value or range of L/L_{Edd} and the local mass density of supermassive black holes. Finally, measurement of a significant number of redshifts for the sources that comprise the hard X-ray and sub-mm background will greatly improve the integral constraint provided by the net luminosity from accretion. Many of these new observations could be realized in the next several years and reduce the need for sweeping assumptions about ϵ or L/L_{Edd}.

The current best estimates or limits for the episodic lifetime are consistent with a wider range of t_Q than the net lifetime, but constraints from the proximity effect in the Lyα forest do point to $t_Q > 10^4$ years. Observations of QSOs when the opacities of neutral helium and hydrogen was much higher suggest much longer episodic lifetimes, although to date are still only based on observations of two QSOs. Constraints such as these on the episodic lifetime provide a valuable complement to the primarily demographic estimates of the net lifetime, as they do not depend on assumptions of accretion physics, but instead on much more straightforward radiation physics.

The next few years offer the hope of tremendous progress in measurement of the QSO lifetime via a variety of techniques. The discovery of additional QSOs associated with He II absorption systems (and hopefully more of these systems), as well as additional QSOs at high redshift $z > 6$, could provide quite strong lower limits to the episodic lifetime, while clustering measurements hold great promise to constrain the net lifetime. As the clustering method does not depend on assumptions about accretion physics, it will provide a valuable complement to the present demographic approaches.

This potential progress suggests that t_Q could be determined to within a factor of 3 in the next three years, which is sufficient precision to address some fundamental questions on the growth of black holes and the physics of the accretion process. One important advance will be to determine if the QSO lifetime determined via clustering is in agreement with that estimated via black hole demographics. As these models employ an independent set of assumptions, these measurements could provide strong evidence that $\epsilon \approx 0.1$ and $L/L_{Edd} \approx 1$ are valid choices.

Measurement of t_Q will also determine if a luminous QSO phase was the dominant growth mechanism for present-day supermassive black holes. If $t_Q > 4 \times 10^7$ yr, then a substantial fraction of the black hole mass in QSO hosts was accreted via this optically luminous mechanism. However, such a long lifetime would also imply that QSOs were quite rare phenomena and only a small fraction of present-day supermassive black holes were created as QSOs. The remainder may have been identifiable as AGN, or were heavily obscured, but would not have been classified as optically luminous QSOs.

If instead t_Q is shorter than the Salpeter time scale, then hosting a short-lived QSO phase was a relatively common occurrence. However, only a small fraction of the present-day black hole mass was accreted during this time and the remainder must have occurred via a less luminous, or less radiatively efficient, mode of accretion, either before or after the QSO epoch. This could then imply that advection-dominated accretion (e.g., Narayan, Mahadevan, & Quataert 1998) or inflow-outflow solutions (Blandford & Begelman 1999) played a significant role in black hole growth.

Acknowledgements. I would like to thank Zoltán Haiman and David Weinberg for helpful comments on this manuscript. I am also grateful to Luis Ho for inviting and encouraging me to prepare this review.

References

Bajtlik, S., Duncan, R. C., & Ostriker, J. P. 1988, ApJ, 327, 570

Barger, A. J., Cowie, L. L., Bautz, M. W., Brandt, W. N., Garmire, G. P., Hornschemeier, A. E., Ivison, R. J., & Owen, F. N. 2001, AJ, 122, 2177

Barth, A. J. 2004, in Carnegie Observatories Astrophysics Series, Vol. 1: Coevolution of Black Holes and Galaxies, ed. L. C. Ho (Cambridge: Cambridge Univ. Press), in press

Begelman, M. C. 2002, ApJ, 568, L97

Bennert, N., Falcke, H., Schulz, H., Wilson, A. S., & Wills, B. J. 2002, ApJ, 574, L105

Bernardi, M., et al. 2003, AJ, 125, 1817

Blandford, R. D., & Begelman, M. C. 1999, MNRAS, 303, L1

Blundell, K. M., Rawlings, S., & Willott, C. J. 1999, AJ, 117, 677

Boyle, B. J, Shanks, T., Croom, S. M., Smith, R. J., Miller, L., Loaring, N., & Heymans, C. 2000, MNRAS, 317, 1014

Brotherton, M. S., et al. 1999, ApJ, 520, L87

Canalizo, G., & Stockton, A. 2001, ApJ, 555, 719

Cavaliere, A., & Padovani, P. 1988, ApJ, 333, L33

——. 1989, ApJ, 340, L5

Cen, R., & McDonald, P. 2002, ApJ, 570, 457

Croom, S. M., Boyle, B. J., Loaring, N. S., Miller, L., Outram, P. J., Shanks, T., & Smith, R. J. 2002, MNRAS, 335, 459

Elvis, M., et al. 1994, ApJS, 95, 1

Fan, X. 2004, in Carnegie Observatories Astrophysics Series, Vol. 1: Coevolution of Black Holes and Galaxies, ed. L. C. Ho (Pasadena: Carnegie Observatories, http://www.ociw.edu/ociw/symposia/series/symposium1/proceedings.html)

Ferrarese, L., & Merritt, D. 2000, ApJ, 539, L9

Gebhardt, K., et al. 2000, ApJ, 539, L13

Haehnelt, M.. & Natarajan, P., & Rees, M. J. 1998, MNRAS, 300, 817

Haehnelt, M., & Rees, M. J. 1993, MNRAS, 263, 168

Haiman, Z., & Cen, R. 2002, ApJ, 578, 702

Haiman, Z., & Hui, L. 2001, ApJ, 547, 27

Haiman, Z., & Loeb, A. 1998, ApJ, 503, 505

——. 2001, ApJ, 552, 459

Heap, S. R., Williger, G. M., Smette, A., Hubeny, I., Sahu, M., Jenkins, E. B., Tripp, T. M., & Winkler, J. N. 2000, ApJ, 534, 69

Hewett, P. C., Foltz, C. B.,, & Chaffee, F. H. 1995, AJ, 109, 1498

Jakobsen, P., Jansen, R. A., Wagner, S., & Reimers, D. 2003, A&A, 397, 891

Kauffmann, G., & Haehnelt, M. 2000, MNRAS, 311, 576

Kawakatu, N., Umemura, M., & Mori, M. 2004, in Carnegie Observatories Astrophysics Series, Vol. 1: Coevolution of Black Holes and Galaxies, ed. L. C. Ho (Pasadena: Carnegie Observatories, http://www.ociw.edu/ociw/symposia/series/symposium1/proceedings.html)

Longair, M. S., & Riley, J. M. 1979, MNRAS, 188, 625

Lynden-Bell, D. 1969, Nature, 223, 690

Martini, P., & Weinberg, D. H. 2001, ApJ, 547, 12

Mo, H. J., & White, S. D. M. 1996, MNRAS, 282, 347

Narayan, R., Mahadevan, R., & Quataert, E. 1998, in The Theory of Black Hole Accretion Discs, ed. M. A. Abramowicz, G. Björnsson, & J. E. Pringle (Cambridge: Cambridge Univ. Press), 148

Norman, C. A., & Scoville, N. Z. 1988, ApJ, 332, 124

Osmer, P. S. 2004, in Carnegie Observatories Astrophysics Series, Vol. 1: Coevolution of Black Holes and Galaxies, ed. L. C. Ho (Cambridge: Cambridge Univ. Press), in press

Press, W. H., & Schechter, P. 1974, ApJ, 187, 425

Rees, M. J. 1984, ARA&A, 22, 471

Richstone, D.O., et al. 1998, Nature, 395, A14

Salpeter, E. E. 1964, ApJ, 140, 796

Salucci, P., Szuszkiewicz, E., Monaco, P., & Danese, L. 1999, MNRAS, 307, 637

Scheuer, P. A. G. 1995, MNRAS, 277, 331

Schneider, D. P., Schmidt, M., & Gunn, J. E. 1994, AJ, 107, 1245

Scott, J., Bechtold, J., Dobrzycki, A., & Kulkarni, V. P. 2000, ApJS, 130, 67

Small, T. A., & Blandford, R. D. 1992, MNRAS, 259, 725

Sołtan, A. 1982, MNRAS, 200, 115

Stephens, A. W., Schneider, D. P., Schmidt, M., Gunn, J. E., & Weinberg, D. H. 1997, AJ, 114, 41

Thorne, K. S. 1974, ApJ, 191, 507

Warren, S. J., Hewett, P. C., & Osmer, P. S. 1994, ApJ, 421, 412

Yu, Q. 2004, in Carnegie Observatories Astrophysics Series, Vol. 1: Coevolution of Black Holes and Galaxies, ed. L. C. Ho (Pasadena: Carnegie Observatories,
  http://www.ociw.edu/ociw/symposia/series/symposium1/proceedings.html)

Yu, Q., & Tremaine, S. 2002, MNRAS, 335, 965

Zel'dovich, Ya. B., & Novikov, I. D. 1964, Sov. Phys. Dokl., 158, 811

12

Fueling gas to the central region of galaxies

KEIICHI WADA
National Astronomical Observatory of Japan, Tokyo

Abstract
Supplying gas to the galactic central regions is one of key ingredients for AGN activity. I will review various fueling mechanisms for a $R \approx 0.1$ kpc region, determined mainly by numerical simulations over the last decade. I will also comment on the bars-within-bars mechanism. Observations suggest that the stellar bar is not a sufficient condition for gas fueling. Moreover, considering the various factors for the onset of gas accretion, stellar bars would not even be a necessary condition. I introduce recent progress obtained through our two- and three-dimensional, high resolution hydrodynamical simulations of the ISM in the central 0.1–1 kpc region of galaxies. Possible structure of the obscuring molecular tori around AGNs is also shown. The nuclear starburst is an important factor in determining the structure of the molecular tori and the mass accretion rate to the nucleus. It is natural that the ISM in the central 100 pc region is a highly inhomogeneous and turbulent structure. As a result, gas accretion to the central parsec region should be time dependent and stochastic. The conventional picture of gas fueling and the AGN unified model may be modified in many respects.

12.1 Conventional Picture of the Fueling Problem

Accretion of gas to the supermassive black hole in the galactic center is the source of all AGN activity. A long-standing issue concerning this gas supply is the "fueling problem"—that is, the question of how to remove the large angular momentum of the gas in a galactic disk and funnel it into the accretion disk in the central AU region. This was one of the main topics at the "Mass-transfer Induced Activity in Galaxies" conference held in Lexington in 1993 (Shlosman 1994). The well known cartoon by E. S. Phinney of a baby being fed by a huge spoon, published in the workshop's proceedings, well represents the essence of the fueling problem. To power an AGN luminosity of $\sim 10^{10-11} L_\odot$, we need a mass accretion rate of $\sim 0.1 M_\odot$ yr^{-1} with a $\sim 10\%$ energy conversion rate. Therefore, to maintain the AGN activity during its lifetime of 10^8 yr, a large amount of the gas, $10^7 M_\odot = 0.1 M_\odot$ yr$^{-1} \times 10^8$ yr, must be funneled into the black hole. The galactic disk is probably a reservoir of the gas, and since the time scale of 10^8 yr is comparable to the rotational time scale of galaxies, it is natural to postulate that the gas is accumulated from the galactic disk. A number of mechanisms for removing the angular momentum of the gas have been proposed. Among them, the use of gravitational torques due to galaxy-galaxy interactions (e.g., major/minor mergers and close encounters) or stellar bars have been considered reasonable means of removing angular momentum of the rotating gas.

© The Observatories of the Carnegie Institution of Washington 2004.

This is the *conventional* picture of the fueling problem. After the comprehensive review paper on this subject by Shlosman, Frank, & Begelman (1990), there were many findings, mainly through numerical simulations. In the next section, I will review various fueling mechanisms on a scale from $R \approx 1$ kpc down to 100 pc, found in the last decade. I will show a revision of the "fueling flowchart" (Shlosman et al. 1990) for cases with and without inner Lindblad resonances. The "bars-within-bars" hypothesis, which has been considered as an important mechanism to connect the large scale and small scale, is discussed in § 12.2.4. In § 12.3, I will introduce our recent work on the dynamics and structure of the gas in the central 100 pc around a supermassive black hole. Finally, I will summarize a new picture for gas fueling and discuss implications from recent observations in § 12.4.

12.2   The Fueling Flowchart and Its Revision

12.2.1   *Fueling Processes with ILRs*

Shlosman et al. (1990) proposed a "flowchart" that describes possible fueling mechanisms from 10 kpc down to the central black hole scale. After the review was published, there was a great deal of theoretical and numerical work on the gas dynamics in a bar potential, especially on the scale from 1 kpc to 100 pc. Shlosman et al.'s fueling flowchart shows that if there are inner Lindblad resonances (ILR), a starburst ring is triggered at $R \approx 1$ kpc, a ringlike, dense region of gas formed due to the resonance-driven mass transfer. However, the gaseous response to the resonances and the final structure of the gas depend not only on the existence of Lindblad resonances, but also on the *type* of Lindblad resonances. Three types of resonance-driven fueling processes were proposed.

(1) If a galaxy has a rigidly rotating central region, two Lindblad resonances are expected to exist around the core radius, depending on the pattern speed of the non-axisymmetric gravitational potential, Ω_p. Owing to the two ILRs, the inner ILR (IILR) and the outer ILR (OILR), an oval gas ring is formed near the two resonances, and if the ring is massive enough, the ring fragments due to gravitational instability. This would cause a ringlike starburst. Wada & Habe (1992), however, showed that if the gas mass is greater than about 10% of the dynamical mass, the fragmented oval gas ring finally collapses. As a result, a large amount of the gas is supplied to the central 100 pc. In this process, the stellar bar removes the angular momentum of the gas, and self-gravity of the gas also plays an essential role. Energy dissipation due to shocks caused by collision of the clumps is also a key physical process in changing the oval gas orbits into inner circular motion (see also Fukunaga & Tosa 1991). Interestingly, similarly elongated gas rings are found in more complicated situations, for instance in numerical simulations of stellar bars with a gas component and galaxy-galaxy encounter systems (Friedli & Benz 1993; Barnes & Hernquist 1996).

(2) Related to the IILR, another fueling mechanism for $R \approx 100$ pc is possible. If the gas disk inside the IILR is massive, there is a more rapid fueling process, by which the gas can fall toward the center well before the ILR ring is formed (Wada & Habe 1995). The gaseous oval orbits near the ILRs are oriented by 45° with respect to the bar potential, and the distortion of the orbits strengthens with time due to the self-gravity of the gas and loss of angular momentum. Eventually, radial shocks are generated along the major axis of the elliptical orbits. When the gas on the elongated orbits rush into the shocked region, their orbits drastically change in the radial direction, and the gas falls toward the center. The gas in the shocked region effectively loses its angular momentum because the torque distribution

exerted by the bar potential is negative in the first and third quadrants (assuming that the bar major axis is located along the x-axis, the gas rotates counter-clockwise), and also because the torque is maximum at an angle of $45°$ to the bar major axis (see Fig. 13 in Wada & Habe 1995).

(3) The third type of bar-driven fueling is related to the so-called nuclear ILR (nILR). The nILR appears when there is a central mass concentration, such as a super massive black hole or a stellar/gaseous core, in a weak bar potential. For such a mass distribution, the linear resonance curve monotonically declines with radius in the region where the central mass dominates, such that $\Omega(R)-\kappa(R)/2=\Omega/2 \propto R^{-3/2}$, where κ is the epicycle frequency. Using two-dimensional, non-self-gravitating SPH (smoothed particle hydrodynamics) simulations, Fukuda, Wada, & Habe (1998) showed that if there is a nILR, offset shocks (ridges) are formed around the nILR. The offset shocks extend to the inner region and connect to a nuclear ring or to a pair of spirals. The offset spirals and ringlike structure of the molecular gas are often observed in the nuclear region of spiral galaxies, for example in IC 342 (Ishizuki et al. 1990) and NGC 4303 (Schinnerer et al. 2002). The famous "twin-peak" structure of CO found in M101, NGC 3351, and NGC 6951 (Kenney et al. 1992) probably corresponds to the inner region of the resonance-driven offset ridges and the ring.

The gas loses its angular momentum as well as energy at the trailing shocks, which are observed as ridges in the molecular gas or dust lanes. After passing through the shocks several times, the gas settles in the nuclear ring, where the orbital energy is in a minimum state for a given angular momentum. The location of the ring is typically a few times smaller than the radius of the nILR, but no further mass inflow beyond the ring is expected, provided that the self-gravity of the gas is not important. Fukuda, Habe, & Wada (2000) showed that the nuclear spirals and ring can be unstable to gravitational instability; the gas ring fragments to many dense clumps, and eventually the clumps fall to the center. Finally, a nuclear dense core, whose size is typically about 50 pc, is formed.

One should note that the offset shocks and ring can be also formed around the OILR, as originally found by Sanders & Tubbes (1980) and van Alvada (1985) (see also Athanassoula 1992; Piner, Stone, & Teuben 1995; Maciejewski et al. 2002). This is because its dynamical character is the same as the nILR. The phase shift of oval orbits near the nILR has a radial dependence similar to that for the OILR; therefore, trailing spirals are formed around both resonances (Wada 1994). This means that the observed offset ridges and rings could be the resonant structure formed around the outer ILR. This is the case when there is a stellar cusp in the galactic center, which is in fact observed in many late-type spiral galaxies (Seigar et al. 2002; Carollo 2004). Recent surveys of molecular gas in the central region of spiral galaxies also suggest that the rotation curves in the central kpc region often show a steep rise (Sofue et al. 1999; Sofue & Rubin 2001).

Considering these facts, it would be natural that the resonant structure driven by the IILR, leading spirals, is not observed, because there is no IILR without a central, rigidly rotating region*. See also Yuan, Lin, & Chen (2004) on the sensitivity of spiral patterns to rotation curves.

---

* Another reason why the leading spirals are not observed is that such spirals are dynamically unstable, and they evolve into an oval ring in a rotational time scale.

12.2.2 *Fueling Processes without ILRs*

As shown in the fueling flowchart by Shlosman et al. (1990), the mass of the gas disk, or its self-gravity, is a key ingredient in determining the fate of gas disks without ILRs. Shlosman, Frank, & Begelman (1989) and Shlosman et al. (1990) suggested the importance of the bar-mode instability of the gas core accumulated by a large-scale stellar bar. The criterion for the bar-mode instability of a rotating disk is $T_{rot}/|W| > t_{crit}$, where T_{rot} is the kinetic energy of rotation, and W is the gravitational energy. I will discuss this bars-within-bars mechanism in § 12.2.4.

Even if the gas disk/core is stable against the global bar-mode instability, the gas disk could be unstable on a local scale, because the radiative cooling is very effective in massive gas disks; thus, the disk can fragment into many clumps on the scale of the Jeans length. The clumps suffer from dynamical friction with the stellar component, and they eventually fall toward the center. Such a process was revealed by three-dimensional *N*-body and sticky-particle simulations by Shlosman & Noguchi (1993) and Heller & Shlosman (1994). They found that when the gas mass fraction is less than about 10%, the gas is channeled toward the galactic center by a growing stellar bar. For higher gas fractions, the gas becomes clumpy. Dynamical friction between gas clumps and stars also contributes to heating of the stellar system. As a result, the growth of the stellar bar is damped.

More recently, high-resolution, hydrodynamical simulations of a massive disk, taking into account self-gravity and radiative cooling below 100 K, revealed that even if the gas disk is globally stable it can be highly inhomogeneous and turbulent on a local scale (Wada & Norman 1999, 2001; Wada, Meurer, & Norman 2002). Wada & Norman (1999, 2001) presented a high-resolution numerical model of the multi-phase ISM in the central 2 kpc region of a disk galaxy with and without energy feedback from massive stars. Using $2048^2 - 4096^2$ grid cells, they found that a globally stable, multi-phase ISM is formed as a natural consequence of the non-linear evolution of thermal and gravitational instabilities in the gas disk. The surface density ranges over 7 orders of magnitude, from 10^{-1} to 10^6 $M_\odot\,pc^{-2}$, and the temperature extends over 5 decades, from 10 to 10^6 K. They also find that, in spite of its very complicated spatial structure, the multi-phase ISM exhibits a one-point probability density function that is a perfect log-normal distribution over 4 decades in density. The log-normal probability density function is very robust even in regions with frequent bursts of supernovae. The radial profile of the turbulent disk changes to a steeper one in a time scale of $\sim 10^8$ yr (Fig. 11 in Wada & Norman 2001). The mass inflow is caused by turbulent viscosity (e.g., Lynden-Bell & Pringle 1974).

The turbulent nature of the self-gravitating gas disk was studied in detail by Wada et al. (2002). They found that the velocity field of the disk in the non-linear phase shows a steady power-law energy spectrum over 3 orders of magnitude in wave number. This implies that the random velocity field can be modeled as fully developed, stationary turbulence. Gravitational and thermal instabilities under the influence of galactic rotation contribute to form the turbulent velocity field. The effective Toomre Q parameter, in the non-linear phase, exhibits a wide range of values, and gravitationally stable and unstable regions are distributed in a patchy manner in the disk. These results suggest that large-scale galactic rotation coupled with the self-gravity of the gas can be the ultimate energy source that maintains the turbulence in the local ISM. Therefore, mass inflow is naturally expected in a rotating gas disk. We just need self-gravity of the dense gas and radiative cooling for fueling.

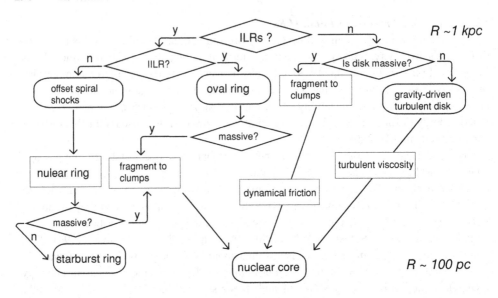

Fig. 12.1. A flowchart of various fueling mechanisms from $R \approx 1$ kpc to 100 pc, which is a partial revision from Shlosman et al. (1990), considering numerical results in the last decade. Shlosman et al. proposed the "bars-within-bars mechanism" to connect the scales from $R \approx 100$ pc to 10 pc.

12.2.3   A Revised Fueling Flowchart

In summary, the fueling flowchart from $R \approx 1$ kpc to 100 pc can be revised as as shown in Figure 12.1*. This shows that there are many possible fueling paths for the $R \approx 100$ pc scale. Dynamical resonances are the key in this diagram, since the redistribution of the gas components seems to be sensitive to them. However, some remarks are necessary. The effects of resonances on the gaseous orbits cannot be ignored, even if there are *no* ILRs, namely when $\Omega_p > \max(\Omega - \kappa/2)$. In this sense, the linear condition of the resonances is not a strict criterion for the response of the gas to the bar potential (see details in Wada 1994). Another point to keep in mind is that the $\Omega - \kappa/2$ diagrams determined from observations can have large errors. Both Ω and κ are determined from rotation curves, but the rotational velocity, especially in the central region of galaxies, is not determined accurately, because of the low spatial resolution and non-circular motion of the gas. One should be especially careful when rotation curves are obtained from position-velocity diagrams of molecular gas in galaxies (Takamiya & Sofue 2002; Koda & Wada 2002). Finally, gas dynamics in a live stellar bar, where there is a back reaction from the gas to the stellar system, could be more complex than that expected from gas dynamics in a fixed bar potential (e.g., Heller & Shlosman 1994).

---

* Note that this flowchart is not complete. For example, Maciejewski et al. (2002) proposed a fueling mechanism due to nuclear spiral shocks in a non-self-gravitating gas disk with a high sound speed. There should be many other factors that determine the fueling processes. One should also realize that the diagram is rather qualitative. The condition "massive or not" depends on the temperature and velocity dispersion of the gas, and it is not a simple function of the gas mass fraction relative to the dynamical mass. All paths are theoretically possible, but it does not say which path is more probable in real galaxies. The time scale for each path is also different. Another point one should keep in mind is that gaseous response to the resonances is not "discrete." Because of the dissipational nature of the gas, resonant structures, such as spirals, can be formed even if the linear resonance condition (e.g., $\Omega_p = \Omega - \kappa/2$) is *not* strictly satisfied (Wada 1994).

The mechanisms described in Figure 12.1 are not the only ones pertaining to angular momentum transfer. Other mechanisms are possible, for example spiral density waves (Goldreich & Lynden-Bell 1965a,b; Lynden-Bell & Kalnajis 1972; Emsellem 2004), galactic shocks (Fujimoto 1968; Roberts 1969), minor mergers (Hernquist & Mihos 1995; Taniguchi & Wada 1996), and the rotational-magnetic instability (Sellwood & Balbus 1999).

By means of one or a combination of these mechanisms, dense gas cores at $R \approx 100$ pc in the galactic center can be formed on a dynamical time scale ($\sim 10^{7-8}$ yr). This seems to be consistent with the fact that many spiral galaxies are molecular gas rich in the central region (Sakamoto et al. 1999). Therefore, a real issue for the fueling problem and AGN activity is *inside* 100 pc. Recall the original fueling flowchart. The only channel toward the galactic center is the "bars-within-bars" mechanism, which was proposed by Shlosman et al. (1989). In the next section, I will review this idea briefly.

12.2.4 *Remarks on the Bars-within-bars Mechanism*

Shlosman et al. (1989) proposed a novel idea, the bars-within-bars mechanism, for fueling AGNs. They used the criterion for bar instability, namely that when the ratio between rotational energy and gravitational energy, $t_{crit} \equiv T_{rot}/|W|$, is larger than some critical value, typically 0.14 for an N-body system or 0.26 for an incompressible gas sphere. They rewrote the criterion as $a_{star}/a_{gas} > C(t_{crit})/g^2$, where a_{star} is the scale length of the stellar system, a_{gas} is the core radius of the gas, $C(t_{crit})$ is a function of t_{crit}, and g is the gas mass fraction relative to the total mass. Using this criterion, Shlosman et al. answered the question: Given g and t_{crit}, how much does the gas disk shrink radially to become unstable? For example, for $g = 0.2$ and $t_{crit} = 0.14$, the disk becomes bar unstable when the gas disk shrinks to 1/10 of the size of the stellar core. This means that if the large-scale stellar bar sweeps the gas inward to about 1/10 of the bar size ($\sim$ a core radius of the stellar disk), then the accumulated gas disk becomes bar unstable. They claimed that the resulting inflow can extend all the way into the inner $\sim$10 pc.

Their argument on the bar instability is reasonable, but they did not actually provide a quantitative discussion on the "resulting inflow." Once the core becomes bar unstable, they expect that some fraction of the gas will falls toward the central region, where viscosity-driven flow would dominate further inflow, i.e. toward $R \approx 10$ pc. However, this has not been theoretically or numerically proven. The redistribution of the mass due to the bar instability can drive only a very small part of the gas into 1/10 of the initial radius. For example, the hydrodynamical simulations of rotating gas spheres by Smith, Houser, & Centrella (1996) show a process of recurrent bar instabilities. As a result of the first bar instability, redistribution of the gas takes place; outer spirals are formed, which transfer angular momentum outward. The second and subsequent instabilities are much weaker than the first one; therefore, the resultant mass distribution is not very different from the initial one. This is because, in order to complete the loop, a large part of the angular momentum must be transferred outward with a small fraction of the mass; otherwise the fraction g becomes too small. Therefore, the gas disk must shrink to a very small radius to satisfy the instability criterion again.

A serious problem here is how to remove a large part of the angular momentum. This would be impossible without external mechanisms of transferring the angular momentum, such as a secondary bar. In this sense, the bars-within-bars mechanism does not solve the

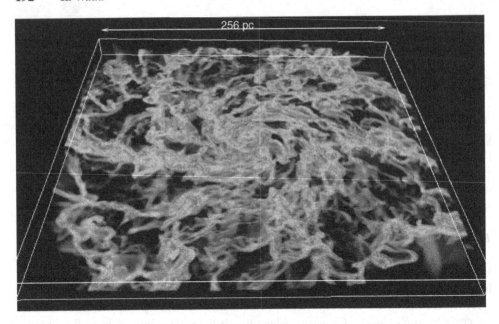

256 pc

Fig. 12.2. Three-dimensional density structure of the ISM in the central region of a galaxy (Wada 2001).

angular momentum problem.* A nested stellar bar on a small scale could work to remove angular momentum of the bar-destabilized gas. However, as mentioned in § 12.2.2, a stellar system can be dynamically heated up due to the interaction between the gas components and the stars, and stellar bars in the nuclear region, where it is especially affected by the dissipation of the gas, can be dissolved when the gas mass fraction is large enough.

12.3   Gas Dynamics in the Inner 100 pc

In order to understand the fueling process from $R \approx 100$ pc to the circumnuclear region, we have to understand, at least, the three-dimensional structure of the ISM around the central massive black hole at sub-pc resolution. Instead of adopting a phenomenological approach for the multi-phase ISM (e.g., Ikeuchi, Habe, & Tanaka 1984; Combes & Gerin 1985), Wada & Norman (2001) tried to obtain relevant numerical models of the ISM by solving time-dependent, non-linear hydrodynamical equations and the Poisson equation that govern the dynamics and structure of the ISM, using a high-accuracy Euler mesh code. They used the advection-upstream splitting method (AUSM; Liou & Steffen 1993), and they achieved third-order spatial accuracy with MUSCL. AUSM is an improvement of the flux-vector splitting scheme, where the advection and pressure terms are separately split at a cell surface. (See details in Wada & Norman 2001.)

Wada & Norman solve the mass, momentum, and energy equations and the Poisson equation numerically in three dimensions to simulate the evolution of a rotating gas disk in a fixed, spherical gravitational potential. The potential of the stars, dark matter, and a

---

* The term "bars-within-bars" is used differently in the literature. In the original Shlosman et al. (1989) paper the term actually means "gas bars-within-stellar bars." Recently many authors use the term to describe "stellar bars-within-stellar bars." See also a recent review paper by Maciejewski (2004a, 2004b) on this subject.

supermassive black hole are assumed to be time independent (no feedback from the gas is considered). The main heating source is supernova explosions. Instead of assuming a "heating efficiency" to evaluate the dynamical effect of the blast waves on the ISM, we explicitly follow the evolution of the blast waves in the inhomogeneous ISM with sub-pc spatial resolution. Radiative cooling is considered not only for the hot gas, but also for gas below 10^4 K, because the pc-scale fine structure of the ISM is mainly determined by cold ($T_g < 100$ K), dense media. Such fine structure is especially important for the fueling processes in the central 100 pc region. We assume a cooling function with solar metallicity for the temperature range between 20 K and 10^8 K.

Figure 12.2 shows the quasi-stable density field of the three-dimensional disk model in a central 256 pc × 256 pc region without a central massive black hole. The plot shows the volume-rendering representation of density, and the greyscale represents relative opacity. As in the two-dimensional models (e.g., Fig. 12 in Wada & Norman 2001), the disk shows a tangled network of many filaments and clumps. Those filaments are formed mainly through tidal interactions between dense clumps. The gas clumps formed by gravitational instability are not rigid bullets; hence, close encounters between them cause tidal tails, and the tails are stretched due to the galactic rotation and local shear motion. The clumps and filaments collide with each other, and this causes the complicated networks. Moreover, supernova explosions are assumed to occur in the model, and their blast waves blow the gas up from the disk plane, enhancing the inhomogeneity.

For the gas around the AGN, the gravitational potential exerted from the central massive black hole gives the structure seen in Figure 12.3, which displays cross sections of the torus. The disk has a complicated internal structure. In this model, an average supernova rate is assumed to be about 1 supernova per year. This corresponds to a star formation rate of $\sim$100 $M_\odot$ yr^{-1}, which would be too high for typical Seyfert 2s with starbursts (Cid Fernandes et al. 2001; Heckman 2004). Note, however, that the scale height of the disk is roughly proportional to $(\text{SFR})^{1/2}r^{1.5}$, as discussed later; therefore, the geometry seen in Figure 12.3 is almost the same for a disk with 2 times larger radius and 1/10 of the star formation rate.

Time evolution of the "torus" shows that the internal motion is not steady, but the global concave geometry is supported by internal turbulence caused by supernova explosions. In Wada & Norman (2002), this is explained using a simple analytic argument, in which the scale height of the thick disk is determined by the balance between the turbulent energy dissipation and the energy feedback from the supernova explosions under the effect of the gravitational potential of the supermassive black hole and the stellar disk. Assuming hydrostatic equilibrium in the vertical direction, we find that the radial dependence of the scale height, $h(r)$, is proportional to $r^{3/2}$ in the region where the black hole potential dominates ($r < r_0$), while it changes as r^{-1} for the outer region ($r > r_0$). Namely,

$$h_1(r) = h_{0,1}(r_0)(r/r_0)^{3/2}, \tag{12.1}$$
$$h_2(r) = h_{0,2}(r_0)(r/r_0)^{-1} \tag{12.2}$$

where $r_0 \approx 60(M_\bullet/10^8 M_\odot)^{1/2}$ pc for a stellar surface density of $10^4 M_\odot$ pc^{-2}, and

$$h_{0,1}(r_0) \approx 35\,\text{SFR}_1^{1/2}r_6^{3/2}M_{g,8}^{-1/2} \text{ pc} \tag{12.3}$$

and

$$h_{0,2}(r_0) \approx 1\,\text{SFR}_1^{2}r_6^{3}M_{g,8}^{-1} \text{ pc}, \tag{12.4}$$

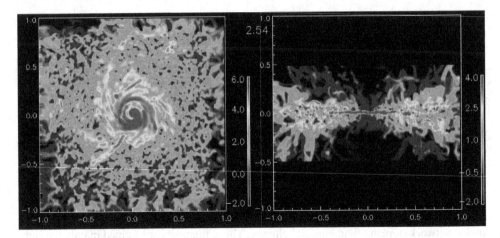

Fig. 12.3. Cross sections of density distribution of the gas disk around a central massive black hole. The boxes are 64 pc across. The greyscale represents log-scaled density ($M_\odot$ pc^{-3}).

where the total gas mass $M_{g,8} \equiv M_g/10^8 M_\odot$, the star formation rate SFR$_1 \equiv 1 M_\odot$ yr^{-1}, and $r_6 \equiv r/60$ pc. The scale heights, $h(r)$, for three star formation rates and for two black hole masses are plotted in Figure 12.4. The solutions have three domains: (I) a stable disk region ($r < 2-5$ pc), (II) a flared disk region [$h(r) \propto r^{3/2}$, $5 < r < 40-60$ pc], and (III) a region where $h(r) \propto r^{-1}$. For a less massive central black hole and larger star formation rate, the disks become thicker. Domain III is more sensitive to the energy input than Domain II; therefore, the scale height of the torus is larger than 1 kpc for SFR$_1 = 100$, which means a "galactic-wind-like" solution. In Domain I, the disk does not fragment to clumps because of the strong shear; thus, no star formation is expected and the disk should be very thin.

The column density toward the nucleus as a function of the viewing angle is plotted in Figure 12.5. A viewing angle of 90° is edge-on. It shows that the viewing angle should be less than about $\pm 40°$ from edge-on to have a large column density ($> 10^{23}$ cm^{-2}), which is suggested in some Seyfert 2s with nuclear starbursts (Levenson, Weaver, & Heckman 2001). However, one should note here that since the internal structure of the torus is very inhomogeneous, the column density for the torus is not a simple function of the viewing angle. There is a large dispersion, ~ 2 orders of magnitude, in the column density for a given viewing angle.

The average mass accretion rate for the $R < 1$ pc region in the above-mentioned model is 0.3 $M_\odot$ yr^{-1}, and it is about twice as large as that for the model without energy feedback. Yamada (1994) suggests a positive correlation between X-ray and CO luminosity in Seyfert galaxies and quasars. If the X-ray and CO luminosity correlate with the mass accretion rate and star formation rate, respectively, the model is qualitatively consistent with the observations. In Figure 12.6, the gas accretion rate to the nucleus is shown. An important point here is that the mass accretion to the nuclear region should not be a steady flow. As seen in the plot, the accretion rate is highly time dependent, with fluctuations over 3 orders of magnitude between 10 and 0.01 $M_\odot$ yr^{-1}. The time scale of the fluctuation is $\sim 10^4 - 10^5$ yr. This is a consequence of the ISM in the inner 100 pc being inhomogeneous.

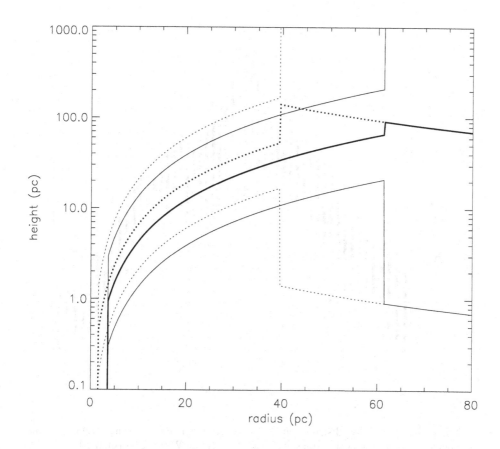

Fig. 12.4. Scale height of the disk around a supermassive black hole for three star formation rates (1, 10, 100 $M_\odot$ yr^{-1}) and two black hole mass (solid lines: $M_\bullet = 1.2 \times 10^8 M_\odot$; dotted lines : $M_\bullet = 5 \times 10^7 M_\odot$). The sharp transitions at $R = 40$ and 60 pc represent the boundaries of Domain II and Domain III. These transitions are expected to be smoother for a realistic mass distribution.

The mass accretion is caused by the kinematic viscosity of the turbulent velocity field, which is maintained by gravitational instability and galactic rotation (the same mechanism mentioned in § 12.2.2); the energy feedback from the supernovae enhances the viscosity. The non-steady inflow for a pc-region would be an important feature of accretion around a supermassive black hole.

Suppose that AGNs generally have circumnuclear gas disks with sizes of several tens pc and that the global structure of the disk is determined by the above-mentioned mechanism. Various types of the AGNs then could be schematically segregated on a plot with three axes, as shown in Figure 12.7. The axes are the total gas mass (or surface density of the gas), the black hole mass, and the star formation rate. Type 2 Seyferts with starbursts would be gas-rich and their star formation rate is high, but the black hole mass would be relatively small. Therefore, the scale height of the disk is large, as shown in Figure 12.3. On the other hand, the gas disks of Seyfert 1s would be thinner than those of Seyfert 2 with starbursts

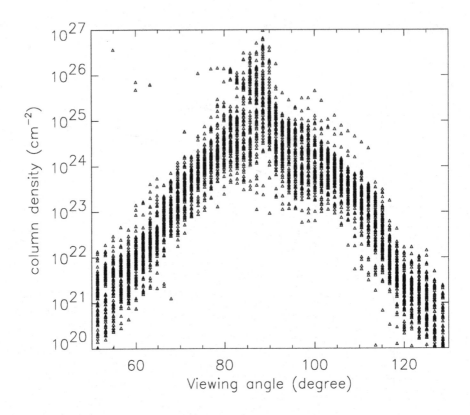

Fig. 12.5. Column density distribution as a function of the viewing angle for the inhomogeneous gas around a supermassive black hole (from Wada & Norman 2002).

because of a lower star formation rate and/or small gas mass. Narrow-line Seyfert 1s might have relatively small black holes. Quasars would have more massive central black holes and a small star formation rate, as a result of which their circumnuclear disks would be very thin, even if they are gas rich. Hence, most quasars are observed as type 1, and type 2 quasars would be observed by chance, only when the gas disks are edge-on. However, there might be a counterpart of type 2 Seyferts/starbursts in the quasar family, in which the nucleus is obscured by the inhomogeneous, dusty thick disk with active star formation. Such obscured quasars might be the sources of the X-ray background radiation (Fabian 2004).

Finally, I should mention an important physical process that has not yet been taken into account for the gas dynamics in the central 100 pc region. Ohsuga & Umemura (2001) explored the formation of dusty gas walls induced by a circumnuclear starburst around an AGN. They found that the radiation force of the circumnuclear starburst works to stabilize optically thick walls surrounding the nucleus. It would be interesting to study the effect of the radiation pressure on the dust in the inhomogeneous, turbulent media found in the simulations discussed above. Another interesting feature to study are the effects of the ionizing radiation from the AGN and the starburst regions. So far, the UV radiation field has been assumed to be uniform, but apparently this is incorrect in the inhomogeneous ISM

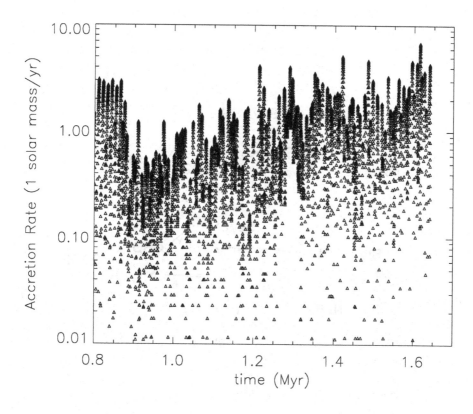

Fig. 12.6. Mass accretion rate for $R < 1$ pc from the model shown in § 12.3.

around an AGN. Solving the radiation field correctly is necessary to compare models with the abundant available observational information on the ionized gas in AGNs. Another observational quantity that could be derived from the simulations is the molecular line intensity. Work is now in progress to incorporate three-dimensional non-LTE radiative transfer for various molecular lines for the molecular tori.

12.4   Summary: Do We Need Triggers for Nuclear Activity?

It has been argued that the fueling problem is essentially equivalent to the questions of how we can remove the large angular momentum of the gas in a galactic disk and how to bring the gas into the small-scale region ($R \approx 1$ pc) during the lifetime of an AGN. Non-axisymmetric perturbation of the gravitational potential, such as that due to stellar bars and companions, has been considered as the most plausible mechanism. However, this conventional picture should be reconsidered.

The majority of recent observations have not shown clear excess of bars or companions in galaxies with AGNs (e.g., Ho, Filippenko, & Sargent 1997, 2003; Mulchaey & Regan 1997; Corbin 2000; Schmitt 2001). Moreover, recent observations of the circumnuclear regions ($R < 100$ pc) by the *Hubble Space Telescope* suggest that no significant differences are found in the structure of the nuclear dust lanes between active and inactive galaxies

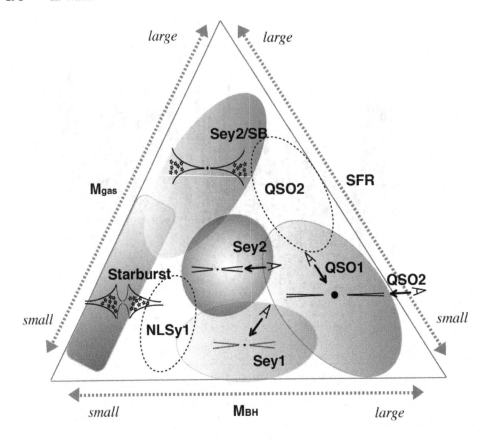

Fig. 12.7. Segregation of various types of AGNs from the point of view of the circumnuclear gas disk.

(Martini et al. 2003a,b). Some research groups claim, on the contrary, that bars are more abundant in Seyfert hosts than non-Seyfert galaxies (Knapen, Shlosman, & Peletier 2000; Laine et al. 2002). Although this is controversial, one should note that the correlation between bars and AGNs is *weak* in any results. For example, Laine et al. (2002) found that 73% (41 out of 56) of Seyfert hosts are barred, while 50% (28 out of 56) are barred in the control sample. In their small sample, the difference between 73% and 50% is statistically subtle. The result also shows that at least 50% of non-barred galaxies host AGNs, and a natural interpretation of this seems to be that AGNs are independent of bars. There is another example: Sakamoto et al. (1999) found that the concentration factors of the CO, $t_{con} \equiv \Sigma_{gas}(R < 500\,\mathrm{pc})/\Sigma_{gas}(R < R_{25})$, where Σ_{gas} is the average surface density of the molecular gas, for SB+SAB galaxies are distributed between 20 and 300, while they are between 10 and 70 for non-barred galaxies (NGC 4414 has an exceptionally low $t_{con} = 0.9$, because CO forms a ring in the galaxy). On average, the degree of gas concentration in the central kpc is higher in barred systems than in unbarred galaxies. One should note again, however, that *more than half* of barred and non-barred galaxies have the same range of the concentration factor, $t_{con} \approx 20-70$. Of course, these samples are still too small to produce a statistically reliable conclusion, but this can be interpreted as indicating that bars have only a weak effect in the concentration of the molecular gas.

Suppose the *weak* correlation between large-scale perturbation and nuclear activity is true. How can we explain this theoretically? One plausible interpretation is that bars or companions are one necessary condition for nuclear activity, but that other conditions related to the pattern speed or strength of the bars, gas mass in a certain radius, or secondary bars (Maciejewski & Sparke 2000) need to be satisfied at the same time in order to trigger gas accretion onto the supermassive black hole. For example, the fraction of the gas mass relative to the dynamical mass may need to exceed ~ 0.1 for fueling through the collapse of the ILR ring (Wada & Habe 1992). The fueling process also depends on the pattern speed of the bar and/or the central rotation curves. Since the ILR ring evolved from leading spirals near the IILR, the ring is not formed either for bars with too fast pattern speeds [i.e., $\Omega_p > \max(\Omega - \kappa/2)$] or for a stellar potential with a central cusp. As mentioned in § 12.2.1, recent studies suggest that most spiral galaxies have a central mass concentration rather than a flat core in the central kpc region. If this is the case, an OILR or nILR is expected. Therefore, the trailing spiral shocks or offset ridges are the probable structures expected in the central parts of barred galaxies. In this case, gas fueling for the nucleus can be stopped at a certain radius inside the OILR or nILR, and its radius depends on the amount of angular momentum and energy loss at the trailing shocks*. Since gravitational instability of the ring is a trigger for further inflow, in this case the gas mass, the strength of the bar, and the pattern speed of the bar are the additional factors needed for the onset of fueling. Fukuda et al. (2000) also suggested that energy feedback from supernovae in the ring enhances gas accretion to the nucleus. As mentioned in § 12.3, energy feedback from star formation in the nuclear region is another important factor in determining the mass accretion rate.

Besides the gravitational torque from a non-axisymmetric potential, dynamical friction between the gas clumps and the stellar system (Shlosman & Noguchi 1993), the viscosity due to clump-clump collisions (Ozernoy, Fridman, & Biermann 1998; Begelman, Frank, & Shlosman 1989), gravity-driven turbulence (Lynden-Bell & Pringle 1974; Paczyński 1978; Wada et al. 2002) and supernova-driven turbulence (von Linden et al. 1993) are important fueling mechanisms that do not involve bars (see Fig. 12.1). In addition to these, radiative avalanche (Umemura, Fukue, & Mineshige 1998; Kawakatsu, Umemura, & Mori 2004; Umemura 2004), magneto-hydrodynamical turbulence (Balbus & Hawley 1991; for a recent three-dimensional global magneto-hydrodynamical simulation of an accretion disk, see Machida, Hayashi, & Matsumoto 2000 and Machida & Matsumoto 2003), and spiral density waves (Goldreich & Lynden-Bell 1965a,b; Lynden-Bell & Kalnajis 1972) may also be important. In these mechanisms, the onset of mass accretion is controlled by many factors, such as the structure of the ISM, stellar mass density, star formation rate, the initial mass function, the dust-to-gas ratio, strength of the magnetic field, ionization fraction, the total gas mass, etc. In light of these various factors for the onset of gas accretion, a stellar bar may not even be a necessary condition for gas fueling.

Finally, I would like to emphasize two important *scales* concerning the fueling problem: the spatial scale and the time scale. AGNs are powered by mass accretion onto supermassive black holes, on scales of $R \approx 10^{-5}$ pc. In order to explain the enormous luminosity of AGNs, we should treat the accretion phenomena on small scales. However, most studies of gas

---

* Maciejewski et al. (2002) showed that for high sound speed a spiral shock propagates to the center, which may be responsible for the fueling.

fueling triggered by galactic-scale phenomena, such as bars, mergers, and interactions, are focused on gas dynamics in regions 100–1000 pc from the galactic center. Apparently, mass accretion to such a region does not mean that the accumulated gas can fall all the way to the accretion disk. In § 12.3, I have shown some new results on the gas dynamics between these two regimes, on scales 1–100 pc. We do not yet have enough observational information on this scale. For example, the structure and dynamics of the molecular gas in the galactic central region have been explored by radio interferometers with $\sim 1''$ resolution, but this is not fine enough to resolve the inhomogeneous structure of the ISM, expected in the numerical simulations, even for nearby galaxies (Wada & Koda 2001). The next generation radio interferometer ALMA (Atacama Large Millimeter/submillimeter Array) will be the instrument for exploring the "missing link" in the fueling problem. The $0.''01$ resolution achievable by ALMA can reveal the sub-pc structure of the molecular gas in the central 100 pc of nearby galaxies (e.g. those in the Virgo cluster).

Time variability is another important feature of AGNs. The average lifetime of AGN activity is $\sim 10^7$ yr (Martini 2004). This does not mean, however, that the mass accretion (e.g., $1 M_\odot$ yr^{-1}), is constant during this lifetime. As mentioned in § 12.2.2, we expect that the turbulence is self-regulated in a dense gas disk (Wada et al. 2002), and the turbulence driven by self-gravity in an inhomogeneous ISM causes stochastic mass accretion. This would be very important as an outer boundary condition for the accretion disk around the supermassive black hole. Our results suggest that the time scale of non-steady mass accretion is $\sim 10^{4-5}$ yr, which would be much shorter than the lifetime of AGNs. If this is the case, any galaxies with a massive gas core and supermassive black hole can be active, and they might be recognized as luminous AGNs only for that short period. Observations with ALMA are again essential to reveal the kinematics of the ISM in the central 100 pc, and to couple this information to the evolution of AGNs.

Acknowledgements. I would like to thank Colin A. Norman, Asao Habe, and Jin Koda for our collaboration. I am also grateful to Luis Ho for organizing this fruitful conference, and to Witold Maciejewski for his valuable comments on the draft. Our numerical simulations were performed on the supercomputer system in the Astronomical Data Analysis Center, National Astronomical Observatory of Japan.

References

Athanassoula, E. 1992, MNRAS, 259, 345

Balbus, S. A., & Hawley, J. F. 1991, ApJ, 376, 214

Barnes, J. E., & Hernquist, L. 1996, ApJ, 471, 115

Begelman, M. C., Frank, J., & Shlosman, I. 1989, in Theory of Accretion Disks, ed. F. Meyer (Dordrecht: Kluwer), 373

Carollo, C. M. 2004, in Carnegie Observatories Astrophysics Series, Vol. 1: Coevolution of Black Holes and Galaxies, ed. L. C. Ho (Cambridge: Cambridge Univ. Press)

Cid Fernandes, R., Jr., Heckman, T. M., Schmitt, H. R., Golzález Delgado, R. M., & Storchi-Bergmann, T. 2001, ApJ, 558, 81

Corbin, M. R. 2000, ApJ, 536, L73

Combes, F., & Gerin, M. 1985, A&A, 150, 327

Emsellem, E. 2004, in Coevolution of Black Holes and Galaxies, ed. L. C. Ho (Pasadena: Carnegie Observatories, http://www.ociw.edu/ociw/symposia/series/symposium1/proceedings.html)

Fabian, A. C. 2004, in Carnegie Observatories Astrophysics Series, Vol. 1: Coevolution of Black Holes and Galaxies, ed. L. C. Ho (Cambridge: Cambridge Univ. Press), in press

Friedli, D., & Benz, W. 1993, A&A, 268, 65

Fujimoto, M., in IAU Symp. 29, Non-stable Phenomena in Galaxies (Yerevan: The Publishing House of the Academy of Sciences of Armenian SSR), 453

Fukuda, H., Habe, A., & Wada, K. 2000, ApJ, 529, 109

Fukuda, H. , Wada, K., & Habe, A. 1998, MNRAS, 295, 463

Fukunaga, M., & Tosa, M. 1991, PASJ, 43, 469

Goldreich, P., & Lynden-Bell, D. 1965a, MNRAS, 130, 97

——. 1965b, MNRAS, 130, 125

Heckman, T. M. 2004, in Carnegie Observatories Astrophysics Series, Vol. 1: Coevolution of Black Holes and Galaxies, ed. L. C. Ho (Cambridge: Cambridge Univ. Press), in press

Heller, C. H., & Shlosman, I. 1994, ApJ, 424, 84

Hernquist, L., & Mihos, J. C. 1995, ApJ, 448, 41

Ho, L. C., Filippenko, A. V., & Sargent, W. L. W. 1997, ApJ, 487, 591

——. 2003, ApJ, 583, 159

Ikeuchi, S., Habe, A., & Tanaka, Y. D. 1984, MNRAS, 207, 909

Ishizuki, S., Kawabe, R., Ishiguro, M., Okumura, S. K., & Morita, K. 1990, Nature, 344, 224

Kawakatsu, N., Umemura, M. & Mori, M. 2004, in Coevolution of Black Holes and Galaxies, ed. L. C. Ho (Pasadena: Carnegie Observatories, http://www.ociw.edu/ociw/symposia/series/symposium1/proceedings.html)

Kenney, J. D. P., Wilson, C. D., Scoville, N. Z., Devereux, N. A., & Young, J. S. 1992, ApJ, 395, L79

Knapen, J. H., Shlosman, I., & Peletier, R. F. 2000, ApJ, 529, 93

Koda, J., & Wada, K. 2002, A&A, 396, 867

Laine, S., Shlosman, I., Knapen, J. H., & Peletier, R. F. 2002, ApJ, 567, 97

Levenson, N. A., Weaver, K. A., & Heckman, T. M. 2001, ApJ, 550, 230

Liou, M.-S., & Steffen, C. J., Jr. 1993, J. Comp. Phys., 107, 23

Lynden-Bell, D., & Kalnajs, A. J. 1972, MNRAS, 157, 1

Lynden-Bell, D., & Pringle, J. E. 1974, MNRAS, 168, 603

Machida, M., Hayashi, M. R., & Matsumoto, R. 2000, ApJ, 532, L67

Machida, M., & Matsumoto, R. 2003, ApJ, 585, 429

Maciejewski, W. 2004a, in Galactic Dynamics, ed. C. Boily et al. (EDP Sciences), in press (astro-ph/0302250)

——. 2004b, in Coevolution of Black Holes and Galaxies, ed. L. C. Ho (Pasadena: Carnegie Observatories, http://www.ociw.edu/ociw/symposia/series/symposium1/proceedings.html)

Maciejewski, W., & Sparke, L. S. 2000, MNRAS, 313, 745

Maciejewski, W., Teuben, P. J., Sparke, L. S., & Stone, J. M. 2002, MNRAS, 329, 502

Martini, P. 2004, in Carnegie Observatories Astrophysics Series, Vol. 1: Coevolution of Black Holes and Galaxies, ed. L. C. Ho (Cambridge: Cambridge Univ. Press), in press

Martini, P., Regan, M. W., Mulchaey, J. S., & Pogge, R. W. 2003a, ApJS, 146, 353

——. 2003b, ApJ, 589, 774

Mulchaey, J. S., & Regan, M. W. 1997, ApJ, 482, L135

Ohsuga, K., & Umemura, M. 2001, A&A, 371, 890

Ozernoy, L. M., Fridman, A. M., & Biermann, P. L. 1998, A&A, 337, 105

Paczyński, B. 1978, Acta Astron., 28, 91

Piner, B. G., Stone, J. M., & Teuben, P. J. 1995, ApJ, 449, 508

Roberts, W. W. 1969, ApJ, 158, 123

Sakamoto, K., Okumura, S. K., Ishizuki, S., & Scoville, N. Z. 1999, ApJS, 124, 403

Sanders, R. H., & Tubbs, A. D. 1980, ApJ, 235, 803

Schinnerer, E., Maciejewski, W., Scoville, N. Z., & Moustakas, L. A. 2002, ApJ, 575, 826

Schmitt, H. R. 2001, AJ, 122, 2243

Seigar, M., Carollo, C. M., Stiavelli, M., de Zeeuw, P. T., & Dejonghe, H. 2002, AJ, 123, 184

Sellwood, J. A., & Balbus, S. A. 1999, ApJ, 511, 660

Shlosman, I. 1994, ed., Mass Transfer Induced Activity in Galaxies (Cambridge: Cambridge Univ. Press)

Shlosman, I., Begelman, M. C., Frank, J. 1990, Nature, 345, 679

Shlosman, I., Frank, J., & Begelman, M. C. 1989, Nature, 338, 45

Shlosman, I., & Noguchi, M. 1993, ApJ, 414, 474

Smith, S. C., Houser, J. L., & Centrella, J. M. 1996, ApJ, 458, 236

Sofue, Y., & Rubin, V. C. 2001, ARA&A, 39, 137

Sofue, Y., Tutui, Y., Honma, M., Tomita, A., Takamiya, T., Koda, J., & Takeda, Y. 1999, ApJ, 523, 136

Takamiya, T., & Sofue, Y. 2002, ApJ, 576, L15

Taniguchi, Y., & Wada, K. 1996, ApJ, 469, 581

Umemura, M., 2004, in Coevolution of Black Holes and Galaxies, ed. L. C. Ho (Pasadena: Carnegie
    Observatories, http://www.ociw.edu/ociw/symposia/series/symposium1/proceedings.html)
Umemura, M., Fukue, J., & Mineshige, S. 1998, MNRAS, 299, 1123
van Albada, G. D. 1985, A&A, 142, 491
von Linden, S., Biermann, P. L., Duschl, W. J., Lesch, H., & Schmutzler, T. 1993, A&A, 280, 468
Wada, K. 1994, PASJ, 46, 165
——. 2001, ApJ, 559, L41
Wada, K., & Habe, A. 1992, MNRAS, 258, 82
——. 1995, MNRAS, 277, 433
Wada, K., & Koda, J. 2001, PASJ, 53, 1163
Wada, K., Meurer, G., & Norman, C. A. 2002, ApJ, 577, 197
Wada, K., & Norman, C. A., 1999, ApJ, 516, L13
——. 2001, ApJ, 546, 172
——. 2002, ApJ, 566, L21
Yamada, T. 1994, ApJ, 423, L27
Yuan, C., Lin, L.-H., & Chen Y.-H. 2004, in Coevolution of Black Holes and Galaxies, ed. L. C. Ho (Pasadena:
    Carnegie Observatories, http://www.ociw.edu/ociw/symposia/series/symposium1/proceedings.html)

13

The AGN-disk dynamics connection

J. A. SELLWOOD and JUNTAI SHEN
Rutgers University

Abstract

Any connection between central activity and the large-scale dynamics of disk galaxies requires an efficient mechanism to remove angular momentum from the orbiting material. The only viable means of achieving inflow from kiloparsec scales is through gravitational stresses created by bars and/or mergers. The inflow of gas in bars today appears to stall at a radius of few hundred parsec, however, forming a nuclear ring. Here we suggest that bars in the early Universe may have avoided this problem, and propose that the progenitors of central supermassive black holes (SMBHs) are created by gas that is driven deep into the centers of galaxies by bars in the early stages of disk formation. The coincidence of the QSO epoch with galaxy formation, the short lifetimes of QSOs, and the existence of SMBHs in the centers of most bright galaxies are all naturally accounted for by disk dynamics in this model. The progenitor SMBHs are the seeds for brighter QSO flares during galaxy mergers. We present a new study of bar weakening by central mass concentrations, which shows that bars are less easily destroyed than previously thought. An extremely massive and compact central mass can, however, dissolve the bar, which creates a pseudo-bulge component in the center of the disk.

13.1 Introduction

It now seems that the masses of supermassive black holes (SMBHs) in the centers of galaxies are strongly correlated with the larger scale dynamical properties of their host galaxies (Kormendy & Richstone 1995; Gebhardt et al. 2000; Ferrarese & Merritt 2000). Yet an utterly insignificant fraction of the mass of a galaxy had low enough angular momentum to create, or accrete directly onto, a SMBH in its center. Why the mass of the SMBH should be so closely related to the properties of its host galaxy is still an open question.

Material orbiting at a galacto-centric radius of a few kiloparsecs, where most of the baryonic galaxy mass resides, must have its angular momentum reduced by several orders of magnitude before it becomes of any relevance to nuclear phenomena. Thus, any connection between galaxy dynamics and nuclear activity requires a mechanism to remove enough angular momentum from the gas to enable it to accrete onto the SMBH. Viscous processes are too slow for gas to sink from orbits at large radii to small, even when augmented by magnetohydrodynamic instabilities (Sellwood & Balbus 1999), and significant radial migration requires gravitational torques. Spiral waves are weak, and generally do more churning of the gas than radial transportation (Sellwood & Binney 2002). Attention has

© The Observatories of the Carnegie Institution of Washington 2004.

therefore focused on the gravitational influence of the strongest non-axisymmetric features: bars in isolated systems and tides during mergers.

Since bars in galaxies today can reduce the angular momentum of gas in the disk by little more than one order of magnitude, other processes would be needed to drive gas originating in the main disk of the galaxy close enough to the nucleus to accrete onto it (see, e.g., Wada 2004). However, the removal of angular momentum by bars could have been somewhat more efficient as galaxies first formed, and we propose a possible connection between disk dynamics and early QSO activity.

It has often been noted that bright galaxies were assembled at about the same time that QSOs flare (e.g., Rees 1997), suggesting a causal connection. In fact, the luminosity function of X-ray selected AGNs seems to track the star formation history of the Universe remarkably closely (Franceschini et al. 1999). Since QSOs are believed to reside in the centers of galaxies (e.g., Bahcall et al. 1997; McLure et al. 1999), it is natural to suppose that they formed there. Many bright galaxies in the local Universe appear to host quiescent SMBHs which are assumed to be the fuel-starved engines of earlier QSO activity (Yu & Tremaine 2002; Ferrarese 2002).

Thus, a convincing model for the formation and evolution central SMBHs should offer answers to at least the following questions:

- Why should QSOs flare during an early stage of galaxy formation?
- Why are the centers of galaxies the preferred sites for QSOs?
- What interrupts the fuel supply to limit QSO lifetimes?
- Why is the mass of the central SMBH related to properties of the host bulge?

Here we outline a model that offers dynamical answers to the first three of these questions, but does not yet answer the fourth. The main ideas are: (1) most large galaxies developed a bar at an early stage of their formation, (2) the central engine is created from gas driven to the center by the bar, and (3) changes to the galaxy potential, caused by mass inflow itself, shut off the fuel supply to the central engine when the mass concentration reaches a small fraction of the galaxy mass. Furthermore, the central mass weakens the bar; we show that complete destruction of the bar creates a (pseudo-)bulge in the stellar distribution but, as yet, we are unable to demonstrate that this is a necessary consequence of SMBH formation. This picture was proposed by Sellwood & Moore (1999); Kormendy, Bender, & Bower (2002) argue for a similar idea with a somewhat different emphasis.

The early sections of this paper review the various ingredients that go into this picture, while the later sections put it together.

13.2   Gas Flow in a Simple Bar

Prendergast (1962) was among the first to realize that a rotating bar in a galaxy would drive large-scale shocks in the interstellar medium, which he identified with the straight, offset dust lanes commonly seen on the leading side of the principal axis of the stellar bar. His insight has been amply confirmed in a host of gas-dynamical calculations reported over the past 40 years.

The gas flow pattern is asymmetric about the axis of the bar, leading to a net torque between the bar and gas. The gas loses angular momentum (to the bar) and energy (in the shocks), which drives it inward. Unfortunately, the inflow rate is not easily predicted from theory or simulations because it depends not only on the mass model for the galaxy and bar, pattern speed, etc., but is particularly sensitive to the effective viscosity (i.e., numerical method and parameters and perhaps also the assumed equation of state). The reason is that the shock is offset farther from the bar major axis as the effective viscosity increases, leading to an increased rate at which the gas loses angular momentum.

It is generally desirable to neglect self-gravity in calculations of the gas flow pattern in a non-axisymmetric potential arising from the more massive stellar component. Self-gravitating, dissipative gas tends to form massive clumps that are, in reality, disrupted by energetic "feedback" from star formation. The wide range of spatial scales makes it impossible for a global simulation of the gas flow to include the small-scale gas dynamics of star formation and feedback in any meaningful way—processes that anyway are not fully understood. Thus, the simulator must include a number of *ad hoc* rules to add energy back to the gas, in addition to calculating its self-gravity, thus making the calculation enormously more expensive in computer time for a questionable improvement in realism.

Self-gravity in the gas should not, of course, be neglected when the gas component is more massive, or when the stellar component is not far from axially symmetric, so that the self-gravity of the gas *creates* the non-axisymmetric structure (e.g., Wada 2004). But this is not the regime of bar flow.

The standard work is by Athanassoula (1992), who shows that the gas builds up in a ring at the inner Lindblad resonance (ILR), if one is present, but is driven in still closer to the center if there is no ILR and the bar is strong. Whether an ILR exists depends on the degree of central concentration in the galactic mass distribution—generally a quite modest bulge component is likely to ensure that an ILR exists.

The conventional definition of the Lindblad resonance is for nearly circular orbits in an axisymmetric potential. The concept can readily be generalized for barred potentials to the region where the orientation of periodic orbits switches from parallel to perpendicular to the bar major axis, which occurs at a radius somewhat interior to that of the naïve definition of the ILR (e.g., Contopoulos & Grosbøl 1989). (The perpendicular orbit family may even disappear in weak bars with little bulge; gas flow without shocks or inflow is possible in such cases.) For simplicity, I loosely use the phrase "ILR ring" to describe the dense ring that forms in the region where gas settles onto non-intersecting orbits in the vicinity of the perpendicular orbit family.

13.3 Nuclear Rings

The general picture of gas inflow down a bar until it stalls at a ring is supported by observation: a gas-rich nuclear ring is seen in many barred galaxies, where an enhanced rate of star formation is observed. Beautiful examples are seen in *Hubble Space Telescope (HST)* images: e.g., NGC 4314 (Benedict et al. 1998) or NGC 1512 and NGC 5248 (Maoz et al. 2001). See also the paper by Carollo (2004).

Furthermore, mm interferometers are mapping the CO emission from the nuclear regions of nearby galaxies at unprecedented resolution (e.g., Sakamoto et al. 1999; Regan et al. 2001; Schinnerer et al. 2002; Sofue et al. 2003). The survey by Regan et al. includes some nice examples of gas accumulating in the centers of barred galaxies and even a partial ring with a central hole can be seen in NGC 4258, although the central hole is not detected in M100. A number of caveats about these data should be noted, however: (1) variations in excitation and optical depth of the CO lines would mean that the observed intensity does not perfectly reflect the CO distribution, (2) estimates of the total gas density depend on the adopted ratio of H_2 to CO, and (3) interferometers frequently detect only a fraction of the CO flux detected by single dishes; this is because they are sensitive only to the inhomogeneous component and are "blind" to smoothly distributed emission.

The very existence of star-forming gas rings requires that any inflow interior to the ring drains the ring more slowly than gas arrives from large radii. Wada (2004) reviews possible mechanisms that can drive inflow inside the ILR ring. We would add that Englmaier & Shlosman (2000) suggest that further mild inflow of gas inside the ILR ring could be achieved through globally driven sound (or pressure) waves. This suggestion is not supported by the *HST* images, however, which often reveal a multi-arm dust distribution that must be caused by other mechanisms. It is likely that spiral features are created by self-gravity in the star-gas mixture and that the behavior inside the ILR, where the quadrupole field of the bar is weak, may not be so different from that in the nuclear regions of unbarred galaxies.

13.4   Double Bars

While not directly relevant to the main theme of this paper, the recent discovery of double bars deserves a mention. These are too striking, and the isophotal twists are too great, to be simply the manifestation of a triaxial ellipsoidal light distribution viewed in projection. Erwin & Sparke (2002) find them to be "surprisingly common"; they are seen in at least 25% (perhaps 40%) of early-type barred galaxies. The random distribution of angles between the inner and outer (or main) bars strongly suggests separately rotating components.

The inner bar probably lies within the ILR of the primary and is some 10% to 15% of length of the primary bar. The origin and dynamics of double bars is not well understood at present, however (see, e.g., Maciejewski & Sparke 2000; Heller, Shlosman, & Englmaier 2001).

It is known that the gas flow pattern is *not* simply a scaled-down version of that in the principal bar; there are no offset dust lanes (Regan & Mulchaey 1999; Shlosman & Heller 2002) and inflow may even be inhibited (e.g., Maciejewski et al. 2002).

13.5   Do Bars Feed AGNs?

Most studies (e.g., Ho, Filippenko, & Sargent 1997) find no significant excess of AGN activity in galaxies with bars over their unbarred counterparts, although enhanced star formation in the circumnuclear environment has long been established (Hawarden et al. 1986). Erwin & Sparke (2002) also find no evidence for excess activity in double barred galaxies.

However, Laine et al. (2002) claim weak evidence ($\sim 2.5\sigma$) for an excess of Seyfert activity in barred galaxies, particularly of later Hubble types. We do not find their result compelling, because it is based on binary binning; there is a continuum both of bar strengths, and of Seyfert activity levels, and the fractions in each bin (Seyfert or non-Seyfert, and barred or not) must depend on where the dividing lines are drawn. Furthermore, visual classification of barred versus unbarred is subjective. It would be better to look for a correlation between a quantitative estimate of bar strength (e.g., Abraham & Merrifield 2000; Buta & Block 2001) and some index of "AGN activity."

13.6   Dissolution of Bars

Many studies (e.g., Pfenniger & Norman 1990) claim that central mass concentrations (CMCs) will dissolve bars. But there have been no previous systematic studies to determine what would be required to dissolve bars, either partially or completely; some (e.g., Friedli 1994; Hozumi & Hernquist 1999) have even claimed that very small CMCs will dissolve bars on a moderate time scale.

Yet bars with CMCs are common, a fact that has led to speculation that the observed bar fraction may indicate the "duty cycle" of bars in galaxies that repeatedly dissolve and form again (e.g., Bournaud & Combes 2002). Since just one cycle of bar formation and destruction leads to a dynamically very hot disk, such a scenario demands prodigious infall of fresh gas before a disk could become responsive enough to form a new bar. If bars really were fragile, this daunting requirement might need to be invoked, but, fortunately, we now know that real bars can survive with realistic CMCs.

We have made the first systematic study of the effect of a CMC on the survival of the bar and our two major findings are illustrated in Figures 13.1 and 13.2. We use the amplitude of the $m = 2$ Fourier component of the particle distribution, relative to the axisymmetric term, as a measure the bar amplitude. (See Shen & Sellwood 2004 for more details.)

Figure 13.1 shows that bars are more robust than some previous studies have suggested. The bar is totally destroyed only when the CMC is very dense with a mass $\gtrsim 4\%$ of the *disk*—a less massive or more diffuse central mass weakens the bar, but does not totally destroy it within ~ 6 Gyr. Figure 13.2 shows the trend in bar amplitude at late times as the radial scale of the Plummer sphere used to model a CMC of fixed mass is varied. Dense CMCs are much more destructive than are diffuse CMCs, with a suggestion that the trend asymptotes to a limit as the size shrinks toward a point mass.

The critical value of $\sim 4\%$ of the *disk* mass needed for rapid bar dissolution by a pointlike CMC is enormously larger than the observed masses of central SMBHs. Gas accumulation at an ILR ring, for example, would also have to be quite unreasonably massive ($\gtrsim 10\%$ of the disk mass) to threaten the survival of a bar. Thus, neither current central SMBHs nor gas concentrations pose a significant threat to the survival of bars today.

Our results are based on very high quality N-body simulations, which have been extensively checked. The results shown are for a regime well clear of significant dependence on the numerical parameters. We have found it essential to pay particular attention to the time step. Figure 13.3 shows that poor orbit integration can cause an erroneous decay of the bar when too long a time step is used in the vicinity of a CMC. We integrate the orbits of particles in the vicinity of the CMC with time steps that are repeatedly halved (as many as nine or ten times) in a set of nested guard annuli around the CMC. It is likely that previous work overestimated the bar decay caused by CMCs because of inadequate care in orbit integration.

Not only does the substantial bar fraction in real galaxies suggest that CMCs pose little threat to bars, but the theoretical picture of scattering of stars on box orbits by SMBHs (e.g., Gerhard & Binney 1985) really does not apply to bars with rapidly tumbling figures where most orbits are loop-type (x_1) that avoid the center. A more likely mechanism for the destruction of bars by massive, dense CMCs is through the breakdown of regular orbits (e.g., Norman, Sellwood, & Hasan 1996). It is perhaps not too surprising that a large, dense mass is required to create a sufficiently extensive chaotic region.

It is very hard to imagine that bar destruction by this mechanism could be achieved more

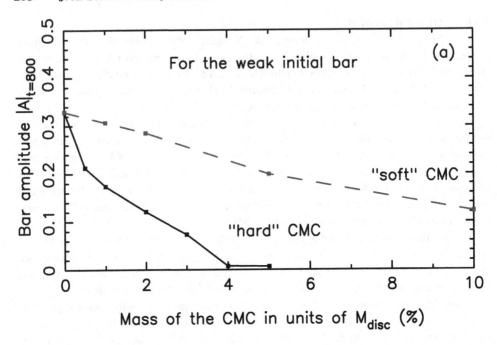

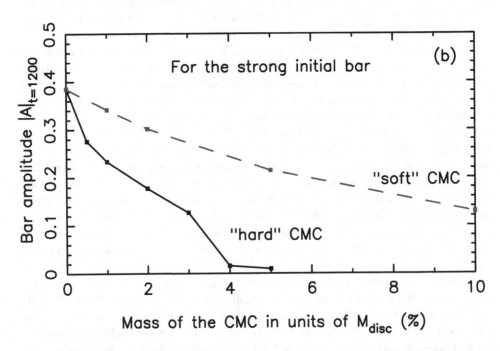

Fig. 13.1. The amplitude of the bar at a fixed time (~ 6 Gyr) after the introduction of CMCs having a range of masses, for both weak (*a*) and strong (*b*) initial bars. The bar is completely destroyed only by massive, dense CMCs.

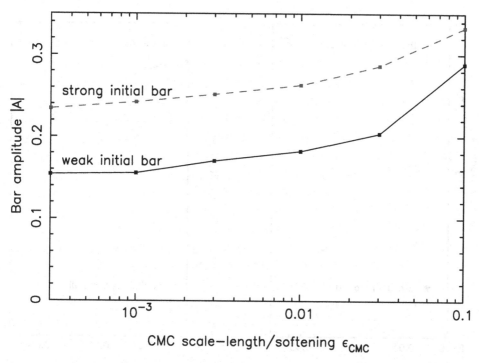

Fig. 13.2. The amplitude of the bar at a fixed time (~ 3 Gyr) after the introduction of CMCs having a range of central concentrations, for both strong and weak initial bars. The softening length ϵ_{CMC} is in units of the exponential scale length of the original disk. Bars are weakened more by dense CMCs than by diffuse ones.

than once in any given galaxy. The formation of a new bar is more difficult, because the disk has become both hot and has acquired a dense center (which inhibits one of the two possible bar forming mechanisms—see next section), but is thought to be possible. However, the dense center would cause the inflow in the new bar to stall at the ILR, preventing the growth of a compact CMC that could threaten its survival.

13.7 Formation of Bars—a Tale of Two Halos

There are two known mechanisms through which a disk galaxy could acquire a bar: (1) a global instability or (2) orbit trapping. The path adopted depends on the mass distribution.

We find a global instability, with no ILR (initially), when the density profile has a large, quasi-uniform core. The global instability occurs on an orbital time scale, and therefore gives rise to a bar immediately in any disk that finds itself in an unstable regime. Such a situation could arise as the mass of the disk increases as primordial gas cools in a protogalactic halo and settles into rotational balance.

Orbit trapping, on the other hand, is the only viable mechanism to form a bar in a galaxy with a steep, inwardly rising density profile. This mechanism is also generally quite fast (see Lynden-Bell 1979 for an alternative) but requires a trigger, such as a mild tidal interaction (e.g., Noguchi 1996) or strong spiral patterns caused by the build up of significant quantities of new, low-velocity dispersion material, in the disk (Sellwood 1989; Sellwood & Moore

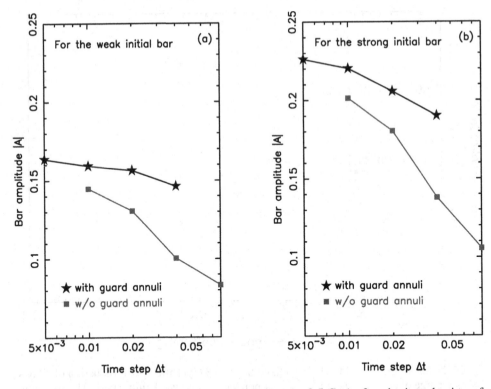

Fig. 13.3. The amplitude of the bar at a fixed time (~ 3.5 Gyr) after the introduction of a 2% disk mass CMC as the time step is varied (lower curves). The upper curves show the amplitude when the time step for orbit integration is sub-divided in nested guard annuli around the CMC.

1999). But isolated disks having dense centers are able to survive for long periods without making bars (Toomre 1981; Sellwood & Evans 2001).

As galaxies form, the mass distribution is dominated by the dark matter halo at first. If the halo has a large, low-density core, we should expect a bar to form once the disk mass begins to dominate in the center. Bars that may develop in halos with density profiles that rise steeply toward the center are formed through orbit trapping in the early, gas-rich disk. The evolution in the two cases is shown in Figure 13.4, and the extensive differences are summarized in Table 13.1.

Figure 13.4a shows that a large bar forms in the disk when the halo has a soft core ($\rho_{halo} \rightarrow$ constant as $r \rightarrow 0$). Soon thereafter it undergoes a collective buckling (aka firehose) instability, caused by the anisotropy of the velocity distribution between the in-plane and vertical velocity dispersions (Toomre 1966; Raha et al. 1991; Merritt & Sellwood 1994). The saturation of this instability converts some of the radial motion into vertical, causing the bar to weaken and to become thicker than the disk from which it formed (see also Combes & Sanders 1981). It seems likely that any gas in the bar region would be driven still closer to the center by this event, although we are unaware of any simulations of gas in a buckling stellar potential.

The bar in the cusped halo ($\rho_{halo} \propto r^{-1}$ for small r), on the other hand, is short at first

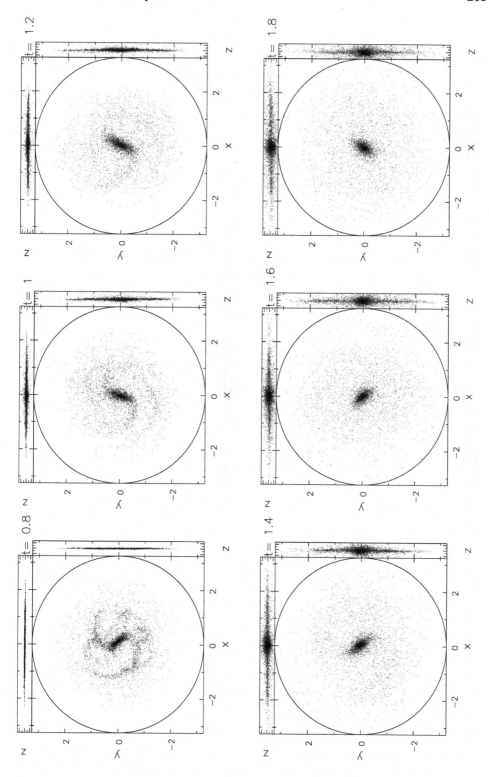

Fig. 13.4. (*a*) The evolution of the disk components in two simulations with different halos. Times are in Gyr and lengths in kpc. A simulation with a soft-core halo profile ($\rho_{\text{halo}} \rightarrow$ constant as $r \rightarrow 0$).

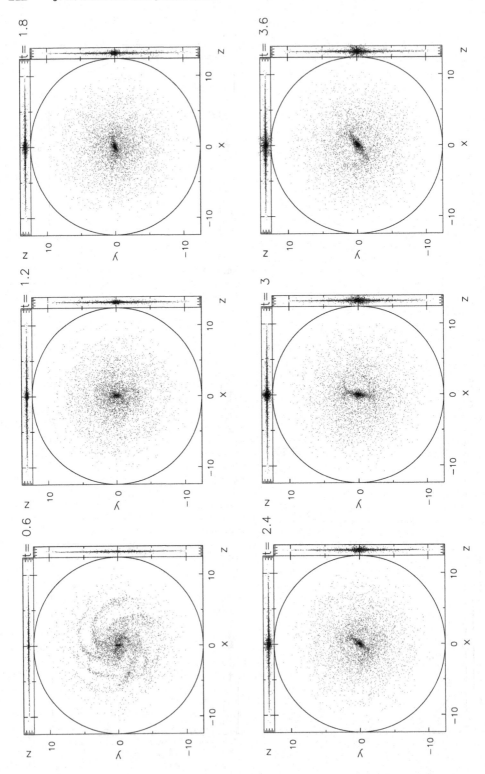

Fig. 13.4. (*b*) As in (*a*), but for a halo with a cusped density profile ($\rho_{halo} \propto r^{-1}$ as $r \to 0$).

Table 13.1. *Differences between the Bars Formed in Two Different Halos*

	Soft core	Cusped halo
Initial bar	large	short
ILR	not initially	yes
Major buckling event	yes	no
Later evolution	smaller and weaker	growing

and grows in length over time (Fig. 13.4*b*). It also thickens somewhat, but because it grows gradually, it does not undergo a severe bending convulsion at any stage.

Gas can be driven deep into the center in the soft-core case, where no ILR is present (initially), and then probably be further compressed by the buckling event. These two successive dynamical instabilities in the gas-rich early stages "naturally" cause a large accumulation of gas in a small volume close to the center on a short time scale. A large concentration of gas is widely believed to be a prerequisite for the growth of a central SMBH (see Shapiro 2004), but some other mechanism may be needed to remove more angular momentum before gas reaches the density required. By contrast, the initial ILR in the cusped-halo case will halt gas inflow at some distance from the center, and the gentle flexing that thickens the bar is unlikely to have much effect on its radial distribution.

The mass of gas accumulated in the center does not have to be very large ($\lesssim 2\%$ of the disk mass) to change the global potential in the soft-core halo by enough to introduce an ILR (Sellwood & Moore 1999). As soon as the ILR is created, further gas supply to the center is shut off.

Thus dynamical evolution of the gas in the soft-core halo case suggests a picture for the origin of QSO activity that has several appealing features: it creates massive concentrations of gas in galaxy centers at the time of galaxy assembly, with a mass perhaps related to the bulge (see below), and a reason the fuel supply is shut off quickly. Note also that the SMBH mass need not be as large as that of the CMC from which it is created, indeed it would be surprising if it were; a larger fraction will form stars and some may be expelled in a wind.

Attractive as it is, such a picture is incomplete for two very obvious reasons: (1) not all SMBHs are in barred galaxies, and (2) it seems to be established that the brightest QSOs are found in merging or elliptical (i.e., post-merger) galaxies (e.g., McLure et al. 1999).

Taking the second point first: It seems likely that at least one galaxy in a merging pair must already host a SMBH in order to make a bright outburst. A bar, and any associated ILR barrier, will be destroyed in the merger, allowing plenty of fresh fuel to be driven inward—this time by the non-axisymmetric forces from the ongoing merger. Since the QSO is reignited, and the mass of the SMBH increases from its previous value, we must expect the brighter QSOs to be found in merging, or post-merger galaxies.

13.8 SMBHs in Non-barred Galaxies

The other problem is the absence of bars in some galaxies with SMBHs. Gas inflow in the early bar creates a CMC from which the SMBH is made. Since there is no initial ILR to stall the inflow, the gas concentration will be compact and the bar will be destroyed quickly if such a CMC exceeds $\sim 4\%$ of the disk mass (Fig. 13.1). However, it is likely that an ILR will form well before the central mass reaches this value, limiting the maximum compact mass that can be achieved.

The bar is weakened substantially by a CMC of $\sim 2\%$ of the disk mass (Fig. 13.1), making it more vulnerable to other destruction mechanisms. Sellwood & Moore (1999) found that ongoing spiral activity in the outer disk, which is not included in our present simulations, could either complete the destruction of the bar or cause it to grow again. Bars can also be destroyed in minor mergers (e.g., Gerin, Combes, & Athanassoula 1990).

Van den Bergh et al. (1996) suggest a deficiency of bars in all galaxies at $z > 0.5$, although the claim is disputed (e.g., Sheth & Regan 2004). There is no doubt that some bars can be missed in blue images (Eskridge et al. 2000; Dickinson 2001), but van den Bergh et al. (2002) vigorously defend the deficiency. New data from the ACS on *HST* will soon settle the question. If the deficiency is real, a second bar must be formed later to account for their present abundance.

13.9   Pseudo-bulges

An appealing by-product of bar dissolution is that the stars which were in the bar form an axisymmetric (pseudo-)bulge component in the galaxy center. Figures 13.5 and 13.6 are made from the particle distribution at a time after a strong bar has been totally destroyed by a 5% central mass. The projected density distribution (Fig. 13.5), which started exclusively in a disk, reveals a distinct central bulge that is only slightly flattened in the inner parts. The projected density profile of the model at this time (Fig. 13.6a) suggests two separate components, even though all particles started in a single disk component. Two distinct components are also seen in the volume density profile (Fig. 13.6b), which also shows the profiles of the rigid central mass and the halo components. (The spherical average used in this plot is appropriate for the bulge, but obviously not for the disk.)

Norman et al. (1996) show that such a pseudo-bulge has a high degree of rotational support, and the velocity field is cylindrically symmetric, as observed (Kormendy 1993).

13.10   Halo Mass Profiles

It would be inappropriate here to review the current controversy over the cosmologically predicted dark matter halo density profiles in galaxies. The solution to the serious discrepancy between the predicted and the observed density profiles is far from clear at present. But we would like to stress that the predicted cuspy mass profiles would force an ILR in every bar when it first formed, which would preclude the formation mechanism for QSOs proposed here.

13.11   Relation to Other Models of SMBH Formation

Following Toomre & Toomre (1972), many workers (e.g., Kauffmann & Haehnelt 2000; Di Matteo et al. 2003; Hatziminaoglou et al. 2003) argue that SMBHs form in mergers, which characterize galaxy formation in cold dark matter (CDM) Universes (e.g., Wechsler et al. 2002). Such an idea has difficulty accounting for SMBHs in galaxies that have long avoided significant mergers; examples include the Milky Way and those possessing pseudo-bulges (Carollo 2004).

The picture proposed here operates in every galaxy in which the disk is massive enough to form a bar through a global instability, and therefore accounts for the existence of SMBHs in every $L \gtrsim L_*$ galaxy. However, the two proposals are not mutually exclusive, and we

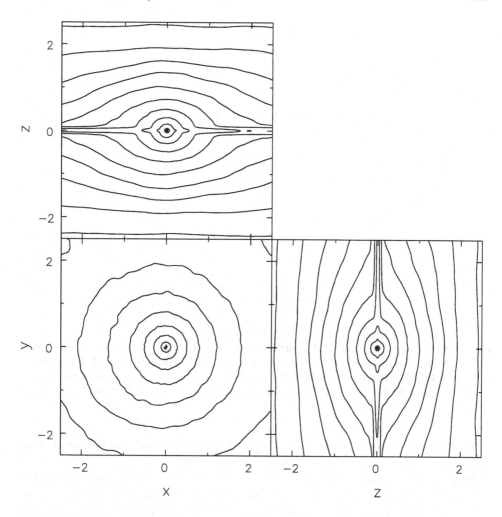

Fig. 13.5. Contours of the projected density of the inner particle distribution in 3 projections after a strong bar is fully destroyed. Distances are reckoned in units of the exponential scale length of the original disk. Notice the bulgelike feature in the edge-on views.

suggest that the seed SMBHs formed by our mechanism are required to be present in the merging fragments in order to produce a bright outburst. The larger SMBHs in giant elliptical galaxies must have grown substantially during the mergers that formed them.

Furthermore, the idea of angular momentum removal by bar formation would be of help in making SMBHs by direct collapse, as discussed by Bromm & Loeb (2003), for example.

13.12 Conclusions

Inflow in barred galaxies today is arrested at a nuclear ring, which is identified as the inner Lindblad resonance (ILR) of the bar. Theoretical predictions of possible further inflow interior to the ring are not yet mature, but the evident nuclear rings in many *HST* images and mm interferometer maps imply that gas drains from the ring at a much slower

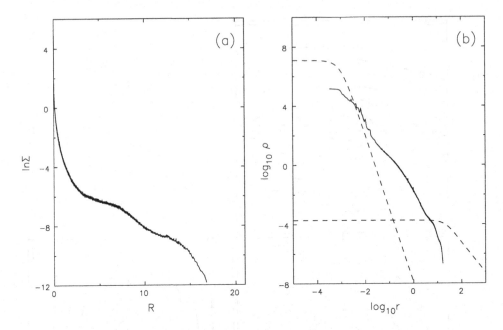

Fig. 13.6. The density profile of the particle distribution (solid curves) after a strong bar is fully destroyed. Panel (*a*) shows the face-on projected surface density of the disk and bulge, while panel (*b*) shows the volume density. The dashed curves in panel (*b*) show the density profiles of the central mass and the halo. The unit of distance is again the scale length of the *original* exponential disk.

rate than it arrives from larger radii. It is likely the inflow rate inside the ILR ring is little different from that in the nuclear regions of unbarred galaxies.

If bars can form without ILRs in the early Universe, then gas can be driven farther into the centers of galaxies. Gas in a bar-unstable, young galaxy disk is driven into the center by two successive instabilities: the usual bar instability, followed by a buckling instability. The gas concentration in these gas-rich early stages is likely to be large, so these instabilities deliver a substantial mass of gas to within $\lesssim$ 100 pc of the galaxy center. While bar flow removes a large fraction of the angular momentum from the gas, other mechanisms, as yet unclear, are required to reduce it still further before an SMBH can form. We simply assume that an SMBH is created in every galaxy in which the disk becomes massive enough to become bar unstable.

The gas inflow itself alters the mass distribution enough to create an ILR where none previously existed, shutting off further gas supply to the nuclear region. It is likely that only a fraction of the gas driven into the center collapses to form the SMBH; probably a larger fraction will form stars, while some might escape in a wind. Any remaining gas that can accrete onto the SMBH will power a low-luminosity QSO for a while.

This picture of SMBH formation accounts for the coincidence of the QSO epoch with that of galaxy formation, the fact that SMBHs are found only in the centers of galaxies, and the short lifetime of QSO activity, because inflow is shut off by the creation of an ILR. The modest SMBHs formed in this way are the seeds for stronger outbursts that must occur when fresh gas is supplied during mergers.

Sellwood & Moore (1999) originally hoped that a further consequence of the initial CMCs that make the SMBHs would be the destruction of the bar to make a small bulge. Unfortunately, we now find that the CMC mass needed to dissolve the bar entirely is larger than could be assembled by gas inflow down the bar; the ILR probably intervenes to shut off the inflow at about half the critical mass for bar destruction. Bars can still be destroyed later, creating a bulgelike component, but this event is decoupled from the initial SMBH formation, which implies that our picture probably cannot hope to offer a simple understanding of SMBH systematics.

Our model of SMBH formation requires the dark matter halos, in which the galaxies form, to have large cores—the instabilities that create the central gas concentration do not occur if the disk forms in a cusped halo. As there are a number of lines of evidence to suggest that the cusped halos predicted in CDM cosmology are not present in real galaxies, we do not regard this requirement to be at all unrealistic. It does, however, preclude predictions of the epoch and rate of QSO activity until whatever is wrong with the CDM prediction is corrected.

Acknowledgements. We would like to thank Laura Ferrarese for a careful read and thoughtful comments on a draft of this paper. This work was supported by NSF grant AST-0098282 and by NASA grant NAG 5-10110.

References

Abraham, R. G., & Merrifield, M. R. 2000, AJ, 120, 2835

Athanassoula, E. 1992, MNRAS, 259, 345

Bahcall, J. N., Kirhakos, S., Saxe, D. H., & Schneider, D. P. 1997, ApJ, 479, 642

Benedict, G. F., Howell, A., Jorgensen, I., Chapell, D., Kenney, J., & Smith, B. J. 1998, STSCI Press Release C98

Bournaud, F., & Combes, F. 2002, A&A, 392, 83

Bromm, V., & Loeb, A. 2003, ApJ, in press (astro-ph/0212400)

Buta, R., & Block, D. L. 2001, ApJ, 550, 243

Carollo, C. M. 2004, in Carnegie Observatories Astrophysics Series, Vol. 1: Coevolution of Black Holes and Galaxies, ed. L. C. Ho (Cambridge: Cambridge Univ. Press), in press

Combes, F., & Sanders, R. H. 1981, A&A, 96, 164

Contopoulos, G., & Grosbøl, P. 1989, A&ARv, 1, 261

Dickinson, M. 2000, Phil. Trans. London A, 358, 2001

Di Matteo, T., Croft, R. A. C., Springel, V., & Hernquist, L. 2003, ApJ, 593, 56

Englmaier, P., & Shlosman, I. 2000, ApJ, 528, 677

Erwin, P., & Sparke, L. S. 2002, AJ, 124, 65

Eskridge, P. B., et al. 2000, AJ, 119, 536

Ferrarese, L. 2002, in Current High-Energy Emission around Black Holes, ed. C.-H. Lee & H.-Y. Chang (Singapore: World Scientific), 3

Ferrarese, L., & Merritt, D. 2000, ApJ, 539, L9

Franceschini, A., Hasinger, G., Takamitsu, M., & Malquori, D. 1999, MNRAS, 310, L5

Friedli, D. 1994, in Mass-Transfer Induced Activity in Galaxies, ed. I. Shlosman (Cambridge: Cambridge Univ. Press), 268

Gebhardt, K., et al. 2000, ApJ, 539, L13

Gerhard, O. E., & Binney, J. 1985, MNRAS, 216, 467

Gerin, M., Combes, F., & Athanassoula, E. 1990, A&A, 230, 37

Hatziminaoglou, E., Mathez, G., Solanes, J-M., Manrique, A., & Salvador-Solé, E. 2003, MNRAS, 343, 692

Hawarden, T. G., Mountain, C. M., Leggett, S. K., & Puxley, P. J. 1986, MNRAS, 221, 41P

Heller, C., Shlosman, I., & Englmaier, P. 2001, ApJ, 553, 661

Ho, L. C., Filippenko, A. V., & Sargent, W. L. W. 1997, ApJ, 487, 591

Hozumi, S., & Hernquist, L. 1999, in Galaxy Dynamics — A Rutgers Symposium, ed. D. Merritt, J. A. Sellwood, & M. Valluri (San Francisco: ASP), 259

Kauffmann, G., & Haehnelt, M. 2000, MNRAS, 311, 576

Kormendy, J. 1993, in Galactic Bulges, ed. H. Dejonghe & H. J. Habing (Dordrecht: Kluwer), 209

Kormendy, J., Bender, R., & Bower, G. 2002, in The Dynamics, Structure, & History of Galaxies, ed. G. S. Da Costa, & H. Jerjen (San Francisco: ASP), 29

Kormendy, J., & Richstone, D. 1995, ARA&A, 33, 581

Laine, S., Shlosman, I., Knapen, J. H., & Peletier, R. F. 2002, ApJ, 567, 97

Lynden-Bell, D. 1979, MNRAS, 187, 101

Maciejewski, W., & Sparke, L. S. 2000, MNRAS, 313, 745

Maciejewski, W., Teuben, P. J., Sparke, L. S., & Stone, J. M. 2002, MNRAS, 329, 502

Maoz, D., Barth, A. J., Ho, L. C., Sternberg, A., & Filippenko, A. V. 2001, AJ, 121, 3048

McLure, R. J., Kukula, M. J., Dunlop, J. S., Baum, S. A., & O'Dea, C. P. 1999, MNRAS, 308, 377

Merritt, D., & Sellwood, J. A. 1994, ApJ, 425, 551

Noguchi, M. 1996, ApJ, 469, 605

Norman, C. A., Sellwood, J. A., & Hasan, H. 1996, ApJ, 462, 114

Pfenniger, D., & Norman, C. 1990, ApJ, 363, 391

Prendergast, K. H. 1962, in Interstellar Matter in Galaxies, ed. L. Woltjer (New York: Benjamin), 217

Raha, N., Sellwood, J. A., James, R. A., & Kahn, F. D. 1991, Nature, 352, 411

Rees, M. J. 1997, RMA, 10, 179

Regan, M. W., & Mulchaey, J. S. 1999, AJ, 117, 2676

Regan, M. W., Thornley, M. D., Helfer, T. T., Sheth, K., Wong, T., Vogel, S. N., Blitz, L., & Bock, D. C.-J. 2001, ApJ, 561, 218

Sakamoto, K., Okamura, S. K., Ishizuki, S., & Scoville, N. Z. 1999, ApJ, 525, 691

Schinnerer, E., Maciejewski, W., Scoville, N. Z., & Moustakas, L. A. 2002, ApJ, 575, 826

Sellwood, J. A. 1989, MNRAS, 238, 115

Sellwood, J. A., & Balbus, S. A., 1999, ApJ, 511, 660

Sellwood, J. A., & Binney, J. J. 2002, MNRAS, 336, 785

Sellwood, J. A., & Evans, N. W. 2001, ApJ, 546, 176

Sellwood, J. A., & Moore, E. M. 1999, ApJ, 510, 125

Shapiro, S. L. 2004, in Carnegie Observatories Astrophysics Series, Vol. 1: Coevolution of Black Holes and Galaxies, ed. L. C. Ho (Cambridge: Cambridge Univ. Press), in press

Shen, J., & Sellwood, J. A. 2004, in preparation

Sheth, K., & Regan, M. W. 2004, in preparation

Shlosman, I., & Heller, C. H. 2002, ApJ, 565, 921

Sofue, Y., Koda, J., Nakanishi, H., & Onodera, S. Kohno, K., Tomita, A., & Okumura, S. K. 2003, PASJ, 55, 17

Toomre, A. 1966, in Geophysical Fluid Dynamics, Notes on the 1966 Summer Study Program at the Woods Hole Oceanographic Institution, ref. no. 66-46

——. 1981, in The Structure and Evolution of Normal Galaxies, ed. S. M. Fall & D. Lynden-Bell (Cambridge: Cambridge Univ. Press), 111

Toomre, A., & Toomre, J. 1972, ApJ, 178, 623

van den Bergh, S., Abraham, R. G., Ellis, R. S., Tanvir, N. R., Santiago, B. X., & Glazebrook, K. G. 1996, AJ, 112, 359

van den Bergh, S., Abraham, R. G., Whyte, L. F., Merrifield, M. R., Eskridge, P. B., Frogel, J. A., & Pogge, R. 2002, 123, 2913

Wada, K. 2004, in Carnegie Observatories Astrophysics Series, Vol. 1: Coevolution of Black Holes and Galaxies, ed. L. C. Ho (Cambridge: Cambridge Univ. Press), in press

Wechsler, R. H., Bullock, J. S., Primack, J. R., Kravtsov, A. V., & Dekel, A. 2002, ApJ, 568, 52

Yu, Q., & Tremaine, S. 2002, MNRAS, 335, 965

14

Black holes and the central structure of early-type galaxies

TOD R. LAUER

National Optical Astronomy Observatory

Abstract

Central massive black holes appear to have a strong role in determining the central structure of elliptical galaxies. The existence of cores in the more luminous ellipticals may be understood as the merger endpoint of two less-luminous galaxies, each harboring a massive black hole. As the two black holes spiral together at the center of the merger product, stars would be ejected from this region, creating a break in the stellar density profile and the shallow cusp in surface brightness characteristic of a core. Further, the resulting binary or merged black hole would preserve the core over subsequent mergers or cannibalism events by disrupting any incoming dense stellar system that would otherwise fill in the core. The effects of black holes are also strongly favored to explain central "double-nuclei," such as that seen in M31, and a recently discovered class of "hollow galaxies," which have local minima in their central stellar density profiles. In short, any discussion of the origin of galaxy morphology must include the ubiquitous presence of massive black holes.

14.1 Introduction

It is now more or less accepted that massive black holes (MBHs) are ubiquitous in the centers of galaxies (Richstone et al. 1998). Indeed, the recent work of Gebhardt et al. (2000) and Ferrarese & Merritt (2000) showing a tight relationship between the masses of black holes and the global dynamics of their hosting galaxies strongly implies that there is a deep relationship between the formation of MBHs and their associated galaxies. Once this premise is accepted, it raises the possibility that some of the characteristic morphological properties of galaxies may in fact be the frank effects of the co-formation of the galaxy with a central MBH.

The creation and preservation of "cores" in luminous elliptical galaxies appears to be one such case where the effects of MBHs on their hosting galaxies are essentially demanded. I will define cores in detail below, but briefly, they are the central region of a galaxy interior to a "break radius," r_b, at which the steep outer envelope brightness profile of the galaxy transitions to a shallow cusp. They are a phenomenon of the more massive elliptical galaxies, which are believed to be the endpoints of possibly several generations of mergers of less-luminous galaxies. A core would be generated during the merging of two galaxies as their respective central black holes spiral into the center of the merger product due to dynamical friction. The black holes form a binary system, which then hardens by ejecting stars from the center of the merged galaxy; this process creates the core structure. The binary or the

© The Observatories of the Carnegie Institution of Washington 2004.

ultimate merged black hole plays an additional role in preserving the core by disrupting any dense stellar systems cannibalized by the new galaxy that would otherwise fill in the core.

Black holes offer a natural explanation for some of the rarer forms of central structure seen in elliptical galaxies in addition to the more common cores. The long-term existence of "double nuclei," such as those seen in M31 (Lauer et al. 1993) and NGC 4486B (Lauer et al 1996) can only be understood as systems where a MBH dominates the central potential, allowing for long-lived and unusual distributions of stellar orbits (Tremaine 1995). Likewise a recently discovered class of "hollow galaxies" (Lauer et al. 2002), where the stellar density profile shows a local *minimum* near the galaxy center, may be cases where the formation of a core through the formation and hardening of a binary MBH has left especially obvious signatures of this mechanism. Overall, MBHs will dominate the centers of their galaxies—it indeed is quite reasonable that they will leave observable signatures in the central stellar density distributions.

14.2   The Cores of Early-type Galaxies

Traditionally, a core was defined to be a central region of a stellar system where the stellar density profile flattened out to a constant value. More specificly, convergence to the central constant value would be quadratic, so as $r \to 0$, $\rho(r) \approx \rho_0(1 + \mathcal{O}r^{-2})$. Cores of this form were seen in some globular clusters—and appeared to be present in elliptical galaxies, based on photographic surface photometry. Schweitzer (1979), however, noted that seeing interacting with an extended object could create the false impression of a "well resolved" core, and questioned the conclusion that cores had been seen in elliptical galaxies. This issue was settled through the use of the then-new CCD cameras to obtain surface photometry, coupled with image processing sufficiently sophisticated to incorporate seeing into the analysis. Both Lauer (1985) and Kormendy (1985) identified several nearby giant ellipticals where the central surface brightness did appear to level off and approach a constant value. Cores did seem to be present, with "core radii," or the half-power point of the brightness profile, r_c, always less than a kiloparsec; however, Lauer (1985) noted that the detailed brightness profiles within the cores flattened out more slowly than quadratic convergence to a flat central surface brightness. At the same time, Lauer and Kormendy both emphasized that seeing did prevent cores from being observed in several, mainly low-luminosity, galaxies—these were especially interesting targets for *HST.*

Once such target was NGC 7457, which was the first galaxy observed with *HST* (Lauer et al. 1991). An extrapolation of the rough relationship between core size and luminosity observed for the galaxies resolved from the ground suggested that NGC 7457 would have a core easily detectable at *HST* resolution. Instead its brightness profile continued into the resolution limit as a steep power law $[I(r) \sim r^{-1}]$; its central stellar density was at least 10^2 higher than had been expected. *HST* observations of M32 (Lauer et al. 1992b) and M87 (Lauer et al. 1992a) also showed these galaxies to have central surface brightness cusps. The case of M87 was of particular interest; from the ground it showed a large core, but one that was strongly "non-isothermal," that is one that did not show a clear zone of constant central surface brightness. *HST* observations showed it to have a shallow cusp in surface brightness that continued unabated into its center; this was significant, as even a shallow cusp in projection implied a steep cusp in luminosity volume density—M87 appeared to have a core, but just as clearly had no extended inner region of constant density.

Subsequent observations of other galaxies with *HST* show that nearly all early-type

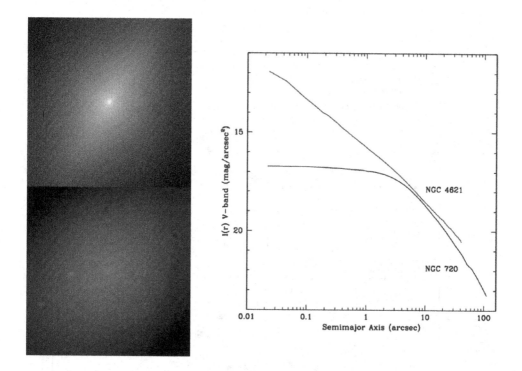

Fig. 14.1. Images and brightness profiles of the central regions of NGC 720 (bottom), a core galaxy, and NGC 4621 (top), a power-law galaxy. The galaxies have same total luminosity.

galaxies have central surface brightness cusps (Crane et al. 1993; Ferrarese et al. 1994; Kormendy et al. 1994; Lauer et al. 1995). None of the galaxies with "cores" seen at ground-based resolution were found to have true constant-density cores. Instead they always had shallow central cusps in surface brightness, which again always implied steep cusps in stellar density. M87, thus, was seen to be typical of galaxies that showed cores at ground-based resolution. The new observations motivated a new formal observational definition of a core as a region where the steep envelope brightness profile of a galaxy "breaks," and transitions to a shallow cusp at smaller radii of the form $I(r) \sim r^{-\gamma}$, with $0 < \gamma \leq 0.3$ (Lauer et al. 1995); a typical core galaxy is shown in Figure 14.1. The transition between the inner cusp and outer envelope is characterized by the "break-radius," r_b, which is analogous to the previous core radius, r_c, parameter; r_b is formally defined as the point of maximum logarithmic second derivative of the brightness profile. In general, r_b is roughly correlated with total galaxy luminosity and is always smaller than 1 kpc. This behavior is neatly described by the empirical "broken power-law" form below, informally known as the "Nuker Law" (Lauer et al. 1995),

$$I(r) = 2^{(\beta-\gamma)/\alpha} I_b \left(\frac{r_b}{r}\right)^{\gamma} \left[1 + \left(\frac{r}{r_b}\right)^{\alpha}\right]^{(\gamma-\beta)/\alpha} , \qquad (14.1)$$

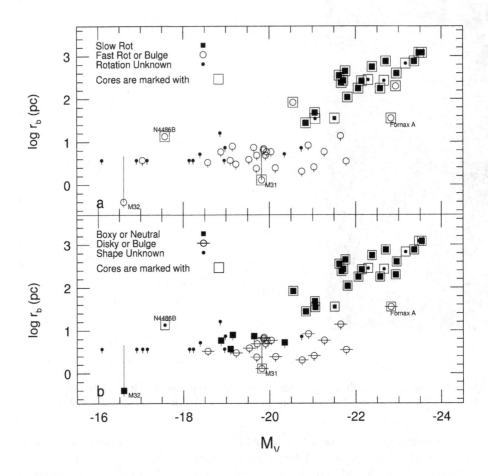

Fig. 14.2. The relationship between break radius, r_b, and total galaxy luminosity (as adopted from Faber et al. 1997). Core galaxies are indicated with an open box. Power-law galaxies are represented as upper limits on r_b. The upper panel additionally classifies the galaxies by the importance of rotation; the bottom panel classifies galaxies by isophote shape. Note that for $M_V < -20$ mag, limits on r_b for power-law galaxies fall below the core-galaxy sequence.

where γ is the inner power-law slope as $r \rightarrow 0$, β is the outer power-law slope as $r \rightarrow \infty$, r_b is the break radius, I_b is the surface brightness at the break radius, and α controls the speed of transition from the inner and outer power laws.

The key point of this revised definition of a core is that it preserves the observational schema that there are central regions of early-type galaxies where there is a physical scale at which the brightness profile becomes much shallower than the envelope, even though it abandons the more abstract idea that this transition must necessarily be associated with a central region of constant luminosity density. We do see cores, and as I will discuss shortly, their existence correlates with important properties of the galaxies in which they are seen.

This is especially clear when we consider the other half of the story presented by the *HST* observations, which is the existence of galaxies in which no core is seen. Just as M87 serves as a prototype of "core" galaxies, NGC 7457 serves as an example of this second class of "power-law" galaxies.

The technical definition of a power-law galaxy is a galaxy for which no inner break in its brightness profile has been resolved; a typical example of a power-law galaxy is shown in Figure 14.1. At first glance this classification may appear to be trivial, since the detection of cores always requires some minimal angular resolution; if power-law galaxies were just core galaxies seen further away, then this definition would be of little use. The power-law classification was motivated instead by the existence of galaxies observed at the same resolution and distance as core galaxies that instead had brightness profiles that continued into the *HST* resolution as steep power-law cusps, or $I(r) \sim r^{-\gamma}$, with $\gamma > 0.5$. These galaxies generally were lower luminosity systems that had no cores detectable at ground-based resolution that now had limits of $r_b < 10$ pc at *HST* resolution. Since r_b for core galaxies roughly correlates with luminosity, it might be assumed for some of the power-law galaxies would have cores appropriate for their luminosity, but which escaped even *HST* resolution. The power-law classification was most interesting, however, for those galaxies that had limits on r_b well below those of core galaxies of the same total luminosity at the same distance. Absence of a core for such systems is thus physically important. It may be that some power-law galaxies may ultimately reveal cores at the few parsec scale or finer when observations that surpass *HST* resolution become available, but there is no reason to presume that this will be the case. Furthermore, the scale of such cores will be discontinuous with the r_b vs. luminosity relationship for core galaxies.

Presence or absence of a core is related to other properties of the galaxies (Faber et al. 1997). Figure 14.2 shows the relationship between core size, as quantified by r_b, and total luminosity. For the most part, core galaxies have $M_V < -21$ mag; for these galaxies $r_b \sim L^{1.2}$, but with significant scatter. Power-law galaxies, as noted above, are in general less luminous, but can be as bright as $M_V \approx -22$ mag. Limits on r_b for power-law galaxies with $M_V < -20$ mag fall well below the r_b vs. L relationship for core galaxies of the same luminosity.

Faber et al. (1997) strongly argue that power-law and core galaxies define two distinct classes of central structure in early-type galaxies. Nearly all early-type galaxies have singular stellar density profiles, but the structural properties of the two types are not continuous. While core galaxies always have $0 < \gamma \leq 0.3$, power-law galaxies always have $\gamma > 0.5$—there is a clear gap between the two structural forms and the distribution of inner cusp slopes is bimodal. This is especially vivid in the comparison of stellar density profiles of both types presented by Gebhardt et al. (1996) shown in Figure 14.3. Recently, Ravindranath et al. (2001) and Rest et al. (2001), who both have studied much larger galaxy samples with *HST*, have identified a few systems with intermediate cusp slopes that fall between the core and power-law division; however, such systems are relatively uncommon, and the overall bimodal distribution in central properties appears to be robust (see Fig. 14.4).

The division of early-type galaxies into core and power-laws is really underscored when the dynamical and morphological properties of the two types are compared (Faber et al. 1997). Core galaxies are slow rotators and have boxy isophotes, while power-law galaxies are marked by rapid rotation and disky isophotes. These differences indeed are especially evident when comparing power-law and core galaxies of the same luminosity. Luminosity is removed as a discriminate, but the galaxies' dynamic and morphological states still match

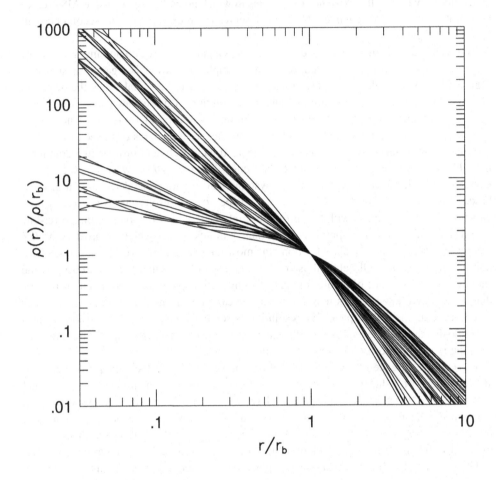

Fig. 14.3. Stellar luminosity density profiles are plotted, showing the dichotomy between core and power-law galaxies. The radial scale is normalized by the break radius; for power-law galaxies, this is taken in this context as the point of maximum logarithmic second derivative in their brightness profile. (Figure adapted from Gebhardt et al. 1996.)

up neatly with the ones that have cores and the ones that do not. These differences plus the overall high luminosities of core galaxies strongly suggest the role of galaxy mergers in creating the core galaxies.

The absence of cores in power-law galaxies is especially important when we compare the stellar *volume density* profiles of core and power-law galaxies. Although both types of galaxies have singular density profiles into the resolution limit, the greatly steeper slopes of the power-law galaxy density profiles means that for $r < 100$ pc these galaxies may have stellar densities that exceed those of core galaxies by several orders of magnitude. As outlined in the next section, this is critically important for understanding the formation of central structure in the context of galaxy formation overall.

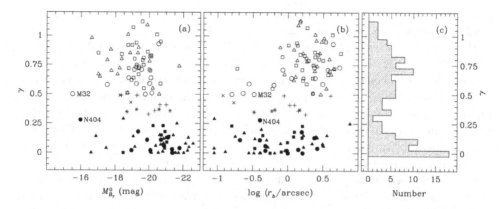

Fig. 14.4. The complete compendium of central cusp slopes as presented in Ravindranath et al. (2001). Dependence of the inner cusp slope on (a) absolute magnitude of the galaxy $(M^0_{B_T})$ and (b) break radius (r_b), showing core galaxies (filled symbols) and power-law galaxies (open symbols). The circles denote the galaxies from Ravindranath et al. (2001), triangles those from Faber et al. (1997), and squares those from Rest et al. (2001). Objects in the sample of Ravindranath et al. that fall in the gap region ($0.3 < \gamma < 0.5$) appear as asterisks, while those from Rest et al. (2001) are shown by plus signs. Two galaxies that fall in this region from Faber et al. (1997) are shown by crosses. (c) Distribution of γ.

14.3 Core Formation and Black Holes

All modern galaxy formation scenarios imply that the most luminous galaxies are the endpoints of a merging hierarchy of less luminous systems. That cores for the most part only appear in the more luminous, nonrotating, and boxy early-type galaxies strongly suggests that cores are not created during the first generation of galaxy formation but arise from mergers at later epochs. Begelman, Blandford, & Rees (1980) were the first to present a specific scenario under which this might happen. Begelman et al. were concerned with the merging of two galaxies, each harboring a central MBH. As the merger progressed the two black holes would sink by dynamical friction into the common center of the merger remnant, creating a binary MBH. Subsequent three-body interactions between the binary and surrounding stars would eject stars from the center of the galaxy and harden the binary. Roughly speaking, a mass exchange would take place at the center of the merger remnant. Coalescence of the binary MBH displaces a similar mass of star, producing a core. Intriguingly, Begelman et al. suggested that enough stars might be ejected from the center that the stellar density profile might actually have a local minimum in the very center. In any case, the density profile would exhibit an abrupt decrease in slope as the center was approached; in projection this would create the appearance of a core.

The Begelman et al. (1980) discussion predated both a general picture of the central structure of early-type galaxies as well as an understanding of the prevalence of MBHs in galaxy centers. More recent work has looked at the creation of cores in the general context of galaxy mergers and again showed that the end state of the merger of two galaxies both with central MBHs created cores. Ebisuzaki, Makino, & Okumura (1991), Makino (1997), and Quinlan (1997) all used N-body simulations of sufficient resolution to elucidate the central structure of mergers produced in this scenario. Ebisuzaki et al. (1991), for example, showed that the core created in the remnant would have a shallow cusp in the density profile, consistent with the *HST* observations that shortly followed after this study.

Milosavljević & Merritt (2001) have recently presented simulations of galaxy mergers with MBHs that have been informed by the modern picture of central galaxy structure provided by *HST* observations. Their starting point is two galaxies with structure appropriate for power-law galaxies. As a MBH binary is created in the merger and hardens, a break appears in the stellar volume profile. The profile interior to the break retains the form of a power law in radius, but with shallower slope than the envelope outside the break. Projection of the merger product not only features a system with a core, but one that has an interior surface brightness cusp, in excellent agreement with the modern picture of core galaxies.

While all the theoretical work quoted above strongly suggests that the creation of a binary black hole in a merger will create a core in the merger remnant, it is mute on the importance of this process for creating cores overall. It is not sufficient just to create a core, however; it must survive until the present epoch against any number of subsequent mechanisms that would deliver gas and stars into central potential of the merged system. That cores have indeed survived over time is the final piece of the puzzle that essentially demands that cores in early-type systems are uniquely tied to MBHs.

The problem of core survival can be understood by considering the existence of cores in the highly luminous brightest cluster galaxies (BCGs), nearly all of which are seen to have cores (Laine et al. 2003). Even if the BCGs were largely created at early times, they continue to cannibalize less luminous cluster galaxies at a significant rate at the present epoch (Merritt 1985; Lauer 1988). Although a modest cannibalism rate may have little effect on the overall structure of these galaxies, this is less true in their centers. The low-luminosity power-law galaxies will be just the systems cannibalized by the BCG, but their high central stellar densities strongly suggest that their central portions should survive partially intact into the cores of the BCGs. In short, how do cores avoid being filled in by cannibalized remnants?

The prevalence of cores at this epoch was raised by Faber et al. (1997) as a strong argument that there must be some disruptive mechanism or guardian that preserves the cores of luminous galaxies over the infall of cannibalized systems. Considering the 14 brightest galaxies in the Lauer et al. (1995) sample, seven of which were BCGs, even a modest cannibalism rate of $0.2L^*$ Gyr^{-1} (Lauer 1988) predicts that *each* of these galaxies should have cannibalized two systems since $z \approx 1$. Given the strong central density contrast, $> 10^3$, between the typical power-law elliptical and BCG, simple tidal arguments suggest that the centers of the power-law galaxies should fall undisrupted into the centers of the BCGs. Kormendy (1984) contemplated this possibility in advance of *HST* observations, suggesting that some luminous galaxies would be found to have a "core within a core"; such systems, however, have never been seen. This scenario was tested in detail by Holley-Bockelmann & Richstone (1999) with *N*-body simulations. If a luminous galaxy *without* a central MBH cannibalizes a less luminous system, above some luminosity difference, the less luminous galaxy will on average be dense enough at its center to survive central accretion into the more massive system. Stripping and tidal disruption of the envelope of the infalling less luminous galaxy will indeed take place, but its nucleus, itself, will indeed settle down unscathed within the core of the massive galaxy.

The hypothesis is that cores, once created, are protected against being filled in by subsequent mergers by the resident MBH acting as a "guardian." The strong tidal forces of an MBH would disrupt any dense stellar system approaching the nucleus. If the incoming nucleus itself contained an MBH, the subsequent formation and hardening of an MBH binary would serve the same purpose.

Again this scenario has been born out by the simulations of Holley-Bockelmann & Richstone (2000), in which they studied the accretion of galaxies with power-law density profiles by more luminous galaxies harboring central MBHs. The strong tidal fields associated with the MBH of typical mass expected based on the Gebhardt et al. (2000) and Ferrarese & Merritt (2000) relationships were more than sufficient to disrupt the power-law galaxies. In short, black holes can serve as a guardian of cores over cannibalism events; there are no viable alternative hypotheses for preserving cores from the accretion of the nuclei of less-luminous galaxies.

One potentially interesting "loose end" to this picture are the small handful of BCGs that have anomalously small cores or steep power-law cusps into the *HST* resolution limit. Faber et al. (1997) identified NGC 1316 (Fornax A) as a galaxy of high luminous with a compact core falling well below the Faber et al. r_b vs. L relationship. Laine et al. (2003) identified three BGCs of similar high luminosity with cores or upper limits on r_b over an order of magnitude smaller than the mean r_b seen at their luminosities. Since the most massive galaxies may under go several generations of mergers or cannibalism, once speculation may be that cores were initially created in these systems, but somehow their MBH were thwarted as guardians later on. One possibility is that the binary MBH created in on merger was ejected in a relativistic three-body interaction with a third MBH introduced by a later cannibalism event. Still later events could fill in the unprotected cores, perhaps also reintroducing MBH of masses more appropriate to the incoming low-luminosity systems. If so the MBH in NGC 1316 that powers its activity may fall below the predictions of the Gebhardt et al. (2000) and Ferrarese & Merritt (2000) relationships.

14.4   Double Nuclei and Hollow Galaxies

The definition of a core given above is entirely in terms of the projected brightness profile and its implicit reflection of the more fundamental stellar density profile. For most core galaxies this is sufficient—the full morphological appearance of the core is completely encapsulated by the one-dimensional density profile, modulo smooth changes in the isophote ellipticity shape or orientation. There is a small handful of systems, however, that have more interesting core structures that may provide unique testimony to the role of black holes in shaping central galaxy structure.

The bare existence of the double nucleus of M31 (Lauer et al. 1993, 1998) initially posed several difficulties. The small, few-parsec separation of the two nuclear components could endure only for an exceptionally short time, yet there was no additional evidence of any recent cannibalism by the M31 bulge. Tremaine (1995) resolved this difficulty by proposing that the bright P1 M31 nuclear component was not a compact star cluster falling into the nucleus, but was rather the projection of stars lingering at the apocenter of an eccentric disk of stars orbiting the $3 \times 10^7 M_\odot$ MBH believed to exist at the center of M31 (Kormendy & Bender 1999). The M31 MBH is thus essentially required as the only mechanism that allows the M31 double nucleus to exist for astronomically long time scales. Identification of NGC 4486B as a second double nucleus system (Lauer et al. 1996), together with the evidence for an especially massive black hole at the center of this system (Kormendy et al. 1997) bolsters, this picture.

Recently, Lauer et al. (2002) identified a handful of core galaxies that had stellar densities

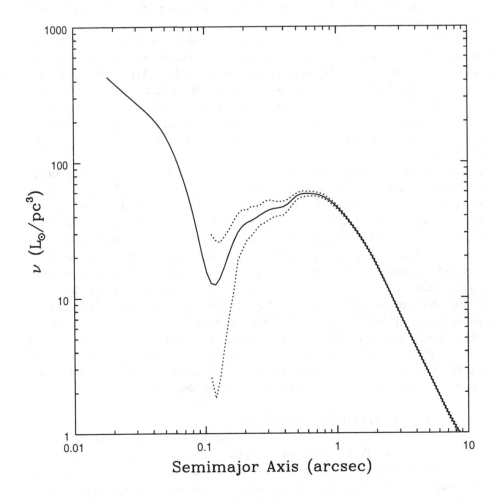

Fig. 14.5. The stellar luminosity density profile for the Virgo elliptical galaxy NGC 4406 is shown. Note the *decrease* in luminosity density interior to one arcsecond. The ultimate central rise appears to be associated with an unresolved compact nuclear source. The bracketing lines give the 1 σ error estimates on the profile.

that actually *decreased* at small radii ($r \ll 100$ pc) moving in toward the galaxy center; an example is shown in Figure 14.5. These systems thus had local *minima* in density interior to their break radii; in a sense their cores could be considered to be partially "hollowed out." These systems may be particularly interesting test cases of the formation of cores by the intercession of MBH binaries. Lauer et al. (2002) advanced two complimentary interpretations of the hollow cores. In one scenario the cores more or less retain a normal stellar density profile, but a diffuse torus of stars has been introduced to the core, creating a local density maximum at the radii it would dominate a normal "background" radial density profile; the local density minimum is thus an illusion generated by the addition of the torus. The *N*-body simulations of Holley-Bockelmann & Richstone (2000) and Zier & Biermann

(2001) indeed show that diffuse stellar tori can be created in a merger as the centers of the two merging galaxies are disrupted by the tidal fields associated with the binary MBH created in the merger.

The second scenario proposed by Lauer et al. (2002) is that the hardening of the binary MBH within the merged core actually is effective enough at transferring energy to stars initially within the core that they are ejected to much larger radii, such that the core really is partially evacuated. As noted above, Begelman et al. (1980) suggested that this mechanism could occur as a binary MBH would harden. The simulations of Makino (1997) indeed appear to show the evacuation of cores in some cases. Comparing the Makino simulations against those of Milosavljević & Merritt (2001), which did not show evacuated cores, raises the possibility that core evacuation depends on the density profiles of the progenitor galaxies. Milosavljević & Merritt merged only power-law galaxies, while Makino merged systems that had pre-existing cores. The binary MBH ejects stars from the center in both cases, but real evacuation of the final merged core is only evident when the initial systems are not very centrally dense in stars to begin with.

14.5   Desiderata

The case that cores are a signature of black holes can be succinctly summarized as follows. *N*-body simulations show that cores can be created in the merger of two galaxies, each possessing its own MBH. The preservation of cores since the epoch of galaxy formation additionally requires MBHs to serve as guardians against destruction of the core by later infall of dense stellar systems. Within this simple picture, however, there remains more work to be done, some of which I outline below.

What are the real starting conditions for building galaxy centers? The discovery of QSOs at $z > 6$ makes it clear the MBHs were forming just as galaxy formation was getting started. Understanding how the construction of galactic nuclei progressed is closely tied to understanding the initial mass function of MBHs plus the properties of the initial stellar systems within which they presumably formed.

How does the onset of core formation fit within the complete context of hierarchical galaxy formation? If cores are formed in mergers, we still have the problem of why they appear so suddenly at $M_V < -21$ mag. Power-law galaxies would have been built in a merging hierarchy as well and also have central MBH. How do they avoid getting cores? Faber et al. (1997) argued that for lower-luminosity systems infall of gas and associated dissipative processes and star formation would have remained important. A complete picture of galaxy formation needs to cleanly explain the transition of power-law to core forms as luminosity increases.

How does the detailed core structure arise? Within the class of core galaxies, systems have diverse inner cusp slopes, a significant spread in core size at any given luminosity, and so on. Are these details interesting diagnostics of the construction history of galaxies?

What other things can massive black holes do to core structure? The notion that creation of an MBH binary is tied to core formation raises any number of issues. As discussed above, interaction of the binary with new MBHs delivered to the nucleus may ultimately cause ejection of all MBHs from the core. The binary may wander within the core at some level. Both of these may have structural implications. Lastly, the central MBH may grow significantly due to central gas accretion. Does the core structure evolve adiabatically as this occurs, the presumption underlying some of the classic inquiries into the effects of MBH on core structure (Young et al. 1978)?

Acknowledgements. I conclude by thanking my collaborators on the Nuker team: Doug Richstone (PI), Ralf Bender, Gary Bower, Alan Dressler, Sandy Faber, Alex Filippenko, Karl Gebhardt, Richard Green, Carl Grillmair, Luis Ho, John Kormendy, John Magorrian, Jason Pinkney, and Scott Tremaine. It has been a pleasure working with them to have at least gotten this far!

References

Begelman, M. C., Blandford, R. D., & Rees, M. J. 1980, Nature, 287, 307

Crane, P., et al. 1993, AJ, 106, 1371

Ebisuzaki, T., Makino, J., & Okumura, S. K. 1991, Nature, 354, 212

Faber, S. M., et al. 1997, AJ, 114, 1771

Ferrarese, L., & Merritt, D. 2000. ApJ, 539, L9

Ferrarese, L., van den Bosch, F. C., Ford, H. C., Jaffe, W., & O'Connell, R. W. 1994, AJ, 108, 1598

Gebhardt, K., et al. 1996, AJ, 112, 105

——. 2000, ApJ, 539, L13

Holley-Bockelmann, K., & Richstone, D. O. 1999, ApJ, 517, 92

——. 2000, ApJ, 531, 232

Kormendy, J. 1984, ApJ, 287, 577

——. 1985, ApJ, 292, L9

Kormendy, J., et al. 1997, ApJ, 473, L91

Kormendy, J., & Bender, R. 1999, ApJ, 522, 772

Kormendy, J., Dressler, A., Byun, Y.-I., Faber, S. M., Grillmair, C., Lauer, T. R., Richstone, D., & Tremaine, S. 1994, in ESO/OHP Workshop on Dwarf Galaxies, ed. G. Meylan & Ph. Prugniel (Garching: ESO), 147

Laine, S., Lauer, T. R., van der Marel, R. P., & Postman, M. 2003, AJ, 125, 478

Lauer, T. R. 1985, ApJ, 292, 104

——. 1988, ApJ, 325, 49

Lauer, T. R., et al. 1991, ApJ, 369, L41

——. 1992a, AJ, 103, 703

——. 1992b, AJ, 104, 552

——. 1993, AJ, 106, 1436

——. 1995, AJ, 110, 2622

——. 1996, ApJ, 471, L79

——. 2002, AJ, 124, 1975

Lauer, T. R., Faber, S. M., Ajhar, E. A., Grillmair, C. J., & Scowen, P. A. 1998, AJ, 116, 2263

Makino, J. 1997, ApJ, 478, 58

Merritt, D. 1985, ApJ, 289, 18

Milosavljević, M., & Merritt, D. 2001, ApJ, 563, 34

Quinlan, G. D. 1997, NewA, 1, 35

Ravindranath, S., Ho, L. C., Peng, C. Y., Filippenko, A. V., & Sargent, W. L. W. 2001, AJ, 122, 653

Rest, A., van den Bosch, F. C., Jaffe, W., Tran, H., Tsvetanov, Z., Ford, H. C., Davies, J., & Schafer, J. 2001, AJ, 121, 2431

Richstone, D., et al. 1998, Nature, 395, A14

Schweizer, F. 1979, 233, 23

Tremaine, S. 1995, AJ, 110, 628

Young, P. J., Westphal, J. A., Kristian, J., Wilson, C. P., & Landauer, F. P. 1978, ApJ, 221, 721

Zier, C., & Biermann, P. L. 2001, A&A, 377, 23

15

The inner properties of late-type galaxies

C. MARCELLA CAROLLO
Eidgenössische Technische Hochschule, CH-8093 Zurich, Switzerland

Abstract

I review some recent results on the inner properties of disk galaxies, and highlight some issues that require either observational or theoretical clarification and that are important for constructing a consistent picture of the formation of the local disk galaxy population.

15.1 Disk Galaxies: Recipe Still Missing

We have not yet achieved a self-consistent theoretical scenario for the formation of the local disk galaxy population, a population which, in crude terms, is a mix of "bulged" and "bulgeless" disks. The origin of the bulges is vigorously debated. In a recent paper in the proceedings of the 1998 workshop on "The Formation of Bulges" (Carollo, Ferguson, & Wyse 1999), Renzini (1999) voices with emphasis the long-standing belief that bulges are nothing more or less than small elliptical galaxies. Basing his argument on the similarity between the magnesium line strength Mg_2 versus absolute r magnitude M_r (and magnesium line strength Mg_2 and velocity dispersion σ; Jablonka, Martin, & Arimoto 1996) relations for bulges and ellipticals, he writes: *"...The close similarity of the $Mg_2 - M_r$ relations argues for spiral bulges and ellipticals sharing a similar star formation history and chemical enrichment. One may argue that origin and evolution have been very different, but differences in age distribution are precisely compensated by differences in the metallicity distributions. This may be difficult to disprove, and I tend to reject this alternative on aesthetic grounds. It requires an unattractive cosmic conspiracy, and I would rather leave to others the burden of defending such a scenario. In conclusion, it appears legitimate to look at bulges as ellipticals that happen to have a prominent disk around them, or to ellipticals as bulges that for some reason have missed the opportunity to acquire or maintain a prominent disk."*

And yet, there is plenty of evidence from observations and numerical experiments that bulges of spiral galaxies may differ significantly from elliptical galaxies. In a pioneering paper, Kormendy (1993; see also Kormendy, Bender, & Bower 2002) reports, for some Sb bulges, V/σ values that are above the oblate line describing the isotropic spheroids in the V/σ-ϵ diagram (with V the maximum velocity, σ the mean velocity dispersion, and ϵ the mean ellipticity of the spheroid; Binney & Tremaine 1987), and makes the point that at least some of the dense structures that are seen inside the disks may actually themselves be disklike systems ("pseudo-bulges"). In numerical simulations, three-dimensional stellar structures result from secular evolution processes that are driven by dynamical instabilities inside the preexisting disks. The fire-hose (or buckling) instability that is seen in simulations of stellar disks can scatter the stars originally in a stellar bar above the plane of the disk, into what resembles a bulgelike structure (Raha et al. 1991). A stellar bar can also drive a high inflow rate of gas toward the center of the disk (Shlosman, Frank, & Begelman 1989); if a mass concentration of the order of $\sim 1\%$ of the total mass is accumulated in the center, this can disrupt the regular orbits supporting the bar and again scatter the stellar orbits above the plane of the disk (Pfenniger & Norman 1990; Norman, Sellwood, & Hasan 1996).

The disks of spiral galaxies also elude us. On large galactic scales, they are a rather homogeneous family, as indicated, for example, by their light profiles, which appear to be exponential over several disk scale lengths (de Jong 1995), the rather common asymptotically flat rotation curves (Persic & Salucci 1995), and the Tully-Fisher relation, which holds over a broad range of surface brightness and mass (Strauss & Willick 1995). On smaller scales, however, where they physically overlap with the bulges (and the rest of the inner structure), disks are not well understood, either observationally or theoretically. Within hierarchical formation schemes, the standard recipe to explain the formation of disks contains three key elements: (1) the angular momentum originates from cosmological torques (Hoyle 1953), (2) the gas and dark matter within virialized systems have initial angular momentum distributions that are identical (Fall & Efstathiou 1980), and (3) the gas conserves its specific angular momentum when cooling (Mestel 1963). These rules are routinely assumed in the (semi-)analytical descriptions of disk galaxies. In contrast, the highest resolution cosmological simulations that include both baryons and cold dark matter (CDM) find significant angular momentum loss for the baryons, especially in the central few kpcs of galaxies (Steinmetz & Navarro 1999). Furthermore, even when disks are assumed to form smoothly and conserving their angular momentum, the resulting disks are more centrally concentrated than single-exponential structures (Bullock et al. 2001; van den Bosch 2001; van den Bosch et al. 2002). Disks with high central densities are seen in the highest-resolution CDM simulations, in which the resulting galaxies have realistic sizes, but a region with low angular momentum and high density is always present at the center (e.g., Governato et al. 2004). It is still a matter of debate whether this is a feature of the structure formation model or is indicative of the lack of a proper treatment of physics (e.g. the effect of supernovae feedback; Springel & Hernquist 2002). Although the CDM simulations still have room for improvement, they could well be correct in their prediction that the central parts of disks might really have quite low angular momentum and high concentration as a result of formation. Although warm dark matter alleviates the angular momentum "catastrophe" (i.e., the loss of angular momentum by the baryons), the angular momentum distributions of warm dark matter halos is identical to that of CDM halos (Knebe, Islam, & Silk 2001; Bullock, Kravtsov, & Colin 2002). Therefore, these halos also predict an excess of low-angular momentum material.

Clearly the central regions of disk galaxies hold important clues to understanding fundamental issues of galaxy formation. Shaping a consistent theory of bulge and disk assembly requires a better understanding of nearby disk galaxies on the nuclear and circumnuclear scales. High-resolution studies of real and simulated disk galaxies are still in their infancy, but have made their first steps in the last few years. Recent reviews on disk galaxies and their subcomponents are presented by Wyse, Gilmore, & Franx (1997) and Carollo et al. (1999). In this review I focus on some recent developments on the central regions of nearby disk galaxies, and discuss some of the related important issues that require future attention. In order to remain faithful to the original studies, and following customary classification schemes, I will often discuss the results maintaining a distinction among systems of early, intermediate, and late types; however, it is this very distinction that I challenge in my concluding remarks.

15.2   First Fact: Complexity is the Rule

All recent high-resolution studies consistently report a large complexity in the inner regions of at least half of the local disk galaxy population of all Hubble types. On scales smaller that 1 kpc, more than half of the galaxies host inner bars (within bars), dust or stellar or gaseous disks, spiral-like dust lanes, star-forming rings, spiral arms, a central cluster (§ 15.5), or simply irregular central emission. (e.g., Carollo et al. 1997c; Carollo, Stiavelli, & Mack 1998; Martini & Pogge 1999; Laine et al. 2002; Böker, Stanek, & van der Marel 2003).

Nuclear bars have received particular attention, as they are claimed to play an important role in feeding gas into the centers of galaxies (Shlosman et al. 1989), potentially building central nuclei and bulges, and fueling nuclear activity.   Intimate links between bars and central starbursts, in particular circumnuclear star-forming rings, are supported by observations (Knapen, Pérez-Ramírez, & Laine 2002). Other studies point out, however, that, on the nuclear scales, stellar rings and inner disks inside large-scale bars of moderately inclined galaxies could be mistaken for secondary bars or even coexist with them, producing erroneous statistics for the occurrence of nuclear bars (Erwin & Sparke 1999). Still, *bona fide* secondary inner bars, typically about 250 pc–1 kpc in size ($\sim$12% the size of their primary bars) appear to be present in as many as 40% of all barred S0–Sa galaxies (Erwin & Sparke 1999). Larger samples of early-type galaxies confirm a high frequency of detection of bars-within-bars (Rest et al. 2001). This high frequency is interpreted to indicate that secondary nuclear bars are relatively long-lived structures. The presence or absence of secondary bars appears to have no significant effect on nuclear activity (as previously reported by, e.g., Regan & Mulchaey 1999). In contrast, nuclear spirals, dusty or star-forming nuclear rings, and off-plane dust are reported to be very often accompanied by LINER or Seyfert nuclei (Erwin & Sparke 1999; Martini & Pogge 1999).

Circumnuclear starburst rings have also been thoroughly investigated (e.g., Maoz et al. 1996, 2001). These rings appear to be a common mode of starbursts in relatively early-type disk galaxies, and are thought to be associated with inner-Lindblad resonances. They contain super star clusters with total luminosities as high as $M_V \approx -15$ mag ($L_V \approx 1.3 \times 10^8 L_{V,\odot}$), radii of the order of a few parsecs, and masses in excess of $10^4 M_\odot$. These clusters are very similar to the super star clusters formed in merging systems (e.g., Whitmore et al. 1999; Hunter et al. 2000): they are bound systems, believed to evolve into stellar structures similar to globular clusters. The starburst rings are thought to be likely associated with bar-driven inflow. Schinnerer et al. (2002) report in the double-barred galaxy NGC 4303, which also hosts a circumnuclear star-forming ring (Colina & Arribas 1999), an extremely good agreement between the observed overall gas geometry and dynamical models for the gas flow in barred galaxies (Englmaier & Shlosman 2000). Observational evidence seems thus to be accumulating in support of the theoretical prediction that disk instabilities on large scales are major drivers of evolution on circumnuclear (and nuclear) galactic scales. Similar to nuclear bars, circumnuclear star-forming rings are found to coexist with AGNs but are not associated one-to-one with AGN activity.

15.3   News on Bulges

It has been known for some time that many bulges have a radial light profile that is not an elliptical-like $r^{1/4}$ law (Andredakis & Sanders 1994; de Jong 1995; Courteau, de Jong, & Broeils 1996); instead, they are reasonably well described by an exponential light profile. Incidentally, the bulge of our own Milky Way also has an exponential light profile (Binney, Gerhard, & Spergel 1997). Recent high-resolution investigations using data from the *Hubble Space Telecope (HST)* have strengthened the evidence for exponential light profiles down to the smallest scales at the end of the spheroids luminosity sequence (Carollo et al. 1998, 2001; Carollo 1999). Several studies have used the generalized surface density profile $I(r) \propto \exp[-(r/r_o)^{(1/n)}]$ introduced by Sérsic (1968) to model the bulge light (Andredakis, Peletier, & Balcells 1995; Graham 2001; MacArthur, Courteau, & Holtzman 2003). These studies report shape-parameter values for bulges of late-type spirals ranging between $n = 0.1$ and 2. Some of the Sérsic analyses attribute a significant meaning to the derived precise values of n (Graham 2001; Balcells et al. 2003). Tests based on simulated data, however, show a large dependence of the derived parameters on, for example, the input parameters; indeed, MacArthur et al. (2003) stress that, on average, the underlying surface density profile for the late-type bulges is adequately described by an exponential distribution. The same studies show the existence of a coupling between bulges and disks that is manifested by an almost-constant scale lengths ratio $h_{bulge}/h_{disk} \approx 0.1$ for late-type spirals, and a similar scaling relation even for earlier-type systems. This is interpreted to indicate a similar origin for bulges of all sizes in hosts of any Hubble type.

For more massive, early-type bulges, ground-based studies using the Sérsic law to describe their light distribution have found values of n close to, or even in excess of the elliptical-like de Vaucouleur's (1948) value of $n = 4$ (Graham 2001). However, the analysis of high-resolution *HST* images for a sample of early-type bulges provides values of the Sérsic shape index n not in excess of ~ 3 (Balcells et al. 2003). The difference in the estimates for n is due to the contribution of photometrically distinct central point sources, which at ground-based resolution are confused for bulge light (see §15.5). Balcells et al. interpret the $n < 3$ Sérsic shape indices in the massive bulges as an indication that even these systems, like the smaller exponential-type bulges, are not the outcome of violent relaxation during collisionless accretion of matter. Both in the ground-based and *HST* analyses, a trend remains between the bulge Sérsic shape parameter n and the bulge luminosity and half-light radius; the trend is in the direction of brighter, bigger bulges having larger n values (Graham et al. 2001; Balcells et al. 2003).

Detailed studies of the integrated stellar populations of bulges of all Hubble types have also been pushed forward by the availability of high-resolution multi-color images from the *HST* (Peletier et al. 1999; Carollo et al. 2001). The independent analyses agree on the basic result that (1) massive early-type bulges have very red colors, unambiguously indicating old ages ($\gtrsim 8$ Gyr) for the average stellar populations of these systems, and (2) the smaller, later-type (almost) exponential bulges have on average significantly bluer colors.

Kinematically, the Sb pseudo-bulges studied by Kormendy (1993) represent the extreme case of a general behavior shown by bulges of any Hubble type and mass: these all appear to have kinematic properties that are closer to disklike structures rather than to elliptical galaxies. Indeed, based on the comparison between minor axis radial velocity dispersions of disks and bulges, Falcón-Barroso et al. (2003) report that even the early-type, massive bulges are actually thickened disks.

The following important considerations on bulges that emerge from the above analyses. (1) The earlier-type bulges form a continuum with the late-type bulges in terms of the shapes of the surface brightness profiles. The smallest bulges are exponential structures, and the largest appear to be intermediate cases between the exponential and the elliptical-galaxy ones. (2) Bulges of spirals are coupled to their host disks in a similar way along the Hubble sequence (with a possible weak trend toward marginally higher h_{bulge}/h_{disk} ratios for early-type bulges). (3) There is a spread in average stellar metallicity and ages amongst bulges, but also a clear trend toward smaller bulges being less enriched and younger stellar structures than the more massive, earlier-type bulges. (4) Bulges of any size show some kinematic features that are typical of disks.

15.4   News on Disks

Recent surveys with the *HST* that have focused on the late-type, allegedly bulgeless Scd–Sm disks, find that only ~30% of these systems have light profiles consistent with being single-exponential structures (Böker et al. 2003). The remaining disks are not well fitted by a single exponential; in particular, the surface brightness in the central few kiloparsecs exceeds the inward extrapolation of the outer exponential disk. The surface brightness profiles of many of these late-type disks are often equally well described either by the sum of two exponential components, or by a single Sérsic profile over the entire radial range with shape parameter n up to a value of 2.5. In the earlier-type systems, a second central exponential component in addition to the outer exponential disk is typically interpreted as a bulge component. Böker et al. (2003) suggest that the frequent detection of such central exponential "excesses" also in systems that, according to the classical classification scheme, should host no bulge component, together with the fact that a single Sérsic profile is often a good alternative to the sum of two exponentials, may indicate that in fact these excesses are not bulges, but rather denser regions of the disks themselves.

A key issue in the context of understanding the nature of the central regions of disk galaxies is one of definitions (see Carollo et al. 1999). Böker et al. distinguish between what they call "the modern theorist" view, assumed to be the correct one, which asserts that a bulge is a kinematically hot component with an extended three-dimensional structure, and the "observers" view, which, in photometric studies, relies on the assumption that disks are exponential structures and that bulges are identifiable as additional light (mass) contributions in the central regions. All the photometric analyses of the local (and distant; see, e.g., Shade et al. 1996) disk galaxy population that are aimed at studying bulge properties indeed assume a constant scale length exponential profile for the disk, and attribute to a bulge any central concentration of light in excess of the inward extrapolation of the outer, constant scale length disk (e.g., Andredakis & Sanders 1994; Andredakis et al. 1995; de Jong 1995; Courteau et al. 1996; Carollo et al. 1998; Balcells et al. 2003; MacArthur et al. 2003). Böker et al. (2003) mention a lack of theoretical support for the disks being exponential, and point out that the assumption that disks remain exponential all the way into the center may not be correct, i.e., that the operational definition of bulges adopted in photometric studies may lead to attributing to a bulge what actually belongs to the disk. It is certainly not an easy task to disentangle into distinct subcomponents the centers of galaxies, where all of these subcomponents are expected to reach their largest densities.

15.5   A Zoom on the Centers: Point Sources and Distinct Nuclei

Photometrically distinct compact nuclei have been known for a while to reside in the centers of bulgeless, weakly active or inactive late-type disks (e.g., Kormendy & McClure 1993; Matthews & Gallagher 1997). Extensive surveys with the *HST* show, however, that distinct pointlike or compact nuclei are the rule rather than the exception in the centers of all sorts of disk galaxies.

An optical (V, WFPC2) and near-infrared (H, J, NICMOS) survey of ~ 100 intermediate-type, Sb-Sc spirals shows that $\sim 70\%$ of these systems host distinct nuclei with visual absolute magnitudes $-8 \gtrsim M_V \gtrsim -16$, comparable at the bright end with young super star clusters in starbursting galaxies (Carollo et al. 1997c, 1998, 2002). Some of these nuclei appear pointlike in the *HST* images. However, many are marginally resolved with half-light radii $\sim 0.\!''1$–$0.\!''2$, corresponding to linear scales of a few to up to ~ 20 pc. These nuclei cover a large range of colors in the range $-0.5\,\mathrm{mag} \lesssim V-H \lesssim 3\,\mathrm{mag}$. Statistically, the distribution of $V-H$ colors is broader for the nuclei than for the surrounding galactic structure; this suggests that star formation, AGN activity, dust reddening, or a combination of these is generally present in the nuclei embedded in intermediate-type hosts. Most of the nuclei, at *HST* resolution, are located at the galaxy centers and appear to be round, star-cluster-like, structures. However, in some cases the nuclei are offset from the isophotal/dynamical centers of the host galaxies or show some degree of elongation (as it is the case for, e.g., the nucleus of M33; Kormendy & McClure 1993; Lauer et al. 1998; Matthews et al. 1999). Both the displacement, which is typically of a few tens of parsecs, and the elongation appear to be uncorrelated with either the luminosity and color of the nucleus, or with the galaxy type. Searches for trends with other galactic subcomponents reveal no clear relationship between the distinct nuclei and nuclear (or even larger-scale) bars: bars are neither ubiquitous nor unusual in nucleated intermediate-type spirals. Some of the nuclei are embedded in exponential-type bulges; actually, every such bulge in this sample hosts a distinct nucleus in its isophotal center or slightly offset from it. These nuclei have low to moderate luminosities, $-8\,\mathrm{mag} \gtrsim M_V \gtrsim -12$ mag. Selection effects may be present. Brighter nuclei are typically embedded in very complex circumnuclear structures that do not allow for the derivation of reliable bulge parameters. Moreover, exponential-type bulges lying under substantial circumnuclear structure but hosting no central nucleus would also, for similar reasons, drop from the sample. The nuclei that have been identified inside the exponential-type bulges have colors compatible with those arising from stellar populations. Under the assumption of no dust reddening, average stellar ages of ~ 1 Gyr are inferred for these relatively faint central star clusters; these ages, in turn, imply masses of about 10^6 to $10^8\ M_\odot$.

Studies of large samples of Scd and later-type disks also find compact, distinct nuclei, identified as "star clusters," in about 75% of the population (Böker et al. 2002). Their distribution of absolute luminosities has a FWHM of about 4 mag and a median value of $M_I = -11.5$ mag. These luminosities, as well as the sizes of these star clusters, are comparable to those of the nuclei embedded in the earlier-type galaxies. For 10 nuclei in the Böker et al. (2002) sample, Walcher et al. (2004) report spectroscopic estimates for the stellar ages that are smaller than about 1 Gyr, and masses of the nuclei estimated from stellar velocity dispersions in the range $10^6 - 10^8\ M_\odot$. These masses are similar to the photometry-based mass estimates derived for the clusters embedded in the exponential-type bulges, and are one order of magnitude higher than expected from stellar synthesis models.

These authors interpret this discrepancy as due to old(er) stellar generations contributing to the total mass of the nuclei. They conclude that the nuclei of late-type disks are grown in multiple star formation events, and suggest that this mechanism may contribute to the formation of Kormendy's pseudo-bulges.

At the other extreme of the Hubble sequence, the large majority of the early-type spiral galaxies that host massive bulges also require a central component, in addition to the bulge and the disk, to reproduce the observed light profiles (Balcells et al. 2003). In most cases, this additional component appears to be a point source at the resolution of NICMOS on the *HST*. Balcells et al. point out that in ground-based data these additional light contributions cannot be disentangled from the bulge light and may produce overestimates of the derived Sérsic *n* parameter (see §15.3).

15.6   Future Challenges

15.6.1   *Disks*

The claim that there is no strong theoretical support for the disks to be exponential (Böker et al. 2003) is partly substantiated by the fact that theories are often tuned so as to reproduce an exponential light profile. While reproducing exponential light profiles is considered to be a test for viable models of disk formation (e.g., Dalcanton, Spergel, & Summers 1997; Silk 2001), it is certainly true that most of the current disk models and simulations do predict large mass fractions with low-angular momentum material that is in excess of the extrapolation of the outer exponential density distribution to the center (see §15.1 and references therein). However, the real problem is to understand whether this excess low-angular momentum material does remain in the disk, or rather forms a three-dimensional bulge.

Furthermore, it is also possible that, rather than being born denser than exponential, the disks may become so during the subsequent galactic evolution. Indeed many processes can occur during the Hubble time that can transform exponential disks into the more complex structures that are observed in late-type disks. In this context, it is worth stressing that a third of the late-type disks remais well described by the simple single-exponential form, showing that this channel of disk formation is indeed accessible to real galaxies.

An obvious example of change in central concentration in disks is the one induced by the formation and subsequent buckling of a stellar bar. To study at unprecedented resolution the effects of secular evolution processes on the central regions of disks, we are conducting state-of-the-art *N*-body and *N*-body+SPH numerical simulations of disk galaxies (details will be published in Debattista et al. 2004 and Mayer et al. 2004). The first *N*-body experiments that I briefly discuss here (from Debattista et al. 2004) consist of live disk components inside frozen halos described either by a spherical logarithmic potential with a central core or a cuspy Hernquist (1990) potential. The initially axisymmetric disks are modeled assuming an exponential profile with a Gaussian thickening; the disks are represented by $(4-7.5) \times 10^6$ equal-mass particles. The spatial resolution that is achieved in the central regions is $\sim$50 pc. The simulations are run on a 3-D cylindrical polar-grid code (described in Sellwood & Valluri 1997). In certain areas of parameter space, the axisymmetric systems are found to be unstable and form bars. Systems in which bars fail to form have only modest heating, indicating that our results are not driven by noise. Every 20 time steps of the disk evolution, we measure the disk velocity dispersions and streaming velocities in annuli, the amplitude

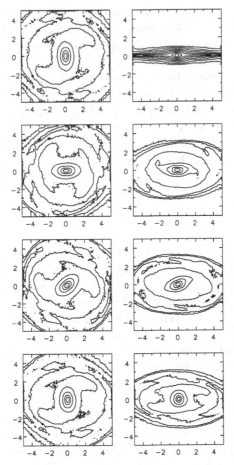

Fig. 15.1. Contour plots for a live-disk/frozen-halo simulation after the development and buckling of a stellar bar. The halo has a Hernquist (1990) profile; the disk initially has a single-exponential profile. Shown are the face-on and edge-on views of the system (upper panels, left and right, respectively), and two different views with disk inclination angles of 30° (left) and 60° (right), respectively. Different panels from second-top to bottom show different orientation angles for the buckled bar of 0°, 45°, and 90°, respectively, with respect to the major axis of the disk. Axes are in units of the initial exponential disk scale length. (From Debattista et al. 2004.)

of the bar from the $m = 2$ Fourier moment, and the amplitude of the buckling from the $m = 2$ Fourier moment of the vertical displacement of particles. We use these quantities to determine when the bar forms, when it buckles, and the evolution of disk properties such as mass density, morphological diagnostics (for any inclination angle of the disk and orientation of the bar), and the V/σ ratio. As an example, Figure 15.1 shows, for various disk inclination angles and tilt angles of the bar with respect to the disk major axis, the isophotal contours of the projected surface density of the system after the bar has formed and buckled. As already pointed out in, for example, Raha et al. (1991), for some projections, including but not uniquely for the edge-on one, the buckled bar looks very much like a normal

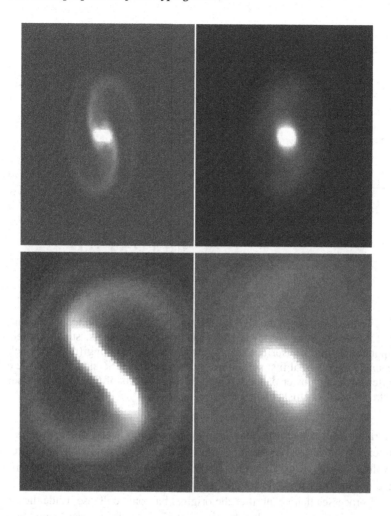

Fig. 15.2. A Hernquist halo simulation (upper panels) and a core halo simulation (lower panels). The images are taken before the buckling (left panels) and after the buckling (right panels) of the bars. The models have inclination angles of 60° and 30°, and bar orientation angles of 90° and 45°, respectively. (From Debattista et al. 2004.)

(three-dimensional) "rounder-than-the-disk" bulge. Figure 15.2 shows, for two different simulations, how the systems would appear on the sky as observed from a specific line-of-sight before and after the buckling of the bars. For the same two models, Figure 15.3 plots the initial surface density profiles (exponential by construction), the surface density profiles after the buckling of the bars, and the post-buckling ellipticity profiles of the systems. Figure 15.3 points out that an initially exponential disk that "nature" makes can be observed to be a more centrally concentrated structure at a later stage, after it has formed and buckled a stellar bar. The final, post-buckling profile in the simulations is well described by the sum of an outer exponential disk and an inner Sérsic component, as observed in real disk galaxies. If the denser-than-exponential profiles in the real late-type spirals were due to the effects

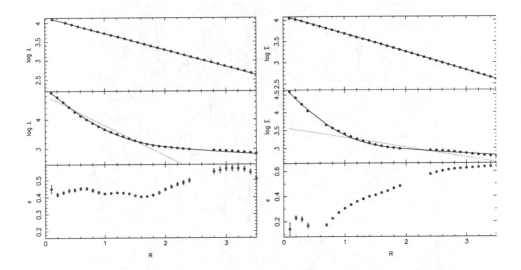

Fig. 15.3. For the same core halo (left) and Hernquist halo (right) models shown in Figure 15.2, we plot here the initial surface density profiles (upper panels: superimposed with the single-exponential fits), the surface density profiles after buckling (middle panels: dotted lines, the single exponential fits; solid lines, exponential plus Sérsic fits), and the ellipticity profiles after buckling (lower panels). The x-axis is in units of the initial exponential disk scale length. (From Debattista et al. 2004.)

of nurture rather than nature, the inner light/mass excesses in the late-type disks would be better associated with "bulge" structures; that is, they would be the "bulges" produced by secular evolution of the disks, which has been extensively discussed in the literature.

Bars are not the only possible solution to increase the central densities in disk galaxies by means of processes that occur after the original baryonic collapse inside the dark halos: viscosity may be important (e.g., Lin & Pringle 1987), and also mergers, satellite accretion, dynamical friction of globulars, etc. Nonetheless, it is fair to conclude that, at this stage, the issue whether the disks are born as denser-than-exponential structures remains open. If this were the case, it will be important to quantify the systematic uncertainties on, for example, bulge scale lengths and luminosities, black hole masses and other galactic properties that are derived assuming that nature, when it makes a disk, makes it exponential.

15.6.2 Bulges

The investigations of the past few years indicate that even the most massive, early-type bulges are not $r^{1/4}$-law systems and have disklike imprints in their kinematics. How do we reconcile, under a common denominator, the differences between bulges and ellipticals with the quoted similarities of stellar population and scaling laws? It is certainly not clear what, for instance, the Mg$_2$ index and the velocity dispersion σ represent in the Mg$_2 - \sigma$ relation. Are the key parameters metallicity, age, or a combination of the two? Are they the depth of potential well, local physics of star formation, or, again, a combination of the two? Local physics imposes thresholds for star formation (e.g., Meurer et al. 1997), which is likely to have an impact on scaling laws such as the $Mg_2 - \sigma$ relation. Indeed, the same $Mg_2 - \sigma$ relation is observed to hold over orders of magnitude of scale lengths, in systems that are very different, ranging from elliptical galaxies to dwarf spheroidals (Bender, Burstein, & Faber 1993). The conclusion is that the $Mg_2 - \sigma$ and similar relations are certainly telling us something important about the formation of stellar systems over a large range of scales, but not necessarily that they all share a similar formation process.

On the other hand, the claims that violent relaxation is not a major player in the formation of bulges, based on the observed Sérsic profiles with $n \lesssim 3$ (Balcells et al. 2003) may also be premature. The consequence of violent relaxation during dissipationless processes such as stellar clumpy collapses (van Albada 1982), mergers of disk galaxies (Barnes 1988), satellite accretion onto disk galaxies (Aguerri, Balcells, & Peletier 2001) is to produce an $r^{1/4}$ profile. However, other studies of violent relaxation in a finite volume show deviations from the $r^{1/4}$ law (Hjorth & Madsen 1995). Furthermore, the same problem of separating nature from nurture may be relevant also in this context. Physical processes may occur during the Hubble time that modify the stellar density profiles in the centers of galaxies, including dynamical friction of globular clusters, dissipative accretion of matter, black hole-driven cusp formation, mergers of black holes (quantitative studies of the latter show that central mass deficits are created from the binding energy liberated by the coalescence of the supermassive binary black holes; see, e.g., Milosavljević et al. 2002, Ravindranath, Ho, & Filippenko 2002, and references therein). Numerical studies of these processes are still rather sketchy and do not explore a vast volume of parameter space; nonetheless, they make the point that the nuclear stellar density profiles may be modified by subsequent evolution. Quantitative work remains to be done to assess whether these or other processes can reproduce the $n \approx 3$ Sérsic profiles typical of the massive bulges and the weak trend between Sérsic shape parameter n and bulge luminosity. The possibility that the disks may not be purely exponential also introduces additional uncertainties on the derived bulge parameters, including the shape index n. If the outer disk can have a Sérsic shape with n values as steep as ~ 2.5, bulge-disk decompositions that use an exponential for the outer disks can systematically offset the bulge parameters. This could even open the question as to whether the observed sequence in n values between the late-type and early-type bulges is a pure bulge sequence, or, rather, at least in part a sequence of different underlying disk profiles.

Concerning support for bulge-building secular evolution processes inside preexisting disks, there is certainly at this point good evidence from high-resolution numerical experiments that the intrinsic evolution of the disks results in transformations of the disks, which can generate three-dimensional structures that resemble bulgelike components. Numerical studies (Pfenniger & Friedli 1991; Zhang & Wyse 2000; Scannapieco & Tisseira 2003; Debattista et al. 2004; see Fig. 15.3) also show that the bulgelike, three-dimensional structures that generally result from the evolution of the disks have the rather low-n Sérsic profiles typical of real bulges. MacArthur et al. (2003) report that simulations by D. Pfenniger (2002, private communication) of self-gravitating disks form bars that may later dissolve into bulgelike components, which show a nearly universal ratio of bulge-to-disk scale lengths, also in agreement with the observed correlations. In the simulations, the universal ratio of bulge-to-disk scale lengths is related to the stellar dynamics of the barred system, for example to the relative position of the vertical to horizontal resonances. There is an additional important ingredient that has been missing so far in the debate concerning the

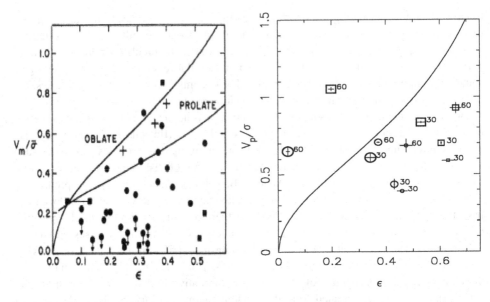

Fig. 15.4. V/σ versus ellipticity ϵ plane. The left panel reproduces Figure 2 of Davies & Illingworth (1983); points are the measurements for observed spheroids, both bulges (crosses), and ellipticals (filled circles and squares). The right panel shows a similar plot derived from our simulations. The oblate-rotator line is also plotted as a solid line. The data points in this panel refer to the end of the simulations, after the bars that have formed in the disks have buckled. The V/σ and ϵ values for the "buckled bars" are derived by averaging the relative profiles inside the half-light radii of the inner Sérsic components that are necessary, in addition to the outer exponential disk components, to obtain good fits to the surface density profiles after buckling. Circles are used for the logarithmic and squares for the Hernquist halo potentials. The size of the symbols refers to the bar orientation angles (90° largest, 45° intermediate, 0° smallest). The disk inclination angle i is explicitly indicated close to symbols. From certain viewing angles and bars orientation angles, buckled bars are indistinguishable from normal spheroids on the V/σ versus ellipticity ϵ plane. (From Debattista et al. 2004.)

possibility that disk secular evolution processes play a substantial role in forming bulgelike structures: namely, the bulges that result from the secular evolution of the disks are, in contrast to what is commonly asserted, not necessarily dynamically cold, "disklike" stellar systems. Due to the fact that eccentric orbits are quickly erased by shocks (Friedli & Benz 1995), the secular evolution of mostly gaseous disks indeed produces cold stellar structures such as the pseudo-bulges discussed by Kormendy (1993); however, the buckling of stellar bars, for example, can produce structures that are, at least from certain viewing angles, indistinguishable from the alleged "normal" bulges in classical diagnostic planes such as the V/σ-ϵ plane. This is shown in Figure 15.4, where the locations on the V/σ-ϵ of a few representative buckled bars from our simulations are shown (right panel) in comparison with what is typically considered the *bona fide* bulge behavior (left panel, figure from Davies & Illingworth 1983).

In summary, from Figures 15.1–15.4, it is evident that, depending on the viewing angle, buckled bars can appear as structures that are simultaneously rounder than the surrounding disks, photometrically identifiable as additional components in excess of outer exponential disks, and kinematically similar to what are considered to be *"bona fide* bulges." In this light, it seems appropriate to question indeed what is a meaningful definition for a *bona fide* bulge. Clearly, the situation is more complex than what is captured in the theorist-versus-observer dichotomy discussed by Böker et al. (§15.4). First, from an observational perspective, even early-type, *bona fide* bulges have been claimed to be thickened disks (Falcón-Barroso et al. 2003). Second, from a theoretical perspective, evolutionary disk processes such as the buckling of progenitor bars inside the disks can produce structures that, in contrast to common belief, are dynamically similar to the *bona fide* bulges that should be the benchmark for the comparison. Thus, as with the photometric classification, even the kinematic classification of bulges is quite fuzzy. Ultimately, this is due to the lack of a proper physical boundary between structures that are forced into different categories by what may be unfolding into an obsolete and confusing classification scheme.

15.6.3 *The Nature and Role of Nuclei and Central Black Holes*

Recent surveys show that central, distinct, compact components, in addition to the disk and the bulge, are present in the large majority of disk galaxies of all Hubble types. Many are clearly star clusters with no AGN contamination. This includes, for example, the "naked" ones in the late-type disks studied by Walcher et al. (2004) and probably the relatively faint population of nuclei embedded in the relatively clean surroundings of the exponential-type bulges (Carollo et al. 1997c, 1998). AGNs are known to be rare in late-type galaxies (Ho, Filippenko, & Sargent 1997; Ulvestad & Ho 2002; Ho 2004). An AGN component may, however, be present in a fraction of the nuclei. This would be statistically consistent with the fact that about 70% of spirals host a distinct nucleus, and about half of them are known to host some form of AGN, even if weak (Ho et al. 1997). Some of the point sources embedded in the early-type bulges of Balcells et al. (2003) may also have an AGN origin or component; pointlike sources associated with AGNs are seen in massive elliptical galaxies (Carollo et al. 1997a, b; Ravindranath et al. 2001).

The young stellar ages plus high velocity dispersions of the central star clusters of late-type disks reported by Walcher et al. (2004) may certainly imply a large spread in stellar population ages, and thus an iterative mass assembly and star formation for the central star clusters, as discussed by the authors. However, the nuclei that are typically selected for the spectroscopic investigations populate the bright end of the luminosity distribution of nuclei. Walcher et al. (2004) stress that in their sample there is no indication that brighter means younger; nevertheless, it is still possible that selection effects are important and that fainter nuclei may have less complex mass assembly and star formation histories. A wide range of star formation histories would be more consistent with a process of growth of central star clusters that is regulated by local physics, for instance by the amount of fuel (either gas or smaller star clusters) available at various epochs in the circumnuclear regions, the angular momentum distribution or orbital structure of this "fuel," and the physical state of the central regions of the disk (e.g., its density or dynamical temperature, in turn determining or originating from the steepness of the gravitational potential, the conditions to develop non-axisymmetric perturbations on small scales, etc.). Furthermore, it is still unknown whether fuel-starved, silent AGN engines—massive black holes—may be present in the central star clusters (e.g., Marconi et al. 2003). The question of whether massive black holes reside in general in the centers of star clusters is far from settled. The case of G1, a globular cluster in

Andromeda in which a central black hole of the mass expected from the linear extrapolation of the relationship reported for the massive spheroids (e.g., Gebhardt et al. 2000) has been detected (Gebhardt, Rich, & Ho 2002), argues for the presence of massive black holes in the centers of star clusters, and supports the suggestion that black holes are ubiquitous and proportionally sized in all spheroids, from mass scale of globular clusters to elliptical galaxies. A small, $\sim 10^{4-5} M_\odot$ black hole is found embedded in the central star cluster of NGC 4395, one of the least luminous and nearest known Type 1 Seyfert galaxies (Filippenko & Ho 2003). On the other hand, the nondetection of a central black hole in the central star cluster of M33 contrasts with the G1 case and argues for the absence of massive black holes in the centers of the distinct nuclei of bulgeless disks. Gebhardt et al. (2001) discuss that, if the mass of a central black hole in the nucleus of M33 was related to its velocity dispersion in the same way that the known supermassive black holes are related to the dispersions of their bulges, then a black hole with mass in the range $\sim 7 \times 10^3 - 6 \times 10^4 M_\odot$ would be expected, well above the measured upper limit of $1500 M_\odot$. Solutions to this inconsistency include those suggested by the authors: the relationship between the mass of the black hole and the velocity dispersion of the host spheroid may be nonlinear; the conditions to make a massive black hole were better in the earlier, denser Universe, when the stars in G1 were made; or M33's young nucleus has not had enough time to create its own black hole. Given the observational uncertainties, other possibilities remain. It could be that G1 is not a star cluster but a harassed spheroidal galaxy [a fact mentioned by Gebhardt et al. (2002) but not considered by the authors as the cause for the discrepancy]. Another possibility is that at least in small-sized spheroids such as star clusters, black holes may not be ubiquitous, or there may not exist a tight correlation between black hole mass and spheroid mass. Or perhaps normal star clusters and the central star clusters in disk galaxies have a different origin.

The case of M33 serves also as a smoking gun in another context. Kormendy & Gebhardt (2001; see also Kormendy et al. 2004) report that the same correlation between the mass of the central black hole and the host luminous spheroid holds for galaxies with both "normal" and kinematically cold, disklike bulges (i.e., the "pseudobulges" discussed by Kormendy 1993). In contrast, M33, a pure disk galaxy with no bulge component of any sort, is indeed found to lack a black hole. Kormendy & Gebhardt (2001) conclude that the basic requirement for making a supermassive central black hole appears to be that the galaxy is capable of forming some kind of dense, bulgelike structure, whatever its nature. Reinterpreting this comment in the light of the bulge/dense-disk conundrum discussed above, the results of Kormendy & Gebhardt (2001) and Gebhardt et al. (2001) may imply that the requirement for making a supermassive central black hole is that the galaxy is capable of reaching sufficiently high central baryonic densities. Either way, from these analyses it appears that black hole masses are not correlated with the total gravitational potential of the disks, and thus of the host dark matter halos. A contrasting report, however, comes from Ferrarese (2002) and Baes et al. (2003), who claim a tight correlation between the circular velocities of galaxies and the masses of their central supermassive black holes, and thus an intimate link between the black holes and the host dark matter halos. Supermassive black holes do form in some pure disk systems, as shown by Filippenko & Ho (2003) for the case of NGC 4395. However, these authors stress that in this galaxy the estimated black hole mass is consistent with the $M_\bullet - \sigma$ relation of Tremaine et al. (2002), if the central cluster is considered in lieu of the bulge. For a $\sigma = 30$ km s^{-1}, a good upper limit for the velocity dispersion of central star cluster in NGC 4395, this relation predicts a $M_\bullet = 6.6 \times 10^4 M_\odot$, consistent with the mass independently estimated from the AGN properties (Filippenko & Ho 2003). Furthermore, it remains a fact that M33, possibly the best candidate to test for the validity of a correlation between the black hole mass and the dark matter halo mass, appears not to support it. As stressed by Gebhardt et al. (2001), if a black hole in M33 were indeed related to the dark matter potential well, then M33 should contain a black hole of mass significantly in excess of $10^6 M_\odot$, which it does not. It may be best to wait for the observational picture to be cleared up before attempting interpretations of the claimed correlation between black hole and dark halo masses.

Finally, given the large frequency of occurrence of nuclei in disk galaxies and the generally accepted idea of hierarchical galaxy assembly, an interesting question is whether the formation and evolution of the nuclei of disk galaxies play any relevant role in the formation of supermassive black holes in the centers of galaxies. More generally, a key question for the future is whether the nearly ubiquitous nuclei are a nuance or rather an important ingredient in the formation process of disk galaxies.

15.7   Concluding Remarks

In summary, at a resolution of typically a few tens to a few hundred parsecs, the local disk galaxy population appears to host bulges that more and more resemble disks, and disks that more and more resemble bulges. Disks may be denser than exponential, and bulges may be less steep than de Vacoluleur's structures. Bulges are claimed to have the kinematics of thickened disks, and buckled bars can be as dynamically hot as the structures claimed to be "true" bulges. The average stellar population properties (i.e., stellar ages and metallicities) remain the only surviving distinction between "massive" and "small" bulges and, more generally, between massive bulges and the centers of disks. This could certainly be an indication of different formation processes, but could also be the result of similar processes occurring at different epochs in the Universe (and thus naturally generating a positive correlation between stellar densities and ages). In my view, what is needed at this point is a shift of the debate from the arena of morphological classifications, where bulges and disks are distinct entities and the question "what is the origin of bulges' is kept distinct from the question "what is the origin of disks," to one where disk galaxies are studied as a whole without the constraints of a rigid classification scheme. The historical focus on morphology, while highlighting many details in the trees, may in fact have hidden the true nature of the forest. The expected outcome will be a renewed concept of the "Hubble sequence" that will be ultimately be based on physical rather than morphological considerations. Clarifying what we really see nearby as the endpoint of the galaxy evolution process is essential in order to meaningfully answer how the distant progenitors, which are seen in the most remote regions of the Universe, transform themselves to become the descendants that populate our own surroundings.

Acknowledgements. I thank L. Ho for the invitation to this very stimulating meeting, and especially for his patience waiting for this manuscript. I am grateful to my collaborators, V. Debattista, L. Mayer and B. Moore for kindly making available some of our results prior to publication. Many thanks to F. van den Bosch and S. Lilly for comments on a previous version of this manuscript.

References

Aguerri, J. A. L., Balcells, M., & Peletier, R. F. 2001, A&A, 367, 428

Andredakis, Y. C., Peletier, R. F., & Balcells, M. 1995, MNRAS, 275, 874

Andredakis, Y. C., & Sanders, R. H. 1994, MNRAS, 267, 283

Baes, M., Buyle, P., Hau, G. K. T., & Dejonghe, H. 2003, MNRAS, 341, L44

Balcells, M., Graham, A. W., Domínguez-Palmero, L., & Peletier, R. F. 2003, ApJ, 582, L79

Barnes, J. E. 1988, ApJ, 331, 699

Bender, R., Burstein, D., & Faber, S. M. 1993, ApJ, 411, 153

Binney, J., Gerhard, O. E., & Spergel, D. N. 1997, MNRAS, 288, 365

Binney, J., & Tremaine, S. 1987, Galactic Dynamics (Princeton: Princeton Univ. Press)

Böker, T., Stanek, R., & van der Marel, R. P. 2003, AJ, 125, 1073

Böker, T., van der Marel, R. P., Laine, S., Rix, H.-W., Sarzi, M., Ho, L. C., & Shields, J. C. 2002, AJ, 123, 1389

Bullock, J. S., Dekel, A., Kolatt, T. S., Kravtsov, A. V., Klypin, A. A., Porciani, C., & Primack, J. R. 2001, ApJ, 555, 240

Bullock, J. S., Kravtsov, A. V., & Colin, P. 2002, ApJ, 564, L1

Carollo, C. M. 1999, ApJ, 523, 566

Carollo, C. M., Danziger, I. J., Rich, R. M., & Chen, X. 1997a, ApJ, 491, 545

Carollo, C. M., Ferguson, H. C., & Wyse, R. F. G., ed. 1999, The Formation of Bulges (Cambridge: Cambridge Univ. Press)

Carollo, C. M., Franx, M., Illingworth, G., & Forbes, D. A. 1997c, ApJ, 481, 710

Carollo, C. M., Stiavelli, M., de Zeeuw, P. T., & Mack, J. 1997c, AJ, 114, 2366

Carollo, C. M., Stiavelli, M., de Zeeuw, P. T., Seigar, M. 2001, ApJ, 546, 216

Carollo, C. M., Stiavelli, M., & Mack, J. 1998, AJ, 116, 68

Carollo, C. M., Stiavelli, M., Seigar, M., de Zeeuw, P. T., & Dejonghe, H. 2002, AJ, 123, 159

Colina, L., & Arribas, S. 1999, ApJ, 514, 637

Courteau, S., de Jong, R. S., & Broeils, A. H. 1996, ApJ, 457, L73

Dalcanto, J. J., Spergel, D. N., & Summers, F. J. 1997, ApJ, 482, 659

Davies, R. L., & Illingworth, G. D. 1983, ApJ, 266, 516

Debattista, V. P., Carollo, C. M., Mayer, L., & Moore, B. 2004, in preparation

de Jong, R. S. 1995, Ph.D Thesis, Univ. Groningen

de Vaucouleurs, G. 1948, Ann. d'Ap., 11, 24

Englmaier, P., & Shlosman, I. 2000, ApJ, 528, 677

Erwin, P., & Sparke, L. S. 1999, ApJ, 521, L37

Falcón-Barroso, J., Balcells, M., Peletier, R. F., & Vazdekis, A. 2003, A&A, 405, 455

Fall, S. M., & Efstathiou, G. 1980, MNRAS, 193, 189

Ferrarese, L. 2002, ApJ, 578, 90

Filippenko, A. V., & Ho, L. C. 2003, ApJ, 588, L13

Friedli, D., & Benz, W. 1995, A&A, 301, 649

Gebhardt, K., et al. 2000, ApJ, 539, L13

——. 2001, AJ 122, 2469

Gebhardt, K., Rich, R. M., & Ho, L. C. 2002, ApJ, 578, L41

Governato, F., et al. 2004, ApJ, submitted (astro-ph/0207044)

Graham, A. 2001, AJ, 121, 820

Hernquist, L. 1990, ApJ, 356, 359

Hjorth, J., & Madsen, J. 1995, ApJ, 445, 55

Ho, L. C. 2004, in Carnegie Observatories Astrophysics Series, Vol. 1: Coevolution of Black Holes and Galaxies, ed. L. C. Ho (Cambridge: Cambridge Univ. Press), in press

Ho, L. C., Filippenko, A. V., & Sargent, W. L. W. 1997, ApJ, 487, 568

Hoyle, F. 1953, ApJ, 118, 513

Hunter, D. A., O'Connell, R. W., Gallagher, III, J. S., & Smecker-Hane, T. A. 2000, AJ, 120, 2383

Jablonka, P., Martin, P., & Arimoto, N. 1996, AJ, 112, 1415

Knapen, J. H., Pérez-Ramírez, D., & Laine, S. 2002, MNRAS, 337, 808

Knebe, A., Islam, R. R., & Silk, J. 2001, MNRAS, 326, 109

Kormendy, J. 1993, in Galactic Bulges, ed. H. Dejonghe & H. J. Habing (Dordrecht: Kluwer), 209

Kormendy, J., et al. 2004, ApJ, submitted

Kormendy, J., Bender, R., & Bower, G. 2002, in The Dynamics, Structure, and History of Galaxies, ed. G. S. Da Costa & H. Jerjen (San Francisco: ASP), 29

Kormendy, J., & Gebhardt, K. 2001, in The 20th Texas Symposium on Relativistic Astrophysics, ed. H. Martel & J. C. Wheeler (Melville: AIP), 363

Kormendy, J., & McClure, R. D. 1993, AJ, 105, 1793

Laine, S., Shlosman, I., Knapen, J. H., & Peletier, R. F. 2002, ApJ, 567, 97

Lauer, T. R., Faber, S. M., Ajhar, E. A., Grillmair, C. J., & Scowen, P. A. 1998, AJ, 116, 2263

Lin, D. C., & Pringle, J. E. 1987, ApJ, 320, L87

MacArthur, L., Courteau, S., & Holtzman, J. A. 2003, ApJ, 582, 689

Maoz, D., Barth, A. J., Ho, L. C., Sternberg, A., & Filippenko, A. V. 2001, AJ, 121, 3048

Maoz, D., Barth, A. J., Sternberg, A., Filippenko, A. V., Ho, L. C., Macchetto, F. D., Rix, H.-W., & Schneider, D. P. 1996, AJ, 111, 2248

Marconi, A., et al. 2003, ApJ, 586, 868

Martini, P., & Pogge, R. W. 1999, AJ, 118, 2646

Matthews, L. D., et al. 1999, AJ 118, 208

Matthews, L. D., & Gallagher, J. S., III 1997, AJ, 114, 1899

Mayer, L., Moore, B., Debattista, V. P., & Carollo, C. M. 2004, in preparation

Mestel, L. 1963, MNRAS, 126, 553

Meurer, G. R., Heckman, T. M., Lehnert, M. D., Leitherer, C., & Lowenthal, J. 1997, AJ, 114, 54

Milosavljević, M., Merritt, D., Rest, A., & van den Bosch, F. C. 2002, MNRAS, 331, L51

Norman, C. A., Sellwood, J. A., & Hasan, H. 1996, ApJ, 462, 114

Peletier, R. F., Balcells, M., Davies, R. L., Andredakis, Y., Vazdekis, A., Burkert, A., & Prada, F. 1999, MNRAS, 310, 703

Persic, M., & Salucci P. 1995, ApJS, 99, 501

Pfenniger, D., & Friedli, D. 1991, A&A, 252, 75

Pfenniger, D., & Norman, C. 1990, ApJ, 363, 391

Raha, N., Sellwood, J. A., James, R. A., & Kahn, F. D. 1991, Nature, 352, 411

Ravindranath, S., Ho, L. C., & Filippenko, A. V. 2002, ApJ, 566, 801

Ravindranath, S., Ho, L. C., Peng, C. Y., Filippenko, A. V., & Sargent, W. L. W. 2001, AJ, 122, 653

Regan, M. W., & Mulchaey, J. S. 1999, AJ, 117, 2676

Renzini, A. 1999, in The Formation of Galactic Bulges, ed. C. M. Carollo, H. C. Ferguson, & R. F. G. Wyse (Cambridge: Cambridge Univ. Press), 1

Rest, A., van den Bosch, F. C., Jaffe, W., Tran, H., Tsvetanov, Z., Ford, H. C., Davies, J., & Schafer, J. 2001, AJ, 121, 2431

Scannapieco, E., & Tissera, P. B. 2003, MNRAS, 338, 880

Schade, D., Lilly, S. J., Le Févre, O., Hammer, F., & Crampton, D. 1996, ApJ, 464, 79

Schinnerer, E., Maciejeweski, W. J., Scoville, N. Z., & Moustakas, L. A. 2002, ApJ, 575, 826

Sellwood, J. A., & Valluri, M. 1997, MNRAS, 287, 124

Sérsic, J. L. 1968, Atlas de Galaxias Australes (Córdoba: Obs. Astron., Univ. Nac. Córdoba)

Shlosman, I., Frank, J., & Begelman, M. C. 1989, Nature, 338, 45

Silk, J. 2001, MNRAS, 324, 313

Springel, V., & Hernquist, L. 2002, MNRAS, 333, 649

Steinmetz, M., & Navarro, J. 1999, ApJ, 513, 555

Strauss, M. A., & Willick, J. A. 1995, Phys. Rep., 261, 271

Tremaine, S., et al. 2002, ApJ, 574, 740

Ulvestad, J. S., & Ho, L. C. 2002, ApJ, 581, 925

van Albada, T. S. 1982, MNRAS, 201, 939

van den Bosch, F. C. 2001, MNRAS, 327, 1334

van den Bosch, F. C., Abel, T., Croft, R. A. C., Hernquist, L., & White, S. D. M. 2002, ApJ, 576, 21

Walcher, C. J., Häring, N., Böker, T., Rix, H.-W., van der Marel, R. P., Gerssen, J., Ho, L. C., & Shields, J. C. 2004, in Carnegie Observatories Astrophysics Series, Vol. 1: Coevolution of Black Holes and Galaxies, ed. L. C. Ho (Pasadena: Carnegie Observatories, http://www.ociw.edu/ociw/symposia/series/symposium1/proceedings.html)

Whitmore, B. C., Zhang, Q., Leitherer, C., Fall, S. M., Schweizer, F., & Miller, B. W. 1999, AJ, 118, 1551

Wyse, R. F. G., Gilmore, G., & Franx, M. 1997, ARA&A, 35, 637

Zhang, B., & Wyse, R. F. G. 2000, MNRAS, 313, 310

16

Influence of black holes on stellar orbits

KARL GEBHARDT

Department of Astronomy, University of Texas at Austin

Abstract

We review the current state of dynamical modeling for galaxies in terms of being able to measure both the central black hole mass and stellar orbital structure. Both of these must be known adequately to measure either property. The current set of dynamical models do provide accurate estimates of the black hole mass *and* the stellar orbital distribution. Generally, these models are able to measure the black hole mass to about 20%–30% accuracy given present observations, and the stellar orbital structure to about 20% accuracy in the radial to tangential dispersions. The stellar orbital structure of the stars near the galaxy center show strong tangential velocity anisotropy for most galaxies studied. Theoretical models that best match this trend are black hole binary/merger models. There is also a strong correlation between black hole mass and the contribution of radial motion at large radii. This correlation may be an important aspect of galaxy evolution.

16.1   Introduction

The first observational evidence that black holes are common in the centers of nearby galaxies is reviewed in Kormendy (1993) and Kormendy & Richstone (1995). The initial studies concentrate mainly on measuring the black hole mass and only somewhat included the effects of different orbital structure. However, it was always apparent that the assumed form for the distribution function has a considerable effect on the measured black hole mass. Thus, the believability in the existence of a central black hole closely paralleled the development of more sophisticated modeling techniques that were designed to be as general as possible.

There are two main aspects for making a general dynamical model. These are the dimensionality of the potential and that of the velocity ellipsoid. For the potential, we know that we have to at least model galaxies as axisymmetric, and, for some, triaxial structure is required (e.g., those with counterrotating cores, polar rings, etc.). While it is important to allow the most freedom for a dynamical model, there is a level of detail that need not be studied (at least at present). For example, we know that no galaxy is exactly symmetric along any axis. Therefore, in order to provide an adequate representation in that case, one cannot use symmetric dynamical models, but instead must rely on *N*-body simulations—similar to what is done when modeling merging systems (Barnes & Hernquist 1992). Using an *N*-body system to model each galaxy is currently not practical, and, furthermore, may not even provide a better understanding of the underlying physics due to the huge parameter space inherent in *N*-body simulations. Thus, the most to gain lies in using general dynamical

© The Observatories of the Carnegie Institution of Washington 2004.

Dimension of Velocity Ellipsoid

parametric
nonparametric

		1D Isotropic	2D r, θ	3D r, θ, φ
1D	**Spherical**	King models Plummer many, many others *Gebhardt & Fischer (95)*	King (66), Michie (63) Merritt (85), Osipkov (79) *Binney & Mamon (82)* *Merritt & Gebhardt (94)* *Rix et al. (97)*	*Richstone & Tremaine (84)*
2D	**Axisymmetric**	Isotropic rotators	Toomre (82), scale–free Richstone (84), scale–free Binney et al. (90) van der Marel (91) Magorrian et al. (98) *Merritt et al. (97), edge–on*	Kuzmin & Kutuzov (62) Dejonghe & de Zeeuw (88) Levison & Richstone (87) *Gebhardt, Richstone (03)* *van der Marel et al. (98)* *Cretton et al. (99)* *Verolme et al. (02)*
3D	**Triaxial**		Stackel Potential de Zeeuw (85), Statler (87) *Zhao (96)* *Verolme et al. (03)*	

Fig. 16.1. The two main assumptions made in dynamical models: the dimension of the potential is along the vertical axis and that of the velocity ellipsoid is along the horizontal axis. Each box includes a few relevant papers for each configuration. These references are not complete and only serve to provide examples. The text type refers to whether the dynamical models assume a parametric (regular) or nonparametric (italics) form for the distribution function.

(Dimension of Potential — label along left vertical axis)

models that may not accurately represent the galaxy, but serve to provide overall trends from which we can infer formational and evolutionary scenarios. In other words, we will always be making some error—no matter which dynamical models we use—but we should be aware of each model's limitations. Below, we first review the dynamical models that have been applied to nearby galaxies, and then summarize the current state-of-the-art and the overall results from these models.

16.2 The Suite of Dynamical Models

Figure 16.1 diagrams the possible dynamical models based on their complexity. The components are the number of symmetry axes for the potential and for the velocity ellipsoid. The potential shapes clearly represent spherical, axisymmetric, and triaxial shapes. The velocity ellipsoid shapes represent isotropic (distribution functions that depend on only one integral of the motion, namely energy), 2-integral, and 3-integral distribution functions. This plot provides the range of possible distribution functions that can be used for dynamical modeling where the system obeys some symmetry axes. Obvious omissions are those systems that obey no symmetry axes.

The goal of the dynamical modeling is to determine the underlying potential of the system as well as the orbital structure. The concern is that, by not using a model that adequately represents the system, the results may be significantly biased. The best way to test for these biases is to model systems with a variety of assumptions and compare the results.

In each grid element are examples from the literature that represent that particular model. This listing is done to provide a few examples each and is not intended to be complete in any way. In fact, a complete listing would take the whole of this proceeding (but see Binney & Tremaine 1987 for a complete discussion). There are two types for the text in each grid: regular text represent analytic models, and italicized text represents nonparametric models. For example, isotropic, spherical models (the upper left grid) encompass an infinite number of density-potential pairs, and only King and Plummer models are listed. Gebhardt & Fischer (1995) present a nonparametric, isotropic, spherical model that determines the potential directly. Isotropic, spherical models have been enormously successful in describing stellar systems, especially for globular clusters (King 1966). For measuring black hole masses, they have done remarkably well; for example, Kormendy (2004) shows the change in the estimated black hole mass for M32 varies little over 20 years of data and a range of model sophistication. However, we are at a level now where the quality of the data is so high that we must use the most general models possible. Furthermore, in order to study the orbital structure one must use nonparametric techniques; otherwise, one restricts the form of the distribution function.

Spherical models are good representations for globular clusters and some of the largest ellipticals (e.g., M87), but we know that most galaxies are not spherical. Tremblay & Merritt (1995) and Khairul Alam & Ryden (2002) argue, based on inversion of the distribution of projected shapes, that, in fact, there are nearly no galaxies that are spherical. We must use, at the least, axisymmetric models. Furthermore, Binney (1978) and Davies et al. (1983) point out that the flattening in galaxies is not consistent with isotropic orbits: i.e., we must also include anisotropy. Thus, there have been a tremendous amount of work in modeling galaxies as 2-integral axisymmetric systems.

Van der Marel (1991) provides one of the first 2-integral studies of a large sample of ellipticals, using the modeling first introduced by Binney, Davies, & Illingworth (1990). From kinematic data taken along the major and minor axis for 37 galaxies, van der Marel finds that 2-integral models have too much motion on the major axis compared to what is seen. The implication is that ellipticals have $\sigma_r > \sigma_\theta$, inconsistent with the 2-integral assumption (where $\sigma_r = \sigma_\theta$). There are multiple ways to cause this inconsistency. For example, galaxies may depend on a third integral of motion; the 2-integral models may be biased by not including a dark halo; galaxies may have significant triaxial shape which also biases axisymmetric models; or the quality of the data may be too poor. We can compare the results of van der Marel to those of Gebhardt et al. (2003), who make 3-integral models for 12 galaxies, including three in common. For half of the sample, $\sigma_r > \sigma_\theta$, consistent with van der Marel. For three galaxies in common (NGC 3379, NGC 4649, and NGC 4697), there is not good agreement. Neither model includes a contribution from a dark halo, which may bias the large radial orbital structure. However, most likely, the differences are due to

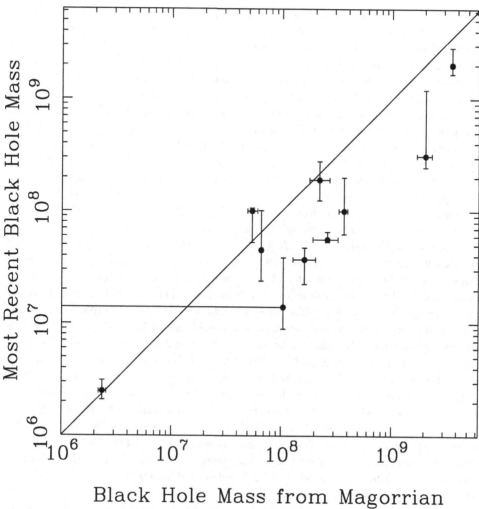

Fig. 16.2. The black hole mass estimate from Magorrian et al. (1998) and those from more recent data and analysis (see Tremaine et al. 2002 for the compilation). The line is one-to-one correspondence. All masses have been corrected to a common distance. The Magorrian et al. (1998) masses are on average 2.4 times larger.

the use of different data sets, and the quality of the data sets can have a significant influence on the results. We now turn to measuring the black hole mass.

The largest sample using 2-integral models to measure black hole masses is that of Magorrian et al. (1998). Magorrian et al. study 36 galaxies with ground-based kinematics and *HST* photometry to provide a systematic estimate of the central black hole mass. Previous black hole studies concentrated on individual cases. These result have been widely used, and also criticized. The major complaint is that the models are still too simplistic (i.e., 2-integral) and that the kinematic data have too low spatial resolution to say anything about the central black hole. Many of the Magorrian et al. galaxies now have *HST* kinematic data and have been modeled with more general models. In Figure 16.2 we compare the black hole mass estimates from Magorrian et al. to these more recent studies. There is a bias in that the 2-integral masses tend to be higher than those from the more recent analysis. The average difference between the two samples is a factor of 2.4. As discussed in Gebhardt et al. (2003), the difference appears to be due to differences in modeling, as opposed to the improved spatial resolution in the kinematics. Clearly, the better kinematics provide a more accurate measurement, but they do not appear to bias the results. Gebhardt et al. (2003) show that the black hole mass is not biased when using only ground-based data compared to using both ground-based and *HST* kinematics.

In order to provide a more accurate estimate of either the black hole mass or the orbital structure, we need to go beyond 2-integral models. Models that allow for three integrals of motion have only recently been applied to dynamical systems. The problem is that the most general form for the third integral is not analytic, and we must rely on numerical approaches. In limiting cases, there are analytic 3-integral models; for example, Dejonghe & de Zeeuw (1988) study 3-integral Kuzmin-Kutuzov (1962) models. However, these models have analytic cores ($d\log\nu/d\log r = 0$ at the center), and since nearly all galaxies have central cusps (Gebhardt et al. 1996; Ravindranath et al. 2001), they will be of limited use. Because the third integral is not analytic, we generally rely on orbit-based, Schwarzschild (1979) codes in order to study them. The first general application of the orbit-based methods is presented in Richstone & Tremaine (1984), applied to spherical systems. They even incorporate rotation in their models to provide one of the first models that include three integrals (energy, E, total angular momentum, L^2, and angular momentum about the pole, L_z), albeit in a spherical system. Rix et al. (1997) extend this analysis to make a detailed orbit-based model of the dark halo around NGC 2434. The first application of an axisymmetric, orbit-based model is that of van der Marel et al. (1998), who measure the black hole mass in M32. A few groups now have axisymmetric, orbit-based codes that have been used to study central black holes. To date, 17 galaxies have been studied with these models, with 14 coming from one code (Gebhardt et al. 2000, 2003; Bower et al. 2001), four from the Leiden group with various codes (van der Marel et al. 1998; Cretton & van den Bosch 1999; Cappellari et al. 2002; Verolme et al. 2002), and one from Emsellem, Dejonghe, & Bacon (1999).

With so few groups using orbit-based codes, we must be certain that the immense freedom allowed by these codes does not bias the results due to some feature of an individual code. The general problem of covering phase space appropriately in these orbit-based codes is tricky. There is a balance that one must obtain between including a large orbit library in order to sample phase space but still maintain a small enough library in order to use a reasonable amount of computer resources. In fact, Valluri, Merritt, & Emsellem (2004) find that there is a large difference when running models using orbit libraries of various size. They have two main results that question the reliability of these models for measuring black hole masses. First, the shape of the χ^2 contours depends on the number of orbits run for a model, using the same data set. Second, for models with large numbers of orbits, there is a degeneracy in black hole mass: the χ^2 contours reach a plateau over a large range of black hole masses. These results are critically important to understand since they may undermine this whole area of study. Fortunately, the other groups involved have done many tests in regards to this degeneracy. We will concentrate on the tests done with the Gebhardt et al. (2003) code.

There are three issues on which we will focus. These are (1) the shape of χ^2 as a function of orbit number, (2) the ability of using the χ^2 contours to measure reliable confidence

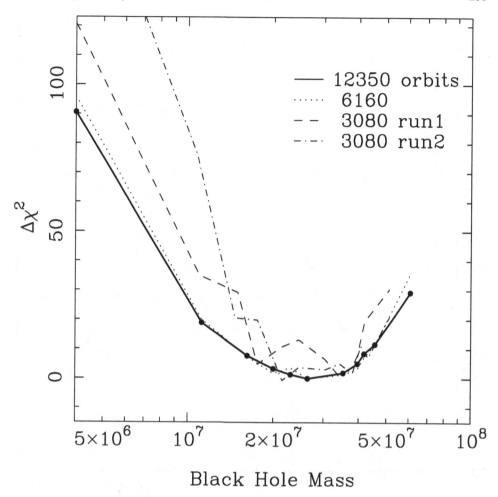

Fig. 16.3. Shape of $\Delta\chi^2$ versus black hole mass for models with different orbit numbers. Each model is a fit to an identical data set, and the only difference is the sampling of phase space. For the two runs with the smallest orbit library, we have run the same number of orbits, but simply sampled phase space differently.

bands, and (3) the dependence on the smoothing parameter. For this last aspect, most groups use regularization for the smoothing while Gebhardt et al. rely on maximizing entropy (Richstone & Tremaine 1988). Regularization imposes smoothing directly in phase space by including a term that represents the noise in the χ^2, typically using the sum of the squared second derivative between phase space elements. Gebhardt et al. (2003) calculate the entropy of each orbit (using entropy equal to $w\log w$, where w is the orbital weight) and use the total entropy as a constraint (see Richstone & Tremaine 1984, 1988 for a complete discussion). Both approaches should provide smooth distribution functions, and there is no obvious desire to use one over the other.

To study the influence of orbit number on the best-fit solution, the obvious test is to run an analytic model where the black hole mass is known and simply increase the orbit number.

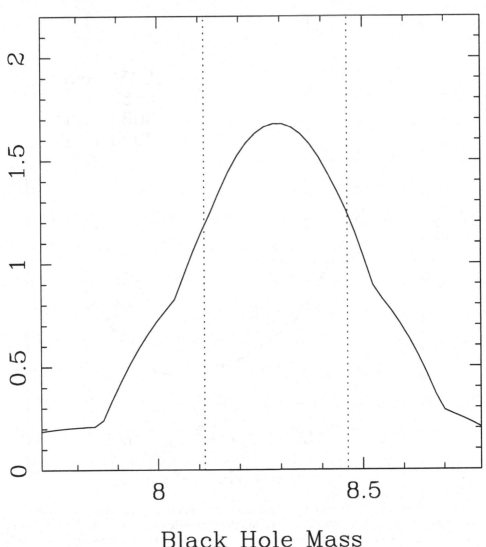

Black Hole Mass

Fig. 16.4. Distribution of black hole masses from Monte Carlo simulations for NGC 3608. The solid line represents the results of changing the input velocity profiles according to the noise in the spectral data (the Monte Carlo approach). The vertical dotted lines represent the 68% confidence limit as measured from the shape of the χ^2 contours. The area inside the dotted lines is close to 68% of the area.

Valluri et al. (2004) have the only paper in which this test has been published. This test, however, has been done by the other groups, but it was never published since nothing was ever seen to be problematic. Figure 16.3 plots this test using the code of Gebhardt et al. (2003). They run four models for the same data set. The total orbit number spans a factor of 4, with the two smallest libraries being run twice but with a different sampling. The two largest libraries show nearly identical χ^2 profiles. The two smallest libraries show a different contour shape, but they have substantial noise, making the comparison difficult. For libraries with an extremely small number of orbits, it is clear that the χ^2 contours must become very noisy since the quality of the fit depends on whether one happens to hit important orbits or not. Thus, having an appropriate number of orbits certainly is important. However, since we see little difference between the two largest libraries that differ by a factor of 2 in orbit number, it appears that the contours do not plateau as a function of black hole mass, as Valluri et al. (2004) find. In fact, even for the small libraries, we see that they tend to trace the true χ^2 contour fairly well, although the noise makes it difficult to follow. The number of orbits in a given library is only useful if one compares it to the number of model grid elements. For published orbit-based models, most have phase space coverage that is adequate to measure the black hole mass. For example, Gebhardt et al. (2003) use about 8000 orbits in each galaxy model with the same number of grid elements shown in Figure. 16.3.

Another issue to understand is whether the uncertainties on the black hole masses are adequately measured. One of the goals of black hole studies is to understand their role in galaxy evolution, and any comparison with galaxy properties must contain accurate uncertainties. All orbit-based models rely on using the shape of the χ^2 to determine their uncertainties. The best method, however, is to run bootstrap simulations on the real data. We have done this for NGC 3608. For each spectrum, we simulate a new realization based on the noise in the spectrum. We then generate 100 realizations. This Monte Carlo method is the same as that used when measuring the uncertainties for the velocity profiles (see Pinkney et al. 2004). We then run the modeling code on each new set of data and estimate the best-fit black hole mass. This procedure is extremely time consuming, and we have only done it for one galaxy so far. Figure. 16.4 plots the distribution of black hole masses obtained by these Monte Carlo simulations. The solid line represents the distribution function using an adaptive kernel estimate of the individual realizations. The dotted lines show the 68% confidence band measured from the shape of the χ^2 contours. The agreement is excellent, as the 68% χ^2 contours are similar to the area that contains 68% of the simulations. The simulations encompass a slightly larger area, but only by a few percent. Thus, it appears that the χ^2 contours can be used to estimate accurately the black hole mass uncertainties. From all of the orbit-based models used to date, the range of black hole mass uncertainties is from 5% to 70%, with an average uncertainty around 20%. Given that the scatter in the $M_\bullet - \sigma$ correlation is less than 30% (Tremaine et al. 2002), we still need to improve the black hole mass uncertainties.

Another important concern is whether the smoothing parameter has an effect on the black hole mass. The choice of this parameter is discussed extensively in Cretton et al. (1999) and Verolme et al. (2002). Their choice of the smoothing parameter is based on comparison with analytic test cases, by finding that smoothing parameter that provides the best match for the phase space distribution function. This cross-validation technique is a standard statistical approach to determine the smoothing parameter. Furthermore, Verolme et al. (2002) have performed tests in which they compare their best-fit mass found with optimal smoothing to that measured when including no smoothing, and find no difference in their black hole mass. Similar results are found in the modeling of Gebhardt et al. (2003). Figure. 16.5 is a plot of χ^2 versus smoothing parameter for many different models of NGC 3608. Each line differs by the mass of the black hole. The final χ^2 versus black hole mass is then obtained by taking the rightmost values in Figure. 16.5. The point of this plot is to show that the best-fit model provides the minimum χ^2 over a large range of smoothing parameters. For the maximum entropy method, the smoothness is employed by increasing the contribution of the entropy

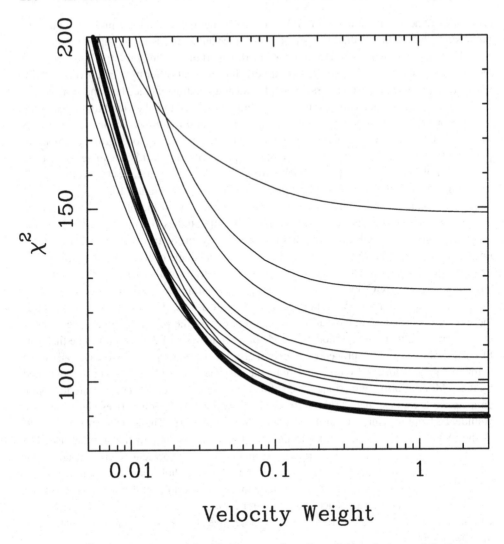

Fig. 16.5. χ^2 between the model and data as a function of the relative weight between entropy and velocity fit for NGC 3608. By increasing the velocity weight, we are decreasing the amount of smoothing. We start each model with maximum smoothing (maximizing entropy only) and then increase the velocity weight until the kinematics are fit as well as possible. The heavy, solid line is the best-fit model when the entropy term has no weight. The best-fit model has the minimum χ^2 over a large range of entropy weights.

term relative to the comparison with the velocities. The velocity weighting is increased until the model provides the best fit to the data and essentially there is no contribution from the entropy term. However, as is seen in Figure. 16.5, the best-fit model provides the minimum χ^2 for a range of 100 in smoothing parameters.

All of the above discussion has focused on measuring the black hole mass and not the stellar orbital structure. The influence of these effects on the orbits is harder to quantify, since the results depend on which aspect of the orbits that concern us. For example, the answer depends on whether one is concerned with the velocity ellipsoid at every position in the galaxy, or whether one wants the radial to tangential components at only two different radii. The former is much harder to measure. We are not at the point where we can study the detailed shape of the velocity ellipsoid throughout the galaxy. The two ingredients required to do this are (1) an understanding of any systematic biases in the orbit-based techniques and (2) having the appropriate data sets to perform this analysis. We will discuss each of these below, but at this point we stress that obtaining a simple measure of radial to tangential motion appears to be robust, and does provide evolutionary constraints. Using the same Monte Carlo simulations discussed above, we can also estimate the distribution of radial to tangential motion from the noise in the spectra. The scatter is remarkably small. Similarly, this ratio has very little dependence on the smoothing parameter. In fact, that ratio changes by a much smaller fraction than the best-fit black hole mass. This quantity is typically measured to around 20% or better. Thus, we are confident that we can use this number to provide good comparison with theoretical predictions.

16.3   Results and Discussion

There are 17 galaxies that have axisymmetric orbit-based models. Figure 16.6 plots the orbital properties of those galaxies against other galaxy properties. We include the black hole mass, the effective dispersion, and the radial to tangential motion at two points in the galaxy—the central region and at 1/4 effective radius. In the central region for each galaxy, the black hole dominates the potential. The $M_\bullet - \sigma$ plot is the most significant correlation. However, there is also a very strong correlation between the black hole mass and the radial motion contribution at large radii (top right plot). There is another correlation of this quantity with effective σ, but this may be secondary to the one with black hole mass. In fact, the correlation with black hole mass is the most significant of all other galaxy properties (total light, total mass, effective radius, etc.). The trend is that those galaxies that have large black hole masses (and hence large σ) have orbits dominated by radial motion at large radii. Tangential motion tends to occur in those galaxies with small black holes. This correlation is one of the strongest for the full set of comparisons in Gebhardt et al. (2003).

The correlation between $M_\bullet$ and σ_r/σ_{te} is likely to be related to the evolutionary history of the galaxy. For the most massive galaxies, at radii near to the effective radius, the orbital distribution is radially biased. This is also the conclusion from Cretton, Rix, & de Zeeuw (2000), who use orbit-based methods to study the giant elliptical NGC 2320; along the major axis, they find strong radial bias in the orbits at large radii. We can compare this radial bias for the most massive galaxies with the N-body simulations of Dubinski (1998). He finds that for the most massive ellipticals, there is an increase in the radial motion from the center (where it is nearly isotropic) to the outer radii (where the merger remnant has $\sigma_r/\sigma_\theta = 1.3$). The most massive galaxies in our sample of 17 approach this amount of radial motion at large radii. For the smaller galaxies, the N-body comparisons are not as developed for measuring the internal orbital structure. However, based on the recent results of N-body simulations (Meza et al. 2003; Samland & Gerhard 2003), we will soon be in a position to compare the internal structure of the smaller galaxies as well. It has long been known that low-luminosity ellipticals rotate rapidly and are often consistent with oblate isotropic rotators, while high-luminosity ellipticals have been thought to be supported by radial anisotropy at large radii (Davies et al. 1983). Since black hole mass correlates with luminosity, the $M_\bullet - \sigma_r/\sigma_{te}$ correlation may then be secondary; however, the radial anisotropy correlates much stronger

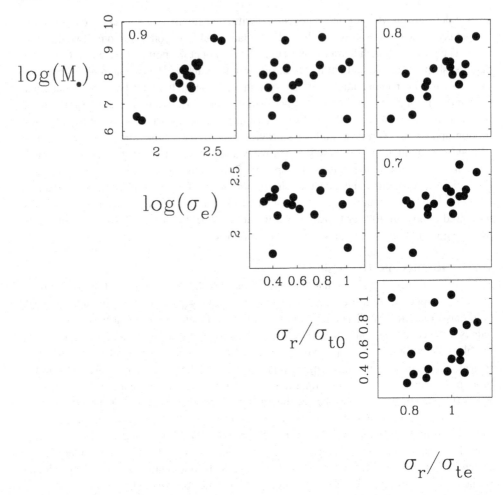

Fig. 16.6. Plots of the orbital properties against various galaxy properties. The properties along the diagonal include the black hole mass, the effective dispersion (σ_e), the ratio of radial to tangential dispersion at the center (σ_r/σ_{t0}), and the ratio of radial to tangential dispersion at 1/4 the effective radius (σ_r/σ_{te}). The number written in the upper left corner of the plot is the Pearson's R correlation coefficient. If the probability from the correlation is below 10%, we do not report R.

with black hole mass than it does with luminosity. There has been a considerable amount of theoretical work in explaining why the black hole mass correlates so well with host galaxy dispersion (see Adams et al. 2003 and references therein for a recent discussion). The correlation may provide additional constraints on the models.

There is also a trend that the galaxies with shallow central density profiles (i.e., the core galaxies) have orbits with the strongest tangential bias near their centers. This correlation has been discussed in Gebhardt et al. (2003). The most likely explanation is that this is caused by binary black hole mergers. We know that the existence of a black hole will leave some amount of tangential anisotropy since it will either eject or accrete those stars that are on radial orbits. This effect has been seen in many *N*-body simulations that consider adiabatic growth of black holes (Quinlan et al. 1995, 1997; Nakano & Makino 1999; Milosavljević & Merritt 2001; Sigurdsson 2004). In all of these case, however, the amount of tangential motion is quite small. In the most detailed study to date, Milosavljević & Merritt (2001) find that the most extreme amount of tangential motion has $\sigma_r/\sigma_\theta = 0.8$. The values that Gebhardt et al. (2003) report are smaller than 0.4. One way to obtain such large amounts of tangential motion is to have a binary black hole that can affect more stars on radial orbits due to its own orbital motion. The binary black hole results from a merger, and we already have seen that binary black holes are one of the best mechanisms to create the division between core and power-law galaxies (Faber et al. 1997; Milosavljević & Merritt 2001; Lauer 2004). However, the N-body simulations that have been studied use fairly restrictive assumptions—most are based on spherical isotropic initial conditions. Once realistic simulations including mergers and central black holes are available, we will be in a much better position to interpret the observational results.

16.4   The Future

There are many aspects of understanding the stellar orbital structure that need improvement—these include the data, analysis, and theoretical comparisons. In regards to the data, with the use of orbit-based models, we can realistically constrain the internal structure of the galaxy. In fact, Verolme et al. (2002) were able to measure with high accuracy the inclination of M32, and thus its intrinsic shape. However, in order to do this they needed two-dimensional kinematic data, which were obtained by the SAURON team (de Zeeuw et al. 2002). Most of the galaxies studied to date with orbit-based models only have limited kinematic data (along 2–4 position angles) and thus cannot be used to study their intrinsic shapes. In fact, as a result of this, most of the models in Gebhardt et al. (2003) are only run as edge-on configurations, and there is a concern that this may bias the results (de Zeeuw 2004). However, for the issues discussed here—the black hole mass and radial to tangential motion—inclined models are unlikely to introduce substantial changes, given the large uncertainties already on these quantities. In any event, significant improvement can be made by using two-dimensional kinematic data. Another area for improvement of the data is to include kinematics at large radii. In the study of Gebhardt et al., they were careful to report only results inside of the effective radii, where the dark halo is unlikely to have any influence. However, any dynamical model needs to include some estimate of the influence of orbits at large radii. Even though the effect of these orbits is expected to be minimal at small radii, they are not ignorable. In order to measure the central black hole and orbital structure, a proper dynamical model should include both high-spatial resolution (i.e., *HST*) and large-radii kinematics. With the advent of integral-field units on many large ground-based telescopes, obtaining this type of data will be feasible. In fact, adaptive optics observations with an integral-field unit will be a tremendous advance to this field of study.

On the data analysis side, while the orbit-based models that have been run offer significant improvement over the previous set of models, there is still a long way to go. For instance, most orbit-based models are axisymmetric and oblate. Prolate and triaxial models need to be included for a proper analysis. As discussed above, even for the oblate models, most include only an edge-on configuration. In addition, many have assumed luminosity density profiles that have constant ellipticity with radius. We know that galaxies have ellipticities and position angles that vary with radius, and so, at some level, the models studied so far incorrectly represent the galaxy light profile. However, at this point, the kinematic uncertainties likely dominate the results, as opposed to assumption biases. One can see this by comparing the inclined models for M32 (Verolme et al. 2002) with the edge-on model of van der Marel et al. (1998). Even there, the difference in the black hole mass is only at the 10% level, and the change in internal orbital structure is even less. Since none of the other black holes are as well measured as M32's (most have uncertainties around 30%), this suggests that the assumption biases will not have a great effect. Yet, once the quality of the data improves, we will have to consider more general models. In fact, triaxial models have already been studied by Verolme et al. (2004). We know that kinematically distinct cores are common in galaxies, and, therefore, axisymmetric models will clearly not provide the best representation. Verolme et al. extend the orbit-based models to include a triaxial distribution function and have successfully reproduced the complicated kinematic structure of NGC 4365. An important step now would be to run both an axisymmetric and triaxial model on the same galaxy to see if any significant differences arise.

The ultimate analysis method includes running an N-body model for each galaxy. We know that at some level there is no galaxy that has perfect symmetry. The question then becomes how significant are the errors one makes when running a model that has some symmetry (spherical, axisymmetric, or triaxial) to an asymmetric galaxy. At least for the black hole mass, the errors are not large. Kormendy (2004) summarizes the changes in black hole mass over time and with different dynamical modeling sophistication. He finds that the change in black hole mass, at least for a few well-studied galaxies, is not very large, considering the enormous change in both data and modeling. The black hole masses measured by Magorrian et al. (1998) using low-quality ground-based data and 2-integral models measured black hole masses to within a factor of 2–3 of the presently accepted values. However, the intrinsic scatter of the $M_\bullet - \sigma$ correlation is consistent with zero, and at most 30% (Tremaine et al. 2002). Furthermore, the correlation of black hole mass with other galaxy properties—concentration index (Graham et al. 2001; Graham 2004) and total mass (Magorrian et al. 1998; McLure & Dunlop 2002)—have a low scatter as well. The fact that the scatter in these correlations is already so low implies that the systematic uncertainties are not terribly measured; otherwise, we would not be able to detect these correlations. In order to better study these correlations, we must have better determined black hole masses, and therefore we must improve the analysis techniques. Hopefully, we will not have to measure black hole masses to much better than 10% to answer the scientifically important questions, since going beyond that will be a challenge in terms of both observations and analysis.

References

Adams, F. C., Graff, D. S., Mbonye, M., & Richstone, D. O. 2003, ApJ, 591, 125
Barnes, J., & Hernquist, L. 1992, ARA&A, 30, 705
Binney, J. 1978, MNRAS, 183, 501
Binney, J., Davies, R. L., & Illingworth, G. D. 1990, ApJ, 361, 78
Binney, J., & Mamon, G. A. 1982, MNRAS, 200, 361
Binney, J., & Tremaine, S. 1987, Galactic Dynamics (Princeton: Princeton Univ. Press)
Bower, G. A., et al. 2001, ApJ, 550, 75
Cappellari, M., Verolme, E. K., van der Marel, R. P., Verdoes Kleijn, G. A., Illingworth, G. D., Franx, M., Carollo, C. M., & de Zeeuw, P. T. 2002, ApJ, 578, 787
Cretton, N., de Zeeuw, P. T., van der Marel, R. P., & Rix, H.-W. 1999, ApJS, 124, 383
Cretton, N., Rix, H.-W., & de Zeeuw, P.T. 2000, ApJ, 536, 319
Cretton, N., & van den Bosch, F. C. 1999, ApJ, 514, 704
Davies, R. L., Efstathiou, G., Fall, S. M., Illingworth, G., & Schechter, P. L. 1983, ApJ, 266, 41
de Zeeuw, P. T. 1985, MNRAS, 216, 273

——. 2004, in Carnegie Observatories Astrophysics Series, Vol. 1: Coevolution of Black Holes and Galaxies, ed. L. C. Ho (Cambridge: Cambridge Univ. Press), in press

de Zeeuw, P. T., et al. 2002, MNRAS, 329, 513

Dejonghe, H., & de Zeeuw, P. T. 1988, ApJ, 333, 90

Dubinski, J. 1998, ApJ, 502, 141

Emsellem, E., Dejonghe, H., & Bacon, R. 1999, MNRAS, 303, 495

Faber, S. M., et al. 1997, AJ, 114, 1771

Gebhardt, K., et al. 2000a, AJ, 119, 1157

——. 2000b, ApJ, 539, L13

——. 2003, ApJ, 583, 92

Gebhardt, K., & Fischer, P. 1995, AJ, 109, 209

Gebhardt, K., Richstone, D., Ajhar, E. A., Kormendy, J., Dressler, A., Faber, S. M., Grillmair, C., & Tremaine, S. 1996, AJ, 112, 105

Graham, A. W. 2004, in Carnegie Observatories Astrophysics Series, Vol. 1: Coevolution of Black Holes and Galaxies, ed. L. C. Ho (Pasadena: Carnegie Observatories,
http://www.ociw.edu/ociw/symposia/series/symposium1/proceedings.html)

Graham, A. W., Erwin, P., Caon, N., & Trujillo, I. 2001, ApJ, 563, L11

Khairul Alam, S. M., & Ryden, B. S. 2002, ApJ, 570, 610

King, I. R. 1966, AJ, 71, 64

Kormendy, J. 1993, in The Nearest Active Galaxies, ed. J. Beckman, L. Colina, & H. Netzer (Madrid: Consejo Superior de Investigaciones Científicas), 197

——. 2004, in Carnegie Observatories Astrophysics Series, Vol. 1: Coevolution of Black Holes and Galaxies, ed. L. C. Ho (Cambridge: Cambridge Univ. Press), in press

Kormendy, J., & Richstone, D. 1995, ARA&A, 33, 581

Kuzmin, G. G., & Kutuzov, S. A. 1962, Bull. Abastumani Ap. Obs., 27, 82

Lauer, T. R. 2004, in Carnegie Observatories Astrophysics Series, Vol. 1: Coevolution of Black Holes and Galaxies, ed. L. C. Ho (Cambridge: Cambridge Univ. Press), in press

Levison, H., & Richstone, D. O. 1987, ApJ, 314, L476

Magorrian, J., et al. 1998, AJ, 115, 2285

McLure, R. J., & Dunlop, J. S. 2002, MNRAS, 331, 795

Merritt, D. 1985, AJ, 90, 1027

Merritt, D., & Gebhardt, K. 1994, in Clusters of Galaxies, Proceedings of the XXIXth Rencontre de Moriond, ed. F. Duret, A. Mazure, & J. Tran Thanh Van (Singapore: Editions Frontiéres), 11

Merritt, D., Meylan, G., & Mayor, M. 1997, AJ, 114, 1074

Meza, A., Navarro, J. F., Steinmetz, M., & Eke, V. R. 2003, ApJ, 590, 619

Michie, R. W. 1963, MNRAS, 125, 127

Milosavljević, M., & Merritt, D. 2001, ApJ, 563, 34

Nakano, T., & Makino, J. 1999, ApJ, 510, 155

Osipkov, L. P. 1979, Pisma Ast. Zh., 5, 77

Pinkney, J., et al. 2004, ApJ, in press

Quinlan, G., & Hernquist, L. 1997, NewA, 2, 533

Quinlan, G., Hernquist, L., & Sigurdsson, S. 1995, ApJ, 440, 554

Ravindranath, S., Ho, L. C., Peng, C. Y., Filippenko, A. V., & Sargent, W. L. W. 2001, AJ, 122, 653

Richstone, D. O. 1984, ApJ, 281, 100

Richstone, D. O., & Tremaine, S. 1984, ApJ, 286, 27

——. 1988, ApJ, 327, 82

Rix, H.-W., de Zeeuw, P. T., Carollo, C. M. C., Cretton, N., & van der Marel, R. P. 1997, ApJ, 488, 702

Samland, M., & Gerhard, O. 2003, A&A, 399, 961

Schwarzschild, M. 1979, ApJ, 232, 236

Sigurdsson, S. 2004, in Carnegie Observatories Astrophysics Series, Vol. 1: Coevolution of Black Holes and Galaxies, ed. L. C. Ho (Cambridge: Cambridge Univ. Press), in press

Statler, T. S. 1987, ApJ, 321, 113

Toomre, A. 1982, ApJ, 259, 535

Tremaine, S., et al. 2002, ApJ, 574, 740

Tremblay, B., & Merritt, D. 1995, AJ, 110, 1039

Valluri, M., Merritt, D., & Emsellem, E. 2004, ApJ, submitted (astro-ph/0210379)

van der Marel, R. P. 1991, MNRAS, 253, 710

van der Marel, R. P., Cretton, N., de Zeeuw, P. T., & Rix, H.-W. 1998, ApJ, 493, 613
Verolme, E. K., et al. 2002, MNRAS, 335, 517
——. 2004, MNRAS, submitted (astro-ph/0301070)
Zhao, H. S. 1996, MNRAS, 283, 149

17

Single and binary black holes and their influence on nuclear structure

DAVID MERRITT

Rutgers University

Abstract

Massive central objects affect both the structure and evolution of galactic nuclei. Adiabatic growth of black holes generates power-law central density profiles with slopes in the range $1.5 \lesssim -d\log\rho/d\log r \lesssim 2.5$, in good agreement with the profiles observed in the nuclei of galaxies fainter than $M_V \approx -20$ mag. However, the shallow nuclear profiles of bright galaxies require a different explanation. Binary black holes are an inevitable result of galactic mergers, and the ejection of stars by a massive binary displaces a mass of order the binary's own mass, creating a core or shallow power-law cusp. This model is at least crudely consistent with core sizes in bright galaxies. Uncertainties remain about the effectiveness of stellar- and gas-dynamical processes at inducing coalescence of binary black holes, and uncoalesced binaries may be common in low-density nuclei. Numerical N-body experiments are not well suited to probing the long-term evolution of black hole binaries due to spurious relaxation.

17.1 Introduction

The effect of a supermassive black hole on its stellar surroundings depends, as so often in stellar dynamics, on how one imagines the system evolved to its present state. Collisional relaxation times are too long to have affected the stellar distribution in all but the densest nuclei (Faber et al. 1997); hence, the structure and kinematics of nuclei are fossil relics of the interactions between stars and black holes. In the simplest scenario, the black hole grows by accreting gas on a time scale long compared with the orbital periods of the surrounding stars (Peebles 1972; Young 1980). This "adiabatic growth" model makes fairly definite predictions about the distribution of stars near the black hole, predictions that are consistent with the steep density cusps observed in faint galaxies but that cannot explain the flatter profiles at the centers of bright galaxies. But galaxies merge, implying the formation of binary black holes (Begelman, Blandford, & Rees 1980) that are efficient at displacing matter as they spiral together. The dynamics of black hole binaries in galactic nuclei are complex; among the unanswered questions are the long-term efficiency of stellar dynamical processes at extracting energy from a binary, and whether decay of black hole binaries ever stalls at separations too great for the efficient emission of gravitational waves. But the binary black hole model is at least crudely consistent with the observed dependence of nuclear structure on galaxy luminosity. This article summarizes theoretical work on the single and binary black hole models and suggests avenues for future progress.

© The Observatories of the Carnegie Institution of Washington 2004.

17.2   Preliminaries

A black hole of mass $M_\bullet$ embedded in a galactic nucleus will strongly affect the motion of stars within a distance $r = r_h$, the "radius of influence." A standard definition for r_h is

$$r_h \equiv \frac{GM_\bullet}{\sigma^2} \approx 10.8 \text{ pc} \left(\frac{M_\bullet}{10^8 M_\odot} \right) \left(\frac{\sigma}{200 \text{ km s}^{-1}} \right)^{-2} \tag{17.1}$$

with σ the 1D velocity dispersion of the stars at $r \gg r_h$. This definition had its origin in the isothermal sphere model for galactic nuclei; σ is independent of r in such a model and r_h is the radius at which the circular velocity around the black hole equals σ. We now know that nuclei are power laws in the stellar density, $\rho \sim r^{-\gamma}$, and that γ can lie anywhere between ~ 0 and ~ 2.5 (Lauer, this volume). The velocity dispersion in a power-law nucleus is only constant if $\gamma = 0$ or 2, hence a definition like equation (17.1) is problematic. One alternative would be to define r_h as the root of $\sigma^2(r) - GM_\bullet/r = 0$. A simpler definition, which will be adopted in this article, is the radius at which the enclosed mass in stars is twice the black hole mass:

$$M_*(r < r_h) = 2M_\bullet. \tag{17.2}$$

This definition is exactly equivalent to equation (17.1) when $\rho(r) = \sigma^2/2\pi G r^2$, the singular isothermal sphere.

17.3   The Adiabatic Growth Model

If a black hole grows at the center of a stellar system through the accretion of gas, the stellar density in the core will also grow as the black hole's gravity pulls in nearby stars (Peebles 1972; Young 1980). The change in the stellar density can be computed straightforwardly if it is assumed that the black hole grows on a time scale long compared with stellar orbital periods. This is reasonable, since even Eddington-limited accretion requires $\sim 10^8$ yr to double the black hole mass, and orbital periods throughout the region dominated by the black hole are $\lesssim 10^6$ yr. Under these assumptions, the adiabatic invariants **J** associated with the stellar orbits are conserved as the black hole grows, and the phase-space density f remains fixed when expressed in terms of the **J**. Computing the final f becomes a simple matter of expressing the final orbital integrals in terms of their initial values under the constraint that the adiabatic invariants remain fixed (Young 1980).

In spherical potentials, the adiabatic invariants are the angular momentum L and the radial action $I = 2 \int_{r_-}^{r_+} \sqrt{2[E - \Phi(r)] - L^2/r^2}$, where $\Phi(r)$ is the gravitational potential and $r_\pm$ are pericenter and apocenter radii. It may be shown (e.g., Lynden-Bell 1963) that orbital shapes remain nearly unchanged when L and I are conserved, implying that an initially isotropic velocity distribution $f(E)$ remains nearly isotropic after the black hole grows (though not exactly isotropic – see below). The final f corresponding to an initially isotropic f is then simply

$$\begin{aligned} f_f(E_f, L) &= f_i(E_i, L) \\ &\approx f_f(E_f) \end{aligned} \tag{17.3}$$

where E_f is related to E_i through the condition $I_f(E_f, L) = I_i(E_i, L)$.

While the form of the nuclear density profile before the black hole appeared is not known, N-body studies of structure formation suggest that power laws are generic (Power et al.

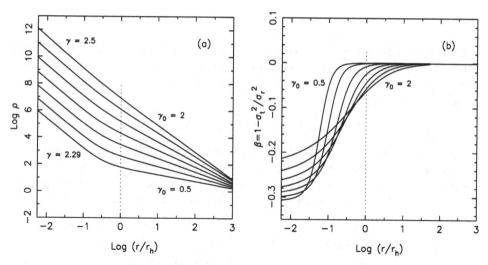

Fig. 17.1. Influence of the adiabatic growth of a black hole on its nuclear environment in a spherical, isotropic galaxy. (*a*) Density profiles after growth of the black hole. Initial profiles were power laws, $\rho_i \propto r^{-\gamma_0}$, with γ_0 increasing upwards in steps of 0.25. The radial scale is normalized to r_h as defined in the initial galaxy (Eq. 17.2). The slope of the final profile at $r < r_h$ is almost independent of the initial slope. (*b*) Velocity anisotropies after growth of the black hole. A slight bias toward circular motions appears at $r < r_h$.

2003 and references therein) and observed nuclear density profiles are often well described as power laws even on scales $r \gg r_h$. Setting $\rho_i \propto r^{-\gamma_0}$, $\Phi_i \propto r^{2-\gamma_0}$ ($0 < \gamma_0 < 2$), the initial distribution function becomes

$$f_i(E_i) \propto E_i^{-\beta}, \quad \beta = \frac{6-\gamma_0}{2(2-\gamma_0)} \quad (0 < \gamma_0 < 2). \tag{17.4}$$

To compute $f_f(E_f)$ we need a relation between E_f and E_i; we restrict attention to the region within the black hole's sphere of influence by setting $E_f = v^2/2 - GM_\bullet/r$. The radial action $I(E,L)$ in the power-law model cannot be computed analytically for every (E,L), but for certain orbits $E_f(E_i)$ has a simple form. For instance, circular orbits have $I = 0$, and conservation of angular momentum implies $r_i M_i(r_i) = r_f M_\bullet$ or $r_f \propto r_i^{4-\gamma_0}$. Thus $E_f \propto -r_f^{-1} \propto -r_i^{\gamma_0-4} \propto -E_i^{(\gamma_0-4)/(2-\gamma_0)}$, or

$$E_i \propto (-E_f)^{-(2-\gamma_0)/(4-\gamma_0)}. \tag{17.5}$$

The same relation turns out to be precisely correct for radial orbits as well and is nearly correct at intermediate eccentricities (Gondolo & Silk 1999). Thus we can write

$$f_f(E_f) = f_i(E_i) \propto E_i^{-\beta} \propto (-E_f)^{\delta}, \quad \delta = \frac{6-\gamma_0}{2(4-\gamma_0)} \quad (0 < \gamma_0 < 2), \tag{17.6}$$

and the final density profile within the sphere of influence of the black hole is

$$\rho_f(r) = \int f_f(v)d^3v \propto \int_{\Phi(r)}^0 (-E)^{\delta}\sqrt{E-\Phi(r)}\,dE$$

$$\propto r^{-\gamma}, \quad \gamma = 2+\frac{1}{4-\gamma_0}. \tag{17.7}$$

For $0 < \gamma_0 < 2$, γ varies only between 2.25 and 2.5; the slope of the final density profile within r_h is almost independent of γ_0.

The form of $\rho_f(r)$ at $r \approx r_h$ must be computed numerically (e. g., Young 1980; Cipollina & Bertin 1994; Cipollina 1995; Quinlan, Hernquist, & Sigurdsson 1995; Gondolo & Silk 1999). Figure 17.1 shows $\rho_f(r)$ when $\rho_i(r) \propto r^{-\gamma_0}$. Defining r_{cusp} to be the radius at which the inner and outer power laws intersect, one finds

$$r_{cusp} = \alpha r_h, \quad 0.19 \lesssim \alpha \lesssim 0.22, \quad 0.5 \leq \gamma_0 \leq 1.5. \tag{17.8}$$

In early treatments of the adiabatic growth model (Peebles 1972; Young 1980), the black hole was assumed to grow inside of a constant-density isothermal core. The index of the power-law cusp that forms from this initial state is $\gamma = 1.5$, compared with the limiting value $\gamma = 2.25$ as $\gamma_0 \to 0$ in the power-law models. This difference can be traced to differences in the central density profile:

$$\rho_i(r) = \rho_0 \times \left(1 + C_1 r + C_2 r^2 + \dots\right). \tag{17.9}$$

The isothermal model has $C_1 = 0$ (an "analytic core") implying a phase-space density that tends to a constant value at low energies. Other sorts of cores have $C_1 \neq 0$ and f diverges at low energies; for instance, the core produced by setting $\gamma = 0$ in $\rho(r) = r^{-\gamma}(1+r)^{-4}$ has $f(E) \to [E - \Phi(0)]^{-1}$. In fact models with finite central ρ's can be found that generate final cusp slopes anywhere in the range $1.5 \leq \gamma \leq 2.25$ (Quinlan et al. 1995). There is probably no way of ruling out an analytic core in the progenitor galaxy on the very small scales that are relevant to the later formation of a cusp, hence the adiabatic growth model is compatible with any final slope in the range $1.5 \lesssim \gamma \lesssim 2.5$. The upper limit could even be extended beyond 2.5 if $\gamma_0 > 2$.

How do these predictions compare with the data? Observed luminosity profiles are well described as power laws at the smallest resolvable radii, and in the case of faint ellipticals, $M_V \gtrsim -20$ mag, the observed range of slopes is $1.5 \lesssim \gamma \lesssim 2.5$ (Lauer, this volume). This is precisely the range in γ predicted by the adiabatic growth model. However, in bright galaxies, γ extends down to ~ 0. A natural interpretation is that the steep cusps in faint galaxies are a result of adiabatic black hole growth, while some additional mechanism, like mergers, has acted to modify the profiles in the brighter galaxies.

Some fine tuning is still required to reproduce the luminosity profiles of galaxies with $1.5 \lesssim \gamma \lesssim 2$. These intermediate slopes require a shallow, core-like initial profile, and if the initial core radius exceeded $\sim r_h$, the final profile will exhibit an upward inflection at $r \approx r_h$ (e.g., Fig. 2 of Young 1980; Fig. 17.1). Such inflections are rarely, if ever, seen; observed profiles have slopes that decrease smoothly inward. A way out is to require that the black hole mass exceed the initial core mass so that its growth obliterates the core; alternatively, all nuclei with $\gamma \lesssim 2$ may have been the products of mergers.

An ingenious attempt to reconcile the adiabatic growth model with both steep and shallow cusps was made by van der Marel (1999). He postulated the existence of isothermal cores in the progenitor galaxies with core masses scaling as $\sim L^{1.5}$, with L the total galaxy luminosity. Since $M_\bullet \sim L$, $M_\bullet/M_{core} \sim L^{-0.5}$ and the black holes in faint galaxies would grow to dominate their cores, producing a cusp profile that approximates the featureless power laws of faint galaxies. In bright galaxies, van der Marel argued that the upward inflection at $r \lesssim r_h$ would be difficult to resolve; hence these galaxies would exhibit nearly unperturbed core profiles. Van der Marel's model is intriguing, although the assumed relation between core size and galaxy luminosity is ad hoc, and the assumption of large, pre-existing cores does not fit naturally into any current model of galaxy formation.

Black hole growth in axisymmetric nuclei with analytic cores was considered by Leeuwin & Athanassoula (2000). They found little dependence of the final cusp slope on the degree of flattening. Merritt & Quinlan (1998) grew black holes of various masses in a triaxial *N*-body model that was formed via gravitational collapse, and found $\rho \sim r^{-2}$ at $r \lesssim r_h$. Triaxial potentials support a wide range of different orbit families and the effects of black hole growth on such models have not been examined in detail.

In the spherical geometry, adiabatic growth of a black hole induces a mild anisotropy in the stellar motions at $r \lesssim r_h$ due to the slightly different ways that circular and eccentric orbits respond to the changing potential; the net effect is a decrease in the average orbital eccentricity (Young 1980; Goodman & Binney 1984; Quinlan et al. 1995; Fig. 17.1). If the progenitor galaxy is rotating, growth of the black hole tends to increase V_{rot} more rapidly than σ (Lee & Goodman 1989; Leeuwin & Athanassoula 2000), although again the effect is slight.

To summarize: the adiabatic growth model is limited in its ability to reproduce the full range of luminosity profiles observed in galactic nuclei. The model predicts power-law profiles at $r \lesssim r_h$ with logarithmic slopes $1.5 \lesssim \gamma \lesssim 2.5$. This nicely brackets the range of slopes observed in the nuclei of galaxies fainter than $M_V \approx -20$ mag. However, slopes less than $\gamma \approx 1.5$ are not naturally produced by the adiabatic growth model, and some fine tuning is required to avoid inflections in the profile at $r \approx r_h$.

17.4   The Binary Black Hole Model

The adiabatic growth model was proposed (Peebles 1972) before the importance of galaxy interactions and mergers was appreciated. We now know that supermassive black holes have been present in at least some galaxies since redshifts of $z \approx 6$ (e.g., Fan et al. 2001), and we believe that most galaxies have experienced at least one major merger since that time; indeed the era of peak quasar activity may coincide with the era of galaxy assembly via mergers (e.g., Cavaliere & Vittorini 2000). If a nucleus forms via the merger of two galaxies containing pre-existing black holes, the net effect on the nuclear density profile is roughly the opposite of what the adiabatic growth model predicts: the black holes *displace* matter as they spiral into the center (Fig. 17.2). This is a natural way to account for the shallow nuclear profiles in bright galaxies. The process may be understood as a sort of dynamical friction, with the "heavy particles" (the black holes) transferring their kinetic energy to the "light particles" (the stars). However, most of the energy transfer takes place after the two black holes have come within each other's sphere of influence, and in this regime, the interaction with the background is dominated by another mechanism, the *gravitational slingshot* (Saslaw, Valtonen, & Aarseth 1974). The massive binary ejects passing stars at high velocities, removing them from the nucleus and simultaneously increasing its binding energy.

We begin by reviewing the dynamics of massive binaries in fixed, homogeneous backgrounds, then consider the more difficult problem of a binary located at the center of an inhomogeneous and evolving galaxy.

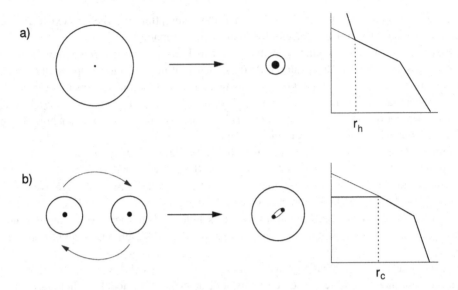

Fig. 17.2. Two schemes for growing black holes at the centers of galaxies. (*a*) The adiabatic growth model: the stellar density within r_h is increased as the black hole pulls in stars. (*b*) The binary black hole model: the inspiraling black holes displace matter within a distance r_c that is roughly the separation between the two black holes when they first form a bound pair.

17.4.1 Dynamics of Massive Binaries

Consider a binary system consisting of two black holes with masses M_1 and M_2, where $p \equiv M_2/M_1 \leq 1$ and $M_{12} \equiv M_1 + M_2$. Let a be the semi-major axis of the Keplerian orbit and e the orbital eccentricity. The binding energy of the binary is

$$|E| = \frac{GM_1M_2}{2a} = \frac{G\mu M_{12}}{2a}, \tag{17.10}$$

with $\mu = M_1M_2/M_{12}$ the reduced mass. The relative velocity of the two black holes, assuming a circular orbit, is

$$V_{bin} = \sqrt{\frac{GM_{12}}{a}} = 658 \text{ km s}^{-1} \left(\frac{M_{12}}{10^8 M_\odot}\right)^{1/2} \left(\frac{a}{1 \text{ pc}}\right)^{-1/2}. \tag{17.11}$$

A binary is called "hard" when its binding energy per unit mass, $|E|/M_{12} = G\mu/2a$, exceeds $\sim \sigma^2$. For concreteness, a binary will here be called hard if its separation falls below a_h, where

$$a_h \equiv \frac{G\mu}{4\sigma^2} \approx 0.27 \text{ pc}\,(1+p)^{-1} \left(\frac{M_2}{10^7 M_\odot}\right) \left(\frac{\sigma}{200 \text{ km s}^{-1}}\right)^{-2}. \tag{17.12}$$

Other definitions of a_h are possible (e. g. Hills 1983; Quinlan 1996).

Stars passing within a distance $\sim 3a$ of the center of mass of a hard binary undergo a complex interaction with the two black holes, followed almost always by ejection at velocity $\sim \sqrt{\mu/M_{12}}V_{bin}$ (Saslaw et al. 1974). Each ejected star carries away energy and angular momentum, causing the semi-major axis, eccentricity and orientation of the binary to change and the local density of stars to drop. If the stellar distribution is assumed fixed far from the binary and if the contribution to the potential from the stars is ignored, the rate at which these changes occur can be computed by carrying out scattering experiments of massless stars against a binary whose orbital elements remains fixed during each interaction.

The results of the scattering experiments can be summarized via a set of dimensionless coefficients $H, J, K, L, ...$ which define the mean rates of change of the parameters characterizing the binary and the stellar background (Hills & Fullerton 1980; Roos 1981; Hills 1983, 1992; Baranov 1984; Mikkola & Valtonen 1992; Quinlan 1996; Merritt 2001, 2002). These coefficients are functions of the binary mass ratio, eccentricity and hardness but are typically independent of a in the limit that the binary is very hard. The hardening rate of the binary is given by

$$\frac{d}{dt}\left(\frac{1}{a}\right) = H\frac{G\rho}{\sigma} \tag{17.13}$$

with ρ the density of stars at infinity. The mass ejection rate is

$$\frac{dM_{ej}}{d\ln(1/a)} = JM_{12} \tag{17.14}$$

with M_{ej} the mass in stars that escape the binary. The rate of change of the binary's orbital eccentricity is

$$\frac{de}{d\ln(1/a)} = K. \tag{17.15}$$

The diffusion coefficient describing changes in the binary's orientation is

$$\langle\Delta\vartheta^2\rangle = L\frac{m_*}{M_{12}}\frac{G\rho a}{\sigma}, \tag{17.16}$$

with m_* the stellar mass. Additional coefficients describe the rate of diffusion of the binary's center of mass, or "Brownian motion" (Merritt 2001).

The binary hardening coefficient H reaches a constant value of ~ 16 in the limit $a \ll a_h$, with a weak dependence on M_2/M_1 (Hills 1983; Mikkola & Valtonen 1992; Quinlan 1996). In a fixed background, equation (17.13) therefore implies that a hard binary hardens at a constant rate:

$$\frac{1}{a(t)} - \frac{1}{a_h} \approx H\frac{G\rho}{\sigma}(t-t_h), \quad t \geq t_h, \quad a(t_h) = a_h. \tag{17.17}$$

If the supply of stars remains steady, hardening continues at this rate until the components of the binary come close enough together that the emission of gravitational radiation is important. In this regime, gravity wave coalescence takes place in a time:

$$\begin{aligned} t_{gr} &= \frac{5}{256F(e)}\frac{c^5}{G^3\mu M_{12}^2}a^4, \\ F(e) &= (1-e^2)^{-7/2}\left(1 + \frac{73}{24}e^2 + \frac{37}{96}e^4\right) \end{aligned} \tag{17.18}$$

(Peters 1964). Coalescence in a time t_{gr} occurs when $a = a_{gr}$, where

$$\frac{a_h}{a_{gr}} \approx 75F^{-1/4}\frac{p^{3/4}}{(1+p)^{3/2}}\left(\frac{\sigma}{200\text{ km s}^{-1}}\right)^{-7/8}\left(\frac{t_{gr}}{10^9\text{yr}}\right)^{-1/4}. \tag{17.19}$$

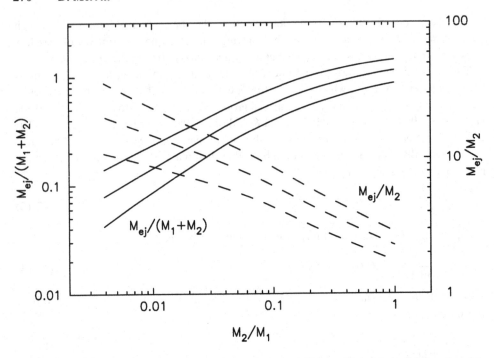

Fig. 17.3. Mass ejected by a decaying binary, in units of $M_{12} = M_1 + M_2$ (solid lines) or M_2 (dashed lines). Curves show mass that must be ejected in order for the binary to reach a separation where the emission of gravity waves causes coalescence on a time scale of 10^{10} yr (lower), 10^9 yr (middle) and 10^8 yr (upper).

The $M_\bullet - \sigma$ relation, $M_{12}/10^8 M_\odot \approx (\sigma/200 \text{ km s}^{-1})^{4.5}$, has been used to express M_{12} in terms of σ. For mass ratios p of order unity and $e \approx 0$, the binary must decay by a factor of $\sim 10^2$ in order for gravitational radiation to induce coalescence in a time shorter than 10^9 yr. Less decay is required if the binary is eccentric or if $M_2 \ll M_1$.

If the binary manages to shrink by such a large factor, the damage done to its stellar surroundings will be considerable. The mass ejected by the binary in decaying from a_h to a_{gr} is given by the integral of equation (17.14):

$$M_{ej} = M_{12} \int_{a_{gr}}^{a_h} \frac{J(a)}{a} da. \tag{17.20}$$

Figure 17.3 shows M_{ej} as a function of the mass ratio M_2/M_1 for $\sigma = 200$ km s^{-1} and various values of t_{gr}. The mass ejected in reaching coalescence is of order M_{12} for equal-mass binaries, and several times M_2 when $M_2 \ll M_1$. A black hole that grew to its current size through a succession of mergers should therefore have displaced a few times its own mass in stars. If this mass came mostly from stars that were originally in the nucleus, the density within r_h would drop drastically and the hardening would slow. Without some way of replenishing the supply of stars, decay could stall at a separation much greater than a_{gr}. This is the "final parsec problem": how to avoid stalling and bring the black holes from their separation of ~ 1 pc, when they first form a hard binary, to $\sim 10^{-2}$ pc, where gravity wave emission is efficient.

Changes in the binary's orbital eccentricity (equation 17.15) are potentially important because the gravity wave coalescence time drops rapidly as $e \rightarrow 1$ (equation 17.18). For a hard binary, scattering experiments give $K(e) \approx K_0 e(1-e^2)$, with $K_0 \approx 0.5$ for an equal-mass binary (Mikkola & Valtonen 1992; Quinlan 1996). The dependence of K on M_2/M_1 is not well understood. The implied changes in e as a binary decays from $a = a_h$ to a_{gr} are modest, $\Delta e \lesssim 0.2$, for all initial eccentricities. The rms change in the orientation of the binary's spin axis (equation 17.16) is $\delta\theta \approx 2(m_*/M_{12})^{1/2}\log^{1/2}(a_h/a)$, which is of order one degree or less if $m_* \approx M_\odot$. Reorientations of the binary's angular momentum affect the spin direction of the coalesced black hole and the direction of any associated jet (Merritt 2002).

17.4.2 Binary Black Holes in Galaxies

The scattering experiments summarized above treat the binary's environment as fixed and homogeneous. In reality, the binary is embedded at the center of an inhomogeneous and evolving galaxy, and the supply of stars that can interact with it is limited.

In a fixed spherical galaxy, stars can interact with the binary only if their pericenters lie within $\sim \mathcal{R} \times a$, where $\mathcal{R}$ is of order unity. Let $L_{lc} = \mathcal{R}a\sqrt{2[E-\Phi(\mathcal{R}a)]} \approx \sqrt{2GM_{12}\mathcal{R}a}$, the angular momentum of a star with pericenter $\mathcal{R}a$. The "loss cone" is the region in phase space defined by $L \leq L_{lc}$. The mass of stars in the loss cone is

$$
\begin{aligned}
M_{lc}(a) &= m_* \int dE \int_0^{L_{lc}} dL\, N(E, L^2) \\
&= m_* \int dE \int_0^{L_{lc}^2} dL^2\, 4\pi^2 f(E, L^2) P(E, L^2) \\
&\approx 8\pi^2 GM_{12}m_*\mathcal{R}a \int dE\, f(E)P_{rad}(E).
\end{aligned}
\tag{17.21}
$$

Here P is the orbital period; in the final line, f is assumed isotropic and P has been approximated by the period of a radial orbit of energy E. An upper limit to the mass that is available to interact with the binary is $\sim M_{lc}(a_h)$, the mass within the loss cone when the binary first becomes hard; this is an upper limit since some stars that are initially within the loss cone will "fall out" as the binary shrinks. Assuming a singular isothermal sphere for the stellar distribution, $\rho \propto r^{-2}$, and taking the lower limit of the energy integral to be $\Phi(a_h)$, equation (17.21) implies

$$
M_{lc}(a_h) \approx 3\mathcal{R}\mu.
\tag{17.22}
$$

We can compute the change in a that would result if the binary interacted with this entire mass, by using the fact the mean energy change of a star interacting with a hard binary is $\sim 3Gm_*\mu/2a$ (Quinlan 1996). Equating the energy carried away by stars with the change in the binary's binding energy gives

$$
\frac{3}{2}\frac{G\mu}{a}dM \approx \frac{GM_1M_2}{2}d\left(\frac{1}{a}\right)
\tag{17.23}
$$

or

$$
\ln\left(\frac{a_h}{a}\right) \approx 3\frac{\Delta M}{M_{12}} \approx \frac{9\mathcal{R}\mu}{M_{12}} \approx 9\mathcal{R}\frac{p}{(1+p)^2}, \qquad p \equiv M_2/M_1,
\tag{17.24}
$$

if ΔM is equated with M_{lc}. Only for very low mass ratios ($M_2 \lesssim 10^{-3} M_1$) is this decay factor large enough to satisfy equation (17.19), but the time required for such a small black hole to reach the nucleus is likely to exceed a Hubble time (Merritt 2000). Hence even under the most favorable assumptions, the binary would not be able to interact with enough mass to reach gravity-wave coalescence.

But the situation is even worse than this, since not all of the mass in the loss cone will find its way into the binary. The time scale for the binary to shrink is comparable with stellar orbital periods, and some of the stars with $r_{peri} \approx a_h$ will only reach the binary after a has fallen below $\sim a_h$. We can account for the changing size of the loss cone by writing

$$\frac{dM}{dt} = \int_{E_0(t)}^{\infty} \frac{1}{P(E)} \frac{dM_{lc}}{dE} dE = 8\pi^2 GM_{12} m_* \mathcal{R} a(t) \int_{E_0(t)}^{\infty} f_i(E) dE, \qquad (17.25)$$

where $M(t)$ is the mass in stars interacting with the binary and $f_i(E)$ is the initial distribution function; setting $P(E_0) = t$ reflects the fact that stars on orbits with periods less than t have already interacted with the binary and been ejected. Combining equations (17.23) and (17.25),

$$\frac{d}{dt}\left(\frac{1}{a}\right) \approx 24\pi^2 \mathcal{R} Gm_* \int_{E_0(t)}^{\infty} f_i(E) dE. \qquad (17.26)$$

Solutions to equation (17.26) show that a binary in a singular isothermal sphere galaxy stalls at $a_h/a \approx 2.5$ for $M_2 = M_1$, compared with $a_h/a \approx 10$ if the full loss cone were depleted (equation 17.24). In galaxies with shallower central cusps, decay of the binary would stall at even greater separations.

It is therefore entirely possible that uncoalesced binaries exist at the centers of many galaxies, particularly galaxies like large ellipticals with low central densities. But there is circumstantial evidence that long-lived black hole binaries are rare. Some radio galaxies show clear evidence of a recent flip in the black hole's spin orientation, as would occur when two black holes coalesce (Dennett-Thorpe et al. 2002), and their numbers are consistent with a coalescence rate that is roughly equal to the galaxy merger rate (Merritt & Ekers 2002). The almost complete lack of correlation between jet directions in Seyfert galaxies and the angular momenta of their disks (Ulvestad & Wilson 1984; Kinney et al. 2000) also suggests that black hole coalescences were common in the past. There are few if any "smoking gun" detections of binary black holes among AGN (e.g., Halpern & Eracleous 2000); the best, but still controversial, case is OJ 287 (Pursimo et al. 2000).

Below are discussed some additional mechanisms that have been proposed for extracting energy from binary black holes and hastening their decay. When the agent interacting with the binary is stars, continued decay of the binary implies continued destruction of the stellar cusp. However, for most of the mechanisms discussed below, the detailed effects on the stellar distribution have yet to be worked out.

17.4.2.1 Scattering of Stars into the Loss Cone

Destruction of a pre-existing stellar cusp generates strong gradients in the phase-space density at $L \approx L_{lc}$, the angular momentum of an orbit that lies just outside the loss cone. A small perturbation can deflect a star on such an orbit into the loss cone. This process has been studied in detail in the context of gravitational scattering of stars into the tidal disruption sphere of a single black hole (Lightman & Shapiro 1977; Cohn & Kulsrud

1978). Once a steady-state flow of stars into the loss cone has been set up, the distribution function near L_{lc} has the form:

$$f(E,L) \approx \frac{1}{\ln(1/R_{lc})} \overline{f}(E) \ln\left(\frac{R}{R_{lc}}\right) \qquad (17.27)$$

(Lightman & Shapiro 1977), where $R \equiv L^2/L_c^2(E)$, $L_c(E)$ is the angular momentum of a circular orbit of energy E, and $\overline{f}$ is the distribution function far from the loss cone, assumed to be isotropic. The mass flow into the central object is $m_* \int \mathcal{F}(E)dE$, where

$$\mathcal{F}(E)dE = 4\pi^2 L_c^2(E) \left\{ \oint \frac{dr}{v_r} \lim_{R \to 0} \frac{\langle(\Delta R)^2\rangle}{2R} \right\} \frac{\overline{f}}{\ln(1/R_{lc})} dE. \qquad (17.28)$$

The quantity in brackets is the orbit-averaged diffusion coefficient in R. Yu (2002) evaluated the contribution of two-body scattering to the decay rate of a massive binary and found that it was usually too small to overcome the stalling that occurs when the loss cone is first emptied.

Standard loss cone theory assumes a quasi-steady-state distribution of stars in phase space near L_{lc}. This assumption is appropriate at the center of a globular cluster, where relaxation times are much shorter than the age of the universe, but is less appropriate for a galactic nucleus, where relaxation times almost always greatly exceed a Hubble time (e.g., Faber et al. 1997). The distribution function $f(E,L)$ immediately following the formation of a hard binary is approximately a step function,

$$f(E,L) \approx \begin{cases} \overline{f}(E), & L > L_{lc} \\ 0, & L < L_{lc}, \end{cases} \qquad (17.29)$$

much steeper than the $\sim \ln L$ dependence in a collisionally relaxed nucleus (equation 17.27). Since the transport rate in phase space is proportional to the gradient of f with respect to L, steep gradients imply an enhanced flux into the loss cone. The total mass in stars consumed by the binary can exceed the predictions of the standard model by factors of a few, implying greater cusp destruction and more rapid decay of the binary (Milosavljević & Merritt 2003b). This time-dependent loss cone refilling might be particularly effective in a nucleus that continues to experience mergers or accretion events, in such a way that the loss cone repeatedly returns to an unrelaxed state with its associated steep gradients.

17.4.2.2 Re-ejection

Unlike the case of tidal disruption of stars by a single black hole, a star that interacts with a massive binary remains inside the galaxy and is available for further interactions. In principle, a single star can interact many times with the binary before being ejected from the galaxy or falling outside the loss cone; each interaction takes additional energy from the binary and hastens its decay. Consider a simple model in which a group of N stars in a spherical galaxy interacts with the binary and receives a mean energy increment of $\langle \Delta E \rangle$. Let the original energy of the stars be E_0. Averaged over a single orbital period $P(E)$, the binary hardens at a rate

$$\frac{d}{dt}\left(\frac{GM_1 M_2}{2a}\right) = m_* \frac{N\langle \Delta E \rangle}{P(E)}. \qquad (17.30)$$

In subsequent interactions, the number of stars that remain inside the loss cone scales as $L_{lc}^2 \propto a$, while the ejection energy scales as $\sim a^{-1}$. Hence $N\langle\Delta E\rangle \propto a^1 a^{-1} \propto a^0$. Assuming the singular isothermal sphere potential for the galaxy, one finds

$$\frac{a_h}{a(t)} \approx 1 + \frac{\mu}{M_{12}} \ln \left[1 + \frac{m_* N\langle\Delta E\rangle}{2\mu\sigma^2} \frac{t-t_h}{P(E_0)} \right] \tag{17.31}$$

(Milosavljević & Merritt 2003b). Hence the binary's binding energy increases as the logarithm of the time, even after all the stars in the loss cone have interacted at least once with the binary. Re-ejection would occur differently in nonspherical galaxies where angular momentum is not conserved and ejected stars could miss the binary on their second passage. However, there will generally exist a subset of orbits defined by a maximum pericenter distance $\lesssim a$, and stars scattered onto such orbits can continue to interact with the binary.

17.4.2.3   Chaotic Loss Cones

Loss cone dynamics are qualitatively different in non-axisymmetric (triaxial or bar-like) potentials, since a much greater number of stars may be on "centrophilic" orbits which take them near to the black hole(s) (Gerhard & Binney 1985). Triaxial models need to be taken seriously given recent demonstrations (Poon & Merritt 2002, 2004) that black hole nuclei can be stably triaxial and that the fraction of mass on centrophilic – typically chaotic – orbits can be large. The frequency of pericenter passages, $r_{peri} < d$, for a chaotic orbit of energy E in a triaxial black hole nucleus is roughly linear in d, $N(r_{peri} < d) \approx d \times A(E)$ (Merritt & Poon 2004). The total rate at which stars pass within a distance $\mathcal{R}a$ of the massive binary is therefore

$$\frac{dM}{dt} \approx \mathcal{R}a \int A(E)M_c(E)dE, \tag{17.32}$$

where $M_c(E)$ is the mass on centrophilic orbits. The implied feeding rate can be comparable to that in a spherical nucleus with a constantly refilled loss cone, and orders of magnitude greater than in a loss cone that is re-supplied via star-star interactions (Merritt & Poon 2004). Even transient departures from axisymmetry, for instance during mergers, might result in substantial loss cone refilling due to this mechanism.

17.4.2.4   Multiple Black Holes

If binary decay stalls, an uncoalesced binary may be present in a nucleus when a third black hole, or a second binary, is deposited there following a subsequent merger. The multiple black hole system that forms will engage in its own gravitational slingshot interactions, eventually ejecting one or more of the black holes from the nucleus (though probably not from the galaxy). This process has been extensively modeled assuming a fixed potential for the galaxy (e.g., Valtaoja et al. 1989; Mikkola & Valtonen 1990; Valtonen et al. 1994). The effect on the stellar distribution of $N \gtrsim 2$ interacting black holes is not well understood, although N-body simulations with $10-20$ massive particles and a "live" background show that the black holes displace ~ 10 times their own mass before being ejected (Merritt & Milosavljević 2003). The separation and eccentricity of the dominant binary can change dramatically during each interaction and this may be an effective way to shorten the gravity-wave coalescence time. In a wide, hierarchical triple, the eccentricity of the dominant binary oscillates through a maximum value of $\sim \sqrt{1-5\cos^2 i/3}$, $|\cos i| < \sqrt{3/5}$, with i the mutual inclination angle (Kozai 1962). Blaes, Lee, & Socrates (2002) estimate that the coalescence times in equal-mass, hierarchical triples can be reduced by factors of ~ 10 via the Kozai mechanism.

17.4.2.5 Gas Dynamics

If the inner ~ 1 pc of the nucleus contains a mass in gas comparable to M_2, torques from the gas will cause the orbit of the smaller black hole to decay in a time of order the gas accretion time (Syer & Clarke 1995; Ivanov, Papaloizou, & Polnarev 1999). Given standard assumptions about accretion disk viscosities, the gas-dynamical decay rate would exceed that from gravity wave emission for $a > a_{acc}$, where

$$\frac{a_{acc}}{a_h} \approx 1 \times 10^{-3} \left(\frac{p}{0.1} \right)^{2/5} \left(\frac{\sigma}{200 \text{ km s}^{-1}} \right)^2 \tag{17.33}$$

(Armitage & Natarajan 2002). If the orbit of the secondary is strongly inclined with respect to the accretion disk around the larger black hole, its passages through the disk could generate periodic outbursts, and this has been suggested as a model for the ~ 12 yr cycle of optical flaring observed in the blazar OJ 287 (Lehto & Valtonen 1996). Gas deposition of the required magnitude almost certainly occurred during the quasar epoch, although it is less clear that this mechanism is effective for galaxies in the current universe.

17.5   *N*-Body Studies

Unless great care is taken, *N*-body studies of binary black hole dynamics are unlikely to give an accurate picture of the evolution expected in real galaxies. This follows from the result (§ 17.4.2) that time scales for two-body scattering of stars into the binary's loss cone are of order the Hubble time or somewhat longer in real galaxies. In *N*-body simulations, relaxation times are shorter by factors of $\sim N/10^{11}$ than in real galaxies, hence the long-term evolution of the binary is likely to be dominated by spurious loss cone refilling, Brownian motion of the black holes and other noise-driven effects (Milosavljević & Merritt 2003b). The stalling that is predicted in the absence of loss cone refilling in real galaxies (Fig. 17.4) can only be reproduced via *N*-body codes if the mean field is artificially smoothed and the black hole binary is "nailed down" (e.g., Quinlan & Hernquist 1997).

N-body studies are most useful at characterizing the early stages of binary formation and decay, or simulating the disruptive effects of a single black hole on an infalling galaxy; indeed scattering experiments (§ 17.4.1) are almost useless in these regimes. Due to algorithmic limitations, most *N*-body studies (e.g., Ebisuzaki, Makino, & Okumura 1991; Makino et al. 1993; Governato, Colpi, & Maraschi 1994; Makino & Ebisuzaki 1996; Makino 1997; Nakano & Makino 1999a,b; Hemsendorf, Sigurdsson, & Spurzem 2002) have been based on galaxy models with unrealistically large cores. The first *N*-body merger simulations using two black holes and realistically dense initial conditions (Merritt & Cruz 2001; see also Merritt et al. 2002) found that the black hole in the larger galaxy was efficient at tidally disrupting the steep cusp in the infalling galaxy, producing a remnant with only slightly higher central density than that of the giant galaxy initially. This result helps explain the absence of dense cusps in bright galaxies (Forbes, Franx, & Illingworth 1995) and the persistence of the "core fundamental plane" in the face of mergers (Holley-Bockelmann & Richstone 1999, 2000).

Quinlan & Hernquist (1997) studied the evolution of a black hole binary inside cuspy models with $\rho \sim r^{-1}$ and $\rho \sim r^{-2}$ and a range of black hole masses and particle numbers,

$N \leq 2 \times 10^5$. Their N-body code was unable to simulate an actual merger and all of their detailed results were derived from initial conditions consisting of a single galaxy into which two "naked" black holes were symmetrically dropped. This artificial starting configuration resulted in substantial evolution of the cusp before the formation of the binary. The late evolution of the binary was found to be strongly dependent on N, due in part to spurious Brownian motion of the black hole particles. The cores that formed were characterized by strong velocity anisotropies, $\sigma_t \gg \sigma_r$, due to the ejection of stars on eccentric orbits.

Milosavljević & Merritt (2001) followed the evolution of cuspy ($\rho \sim r^{-2}$) galaxy models containing black holes, starting from pre-merger initial conditions and continuing until the binary separation had decayed a factor of ~ 10 below a_h. The initially steep nuclear cusps were converted to shallower, $\rho \sim r^{-1}$ profiles shortly after the black holes had formed a hard binary; thereafter the nuclear profile evolved slowly toward even shallower slopes. The decay rate of the binary was found not to be strongly dependent on N, probably due to the fact that the loss cone was continuously refilled by two-body scattering in these collisional simulations (Milosavljević & Merritt 2003b). The velocity anisotropies created during formation of the core were much milder than in the simulations of Quinlan & Hernquist (1997), and similar in magnitude to the anisotropies predicted by the adiabatic growth model (Fig. 17.1).

17.6   Observational Evidence for the Binary Black Hole Model

Figure 17.3 shows that a massive binary must eject of order its own mass in reaching a separation of a_{gr} if $M_2 \approx M_1$, or several times M_2 if $M_2 \ll M_1$. These numbers should be interpreted with caution since: (1) binaries might not decay as far as a_{gr}, or the final stages of decay might be driven by some process other than energy exchange with stars; (2) the definition of "ejection" used in Figure 17.3 is escape of a star from an isolated binary, and does not take into account the confining effect of the nuclear potential; (3) the effect of repeated mergers on nuclear density profiles, particularly mergers involving very unequal-mass binaries, is poorly understood. Nevertheless, a clear prediction of the binary black hole model is that galactic mergers should result in the removal of a mass of order M_{12} from the nucleus. As in the adiabatic growth model, we are handicapped in testing the theory by lack of knowledge of the primordial nuclear profiles. A reasonable guess is that all galaxies originally had steep, $\rho \sim r^{-2}$ density cusps, since these are generic in the low-luminosity ellipticals which are least likely to have been influenced by mergers, and since the adiabatic growth model predicts $\rho \sim r^{-2}$ (Fig. 17.1).

The "mass deficit" is defined as the difference in integrated mass between the observed density profile, and a $\rho \propto r^{-\gamma_0}$ profile extrapolated inward from the break radius r_b:

$$M_{def} \equiv 4\pi \int_0^{r_b} \left[\rho(r_b) \left(\frac{r}{r_b} \right)^{-\gamma_0} - \rho(r) \right] r^2 dr. \tag{17.34}$$

Milosavljević et al. (2002) and Ravindranath, Ho, & Filippenko (2002) computed M_{def} in samples of "core"-profile elliptical galaxies. The former authors found a good proportionality between M_{def} and $M_\bullet$, with $\langle M_{def}/M_\bullet \rangle \approx 1$ for $\gamma_0 = 1.5$ and $\langle M_{def}/M_\bullet \rangle \approx 10$ for $\gamma_0 = 2$ (Fig. 17.4). These numbers are within the range predicted by the binary black hole model, given the uncertainties associated with multiple mergers. Ravindranath et al. (2002) computed black hole masses using a shallower assumed slope for the $M_\bullet - \sigma$ relation, $M_\bullet \propto \sigma^{3.75}$, and found a steeper, nonlinear dependence of M_{def} on $M_\bullet$.

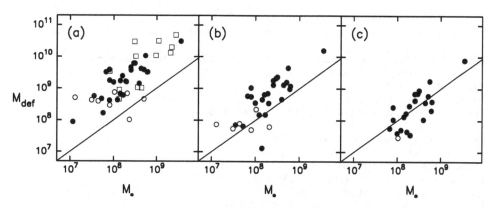

Fig. 17.4. Mass deficit versus black hole mass for three different assumed values of γ_0, the logarithmic slope of the density cusp before energy input from the black holes: (*a*) $\gamma_0 = 2$, (*b*) $\gamma_0 = 1.75$, and (*c*) $\gamma_0 = 1.5$. Units are solar masses. Solid lines are $M_{def} = M_\bullet$. (Adapted from Milosavljević et al. 2002).

More rigorous tests of the binary black hole model will require a better understanding of the expected form of $\rho(r)$. As discussed above, while the best current N-body simulations suggest $\rho \sim r^{-1}$ following binary formation (Milosavljević & Merritt 2001), the simulations are dominated by noise over the long term. If the decay stalls, the predicted density profile can be very different: a "hole" forms inside of $\sim 3a_{stall}$ (e.g., Fig. 1 of Zier & Biermann 2001). Central minima may in fact have been seen in the luminosity profiles of a few galaxies (Lauer et al. 2002).

Other processes could result in energy exchange between binary black holes and stars. If the binary eventually coalesces, the gravitational radiation carries a *linear* momentum leading to a recoil of the coalesced black hole at a velocity $v_{recoil} \sim 10^2 - 10^3$ km s^{-1} (Bekenstein 1973; Fitchett 1983; Eardley 1983), and possibly even higher if the black holes were rapidly spinning prior to coalescence (Redmount & Rees 1989). Recoil velocities of this order would eject the black hole from the center of the nucleus and its subsequent infall would displace stars. Quantitative evaluation of this effect will require more accurate estimates of v_{recoil} based on fully general-relativistic calculations of black hole coalescence.

A major focus of future work should be to calculate the evolution of $\rho(r)$ as predicted by the various scenarios for binary decay discussed in this article.

Acknowledgements. I thank M. Milosavljević and L. Ho for their detailed reading of the manuscript. This work was supported by NSF grants AST 00-71099 and AST 02-06031, by NASA grants NAG5-6037 and NAG5-9046, and by grant HST-AR-08759 from STScI.

References

Armitage, P. J., & Natarajan, P. 2002, ApJ, 567, L9
Baranov, A. S. 1984, Astron. Zh., 61, 1098
Begelman, M. C., Blandford, R. D., & Rees, M. J. 1980, Nature, 287, 307
Bekenstein, J. 1973, ApJ, 183, 675
Blaes, O., Lee, M. H., & Socrates, A. 2002, ApJ, 578, 775
Cavaliere, A., & Vittorini, V. 2000, ApJ, 543, 599
Cipollina, M. 1995, A&AS, 110, 155
Cipollina, M., & Bertin, G. 1994, A&A, 288, 43

Cohn, H., & Kulsrud, R. 1978, ApJ, 226, 1087

Dennett-Thorpe, J., Scheuer, P. A. G., Laing, R. A., Bridle, A. H., Pooley, G. G., & Reich, W. 2002, MNRAS, 330, 609

Eardley, D. M. 1985, in Gravitational Radiation, ed. N. Deruelle & T. Piran (Amsterdam: North-Holland), 257

Ebisuzaki, T., Makino, J., & Okumura, S. K. 1991, Nature, 354, 212

Faber, S. M., et al. 1997, AJ, 114, 1771

Fan, X., et al. 2001, AJ, 122, 2833

Forbes, D. A., Franx, M., & Illingworth, G. D. 1995, AJ, 109, 1988

Fitchett, M. 1983, MNRAS, 203, 1049

Gerhard, O. E., & Binney, J. 1985, MNRAS, 216, 467

Gondolo, P., & Silk, J. 1999, Phys. Rev. Lett., 83, 1719

Goodman, J., & Binney, J. 1984, MNRAS, 207, 511

Governato, F., Colpi, M., & Maraschi, L. 1994, MNRAS, 271, 317

Halpern, J. P., & Eracleous, M. 2000, ApJ, 531, 647

Hemsendorf, M., Sigurdsson, S., & Spurzem, R. 2002, ApJ, 581, 1256

Hills, J. G. 1983, AJ, 88, 1269

——. 1992, AJ, 103, 1955

Hills, J. G., & Fullerton, L. W. 1980, AJ, 85, 1281

Holley-Bockelmann, K., & Richstone, D. 1999, ApJ, 517, 92

——. 2000, ApJ, 531, 232

Ivanov, P. B., Papaloizou, J. C. B., & Polnarev, A. G. 1999, MNRAS, 307, 79

Kinney, A. L., Schmitt, H. R., Clarke, C. J., Pringle, J. E., Ulvestad, J. S., & Antonucci, R. R. J. 2000, ApJ, 537, 152

Kozai, Y. 1962, AJ, 67, 591

Lauer, T. R., et al. 2002, AJ, 124, 1975

Lee, M. H., & Goodman, J. 1989, ApJ, 343, 594

Leeuwin, F., & Athanassoula, E. 2000, MNRAS, 317, 79

Lehto, H. J., & Valtonen, M. J. 1996, ApJ, 460, 207

Lightman, A. P., & Shapiro, S. L. 1977, ApJ, 211, 244

Lynden-Bell, D. 1963, The Observatory, 83, 23

Makino, J. 1997, ApJ, 478, 58

Makino, J., & Ebisuzaki, T. 1996, ApJ, 465, 527

Makino, J., Fukushige, T., Okumura, S. K., & Ebisuzaki, T. 1993, PASJ, 45, 303

Merritt, D. 2000, in Dynamics of Galaxies, ed. F. Combes, G. A. Mamon, & V. Charmandaris, (San Francisco: ASP), 221

——. 2001, ApJ, 556, 245

——. 2002, ApJ, 568, 998

Merritt, D., & Cruz, F. 2001, ApJ, 551, L41

Merritt, D., & Ekers, R. D. 2002, Science, 297, 1310

Merritt, D., & Milosavljević, M. 2003, DARK 2002: 4th International Heidelberg Conference on Dark Matter in Astro and Particle Physics, ed. H. V. Klapdor-Kleingrothaus & R. Viollier (astro-ph/0205140)

Merritt, D, Milosavljević, M., Verde, L., & Jimenez, R. 2002, Phys. Rev. Lett., 88, 191301

Merritt, D., & Poon, M. Y. 2004, in preparation

Merritt, D., & Quinlan, G. D. 1998, ApJ, 498, 625

Mikkola, S., & Valtonen, M. J. 1990, ApJ, 348, 412

——. 1992, MNRAS, 259, 115

Milosavljević, M., & Merritt, D. 2001, ApJ, 563, 34

——. 2003a, to appear in Proceedings of the 4th LISA Symposium (astro-ph/0212270)

——. 2003b, ApJ, in press (astro-ph/0212459)

Milosavljević, M., Merritt, D., Rest, A., & van den Bosch, F. C. 2002, MNRAS, 331, L51

Nakano, T., & Makino, J. 1999a, ApJ, 510, 155

——. 1999b, ApJ, 525, L77

Peebles, P. J. E. 1972, Gen. Rel. Grav., 3, 61

Peters, P. C. 1964, Phys. Rev. B, 136, 1224

Poon, M., & Merritt, D. 2002, ApJ, 568, L89

——. 2004, ApJ, in press (astro-ph/0212581)

Power, C., Navarro, J. F., Jenkins, A., Frenk, C. S., White, S. D. M., Springel, V., Stadel, J., & Quinn, T. 2003, MNRAS, 338, 14

Pursimo, T., et al. 2000, A&AS, 146, 141

Quinlan, G. D. 1996, New Astron., 1, 35

Quinlan, G. D., & Hernquist, L. 1997, New Astron., 2, 533

Quinlan, G. D., Hernquist, L., & Sigurdsson, S. 1995, ApJ, 440, 554

Ravindranath, S., Ho, L. C., & Filippenko, A. V. 2002, ApJ, 566, 801

Redmount, I. H., & Rees, M. J. 1989, Comm. Astrophys., 14, 165.

Roos, N. 1981, A&A, 104, 218

Saslaw, W. C., Valtonen, M. J., & Aarseth, S. J. 1974, ApJ, 190, 253

Syer, D., & Clarke, C. J. 1995, MNRAS, 277, 758

Ulvestad, J. S., & Wilson, A. S. 1984, ApJ, 285, 439

Valtaoja, L., Valtonen, M. J., & Byrd, G. G. 1989, ApJ, 343, 47

Valtonen, M. J., Mikkola, S., Heinamaki, P., & Valtonen, H. 1994, ApJS, 95, 69

van der Marel, R. P. 1999, AJ, 117, 744

Young, P. J. 1980, ApJ, 242, 1232

Yu, Q. 2002, MNRAS, 331, 935

Zier, C., & Biermann, P. L. 2001, A&A, 377, 23

18

Supermassive black holes: demographics and implications

DOUGLAS RICHSTONE
Department of Astronomy, University of Michigan

Abstract

The central fact in the demographics of supermassive black holes is the correlation between black hole mass and the mass, luminosity or velocity dispersion of the bulge of the host galaxy. We are not aware of a "second parameter" or of any systematic departures (such as might be due to measurement technique) from these fundamental relationships. The cosmic density of supermassive black holes at zero redshift is $4.8 \times 10^5 h^2 M_\odot \mathrm{Mpc}^{-3}$ or $\Omega_\bullet = 1.7 \times 10^{-6}$. This is consistent with the quasar "background" light, which has an *emitted* energy density of about $1.3 \times 10^{-15} \mathrm{erg} \, \mathrm{cm}^{-3}$.

The near equality of the cosmic supermassive black hole density with the quasar background energy (corrected by a suitable radiative efficiency) implies that the quasars are a by-product of the growth of black holes. We review a sampling of the wide variety of proposed theories for the growth of black holes and the coupling between black hole formation and galaxy bulge formation.

18.1   Introduction

The existence of this volume, reflecting a major meeting on supermassive black holes (BHs), testifies to the rapid evolution of this subject. As recently as 15 years ago, skepticism about the existence of BHs and their connection to active galactic nuclei (AGNs) was respectable. At this point BHs are the most conservative explanation for the dark mass concentrations in galactic centers. Elsewhere in this volume, Barth (2004), Gebhardt (2004), and Kormendy (2004) provide considerable information about the kinds of analysis that have been carried out, and the current state of our understanding of specific objects. In this chapter I focus on the inferences that can be drawn from these objects viewed together.

18.2   Correlations of BH Mass with Main Body Galaxy Properties

The central fact in discussions of BH demography is the relationship between BH mass and properties of the main body of the bulge of the host galaxy. Considerable attention has been drawn to the relationship between the BH mass and the bulge light (for example, in Kormendy & Richstone 1995; Kormendy & Gebhardt 2001), and the $M_\bullet - \sigma$ relation between the BH mass and rms velocities of bulge stars (Gebhardt et al. 2000; Ferrarese & Merritt 2000). We will rely heavily on the $M_\bullet - \sigma$ relation in this article. We illustrate both relationships in Figure 18.1. We note that other authors have argued that the galaxy luminosity profile also predicts the BH mass (see Erwin, Graham, & Caon 2004) We will not comment on this.

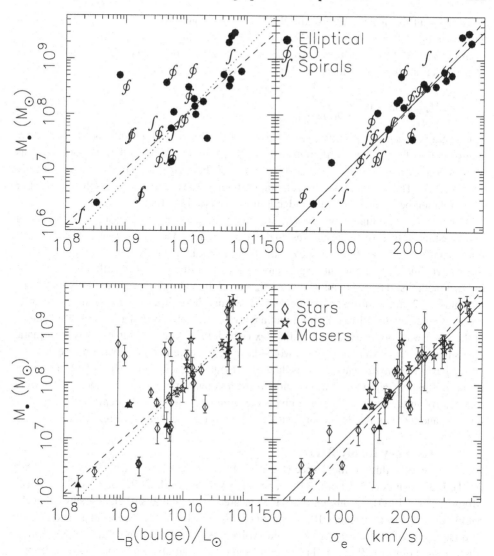

Fig. 18.1. Dynamically measured BH masses from Tremaine et al. (2002; except as noted in the text) are shown versus the host galaxy B-band bulge luminosity L_B (left-hand panels; from Kormendy & Gebhardt 2001) and, versus the slit-averaged bulge rms velocities out to half an effective radius (right-hand panels; from the tabulation in Tremaine et al.). In the top panels the points are marked by Hubble type, while in the lower panels they are designated by the method used to determine the BH mass. In the left-hand panel the dashed line represents $M_\bullet = 0.01 L_B$ and the dotted line represents $M_\bullet = 2.0 \times 10^{-3} \mathcal{M}_{\text{bulge}}$, where $\mathcal{M}_{\text{bulge}}$ is the bulge mass and we also assume that the bulge mass-to-light ratio is 5 at $L_B = 10^{10} L_\odot$ and that it varies as $\mathcal{M}_{\text{bulge}}/L \propto L^{0.2}$. In the right-hand panel the solid line represents the Tremaine et al. $M_\bullet - \sigma$ relation, and the dashed line represents the Merritt & Ferrarese (2001b) relation.

Throughout this article, we use the BH masses tabulated in Tremaine et al. (2002) with the exception of the mass of the BH in the Galaxy, for which we adopt $M_\bullet = 3.5 \times 10^6 M_\odot$ from Ghez (2004). In that paper, Tremaine et al. (2002) did a very careful review of the $M_\bullet - \sigma$ relation. In what follows (see Fig. 18.1), we use their result,

$$M_\bullet = 1.35 \times 10^8 \left(\frac{\sigma}{200 \text{ km s}^{-1}} \right)^{4.02}. \tag{18.1}$$

This result is consistent with the original result of Gebhardt et al. (2000) and is about 1.5σ shallower than the shallowest fit published by Merritt & Ferrarese (2001b). Possible reasons for this disagreement are discussed in Tremaine et al. but at this writing we still do not fully understand it. The scatter about this relation is about 0.3 dex, noticeably less than the scatter in the relation between $M_\bullet$ and the bulge luminosity (see Fig. 18.1).

Although the residuals from this fit are barely larger than the measurement errors, it is interesting is to investigate the residuals as a function of measurement method or as a function of galaxy type. In the first case, the BH masses are measured by stellar dynamics or by gas dynamics, and we would expect any systematic effects to have a different impact on the two methods. The residuals from the $M_\bullet - \sigma$ relation are plotted for these two methods in Figure 18.2. A t-test of sample means verifies that these distributions are insignificantly different. We conclude from this plot that any systematic effects either affect both methods in the same way, or are smaller than 0.1 dex (which would be detectable). Any systematic errors in these two different methods would have to conspire to create this agreement.

We have also investigated the residuals for evidence of an obvious second parameter in the BH-galaxy relation. We have not found a convincing second parameter. We were particularly interested in whether Hubble type might correlate with departures from this relation, and were unable to see any evidence of that in these data (see Fig. 18.2).

18.3   Are They Really BHs?

We note three critical special cases. In the case of our own Galaxy (see Ghez 2004), M32 (van der Marel et al. 1998; see also Verome et al. 2002) and NGC 4258 (see Miyoshi et al. 1995), an unseen mass is measured in a volume sufficiently small that the enclosed mass density exceeds $10^9 M_\odot \text{pc}^{-3}$ ($10^8 M_\odot \text{pc}^{-3}$ for M32). At this mass and density a single mass will shine very brightly and evolve rapidly, while an aggregate of any *known* astrophysical object will either (1) evaporate and evolve rapidly or (2) collapse to a BH in a time much less than the probable ages of the stars in these galaxies (see Maoz 1995, 1998). We summarize Maoz's argument below. It has two strands. If we treat the aggregate as point masses, then two-body relaxation sets an upper limit to the typical mass of the constituent particles. As this mass drops below the neutron star limit, the size of the objects is limited from below by ordinary electron degeneracy to the white dwarf mass-radius relation (in the solar mass range) or to the brown dwarf and planetary mass-radius relation (as one drops below $0.1 M_\odot$). These finite-sized objects collide in much less time than their likely age, regardless of their mass (at densities above the $10^9 M_\odot \text{pc}^{-3}$). Hence, these objects are not long-lived dynamical equilibria. One is free to construct collections of dark matter for the purpose of evading these constraints, but not with *known* astrophysical objects except for BHs of mass $\leq 0.2 M_\odot$.

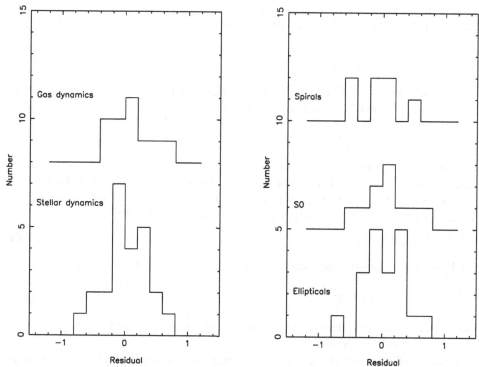

Fig. 18.2. Logarithmic residuals from the Tremaine et al. (2002) $M_\bullet - \sigma$ relation shown in Fig. 18.1. In the left hand-panel the residuals have been sorted by the method used to dynamically determine the BH mass. In the right-hand panel they are segregated by Hubble type. We see no statistically significant difference within either set of subsamples.

18.4 The Density of BHs and of the AGN Background

We can use the $M_\bullet - \sigma$ relationship to compute the mass density of BHs directly. Aller & Richstone (2002) do this by converting the galaxy luminosity function for different Hubble types, and the Faber-Jackson relation and its analog for bulges, to a "velocity dispersion function," and combine this with the $M_\bullet - \sigma$ relation to get a BH mass function from two different galaxy surveys (Marzke et al. 1994a; Marzke, Huchra, & Geller 1994b; Madgwick et al. 2002). The integrated density in BH mass at zero redshift is about

$$\rho_\bullet = 4.8 \times 10^5 h^2 \ M_\odot \mathrm{Mpc}^{-3}. \tag{18.2}$$

Yu & Tremaine (2002) used the Sloan Digital Sky Survey result giving the number of galaxies as a function of velocity dispersion (Bernardi et al. 2003) to obtain

$$\rho_\bullet = 2.5 \pm 0.4 \times 10^5 \ M_\odot \mathrm{Mpc}^{-3}, \tag{18.3}$$

for $h = 0.65$. These two numbers are in excellent agreement. An earlier estimate obtained by Merritt & Ferrarese (2001a) using the $M_\bullet - \sigma$ relation to recalibrate an $M_\bullet - L_{\mathrm{bulge}}$ relation gave $5 \times 10^5 M_\odot \mathrm{Mpc}^{-3}$ and was later adjusted downward to about $3 \times 10^5 M_\odot \mathrm{Mpc}^{-3}$ (Merritt & Ferrarese 2001c).

It is also useful to compare the cosmic BH mass density at $z = 0$ from Equation (18.2) or (18.3) to the luminous density of galaxies (this is a BH mass-to-galaxy light ratio). Marzke et al. (1994a, b) and Madgwick et al. (2002) both report a total luminous density in galaxies of

$$\rho_L = 2.0 \times 10^8 h \ L_\odot \mathrm{Mpc}^{-3}, \tag{18.4}$$

a value a factor of 2 larger than Loveday et al. (1992). Using Equations (18.2) and (1.4), we obtain a ratio of BH mass density to galaxy light density of

$$\frac{\rho_\bullet}{\rho_L} = 2.4 \times 10^{-3} h \ L_\odot \mathrm{Mpc}^{-3}. \tag{18.5}$$

Note that this is about 4 times smaller than the mean BH mass-to-bulge light ratio Υ illustrated in Figure 18.1, which is about $\Upsilon = 10^{-2}$, consistent with the fact that about 1/4 of the luminosity in galaxies at $z = 0$ is in bulges or ellipticals (Simien & de Vaucouleurs 1986). We can compute the fraction of the critical density carried by this BH mass,

$$\Omega_\bullet = 1.7 \times 10^{-6}, \tag{18.6}$$

and the BH mass fraction of baryonic matter (taking $\Omega_b h^2 = 0.0224$; Spergel et al. 2003)

$$\frac{\Omega_\bullet}{\Omega_b} = 0.76 \times 10^{-4} h^2. \tag{18.7}$$

This shows that the mass density in (supermassive) BHs is small compared to the cosmic baryon density (or the critical density).

The cosmic density of BHs can more interestingly be compared to the energy density in the radiation background due to quasars. Since quasars are thought to be accreting BHs we expect that the cosmic BH density is at least as large as the quasars would predict. We believe that Sołtan (1982) was the first to note the possibility of making this comparison and to use it to predict the masses of BHs in galactic nuclei today. Chokshi & Turner (1992) reexamined that result, and we follow their notation. They argue that the integrated comoving energy density in quasar light is

$$u_e = \int_0^\infty \int_0^\infty L\Phi(L|z)dL \frac{dt}{dz} dz = 1.3 \times 10^{-15} (C_B/16.5) \ \mathrm{erg \ cm}^{-3}, \tag{18.8}$$

where C_B is the bolometric correction at B band. This emitted energy density is related to the present-day energy density u_0 by

$$u_0 = u_e (1+\bar{z})^{-1} \approx 4.3 \times 10^{-16} (C_B/16.5) \ \mathrm{erg \ cm}^{-3}, \tag{18.9}$$

where $\bar{z}$ is the mean redshift of quasar emission, which we took to be $\bar{z} = 2$ for the estimate shown. The redshift dependence appears because, like the comoving mass density, the *comoving* photon density does not change with z, but the energy of any individual photon (and therefore the comoving radiation energy density) varies as $(1+z)^{-1}$ The local comoving energy density is, in principle, a direct observable given by $u_0 = 4\pi I/c$, where I is the frequency-integrated surface brightness of quasar light on the sky, and the explicit integral is over solid angle.

We define the "Sołtan number" S as the local mass density of the emitted quasar light corrected for the likely radiative efficiency of 10%. The next step in this procedure, using Equation (18.8), is to relate u_e to the mass in BHs.

Table 18.1. *Selected Backgrounds*

Name	Intensity (nW m^{-2} sr^{-1})	Energy Density (erg cm^{-3})
CMB	990	4.17×10^{-13}
IR	50.3	2.11×10^{-14}
UV–Opt	73.	3.1×10^{-14}
X-ray	0.35	1.45×10^{-16}
Quasars[a]		4.3×10^{-16}

[a] Present-day energy density u_0.

$$S = u_e \frac{1-\epsilon}{\epsilon c^2} = 2.2 \times 10^5 [(1-\epsilon)\epsilon_{0.1}^{-1}(C_B/16.5)] \ M_\odot \, \text{Mpc}^{-3}, \tag{18.10}$$

where ϵ is the radiative efficiency of accretion (the ratio of luminosity to $\dot{M} c^2$ in the accretion flow), and $\epsilon_{0.1} = \epsilon/0.1$. We parameterize most of our results in terms of $\epsilon_{0.1}$ since $\epsilon = 0.1$ is a fairly generic accretion efficiency for optically thick, geometrically thin accretion disks in Schwarzschild geometries (nonrotating BHs).

Yu & Tremaine (2002) also use recent (Boyle et al. 2000) estimates of the quasar distribution in luminosity and redshift to revisit the Chokshi & Turner estimate of the Sołtan number. They find

$$S = 2.1 \times 10^5 (C_B/11.8)[0.1(1-\epsilon)/\epsilon)] \ M_\odot \, \text{Mpc}^{-3}, \tag{18.11}$$

in excellent agreement with Chokshi & Turner. However, Yu & Tremaine use a bolometric correction of 11.8 (reflected in Eq. 18.11) based on work by Elvis et al. (1994).

The discussion above draws attention to the relationship of S to a background energy density for two reasons. First, it emphasizes that S is nearly independent of cosmological parameters like h, Ω, or Λ. Second, it emphasizes the relationship of the QSO radiation background to the X-ray background. We emphasize, however, that the calculation, as usually performed, renders S critically dependent on an assumed bolometric correction and further corrections (often embodied in the bolometric correction) for obscured or undercounted objects.

It is widely believed that all of the X-ray background is produced by accretion onto BHs. It is therefore of considerable interest to compare u_0 to the X-ray background. We illustrate backgrounds in many parts of the electromagnetic spectrum in Figure 18.3 (Hauser & Dwek 2001).

We can compare the quasar energy from Equation (18.9) with the X-ray background. The four indicated backgrounds are integrated over their wavelength range and given in Table 18.1. The energy density in the X-ray background is $u_0 = 1.45 \times 10^{-16}$ erg cm^{-3}—about a factor of 10 less than the emitted quasar energy in Equation (18.8).

Fabian & Iwasawa (1999), correcting for soft X-ray absorption and for completely buried X-ray emitters, derive a local energy density of the X-ray background of $u_0 = 1.3 \times 10^{-15}$ erg cm^{-3}. Assuming a redshift distribution of unseen objects that is comparable to bright quasars, they then obtain a rest-frame comoving energy density of $u_e = 4.1 \times$

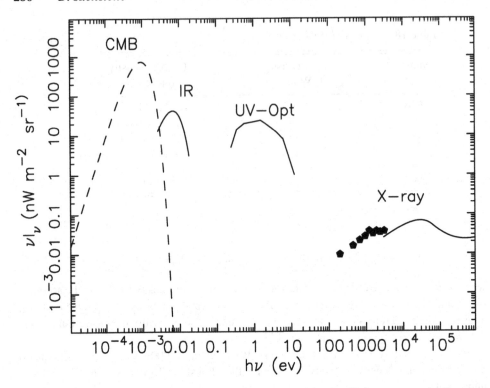

Fig. 18.3. The background radiation actually observed in the Universe at redshift zero. (Adapted from Hauser & Dwek 2001.)

10^{-15} erg cm^{-3}, about 3 times larger than the Chokshi & Turner estimate in Equation (18.8). This represents an extrapolation of a factor of 30 over the actual observed X-ray background. A factor of 10 came from estimates of the absorption of X-ray emission (or by fully obscured sources) and a factor of 3 came from an assumption that much of the absorption came from $z \approx 2$ as in the case of the bright optical quasars. Fabian (2004) discusses a downward revision of the X-ray background to an energy density corresponding to u_e in Equation (18.8) above.

We note that the infrared background shown above exceeds the quasar "background" by about an order of magnitude. Bernstein, Freedman, & Madore (2002, and references therein) have argued that consistency with the production of heavy elements requires that most of the infrared background be produced by stars and not quasars.

Supermassive BH masses range from about $10^6 M_\odot$ to about $3 \times 10^9 M_\odot$, with a typical mass for L^* galaxies of about $10^8 M_\odot$. The Eddington luminosity (above which the outward radiation force on free electrons exceeds the gravitational attraction exerted on protons) is

$$L_E = 1.3 \times 10^{46} (M/10^8 M_\odot) \text{ erg s}^{-1}, \tag{18.12}$$

comparable to the luminosity of bright quasars.

Since the mass density and the masses of the local BH population are consistent with the integrated quasar luminosity and the individual quasar luminosities, it is reasonable to suppose that the numbers of quasars are consistent with the numbers of BHs locally. Small

& Blandford (1992) suggest that this is not the case. Richstone et al. (1998) obtained a similar result by comparing the number of quasars—about $10^{-6} \mathrm{Mpc}^{-3}$—brighter than $6 \times 10^{46} \mathrm{erg\, s}^{-1}$ at the peak of the quasar epoch, to the number of bulges with BH of masses of at least $4 \times 10^8 M_\odot$. The brightest quasars are likely to be emitting near their Eddington limit, and hence their engines have BHs of about $4 \times 10^8 M_\odot$. Bulges with BHs this massive have a local density of $10^{-3} \mathrm{Mpc}^{-3}$. Why then, are there so few quasars? Since the number seen at any time is the duty fraction η times the total number, it may simply be that the duty fraction is tiny. The Richstone et al. analysis implies $\eta = 10^{-3}$.

One possible problem raised by such a small duty fraction is that it suggests a lifetime of 10^6 yr for the observed quasars. The ratio of this lifetime to the time scale for mass e-folding during Eddington-limited accretion, the "Salpeter time,"

$$t_S = M/\dot{M} = 4 \times 10^7 \, \epsilon_{0.1} \text{ yr,} \tag{18.13}$$

is about 0.025, implying that BHs acquire only a small fraction of their mass during the bright quasar phase. This would render the rough equality of the BH mass density and the Sołtan number a coincidence. The Small & Blandford (1992) study is consistent with a less extreme duty fraction of about 10^{-2}, which weakens this argument considerably.

18.5   Supermassive BH Formation and Evolution

18.5.1   Some General Inferences

Setting aside the dilemma posed by the discrepancy between the number of the biggest BHs and the number of the brightest quasars, the approximate equality of the Sołtan number and the BH mass density leads to two important inferences: (1) supermassive BHs are the relics of the quasars (they have the correct cosmic density and individual masses); (2) much of their mass was acquired during the quasar epoch and this accretion is electromagnetically visible. These two points suggest that the bulk of BH accretion is by accreting gas, not whole stars, dark matter, or other BHs. Otherwise, the equality of the cosmic density of BHs and the Sołtan number is a coincidence.

The luminosity of bright quasars suggests that during the quasar epoch the BHs are already massive, because the brightest are almost surely limited by their Eddington luminosity (Eq. 18.12) and require masses greater than $10^8 M_\odot$. Moreover, reaching this mass by Eddington-limited accretion is a challenge, partly because of the duty fraction dilemma, but also because the age of an $\Omega_\mathrm{m} = 0.3, \Omega_\Lambda = 0.7$ Universe at $z = 6.4$ (when the first quasars are known to exist; see Osmer 2004) is $0.8 \mathrm{Gyr} \approx 20 t_S$ (see Efstathiou & Rees 1988 and Turner 1991 for somewhat divergent views of this issue). It is conceivable that seed BHs of significantly lower mass formed well before $z = 6.4$.

Perhaps the most important inference concerns the connection between BH growth and bulge formation. Because the velocity dispersion at the effective radius is closely related to the binding energy per unit mass of the bulge (which is, inside that radius, dominated by visible matter), and is set during bulge formation or subsequent mergers, and because the BHs have already formed by redshifts $z \approx 3$, the growth of the BH and the formation of its host bulge are likely physically coupled.

18.5.2   A Taxonomy of Models for the $M_\bullet - \sigma$ Relation

It is interesting to consider how the $M_\bullet - \sigma$ relation might have been created. There are a wide array of theories relying on rather different physics. One can classify these theories in at least three dimensions. First, there is the question of what is accreted (gas, stars, other BHs, dark matter). Second, there is the question of whether BH formation (or, more correctly, most of the growth of the BHs) precedes bulge formation or *vice versa*, or whether the two are essentially synchronous. Finally, there is the question of what physical process produces or regulates the $M_\bullet - \sigma$ relation.

One class of models depends on the competition between star formation and BH accretion for gas in the protogalactic bulge (Burkert & Silk 2001). They suggest that the mass accreted by a BH in a newly formed bulge is limited by competition with star formation, and that both processes proceed on a dynamical time (for different reasons). They argue that the mass accreted by the BHs therefore scales as $M_\bullet \propto f_g \sigma^3 (1+z)^{-3/2}$, where f_g is the gas mass fraction in the bulge. This model eventually leads to $M_\bullet \propto \sigma^3$, somewhat shallower than observed. The virtues of models of this sort are (1) inevitability (competition between BH accretion and star formation for gas in the bulge seems unavoidable, although it may not be regulated in a manner conceived by the authors); (2) connection to quasars (most BH growth is due to gas accretion and will therefore produce the required radiation, and is likely to occur during the epoch of bulge formation, which probably corresponds to the epoch of BH formation); (3) connection to popular views of galaxy formation [the model offers a possible physical mechanism to produce the initial $M_\bullet - \sigma$ relation assumed in the semi-analytic model of Haehnelt & Kauffmann (2000)].

Silk & Rees (1998) and Blandford (1999) have suggested that the mechanical force exerted on the gas by the quasar (perhaps through its jets) limits its growth by accretion. This is an energy-limited accretion process. This has many of the same virtues of the Burkert & Silk model, but predicts a steeper relationship in the form of an inequality $M_\bullet \leq (f_g/f_w)\sigma^5$, where f_w is the fraction of the quasar luminosity output as mechanical energy. Another energy-limiting model due to Ciotti & Ostriker (1997, 2001) regulates the accretion rate of the BH by heating the accreting gas.

Direct infall from a cold mass distribution with $\rho \propto r^{-2}$ and constant angular velocity onto a seed BH (Adams, Graff, & Richstone 2001, 2003) can also produce a BH from either noninteracting or baryonic matter (so long as it is cold). The mass not captured by the BH forms the bulge in this model. An important limitation of this model is its fragility against modification of the initial conditions, as it depends very sensitively on the angular momentum distribution in the protogalactic (or galactic) material.

Two other models provide very interesting foils. Ostriker (2000) has proposed a model in which self-interacting dark matter is accreted by a seed BH following the Bondi (1952) formula, and shows that the model leads to $M_\bullet \propto \sigma^4$, provided that the BH is in the center of a spherical, singular isothermal mass distribution ($\rho \propto r^{-2}$).

A second interesting foil is one in which two-body relaxation drives the bulge to core collapse and creates a supermassive BH via the merger of degenerate or collapsed objects. In globular clusters, the binding energy of a single contact, nondegenerate binary star is comparable to the entire binding energy of the cluster, so the extraction of energy from hard binaries can halt the collapse. Bulge masses are much greater, so binary formation will not necessarily halt their collapse. A model of bulge core collapse has been proposed by Ebisuzaki et al. (2001). An important feature of relaxation is conservation of energy. As long as the core collapse and BH formation are driven by conservative two-body relaxation, the flow of energy proceeds from the most tightly bound part of the cluster to the evaporating stars, which carry away very little energy. In order to make a BH, energy of the order of $M_\bullet c^2$

must be extracted from the stars that will form the BH and deposited in the rest of the mass reservoir, which cannot absorb much more than its binding energy $M_t\sigma^2$. Hence,

$$\frac{M_\bullet}{M_t} \le \frac{\sigma^2}{c^2} \approx 0.5 \times 10^{-6}, \tag{18.14}$$

where the numerical result is for a bulge of dispersion $200\ \mathrm{km\ s^{-1}}$. Ebisuzaki et al. extract some of the energy in their model by gravitational radiation, but in order to reach $M_\bullet/M_t \approx 10^{-3}$ (see Eq. 18.7), they need to rely almost entirely on gravitational radiation or some other form of dissipation.

The Ostriker and Ebisuzaki et al. models have the property that the accretion is electromagnetically invisible. Hence, if these processes dominate the growth of BHs we need to seek an alternate source for the energy radiated by quasars, and the approximate equality of the Soltan number and the cosmic density of BHs is a false clue to the dominant accretion mechanism.

18.5.3 Coevolution of Galaxies and BHs

Once produced, the supermassive BHs coevolve with their host galaxies. The galaxy controls any further fueling of the BH. However, there are (at least) three ways that the BH can influence the galaxy (Richstone 2004).

First, the luminosity of an accreting BH can be compared to the stars in the galaxy by

$$\frac{L_{\mathrm{BH}}}{L_{\mathrm{stars}}} = 3 \times 10^4 \left(\frac{M_\bullet}{\mathcal{M}_{\mathrm{bulge}}}\right) \Upsilon_{\mathrm{bulge}} = 30\ \Upsilon_{\mathrm{bulge}}, \tag{18.15}$$

where the $\Upsilon_{\mathrm{bulge}}$ is the usual bulge mass-to-light ratio. Thus, during the quasar epoch, the luminous output of the quasar can dominate the starlight in the galaxy (especially in X-rays) and thereby dominate the heat input to the interstellar medium. Second, the BH dominates the mass and gravitational field in the inner few parsecs of the galaxy, and may (or may not) destroy triaxial structure (see Holley-Bockelmann et al. 2001; Poon & Merritt 2001; and references therein). Finally, the BH can play a major role during galactic and protogalactic mergers. During these mergers, the BH can scatter stars, altering the stellar distribution function and reducing the density of stars near the center (see, for example, Milosavljević et al. 2002; Ravindranath, Ho, & Filippenko 2002; Merritt 2004).

Acknowledgements. I thank Rebecca Bernstein and Kelly Holley-Bockelmann for helpful comments and Michael Hauser and Eli Dwek for helping me to adapt Figure 18.3 from their review. I have enjoyed valuable conversations with the Nuker team for 15 years. I thank the editor of these proceedings, Luis Ho, for gently and persistently getting me to write this review.

References

Adams, F. C., Graff, D. S., & Richstone, D. O. 2001, ApJ, 551, L31
——. 2003, ApJ, 591, 125
Aller, M. C., & Richstone, D. 2002, AJ, 124, 3035
Barth, A. J. 2004, in Carnegie Observatories Astrophysics Series, Vol. 1: Coevolution of Black Holes and
     Galaxies, ed. L. C. Ho (Cambridge: Cambridge Univ. Press), in press
Bernardi, M., et al. 2003, AJ, 125, 1817
Bernstein, R. A., Freedman, W. L., & Madore, B. F. 2000, ApJ, 571, 107

Blandford, R. D. 1999, in Galaxy Dynamics, ed. D. R. Merritt, J. A. Sellwood, & M. Valluri (San Francisco: ASP), 87

Bondi, H. 1952, MNRAS, 112, 195

Boyle, B. J, Shanks, T., Croom, S. M., Smith, R. J., Miller, L., Loaring, N., & Heymans, C. 2000, MNRAS, 317, 1014

Burkert, A., & Silk, J. 2001, ApJ, 554, L151

Chokshi A., & Turner, E. L. 1992, MNRAS, 259, 421

Ciotti, L., & Ostriker, J. P. 1997, ApJ, 487, L105

——. 2001, ApJ, 551, 131

Ebisuzaki, T., et al. 2001, ApJ, 562, L19

Efstathiou, G., & Rees, M. J. 1988, MNRAS, 230, 5P

Elvis, M., et al. 1994, ApJS, 95, 1

Erwin, P., Graham, A. W., & Caon, N. 2004, in Carnegie Observatories Astrophysics Series, Vol. 1: Coevolution of Black Holes and Galaxies, ed. L. C. Ho (Pasadena: Carnegie Observatories, http://www.ociw.edu/ociw/symposia/series/symposium1/proceedings.html)

Fabian, A. C. 2004, in Carnegie Observatories Astrophysics Series, Vol. 1: Coevolution of Black Holes and Galaxies, ed. L. C. Ho (Cambridge: Cambridge Univ. Press), in press

Fabian, A. C., & Iwasawa, K. 1999, MNRAS, 303, L34

Ferrarese, L., & Merritt, D. 2000, ApJ, 539, L9

Gebhardt, K. 2004, in Carnegie Observatories Astrophysics Series, Vol. 1: Coevolution of Black Holes and Galaxies, ed. L. C. Ho (Cambridge: Cambridge Univ. Press), in press

Gebhardt, K., et al. 2000, ApJ, 539, L13

——. 2001, AJ 122, 2469

Ghez, A. M. 2004, in Carnegie Observatories Astrophysics Series, Vol. 1: Coevolution of Black Holes and Galaxies, ed. L. C. Ho (Cambridge: Cambridge Univ. Press), in press

Haehnelt, M. G., & Kauffmann, G. 2000, MNRAS, 318, L35

Hauser, M. G., & Dwek, E. 2001, ARA&A, 39, 249

Holley-Bockelmann, K., Mihos, J. C., Sigurdsson, S., & Hernquist, L. 2001, ApJ, 549, 862

Kormendy, J. 2004, in Carnegie Observatories Astrophysics Series, Vol. 1: Coevolution of Black Holes and Galaxies, ed. L. C. Ho (Cambridge: Cambridge Univ. Press), in press

Kormendy, J., & Gebhardt, K. 2001, in The 20th Texas Symposium on Relativistic Astrophysics, ed. H. Martel & J. C. Wheeler (Melville: AIP), 363

Kormendy, J., & Richstone, D. O. 1995, ARA&A, 33, 581

Loveday, J., Peterson, B. A., Efstathiou, G., & Maddox, S. J. 1992, ApJ, 390, 338

Madgwick, D. S., et al. 2002, MNRAS, 333, 133

Maoz, E. 1995, ApJ, 447, L91

——. 1998, ApJ, 494, L181

Marzke, R. O., Geller, M. J., Huchra, J. P., & Corwin, H. G. 1994a, AJ, 108, 437

Marzke, R. O., Huchra, J. P., & Geller, M. J. 1994b, ApJ, 428, 43

Merritt, D. 2004, in Carnegie Observatories Astrophysics Series, Vol. 1: Coevolution of Black Holes and Galaxies, ed. L. C. Ho (Cambridge: Cambridge Univ. Press), in press

Merritt, D., & Ferrarese, L. 2001a, MNRAS, 320, L30

——. 2001b, ApJ, 547, 140

——. 2001c, in The Central Kpc of Starbursts and AGN: The La Palma Connection, ed. J. H. Knapen et al. (San Francisco: ASP), 335

Milosavljević, M., Merritt, D., Rest, A., & van den Bosch, F. C. 2002, MNRAS, 331, L51

Miyoshi, M., Moran, J., Herrnstein, J., Greenhill, L., Nakai, N., Diamond, P., & Inoue, M. 1995, Nature, 373, 127

Osmer, P. S. 2004, in Carnegie Observatories Astrophysics Series, Vol. 1: Coevolution of Black Holes and Galaxies, ed. L. C. Ho (Cambridge: Cambridge Univ. Press), in press

Ostriker, J. P. 2000, Phys. Rev. Lett., 84, 5258

Poon, M. Y., & Merritt, D. 2001, ApJ, 549, 192

Ravindranath, S., Ho, L. C., & Filippenko, A. V. 2002, ApJ, 566, 801

Richstone, D. 1998, in Laser Interferometer Space Antenna, ed. W. M. Faulkner (New York: AIP), 41

——. 2004, in IAU Symp. 208, Astrophysical Supercomputing Using Particles, ed. J. Makino & P. Hut (San Francisco: ASP), in press

Sarzi, M., Rix, H.-W., Shields, J. C., Rudnick, G., Ho, L. C., McIntosh, D. H., Filippenko, A. V., & Sargent, W. L. W. 2001, ApJ, 550, 65

Silk, J., & Rees, M. J. 1998, A&A, 331, L1
Simien, F., & de Vaucouleurs, G. 1986, ApJ, 302, 564
Small, T. A., & Blandford, R. D. 1992, MNRAS, 259, 725
Sołtan, A. 1982, MNRAS, 200, 115
Spergel, D. N., et al. 2003, ApJS, 149, 175
Tremaine, S., et al. 2002, ApJ, 574, 740
Turner, E. L. 1991, AJ, 101, 5
van der Marel, R. P., Cretton, N., de Zeeuw, P. T., & Rix, H.-W. 1998, ApJ, 493, 613
Verolme, E. K., et al. 2002, MNRAS, 335, 517
Yu, Q., & Tremaine, S. 2002, MNRAS, 335, 965

19

Black hole demography from nearby active galactic nuclei

LUIS C. HO

The Observatories of the Carnegie Institution of Washington

Abstract

A significant fraction of local galaxies show evidence of nuclear activity. I argue that the bulk of this activity, while energetically not remarkable, derives from accretion onto a central massive black hole. The statistics of nearby active galactic nuclei thus provide an effective probe of black hole demography. Consistent with the picture emerging from direct dynamical studies, the local census of nuclear activity strongly suggests that most, perhaps all, galaxies with a significant bulge component contain a central massive black hole. Although late-type galaxies appear to be generally deficient in nuclear black holes, there are important exceptions to this rule. I highlight two examples of dwarf, late-type galaxies that contain active nuclei powered by intermediate-mass black holes.

19.1 Introduction

The search for massive black holes (BHs) has recently enjoyed dramatic progress, to the point that the statistics of BH detections have begun to yield useful clues on the connection between BHs and their host galaxies, the central theme of this Symposium. Lest one becomes complacent, however, we should recognize that our knowledge of the demographics of BHs in nearby galaxies—on which much of the astrophysical inferences depends—remains highly incomplete. Direct measurements of BH masses based on resolved gas or stellar kinematics, while increasingly robust, are still far from routine and presently are available only for a limited number of galaxies (see Barth 2004 and Kormendy 2004). Certainly nothing approaching a "complete" sample exists yet. More importantly, it is far from obvious that the current statistics are unbiased. As discussed by Barth (2004), most nearby galaxies possess chaotic nuclear rotation curves that defy simple analysis. Stellar kinematics provide a powerful alternative to the gas-based method, but in practice this technique thus far has been limited to relatively dust-free systems and, for practical reasons, to galaxies of relatively high central surface brightness. The latter restriction selects against luminous, giant ellipticals. Lastly, current surveys severely underrepresent disk-dominated (Sbc and later) galaxies, because the bulge component in these systems is inconspicuous and star formation tends to perturb the velocity field of the gas.

Given the above limitations, it would be important to consider alternative constraints on BH demography. This contribution discusses the role that active galactic nuclei (AGNs) can play in this regard. The commonly held, but by now well-substantiated premise that AGNs derive their energy output from BH accretion implies that an AGN signifies the presence of a central BH in a galaxy. The AGN signature in and of itself provides no direct

© The Observatories of the Carnegie Institution of Washington 2004.

information on BH masses, but AGN statistics can inform us, effectively and efficiently, some key aspects of BH demography. For example, what fraction of all galaxies contain BHs? Do BHs exist preferentially in galaxies of certain types? Does environment matter? Under what conditions do BHs light up as AGNs and how long does the active phase last? What is their history of mass build-up? These and many other related issues are inextricably linked with the statistical properties of AGNs as a function of cosmological epoch. This contribution concentrates on the local ($z \approx 0$) AGNs; Osmer (2004) considers the high-redshift population.

This review is structured as follows. I begin with an overview of the basic methodology of the spectral classification of emission-line nuclei (§19.2) by describing the currently adopted system, its physical motivation, the complications of starlight subtraction, and some practical examples. Section 19.3 briefly summarizes past and current spectroscopic surveys and introduces the Palomar survey. The demographics of nearby AGNs is the subject of §19.4, covering detection rates based on optical surveys, detection rates based on radio work, the detection of weak broad emission lines, issues of robustness and completeless in current surveys, the local AGN luminosity function, the statistics of accretion luminosities, host galaxy properties and environmental effects, and intermediate-mass black holes. No discussion on nearby AGNs would be complete without a proper treatment on LINERs (§19.5). I focus on what I believe are the three most important topics, namely the current evidence that the majority of LINERs are indeed powered by accretion, AGN photoionization as their dominant excitation mechanism and the demise of competing alternatives, and the largely still-unresolved nature of the so-called transition objects. Section 19.6 gives a synopsis of the main points.

19.2 Spectral Classification of Galactic Nuclei

19.2.1 *Physical Motivation*

AGNs can be identified by a variety of methods. Most AGN surveys rely on some aspect of the distinctive AGN spectrum, such as the presence of strong or broad emission lines, an unusually blue continuum, or strong radio or X-ray emission. While all of these techniques are effective, none is free from selection effects. To search for AGNs in nearby galaxies, where the nonstellar signal of the nucleus is expected to be weak relative to the host galaxy, the most effective and least biased method is to conduct a spectroscopic survey of a complete, optical-flux limited sample of galaxies. To be sensitive to weak emission lines, the survey must be deep and of sufficient spectral resolution. To obtain reliable line intensity ratios, on which the principal nuclear classifications are based, the data must have accurate relative flux calibration, and one must devise a robust scheme to correct for the starlight contamination. These issues are discussed below. But first, I must cover some basic material on spectral classification.

The most widely used system of spectral classification of emission-line nuclei follows the method outlined by Baldwin, Phillips, & Terlevich (1981), and later modified by Veilleux & Osterbrock (1987). The basic idea is that the relative strengths of certain prominent emission lines can be used to probe the nebular conditions of a source. In the context of the present discussion, the most important diagnostic is the source of excitation, which broadly falls into two categories: stellar photoionization or photoionization by a centrally located, spectrally hard radiation field, such as that produced by the accretion disk of a massive BH. The latter class of sources are generically called AGNs, which are most relevant to issues of BH demography.

How does one distinguish stellar from nonstellar photoionization? The forbidden lines of the doublet [O I] $\lambda\lambda6300$, 6364 rise from collisional excitation of O^0 by hot electrons. Since the ionization potential of O^0 (13.6 eV) is nearly identical to that of hydrogen, in an ionization-bounded nebula [O I] is produced predominantly in the "partially ionized zone," wherein both neutral oxygen and free electrons coexist. In addition to O^0, the conditions of the partially ionized zone are also favorable for S^+ and N^+, whose ionization potentials are 23.3 eV and 29.6 eV, respectively. Hence, in the absence of abundance anomalies, [N II] $\lambda\lambda6548$, 6583 and [S II] $\lambda\lambda6716$, 6731 are strong (relative to, say, Hα) whenever [O I] is strong, and *vice versa*.

In a nebula photoionized by young, massive stars, the partially ionized zone is very thin because the ionizing spectrum of OB stars contains few photons with energies greater than 1 Rydberg. Hence, in the optical spectra of H II regions and starburst nuclei (hereinafter H II nuclei*) the low-ionization transitions [N II], [S II], and especially [O I] are very weak. By contrast, a harder radiation field, such as that of an AGN power-law continuum that extends into the extreme-ultraviolet (UV) and X-rays, penetrates much deeper into an optically thick cloud. X-ray photoionization and Auger processes release copious hot electrons in this predominantly neutral region, creating an extensive partially ionized zone. The spectra of AGNs, therefore, exhibit relatively strong low-ionization forbidden lines.

19.2.2   Sample Spectra

The spectra shown in Figure 19.1 illustrate the empirical distinction between AGNs and H II nuclei. In NGC 7741, which has a well-known starburst nucleus (Weedman et al. 1981), [O I], [N II], and [S II] are weak relative to Hα. The [O III] $\lambda\lambda4959$, 5007 doublet is quite strong compared to [O II] $\lambda3727$ or Hβ because the metal abundance of NGC 7741's nucleus is rather low, although the ionization level of H II nuclei can span a wide range, depending on metallicity (Ho, Filippenko, & Sargent 1997c). On the other hand, the low-ionization lines are markedly stronger in the other two objects shown, both of which qualify as AGNs. NGC 1358 is an example of a galaxy with a "high-ionization" AGN or "Seyfert" nucleus. NGC 1052 is the prototype of the class known as "low-ionization nuclear emission-line regions" or "LINERs." The ionization level can be judged by the relative strengths of the oxygen lines, but in practice is most easily gauged by the [O III]/Hβ ratio. In the commonly adopted system of Veilleux & Osterbrock (1987), the division between Seyferts and LINERs occurs at [O III] $\lambda5007/$H$\beta = 3.0$. Ho, Filippenko, & Sargent (2003) stress, however, that this boundary has no strict physical significance. The ionization level of the narrow-line region (NLR) in large, homogeneous samples of AGNs spans a wide and apparently continuous range; contrary to the claims of some studies (e.g., Véron-Cetty & Véron 2000) there is no evidence for any clear-cut transition between Seyferts and LINERs (Ho et al. 2003; Heckman 2004).

The classification system discussed above makes no reference to the profiles of the emission lines. Luminous AGNs such as quasars and many "classical" Seyfert galaxies

---

* As originally defined (Weedman et al. 1981; Balzano 1983), a star*burst* nucleus is one whose current star formation rate is much higher than its past average rate. This terminology presupposes knowledge of the star formation history of the system. Since this information is usually not available for any individual object, I will adopt the more general designation of "H II nucleus."

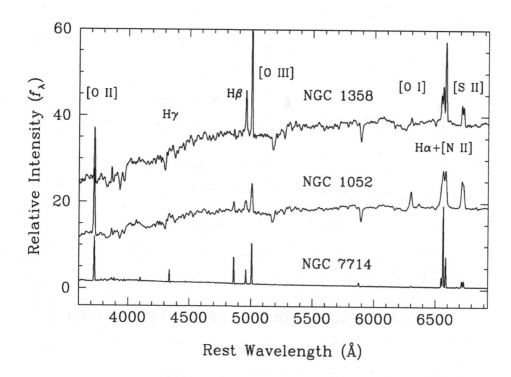

Fig. 19.1. Sample optical spectra of the various classes of emission-line nuclei. NGC 1358 = Seyfert; NGC 1052 = LINER; NGC 7714 = H II. The prominent emission lines are identified. (Based on Ho et al. 1993a and unpublished data.)

exhibit permitted lines with a characteristically broad component, with FWHM widths of $\sim 1000 - 10{,}000$ km s^{-1}. This component arises from the broad-line region (BLR), which is thought to be physically distinct from the NLR responsible for the narrow lines. Following Khachikian & Weedman (1974), it is customary to refer to Seyferts with and without (directly) detectable broad lines as "type 1" and "type 2" sources, respectively. As discussed in § 19.4.3, this nomenclature can also be extended to include LINERs, which also contain broad emission lines. Figure 19.2 gives some examples. The spectrum of the bright Seyfert galaxy NGC 4151 is familiar to all: strong, broad permitted lines superposed on an unambiguous featureless, nonstellar blue continuuum. But this object is not typical. Even within the Seyfert class, most objects resemble NGC 5273, where the broad component is easily visible only for Hα and the featureless continuum is heavily diluted by the host galaxy light. The same applies to LINERs (e.g., NGC 3998), where the host galaxy dilution is even more extreme; nonetheless, with careful starlight subtraction (§ 19.2.4) and profile modeling (§ 19.4.3), one can detect broad Hα emission in many LINERs.

19.2.3 *Diagnostic Diagrams*

The classification system of Veilleux & Osterbrock (1987), which I adopt throughout this paper, is based on two-dimensional line-intensity ratios constructed from

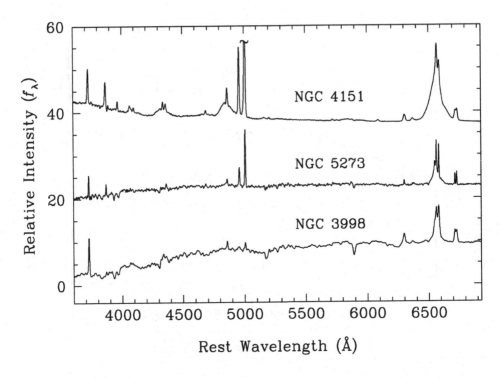

Fig. 19.2. Sample optical spectra of broad-line AGNs. NGC 4151 = "classical" Seyfert 1; NGC 5273 = typical low-luminosity Seyfert 1; NGC 3998 = LINER 1. (Based on Ho et al. 1993a and unpublished data.)

[O III] $\lambda5007$, Hβ $\lambda4861$, [O I] $\lambda6300$, Hα $\lambda6563$, [N II] $\lambda6583$, and [S II] $\lambda\lambda6716, 6731$ (here Hβ and Hα refer only to the narrow component of the line). The main virtues of this system, shown in Figure 19.3, are (1) that it uses relatively strong lines, (2) that the lines lie in an easily accessible region of the optical spectrum, and (3) that the line ratios are relatively insensitive to reddening corrections because of the close separation of the lines. The definitions of the various classes of emission-line objects are given in Ho, Filippenko, & Sargent (1997a)*. In addition to the three main classes discussed thus far—H II nuclei, Seyferts, and LINERs—Ho, Filippenko, & Sargent (1993a) identified a group of "transition objects" whose [O I] strengths are intermediate between those of H II nuclei and LINERs. Since they tend to emit weaker [O I] emission than classical LINERs, previous authors have called them "weak-[O I] LINERs" (Filippenko & Terlevich 1992; Shields 1992; Ho & Filippenko 1993). Ho et al. (1993a) postulated that transition objects are composite systems having both an H II region and a LINER component; I will return to the nature of these sources in § 19.5.3.

I note that my definition of LINERs differs from that originally proposed by Heckman (1980b), who used solely the oxygen lines: [O II] $\lambda3727 >$ [O III] $\lambda5007$ and [O I] $\lambda6300$

---

* The classification criteria adopted here differ slightly, but not appreciably, from those proposed by Kewley et al. (2001) based on theoretical models.

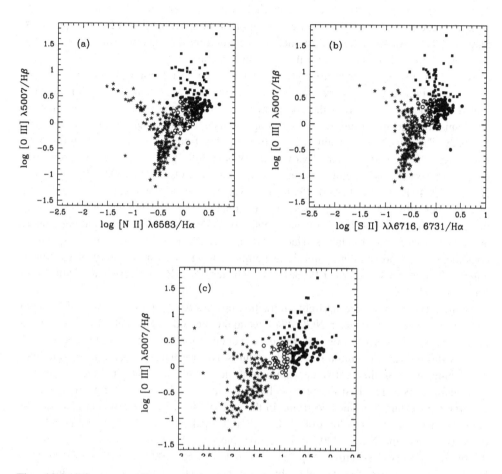

Fig. 19.3. Diagnostic diagrams plotting (*a*) log [O III] λ5007/Hβ versus log [N II] λ6583/Hα, (*b*) log [O III] λ5007/Hβ versus log [S II] λλ6716, 6731/Hα, and (*c*) log [O III] λ5007/Hβ versus log [O I] λ6300/Hα. The nuclear spectral classes shown are H II nuclei (*asterisks*), Seyfert nuclei (*squares*), LINERs (*solid circles*), and transition objects (*open circles*). (Adapted from Ho et al. 1997a.)

> 0.33 [O III] λ5007. The two definitions, however, are nearly equivalent. Inspection of the full optical spectra of Ho et al. (1993a), for example, reveals that emission-line nuclei classified as LINERs based on the Veilleux & Osterbrock diagrams almost invariably also satisfy Heckman's criteria. This is a consequence of the inverse correlation between [O III]/Hβ and [O II]/[O III] in photoionized gas with fairly low excitation ([O III]/Hβ ≲3; see Fig. 2 in Baldwin et al. 1981).

19.2.4   *Starlight Subtraction*

The scheme outlined above, while conceptionally simple, overlooks one key practical complication. The integrated spectra of galactic nuclei include emission from stars, which in most nearby systems overwhelms the nebular line emission. This can be seen in Figure 19.1, or from a cursory examination of the spectral atlas of Ho, Filippenko, & Sargent (1995). Any reliable measurement of the emission-line spectrum of galactic nuclei, therefore, *must* properly account for the starlight contamination.

An effective strategy for removing the starlight from an integrated spectrum is that of "template subtraction," whereby a template spectrum devoid of emission lines is suitably scaled to and subtracted from the spectrum of interest to yield a continuum-subtracted, pure emission-line spectrum. A number of approaches have been adopted to construct the template. These include (1) using the spectrum of an off-nuclear position within the same galaxy (e.g., Storchi-Bergmann, Baldwin, & Wilson 1993); (2) using the spectrum of a different galaxy devoid of emission lines (e.g., Costero & Osterbrock 1977; Filippenko & Halpern 1984; Ho et al. 1993a); (3) using a weighted linear combination of the spectra of a number different galaxies, chosen to best match the stellar population and velocity dispersion (Ho et al. 1997a); (4) using the spectrum derived from a principal-component analysis of a large set of galaxies (Hao & Strauss 2004); and (5) using a model spectrum constructed from population synthesis techniques, using as input a library of spectra of either individual stars (e.g., Keel 1983a) or star clusters (e.g., Bonatto, Bica, & Alloin 1989; Raimann et al. 2001).

Figure 19.4 illustrates the starlight subtraction process for the H II nucleus in NGC 3596 and for the Seyfert 2 nucleus in NGC 7743, using the method of Ho et al. (1997a). Given a list of input spectra derived from galaxies devoid of emission lines and an initial guess of the velocity dispersion, a χ^2-minimization algorithm solves for the systemic velocity, the line-broadening function, the relative contribution of the various input spectra, and the general continuum shape. The best-fitting model is then subtracted from the original spectrum, yielding a pure emission-line spectrum. In the case of NGC 3596, the model consisted of the combination of the spectrum of NGC 205, a dE5 galaxy with a substantial population of A-type stars, and NGC 4339, an E0 galaxy having a K-giant spectrum. Note that in the original observed spectrum (top), Hγ, [O III] $\lambda\lambda 4959$, 5007, and [O I] $\lambda 6300$ were hardly visible, whereas after starlight subtraction (bottom) they can be easily measured. The intensities of both Hβ and Hα have been modified substantially, and the ratio of the two [S II] $\lambda\lambda 6716$, 6731 lines changed. The effective template for NGC 7743 made use of NGC 205, NGC 4339, and NGC 628, an Sc galaxy with a nucleus dominated by A and F stars.

Some studies (e.g., Kim et al. 1995) implicitly assume that only the hydrogen Balmer lines are contaminated by starlight, and that the absorption-line component can be removed by subtracting a constant equivalent width (2–3 Å). This procedure is inadequate for a number of reasons. First, the stellar population of nearby galactic nuclei, although relatively uniform, is by no means invariant (Ho et al. 2003). Second, the equivalent widths of the different Balmer absorption lines within each galaxy are generally not constant. Third, the Balmer absorption lines affect not only the strength but also the shape of the Balmer emission lines. And finally, as the above examples show, starlight contaminates lines other than just the Balmer lines.

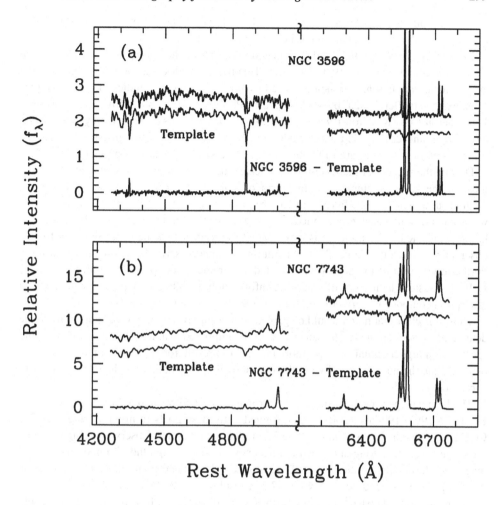

Fig. 19.4. Illustration of the method of starlight subtraction. In each panel, the top plot shows the observed spectrum, the middle plot the best-fitting "template" used to match the stellar component, and the bottom plot the difference between the object spectrum and the template. In the case of NGC 3596 (*a*), the model was constructed from NGC 205 and NGC 4339, while for NGC 7743 (*b*), the model was derived from a linear combination of NGC 205, NGC 4339, and NGC 628. (Adapted from Ho et al. 1997a.)

19.3 Spectroscopic Surveys of Nearby Galactic Nuclei

It was apparent from some of the earliest redshift surveys that the central regions of galaxies often show evidence of strong emission lines (e.g., Humason, Mayall, & Sandage 1956). A number of studies also indicated that in many instances the spectra revealed abnormal line-intensity ratios, most notably the unusually great strength of [N II] relative to Hα (Burbidge & Burbidge 1962, 1965; Rubin & Ford 1971). That the optical emission-line spectra of some nuclei show patterns of low ionization was noticed from time to time, primarily by Osterbrock and his colleagues (e.g., Osterbrock & Miller 1975; Koski & Osterbrock 1976; Costero & Osterbrock 1977; Grandi & Osterbrock 1978; Phillips 1979), but also by others (e.g., Disney & Cromwell 1971; Danziger, Fosbury, & Penston 1977; Fosbury et al. 1977, 1978; Penston & Fosbury 1978; Stauffer & Spinrad 1979).

Most of the activity in this field culminated in the 1980s, beginning with the recognition (Heckman, Balick, & Crane 1980; Heckman 1980b) of LINERs as a major constituent of the extragalactic population, and then followed by further systematic studies of larger samples of galaxies (Stauffer 1982a, b; Keel 1983a, b; Phillips et al. 1986; Véron & Véron-Cetty 1986; Véron-Cetty & Véron 1986; see Ho 1996 for more details). These surveys established three important results: (1) a large fraction of local galaxies contain emission-line nuclei; (2) many of these sources are LINERs; and (3) LINERs may be accretion-powered systems.

Despite the success of these seminal studies, there was room for improvement. Although most of the surveys attempted some form of starlight subtraction, the accuracy of the methods used tended to be fairly limited (see discussion in Ho et al. 1997a), the procedure was sometimes inconsistently applied, and in some of the surveys starlight subtraction was largely neglected. The problem is exacerbated by the fact that the apertures used for the observations were quite large, thereby admitting an unnecessarily large amount of starlight. Furthermore, most of the data were collected with rather poor spectral resolution (FWHM ≈ 10 Å). Besides losing useful kinematic information, blending between the emission and absorption components further compromises the ability to separate the two.

Thus, it is clear that much would be gained from a survey having greater sensitivity to the detection of emission lines. The sensitivity can be improved in at least four ways—by taking spectra with higher signal-to-noise ratio and spectral resolution, by using a narrower slit to better isolate the nucleus, and by employing more effective methods to handle the starlight correction.

The Palomar spectroscopic survey of nearby galaxies (Filippenko & Sargent 1985, 1986; Ho et al. 1995, 1997a–e, 2003) was designed with these goals in mind. Using a double CCD spectrograph mounted on the Hale 5-m reflector at Palomar Observatory, high-quality, moderate-resolution, long-slit spectra were obtained for a magnitude-limited ($B_T \leq 12.5$ mag) sample of 486 northern ($\delta > 0°$) galaxies. The spectra simultaneously cover the wavelength ranges 6210–6860 Å with $\sim$2.5 Å resolution (FWHM) and 4230–5110 Å with $\sim$4 Å resolution. Most of the observations were obtained with a narrow slit (generally $2''$, and occasionally $1''$), and the exposure times were suitably long (up to 1 hr or more for some objects with low central surface brightness) to secure data of high signal-to-noise ratio. This survey contains the largest database to date of homogeneous and high-quality optical spectra of nearby galaxies. It is also the most sensitive; the detection limit for emission lines is $\sim$0.25 Å, roughly an order-of-magnitude improvement compared to previous work. The selection criteria of the survey ensure that the sample gives a fair representation of the local ($z \approx 0$) galaxy population, and the proximity of the objects (median distance = 17 Mpc) enables relatively good spatial resolution to be achieved (typically $\lesssim$200 pc). These properties of the Palomar survey make it ideally suited to address issues on the demographics and physical properties of nearby, and especially low-luminosity, AGNs. Unless otherwise noted, the main results presented in the rest of this paper will be taken from the Palomar survey.

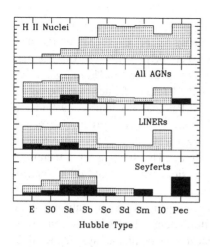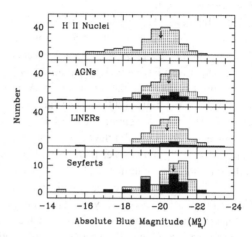

Fig. 19.5. (*Left*) Percentage of galaxies with the various classes of emission-line nuclei detected as a function of Hubble type. (*Right*) Distribution of the classes of emission-line nuclei as a function of the absolute *B* magnitude of the host galaxy. The arrow in each panel marks the median of the distribution. The solid and hatched histograms denote type 1 and type 1 + type 2 sources, respectively. (Adapted from Ho et al. 1997b.)

19.4   Demographics of Nearby AGNs

19.4.1   Detection Rates

In qualitative agreement with previous surveys, the Palomar survey finds that a substantial fraction (86%) of all galaxies contain detectable emission-line nuclei (Ho et al. 1997b). The detection rate is essentially 100% for all disk (S0 and spiral) galaxies, and over 50% for elliptical galaxies. One of the most surprising results is the large fraction of objects classified as AGNs or AGN candidates. Summed over all Hubble types, 43% of all galaxies that fall in the survey limits can be considered "active" (Fig. 19.5). This percentage becomes even more remarkable for galaxies with an obvious bulge component, rising to ~50%–70% for Hubble types E–Sbc. By contrast, the detection rate of AGNs drops dramatically toward later Hubble types (Sc and later), which almost invariably (80%) host H II nuclei. This strong dependence of nuclear spectral class on Hubble type has been noticed in earlier studies (Heckman 1980a; Keel 1983a; Terlevich, Melnick, & Moles 1987).

Among the active sources, 11% have Seyfert nuclei, at least doubling older estimates (Stauffer 1982b; Keel 1983b; Phillips, Charles, & Baldwin 1983; Huchra & Burg 1992). LINERs constitute the dominant population of AGNs. "Pure" LINERs are present in ~20% of all galaxies, whereas transition objects, which by assumption also contain a LINER component, account for another ~13%. Thus, if all LINERs can be regarded as genuine AGNs (see § 19.5), they truly are the most populous constituents—they make up 1/3 of all galaxies and 2/3 of the AGN population (here taken to mean all objects classified as Seyferts, LINERs, and transition objects).

The sample of nearby AGNs emerging from the Sloan Digital Sky Survey (SDSS) (Kauffmann et al. 2003; Hao & Strauss 2004; Heckman 2004) far surpasses that of the Palomar survey in number. Within the magnitude range $14.5 < r < 17.7$, Kauffmann et al. (2003) report an overall AGN fraction (for narrow-line sources) of ~40%, of which ~10% are Seyferts, the rest LINERs and transition objects. Using a different method of starlight subtraction, Hao & Strauss (2004) obtain very similar statistics for their sample of Seyfert galaxies. Although these detection rates broadly resemble those of the Palomar survey, one should recognize important differences between the two surveys. The Palomar objects extend much farther down the luminosity function than the SDSS. The emission-line detection limit of the Palomar survey, 0.25 Å, is roughly 10 times fainter than the cutoff chosen by Hao & Strauss (2004). The faint end of the Palomar Hα luminosity function reaches 1×10^{38} erg s^{-1}, again a factor of 10 lower than the SDSS counterpart. Moreover, the 3″-diameter fibers used in the SDSS subtend a physical scale of ~5.5 kpc at the typical redshift $z \approx 0.1$, 30 times larger than in the Palomar survey. The SDSS spectra, therefore, include substantial contamination from off-nuclear emission, which would dilute, and in some cases inevitably confuse, the signal from the nucleus.

Contamination by host galaxy emission has two consequences. First, only relatively bright nuclei have enough contrast to be detected; this is consistent with the sensitivity difference described above. But second, it can introduce a more pernicious systematic effect that can be hard to quantify. Apart from normal H II regions, galactic disks are known to contain emission-line regions that exhibit low-ionization, LINER-like spectra, which can be confused with genuine *nuclear* LINERs. Examples include gas shocked by supernova remnants (e.g., Dopita & Sutherland 1995), ejecta from starburst-driven winds (Armus, Heckman, & Miley 1990), and diffuse ionized plasma (e.g., Lehnert & Heckman 1994; Collins & Rand 2001). Massive, early-type galaxies, though generally lacking in ongoing star formation, do often possess X-ray emitting atmospheres that exhibit extended, low-ionization emission-line nebulae (e.g., Fabian et al. 1986; Heckman et al. 1989). These physical processes, while interesting in their own right, are not directly related, and thus irrelevant, to the AGN phenomenon. Thus, "LINERs" selected from samples of distant galaxies should be regarded with considerable caution.

19.4.2   Statistics from Radio Surveys

The prevalence of weak AGNs in nearby galaxies is corroborated by high-resolution radio continuum surveys. Sadler, Jenkins, & Kotanyi (1989) and Wrobel & Heeschen (1991) report a relatively high incidence (~50%) of compact radio cores in complete, optical flux-limited samples of elliptical and S0 galaxies. The radio powers are quite modest, generally in the range of $10^{19} - 10^{21}$ W Hz^{-1} at 5 GHz. When available, the spectral indices tend to be relatively flat (e.g., Slee et al. 1994). The optical counterparts of the radio cores are usually spectroscopically classified as LINERs (Phillips et al. 1986; Ho 1999a).

19.4.3   Broad Emission Lines

Broad emission lines, a defining attribute of classical Seyferts and quasars, are also found in nuclei of much lower luminosities. The well-known case of the nucleus of M81 (Peimbert & Torres-Peimbert 1981; Filippenko & Sargent 1988), for example, has a broad (FWHM $\approx$ 3000 km s^{-1}) Hα line with a luminosity of only 2×10^{39} erg s^{-1} (Ho, Filippenko, & Sargent 1996), and many other less conspicuous cases have been discovered in the Palomar survey (Ho et al. 1997e; Fig. 19.6). Searching for broad Hα emission in nearby nuclei is nontrivial, because it entails measurement of a (generally) weak, low-

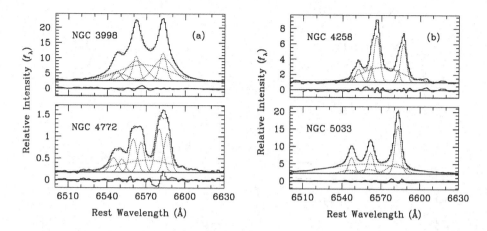

Fig. 19.6. Examples of (*a*) LINERs and (*b*) Seyferts with broad Hα emission. [N II] λλ6548, 6583 and the narrow component of Hα are assumed to have the same shape as [S II] λλ6716, 6731, and the broad component of Hα is modeled as a single Gaussian. Residuals of the fit are shown on the bottom of each panel. (Adapted from Ho et al. 1997e.)

contrast, broad emission feature superposed on a complicated stellar background. Thus, the importance of careful starlight subtraction cannot be overemphasized. Moreover, even if one were able to perfectly remove the starlight, one still has to contend with deblending the Hα + [N II] λλ6548, 6583 complex. The narrow lines in this complex are often heavily blended together, and rarely do the lines have simple profiles. The strategy adopted by Ho et al. (1997e) is to use the empirical line profile of the [S II] lines to model [N II] and the narrow component of Hα.

Of the 221 emission-line nuclei in the Palomar survey classified as LINERs, transition objects, and Seyferts, 33 (15%) definitely have broad Hα, and an additional 16 (7%) probably do. Questionable detections were found in another 8 objects (4%). Thus, approximately 20%–25% of all nearby AGNs are type 1 sources. These numbers, of course, should be regarded as lower limits, since undoubtedly there must exist AGNs with even weaker broad-line emission that fall below the detection threshold.

It is illuminating to consider the incidence of broad Hα emission as a function of spectral class. Among objects formally classified as Seyferts (according to their narrow-line spectrum), approximately 40% are Seyfert 1s. The implied ratio of Seyfert 1s to Seyfert 2s (1:1.6) has important consequences for some models concerning the evolution and small-scale geometry of AGNs (e.g., Osterbrock & Shaw 1988; Lawrence 1991). Despite claims to the contrary (Krolik 1998; Sulentic, Marziani, & Dultzin-Hacyan 2000), broad emission lines emphatically are *not* exclusively confined to Seyfert nuclei. Within the Palomar sample, nearly 25% of the "pure" LINERs have detectable broad Hα emission. By direct analogy with the familiar nomenclature established for Seyferts, LINERs can be divided into "type 1" and "type 2" sources according to the presence or absence of broad-line emission, respectively (Ho et al. 1997a, 1997e). The detection rate of broad Hα, however, drops drastically for transition objects. The cause for this dramatic change is unclear, but a

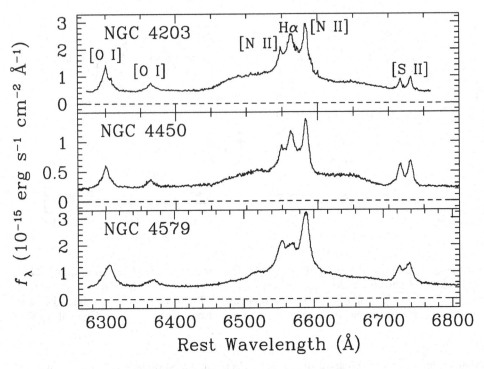

Fig. 19.7. Examples of LINERs with broad, double-peaked Hα emission discovered with *HST*. (Adapted from Ho et al. 2000, Shields et al. 2000, and Barth et al. 2001.)

possible explanation is that the broad-line component is simply too weak to be detected in the presence of substantial contamination from the H II region component.

A subset of LINERs contain broad lines with *double-peaked* profiles (Fig. 19.7), analogous to those seen in a minority of radio galaxies (Eracleous & Halpern 1994), where they are often interpreted as a kinematic signature of a relativistically broadened accretion disk (Chen & Halpern 1989). Most of the nearby cases have been discovered serendipitously, either as a result of the broad component being variable (e.g., Storchi-Bergmann et al. 1993) or because of the increased sensitivity to weak, broad features afforded by small-aperture measurements made with the *Hubble Space Telescope (HST)* (Shields et al. 2000; Ho et al. 2000, and references therein).

19.4.4 *Robustness and Completeness*

To gain confidence in the current AGN statistics, one must have some handle on whether the existing AGN detections are trustworthy and whether there are many AGNs that have been missed. The robustness issue hinges on the question of whether the weak, nearby sources classified as AGNs are truly accretion-powered. As summarized in § 19.5, this appears to be largely the case. The completeness issue can be examined in two regimes. Among bulged (Sbc and earlier) galaxies, for which the spectroscopic AGN fractions are already very high (~50%–75%; Fig. 19.5), there is no room for a large fraction of missing AGNs. The same does not necessarily hold for galaxies of Hubble types Sc and later. While the majority of these systems are spectroscopically classified as H II nuclei, one must be wary that weak AGNs, if present, may be masked by brighter off-nuclear H II regions or H II regions projected along the line of sight. After all, some very late-type galaxies *do* host *bona fide* AGNs (see §19.4.8).

The AGN content of late-type galaxies can be independently assessed by using a diagnostic less prone to confusion by star-forming regions. The presence of a compact, nuclear radio or X-ray core turns out to be a useful AGN filter, since genuine AGNs almost always possess compact emission in these bands. Because of the expected weakness of the nuclei, however, any search for core emission must be conducted at relatively high sensitivity and angular resolution ($\lesssim 1''$). In practice, this requires *Chandra* for the X-rays and an interferometer such as the Very Large Array (VLA) for the radio.

Ulvestad & Ho (2002) have performed a VLA survey for radio cores in a distance-limited sample of 40 Palomar Sc galaxies classified as hosting H II nuclei. To a sensitivity limit of $P_{\mathrm{6cm}} \approx 10^{18} - 10^{20}$ W Hz^{-1} at $\sim 1''$ resolution, they found that *none* of the galaxies contain radio cores. They detected nuclear emission in three galaxies, but in all cases the morphology was diffuse, consistent with that seen in nearby circumnuclear starbursts such as NGC 253. The VLA study of Filho, Barthel, & Ho (2000) also failed to detect radio cores in a more heterogeneous sample of 12 H II nuclei.

Information on nuclear X-ray cores in late-type galaxies is much more limited because to date there has been no systematic investigation of these systems with *Chandra*. A few studies, however, have exploited the High Resolution Imager (HRI) on *ROSAT* to resolve the soft X-ray (0.5–2 keV) emission in late-type galaxies (Colbert & Mushotzky 1999; Lira, Lawrence, & Johnson 2000; Roberts & Warwick 2000). Although the resolution of the HRI ($\sim 5''$) is not ideal, it is nonetheless quite effective for identifying point sources given the relatively diffuse morphologies of late-type galaxies. Compact X-ray sources, often quite luminous ($\gtrsim 10^{38}$ erg s^{-1}), are frequently found, but generally they do *not* coincide with the galaxy nucleus; the nature of these "ultraluminous X-ray sources" is discussed by van der Marel (2004).

To summarize: unless H II nuclei in late-type galaxies contain radio and X-ray cores far weaker than the current survey limits—a possibility worth exploring—they do not appear to conceal a significant population of undetected AGNs.

19.4.5　The $z \approx 0$ AGN Luminosity Function

Many astrophysical applications of AGN demographics benefit from knowing the AGN luminosity function, $\Phi(L, z)^*$. Whereas $\Phi(L, z)$ has been reasonably well charted for high L and high z using quasars (Osmer 2004), it is very poorly known at low L and low z. Indeed, until very recently there has been no reliable determination of $\Phi(L, 0)$.

The difficulty in determining $\Phi(L, 0)$ can be ascribed to a number of factors, as discussed in Huchra & Burg (1992). First and foremost is the challenge of securing a reliable, spectroscopically selected sample, as discussed in §§ 19.2–19.4. Since nearby AGNs are expected to be faint relative to their host galaxies, most of the traditional techniques used to identify quasars cannot be applied without introducing large biases. The faintness of nearby AGNs presents another obstacle, namely how to disentangle the nuclear emission—the only component relevant to the AGN—from the usually much brighter contribution from the host galaxy. Finally, most optical luminosity functions of bright, more distant AGNs are specified in terms of the nonstellar optical continuum (usually the *B* band), whereas spectroscopic

---

* For the purposes of this paper, I will only consider the optical luminosity function.

surveys of nearby galaxies generally only reliably measure optical line emission (e.g., Hα) because the featureless nuclear continuum is often impossible to detect in ground-based, seeing-limited apertures.

Huchra & Burg (1992; see also Osterbrock & Martel 1993) presented the first optical luminosity function of nearby Seyfert galaxies, based on the sample of AGNs selected from the CfA redshift survey. They also calculated the luminosity function of LINERs, but it was known to be highly incomplete. Huchra & Burg, however, did not have access to true nuclear luminosities for their sample; their luminosity function was based on *total* (nucleus plus host galaxy) magnitudes.

A different strategy can be explored by taking advantage of the fact that Hα luminosities are now available for nearly all of the AGNs in the Palomar survey (Ho et al. 1997a, 2003). Ho et al. (2004) begin by calculating the nuclear Hα luminosity function using Schmidt's (1968) V/V_{max} method, where for each source V is the volume it occupies given its distance and V_{max} is the maximum volume it could occupy were it to lie within the flux limit of the survey, taken to be the larger of the two volumes as calculated from the total optical magnitude limit of the survey ($B_T = 12.5$ mag) and the flux limit for detecting emission lines. Next, Ho et al. (2004) exploit the fact that luminous AGNs obey a tight correlation between the luminosity of the optical, nonstellar continuum and the luminosity of the hydrogen Balmer lines, a relation that follows naturally from simple photoionization arguments (Searle & Sargent 1968; Weedman 1976; Yee 1980; Shuder 1981). Using *nuclear* continuum magnitudes extracted from high-resolution *HST* images, Ho & Peng (2001) showed that low-luminosity AGNs, too, obey the correlation established by the more luminous sources, albeit with somewhat greater scatter.

Figure 19.8 presents the *B*-band nuclear luminosity function for the Palomar AGNs, computed by translating the extinction-corrected Hα luminosities to *B*-band absolute magnitudes with the aid of the empirical calibration between Hβ luminosity and M_B of Ho & Peng (2001) and an assumed Hα/Hβ ratio of 3.1. Two versions are shown, each representing an extreme view of what kind of sources should be regarded as *bona fide* AGNs. The open circles include only type 1 nuclei, sources in which broad Hα emission was detected and hence whose AGN status is incontrovertible. This may be regarded as the most conservative assumption and a lower bound, since we know that genuine narrow-lined AGNs do exist (e.g., M104 or NGC 4261). The solid circles lump together all sources classified as LINERs, transition objects, or Seyferts, both type 1 and type 2. This represents the most optimistic view and an upper bound, since undoubtedly *some* narrow-lined sources must be stellar in origin but masquerading as AGNs. The true space density of local AGNs most likely lies between these two possibilities. In either case, the differential luminosity function is reasonably well approximated by a single power law from $M_B \approx -5$ to -18 mag, roughly of the form $\Phi \propto L^{-1.2\pm0.2}$. The slope may flatten for $M_B \gtrsim -7$ mag, but the luminosity function is highly uncertain at the faint end because of density fluctuations in our local volume.

For comparison, I have overlaid the luminosity function of $z \lesssim 0.3$ quasars and Seyfert 1 nuclei as determined by Köhler et al. (1997) from the Hamburg/ESO UV-excess survey*. This sample extends the luminosity function from $M_B \approx -18$ to -26 mag. Although the two samples do not strictly overlap in luminosity, it is apparent the two samples roughly merge, and that the break in the combined luminosity function most likely falls near $M_B^* \approx -19$ mag,

---

* Scaled to our adopted cosmological parameters of $H_0 = 75$ km s^{-1} Mpc^{-1}, $\Omega_m = 0.3$, and $\Omega_\Lambda = 0.7$.

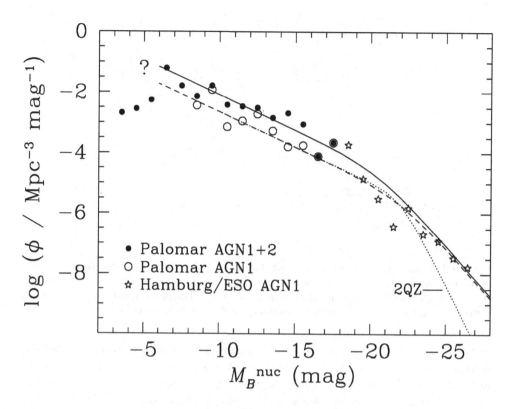

Fig. 19.8. The *B*-band nuclear luminosity function of nearby AGNs derived from the Palomar survey. The filled circles include all (type 1 + type 2) sources, while the open circles include only type 1 sources. The sample of luminous Seyfert 1s and quasars from the Hamburg/ESO survey of Köhler et al. (1997) is shown as stars. A double power-law fit to the Palomar and Hamburg/ESO samples is shown as a solid (type 1 + type 2) and dashed (type 1) curve. The dotted curve represents the quasar luminosity function derived from the 2dF quasar redshift survey (2QZ), shifted to $z = 0$ according to the luminosity evolution model of Boyle et al. (2000). (Adapted from Ho et al. 2004.)

where the space density $\phi \approx 1 \times 10^{-4}$ Mpc^{-3} mag^{-1}. I also plotted the quasar luminosity function obtained from the 2dF quasar redshift survey (2QZ), after evolving it to $z = 0$ following the luminosity evolution prescription of Boyle et al. (2000). The faint-end slope matches that of the local value quite well, but the break the 2QZ luminosity function drops much more sharply than the local sample.

19.4.6 *Bolometric Luminosities and Eddington Ratios*

To gain further insight into the physical nature of nearby AGNs, it is more instructive to examine their bolometric luminosities, rather than their luminosities in a specific band or emission line. Because AGNs emit a very broad spectrum, their bolometric luminosities ideally should be measured directly from their full spectral energy distributions (SEDs). In practice, however, complete SEDs are not readily available for most AGNs,

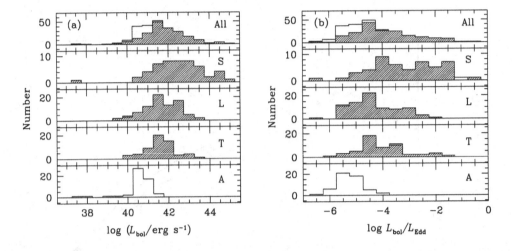

Fig. 19.9. Distribution of (*a*) bolometric luminosity, L_{bol}, and (*b*) ratio of bolometric luminosity to the Eddington luminosity, L_{bol}/L_{Edd}, for all objects, Seyferts (L), LINERs (L), transition nuclei (T), and absorption-line nuclei (A). The hatched and open histograms denote detections and upper limits, respectively. (Adapted from Ho 2004.)

and one commonly estimates the bolometric luminosity by applying bolometric corrections derived from a set of well-observed calibrators. As discussed in more detail in §19.5, the SEDs of low-luminosity AGNs differ quite markedly from those of conventionally studied AGNs. Nonetheless, they do exhibit a characteristic shape, which enables bolometric corrections to be calculated. AGN researchers customarily choose as the reference point either the optical *B* band or an X-ray band. While the same strategy may be used for low-luminosity AGNs (e.g., Ho et al. 2000), it cannot yet be widely employed because nuclear optical or X-ray fluxes are not yet available for large samples. What is available, by selection, is nuclear emission-line fluxes, and upper limits thereof. Although the Hα luminosity comprises only a small percentage of the total power, its fractional contribution to the bolometric luminosity, as Ho (2003, 2004) notes, turns out to be fairly well defined.

Figure 19.9 shows the distributions of bolometric luminosities and their values normalized with respect to the Eddington luminosity for Palomar galaxies with measurements of Hα luminosity and central stellar velocity dispersions. The $M_\bullet - \sigma$ relation of Tremaine et al. (2002) was used to obtain L_{Edd}. Whereas LINER and transition nuclei both have a median $L_{bol} \approx 2 \times 10^{41}$ erg s^{-1}, Seyfert nuclei are typically an order of magnitude more luminous (median $L_{bol} \approx 2 \times 10^{42}$ erg s^{-1}). The upper limits for the objects lacking any detectable line emission (absorption-line nuclei) cluster near $L_{bol} \approx 3 \times 10^{40}$ erg s^{-1}. These systematic trends persist when I consider the Eddington ratios. One again, the distribution of L_{bol}/L_{Edd} for LINERs is rather similar to that of transition objects (median $L_{bol}/L_{Edd} \approx 2 \times 10^{-5}$ and 3×10^{-5}, respectively), but both are quite distinct from Seyferts (median $L_{bol}/L_{Edd} \approx 4 \times 10^{-4}$). Notably, the vast majority of nearby nuclei have highly sub-Eddington luminosities.

19.4.7   Host Galaxy Properties

The near dichotomy in the distribution of Hubble types for galaxies hosting active versus inactive nuclei (Fig. 19.5) leads to the expectation that the two populations ought to have fairly distinctive global, and perhaps even nuclear, properties. Moreover, a detailed examination of the host galaxies of AGNs may shed light on the origin of their spectral diversity. These issues were recently examined by Ho et al. (2003) using the database from the Palomar survey. The main results are summarized here.

The host galaxies of Seyferts, LINERs, and transition objects display a remarkable degree of homogeneity in their large-scale properties. After factoring out spurious differences arising from slight mismatches in Hubble type distribution, all three classes have essentially identical total luminosities ($\sim L^*$), bulge luminosities, sizes, and neutral gas content. The only exception is that, relative to LINERs, transition objects may show a mild enhancement in the level of star formation, and they may be preferentially more inclined. This is consistent with the hypothesis that the transition class arises from spatial blending of emission from a LINER and H II regions.

Theoretical studies (e.g., Heller & Shlosman 1994) suggest that large-scale stellar bars can be highly effective in delivering gas to the central few hundred parsecs of a spiral galaxy, thereby potentially leading to rapid star formation. Further instabilities result in additional inflow to smaller scales. Thus, provided that an adequate reservoir of gas exists, the presence of a bar might be expected to influence the BH fueling rate, and hence the level of nonstellar activity. The Palomar sample is ideally suited for statistical tests of this nature, which depend delicately on issues of sample selection effects and completeness. Ho et al. (1997d) find that while the presence of a bar indeed does enhance both the probability and rate of star formation in galaxy nuclei, it appears to have no impact on either the frequency or strength of AGN activity. Bearing in mind the substantial uncertainties introduced by sample selection (see discussion in Appendix B of Ho & Ulvestad 2001), other studies broadly come to a similar conclusion (see review by Combes 2003).

In the same vein, dynamical interactions with neighboring companions should lead to gas dissipation, enhanced nuclear star formation, and perhaps central fueling (e.g., Hernquist 1989). Schmitt (2001) and Ho et al. (2003) studied this issue using the Palomar data, parameterizing the nearby environment of each object by its local galaxy density and the distance to its nearest sizable neighbor. After accounting for the well-known morphology-density relation, it was found that the local environment, like bars, has little impact on AGNs.

The uniformity in host galaxy properties extends even to small, nuclear (~ 200 pc) scales, in two important respects. First, the velocity field of the ionized gas, as measured by the width and asymmetry of the narrow emission lines, appears to be crudely similar among the three classes, an observation that argues against the proposition that fast shocks primarily drive the spectral variations observed in nearby galactic nuclei. Second, the homogeneity among the three AGN classes is seen in their nuclear stellar content, which nearly always appears evolved. The general dearth of young or intermediate-age stars presents a serious challenge to proposals that seek to account for the excitation of the emission lines in terms of starburst or post-starburst models.

19.4.8   Do Intermediate-mass Black Holes Exist?

As summarized by Barth (2004) and Kormendy (2004), the observational evidence for supermassive BHs in the mass range $M_\bullet \approx 10^6 - 10^{9.5} \, M_\odot$ has become quite secure, to the

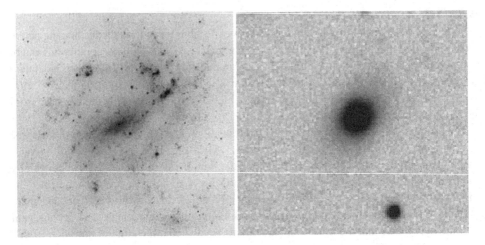

Fig. 19.10. Two examples of AGNs in late-type galaxies. The *left* panel shows an optical image of NGC 4395, adapted from the Carnegie Atlas of Galaxies (Sandage & Bedke 1994); the image is $\sim15'$ (17 kpc) on a side. The *right* panel shows an R-band image of POX 52, adapted from Barth et al. (2004); the image is $\sim25''$ (11 kpc) on a side.

point that important inferences on their demographics can be drawn (Richstone 2004). Do central (nonstellar) BHs with masses below $10^6 \, M_\odot$ exist? The current low end of the mass scale may reflect observational limitations rather than a true physical limit. There is certainly no compelling theory that prohibits the existence of BHs in the large gap between the stellar regime of $10 \, M_\odot$ and $10^6 \, M_\odot$. The recent dynamical studies of Gerssen et al. (2002) and Gebhardt, Rich, & Ho (2002) suggest that some massive star clusters may contain BHs in the mass range $10^3 < M_\bullet < 10^4 \, M_\odot$ (see review by van der Marel 2004).

The existence of intermediate-mass BHs is further supported by the detection of AGNs in at least some very late-type and dwarf galaxies. Two examples are particularly noteworthy (Fig. 19.10). The nearby ($\sim$4 Mpc) galaxy NGC 4395 contains all the usual attributes of a respectable AGN: broad optical and UV emission lines (Filippenko & Sargent 1989; Filippenko, Ho, & Sargent 1993), a compact radio core (Ho & Ulvestad 2001; Wrobel, Fassnacht, & Ho 2001), and rapidly variable hard X-ray emission (Moran et al. 1999, 2004; Shih, Iwasawa, & Fabian 2003). Contrary to expectations, however, NGC 4395 is a bulgeless, extremely late-type (Sdm) spiral, whose central stellar velocity dispersion does not exceed ~ 30 km s^{-1} (Filippenko & Ho 2003). If NGC 4395 obeys the $M_\bullet - \sigma$ relation, its central BH should have a mass $\lesssim 10^5 \, M_\odot$. This limit agrees surprisingly well with the value of $M_\bullet$ estimated from its broad Hβ line width and X-ray variability properties ($\sim 10^4 - 10^5 \, M_\odot$; Filippenko & Ho 2003).

As first noted by Kunth, Sargent, & Bothun (1987), the presence of a Seyfert-like nucleus in POX 52 is unusual because of the low luminosity of the host galaxy. Barth et al. (2004) show that POX 52 bears a close spectroscopic resemblance to NGC 4395. Based on the broad profile of Hβ, these authors derive a virial BH mass of $1.6 \times 10^5 \, M_\odot$ for POX 52, again remarkably close to the value of $1.3 \times 10^5 \, M_\odot$ predicted from the $M_\bullet - \sigma$ relation given the measured central stellar velocity dispersion of 35 km s^{-1}. Deep images reveal POX 52 to be most akin to a dwarf elliptical galaxy, to date an unprecedented morphology for an AGN host galaxy.

The two objects highlighted above demonstrate that the mass spectrum of nuclear BHs very likely extends below $10^6 \, M_\odot$. Furthermore, they provide great leverage for anchoring the $M_\bullet - \sigma$ at the low end. But how common are such objects? AGNs hosted in high-surface brightness, late-type spirals appear to be quite rare in the nearby Universe. Within the comprehensive Palomar survey, NGC 4395 emerges as a unique case, and as argued in §19.4, the late-type galaxy population probably does not conceal a large number of misidentified AGNs. The majority of late-type spirals do possess compact, photometrically distinct nuclei (Böker et al. 2002), morphologically not dissimilar from NGC 4395, but these nuclei are compact star clusters, not AGNs (Walcher et al. 2004). The true incidence of AGNs like POX 52 is more difficult to assess. Most spectroscopic surveys of blue-selected emission-line objects do not have sufficient signal-to-noise ratio to identify weak, broad emission lines. On the other hand, Greene & Ho (2004), in a preliminary analysis of the first data release from SDSS, have uncovered a number of broad-line AGNs in low-luminosity, presumably low-mass, host galaxies. These appear to be excellent candidates for late-type galaxies with intermediate-mass BHs.

19.5   The Nature of LINERs

The recognition and definition of LINERs is based on their spectroscopic properties at optical wavelengths. In addition to the AGN scenario, the optical spectra of LINERs unfortunately can be interpreted in several other ways that do not require an exotic energy source. The principal alternatives are shocks and hot stars (see reviews by Barth 2002 and Filippenko 2003). As a consequence, it has often been suggested that LINERs may be a mixed-bag, heterogeneous collection of objects. While the nonstellar nature of some well-studied LINERs is incontrovertible (e.g., M81, M87), the AGN content in the majority of LINERs continues to be debated. Determining the physical origin of LINERs is more than of mere phenomenological interest. Because LINERs are so numerous, they have repercussions on many issues related to AGN demographics.

19.5.1   *Evidence for an AGN Origin*

A number of recent developments provide considerable new insight into the origin of LINERs. I outline these below, and I use them to advance the proposition that *most* LINERs are truly AGNs[*]. The discussion first focuses on "pure" LINERs, returning later to the class of transition objects, whose nature remains more obscure.

(1) *Detection of broad emission lines:* Luminous, unobscured AGNs distinguish themselves unambiguously by their characteristic broad permitted lines. The detection of broad Hα emission in ∼25% of LINERs (Ho et al. 1997e) thus constitutes strong evidence in favor of the AGN interpretation of these sources. LINERs, like Seyferts, evidently come in two flavors—some have a visible BLR (type 1), and others do not (type 2). The broad component becomes progressively more difficult to detect in ground-based spectra for permitted lines

---

[*] Note that this paper is concerned only with compact, *nuclear* LINERs ($r \lesssim 100$ pc), which are most relevant to the AGN issue. I do not consider LINER-like extended nebulae such as those associated with cooling flows, nuclear outflows, starburst-driven winds, and circumnuclear disks. LINERs selected from samples of distant, interacting, or infrared-bright galaxies are particularly vulnerable to confusion from these extended sources. (See also discussion in § 19.4.1.)

weaker than Hα. Where available, however, broad, higher-order Balmer lines as well as UV lines such as Lyα, C IV λ1549, Mg II λ2800, and Fe II multiplets can be seen in *HST* spectra (Barth et al. 1996; Ho et al. 1996). I further recall the double-peaked broad lines mentioned in § 19.5.3.

(2) *Detection of hidden BLRs:* An outstanding question, however, is what fraction of the more numerous LINER 2s are AGNs. By analogy with the Seyfert 2 class, surely *some* LINER 2s must be genuine AGNs—that is, LINERs whose BLR is obscured along the line of sight to the viewer. There is no *a priori* reason why the unification model, which has enjoyed such success in the context of Seyfert galaxies, should not apply equally to LINERs. The existence of an obscuring torus does not obviously depend on the ionization level of the NLR. If we suppose that the ratio of LINER 2s to LINER 1s is the same as the ratio of Seyfert 2s to Seyfert 1s, that ratio being 1.6:1 in the Palomar survey, we can reasonably surmise that the AGN fraction in LINERs may be as high as ∼60%. That at least some LINER 2s *do* contain a hidden BLR was demonstrated by the spectropolarimetric observations of Barth, Filippenko, & Moran (1999a, b).

(3) *Naked LINER 2s:* The BLR in some LINER 2s may be intrinsically absent, not obscured. If BLR clouds arise from condensations in a radiation-driven, outflowing wind (e.g., Murray & Chiang 1995), a viewpoint now much espoused, then it is reasonable to expect that very low-luminosity sources would be incapable of generating a wind, and hence of sustaining a BLR. A good example of such a case is NGC 4594 (the "Sombrero" galaxy). Its nucleus, although clearly an AGN, shows no trace of a broad-line component, neither in direct light (Ho et al. 1997e), not even when very well isolated with a small *HST* aperture (Nicholson et al. 1998), nor in polarized light (Barth et al. 1999b). Its Balmer decrement indicates little reddening to the NLR. For all practical purposes, the continuum emission from the nucleus looks unobscured: it is detected in the UV (Maoz et al. 1998) and in the soft and hard X-rays (Fabbiano & Juda 1997; Nicholson et al. 1998; Ho et al. 2001; Pellegrini et al. 2002; Terashima et al. 2002), with evidence for only moderate intrinsic absorption (Pellegrini et al. 2003). In short, there is no sign of anything being hidden or much doing the hiding. So where is the BLR? It is not there. A number of authors have also emphasized the existence of unabsorbed Seyfert 2 nuclei (e.g., Panessa & Bassani 2002; Gliozzi et al. 2004). These considerations lead to the conclusion that the mere absence of a BLR does not constitute evidence against the AGN pedigree of an object.

(4) *Compact cores:* AGNs, at least when unobscured, traditionally reveal themselves as pointlike nuclear sources at virtually all wavelengths. This fact can be exploited by searching for compact cores in LINERs. To overcome the contrast problem, imaging observations of this kind are only meaningful for sufficiently high angular resolution. Moreover, in practice certain spectral windows are more advantageous than others. Nuclear point sources turn out to be surprisingly difficult to extract from optical and near-infrared images of nearby galaxies, even at the 0."1 resolution of *HST* (see, e.g., Ho & Peng 2001; Ravindranath et al. 2001; Peng et al. 2002). This is due to a number of factors, chiefly the dominance of the host galaxy bulge and the complexity of dust structures in nuclear regions at these wavelengths. The bulge light largely disappears in the UV, but the detection rate of nuclear cores in the UV is only ∼25% for LINERs (Maoz et al. 1995; Barth et al. 1998). This band is especially hard to work with because it is particularly susceptible to dust extinction, to confusion from young stars when present (Maoz et al. 1998), and to intrinsic variations due to the form of the SEDs of low-luminosity AGNs (Ho 1999b; see point 5 below).

Compact nuclei can be detected most cleanly in the X-rays and radio (see also § 19.4.4). These regions are least sensitive to obscuration and provide the highest contrast between nonstellar and stellar emission. There have been a number of attempts to systematically investigate LINERs using X-ray data obtained from *ROSAT* (Koratkar et al. 1995; Komossa, Böhringer, & Huchra 1999; Roberts & Warwick 2000; Halderson et al. 2001; Roberts, Schurch, & Warwick 2001), *ASCA* (Ptak et al. 1999; Terashima, Ho, & Ptak 2000; Terashima et al. 2000, 2002), and *BeppoSAX* (Georgantopoulos et al. 2002). While these efforts have been enormously useful in delineating the basic X-ray properties of LINERs, particularly in the spectral domain, they suffer from two crucial limitations. First, the angular resolutions of all the above X-ray facilities, with the possible exception of the *ROSAT*/HRI under some circumstances (§19.4.4), are grossly incapable of properly isolating any but the brightest nuclei. And second, the faintness of the typical targets compels most investigators to study only limited, inevitably biased, samples.

The advent of *Chandra* has dramatically improved this situation. The high angular resolution of the telescope ($\sim 1''$) and the low background noise of the CCD detectors allow faint point sources to be detected with brief (few ks) exposures (Ho et al. 2001; Terashima & Wilson 2003). This makes feasible, for the first time, X-ray surveys of large samples of galaxies selected at non-X-ray wavelengths. Ho et al. (2001) used the ACIS camera to image a distance-limited sample of Palomar AGNs. Their analysis of a preliminary subset indicates that X-ray cores, some as faint as $\sim 10^{38}$ erg s^{-1} in the 2–10 keV band, are found in $\sim 75\%$ of LINERs; the detection rate is roughly similar for LINER 1s and 2s.

As in the X-rays, AGNs nearly universally emit at some level in the radio. Although most AGNs are radio quiet, they are seldom radio silent when observed with sufficient sensitivity and angular resolution. For example, Seyfert galaxies generally contain radio cores, which are often accompanied by linear, jetlike features (e.g., Ulvestad & Wilson 1989; Kukula et al. 1995; Thean et al. 2000; Ho & Ulvestad 2001; Schmitt et al. 2001). The complete sample of Seyferts selected from the Palomar survey shows a detection rate of $\sim 80\%$ at 5 GHz and $1''$ resolution (Ho & Ulvestad 2001; Ulvestad & Ho 2001a); their radio powers span $10^{18} - 10^{21}$ W Hz^{-1}.

LINERs have been surveyed somewhat less extensively than Seyferts. As mentioned in § 19.4.2, radio interferometric surveys of nearby elliptical and S0 galaxies detect a high fraction of low-power cores, most of which are optically classified as LINERs. VLA studies of well-defined subsamples of LINERs chosen from the Palomar survey show qualitatively similar trends (Van Dyk & Ho 1997; Nagar et al. 2000, 2002). At 5 and 8 GHz, where the sensitivity is highest, $\sim 60\%$–80% of LINERs, independent of type, contain radio cores. VLBI observations of the brighter, VLA-detected sources generally reveal brightness temperatures $\gtrsim 10^{6-8}$ K (Falcke et al. 2000; Ulvestad & Ho 2001b; Filho, Barthel, & Ho 2002b; Anderson, Ulvestad, & Ho 2004).

To summarize: the majority of LINERs, both type 1 and 2, photometrically resemble AGNs insofar as they emit compact, pointlike hard X-ray and radio emission.

(5) *Spectral energy distributions:* The broad-band spectrum of luminous, unabsorbed AGNs follows a fairly universal shape (e.g., Elvis et al. 1994). The SED from the infrared to the X-rays can be roughly represented as the sum of an underlying power law ($L_\nu \propto \nu^{-1}$) and a few distinct components, the most prominent of which is the "big blue bump" usually attributed to thermal emission from an optically thick, geometrically thin accretion disk (Shields 1978; Malkan & Sargent 1982). The SEDs of LINERs deviate markedly from the standard form of high-luminosity AGNs (Ho 1999b, 2002b; Ho et al. 2000), as shown in Figure 19.11. The most conspicuous difference can be seen in the apparent absence of a UV excess. The SEDs of these sources also tend to be generically "radio loud," defined here by the convention that the radio-to-optical luminosity ratio exceeds a value of 10. In fact, radio loudness seems

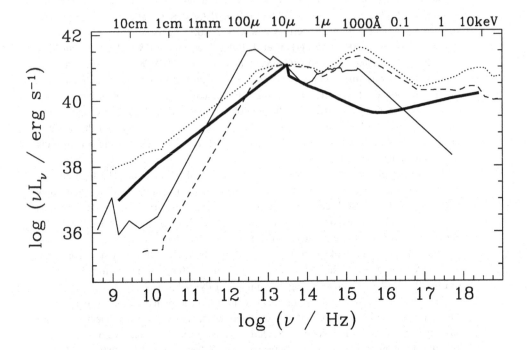

Fig. 19.11. The average SED of low-luminosity AGNs (*heavy solid line*), adapted from Ho (1999b). Overplotted for comparison are the average SEDs of powerful radio-loud (*dotted line*) and radio-quiet (*dashed line*) AGNs (Elvis et al. 1994), and of low-extinction starburst galaxies (*light solid line*; Schmitt et al. 1997). The curves have been arbitrarily normalized to the luminosity at 10 μm.

to be a property common to essentially *all* nearby weakly active nuclei (Ho 2002a) and a substantial fraction of Seyfert nuclei (Ho & Peng 2001). Using a definition of radio loudness based on the relative strength of the X-ray and radio emission, Terashima & Wilson (2003) also find that LINERs tend to be radio loud.

While the SEDs of LINERs differ from those of those of traditional AGNs, it is important to emphasize that they *do* approximate the SEDs predicted for radiatively inefficient accretion flows onto BHs (e.g., Quataert et al. 1999; Ptak et al. 2004). At the same time, they definitely bear little resemblance to SEDs characteristic of "normal" stellar systems (see, e.g., Schmitt et al. 1997). Inactive galaxies or starburst systems not strongly affected by dust extinction emit the bulk of their radiation in the optical–UV and in the thermal infrared regions, with only an energetically miniscule contribution from X-rays (Fig. 19.11). Indeed, the decidedly *nonstellar* nature of the SEDs of LINERs can be regarded as compelling evidence that LINERs are accretion-powered sources, albeit of an unusual ilk.

(6) *Host galaxies:* As discussed in § 19.4.7, LINERs and Seyferts live in virtually identical host galaxies. To the extent that Seyferts are regarded as AGNs, the close similarity of their hosts with those of LINERs lends supporting, if albeit indirect, evidence that the two classes share a common origin.

(7) *Detection of massive BHs:* Finally, I note the obvious fact that a significant fraction of the

galaxies with detected BHs are, in fact, well-known LINERs. These include M81, M84, M87, NGC 4261, and NGC 4594. Although certainly no statistical conclusions can yet be drawn from such meager statistics, these examples nevertheless illustrate that at least some LINERs seem to be directly connected with BH accretion.

19.5.2   Excitation Mechanisms

The above arguments lend credence to the hypothesis that a sizable fraction of LINERs—indeed the *majority*—are directly related to AGNs. In this context, photoionization by a central AGN surfaces as the most natural candidate for the primary excitation mechanism of LINERs. The optical spectra of LINERs can be readily reproduced in AGN photoionization calculations by adjusting the ionization parameter, U, defined as the ratio of the density of ionizing photons to the density of nucleons at the illuminated face of a cloud. Whereas the NLR spectrum of Seyferts can be well fitted with $\log U \approx -2.5 \pm 0.5$ (e.g., Ferland & Netzer 1983; Stasińska 1984; Ho, Shields, & Filippenko 1993b), that of LINERs requires $\log U \approx -3.5 \pm 1.0$ (Ferland & Netzer 1983; Halpern & Steiner 1983; Péquignot 1984; Binette 1985; Ho et al. 1993a). What factors contribute to the lower ionization parameters in LINERs? Ho et al. (2003) identify the central luminosity and gas density as two relevant factors. According to the statistics from the Palomar survey, LINERs on average have lower luminosities and lower gas densities than Seyferts. For a given volume filling factor, this leads to lower ionization parameters in LINERs compared to Seyferts, although not at a level sufficient to account for the full difference between the two classes.

Despite the natural appeal of AGN photoionization, alternative excitation mechanisms for LINERs have been advanced. Collisional ionization by shocks has been a popular contender from the outset (Koski & Osterbrock 1976; Fosbury et al. 1978; Heckman 1980b; Dopita & Sutherland 1995; Alonso-Herrero et al. 2000; Sugai & Malkan 2000). Dopita & Sutherland (1995) showed that the diffuse radiation field generated by fast ($v \approx 150$–500 km s^{-1}) shocks can reproduce the optical narrow emission lines seen in both LINERs and Seyferts. In their models, LINER-like spectra are realized under conditions in which the precursor H II region of the shock is absent, as might be the case in gas-poor environments. The postshock cooling zone attains a much higher equilibrium electron temperature than a photoionized plasma; consequently, a robust prediction of the shock model is that shocked gas should produce a higher excitation spectrum, most readily discernible in the UV, than photoionized gas. In all the cases studied so far, however, the UV spectra are inconsistent with the fast-shock scenario because the observed intensities of the high-excitation lines such as C IV $\lambda 1549$ and He II $\lambda 1640$ are much weaker than predicted (Barth et al. 1996, 1997; Maoz et al. 1998; Nicholson et al. 1998; Gabel et al. 2000). Dopita et al. (1997) used the spectrum of the circumnuclear *disk* of M87 to advance the view that LINERs are shock excited. This argument is misleading because their analysis deliberately avoids the nucleus. Sabra et al. (2003) demonstrate that the UV–optical spectrum of the *nucleus* of M87 is best explained by a multi-component photoionization model.

A recent analysis of the emission-line profiles of the Palomar nuclei further casts doubt on the viability of the shock scenario (Ho et al. 2003). The velocity dispersions of the nuclear gas generally fall short of the values required for shock excitation to be important. Furthermore, the close similarity between the velocity field of LINERs and Seyferts, as deduced from their line profiles, contradicts the basic premise that shocks are primarily responsible for the spectral differences between the two classes of objects.

Another widely discussed class of models invokes hot stars, formed in a short-duration burst of star formation, to supply the primary ionizing photons. Ordinary O-type stars with effective temperatures typical of those inferred in giant H II regions in galactic disks do not produce sufficiently strong low-ionization lines to account for the spectra of LINERs. The physical conditions in galactic nuclei, on the other hand, may be more favorable for generating LINER-like spectra. For example, Terlevich & Melnick (1985) postulate that the high-metallicity environment of galactic nuclei may be particularly conducive to forming very hot [$T \approx (1-2) \times 10^5$ K], luminous Wolf-Rayet stars, whose ionizing spectrum would effectively mimic the power-law continuum of an AGN. The models of Filippenko & Terlevich (1992) and Shields (1992) appeal to less extreme conditions. These authors show that photoionization by ordinary O stars embedded in an environment with high density and low ionization parameter can explain the spectral properties of transition objects. Barth & Shields (2000) extended this work by modeling the ionizing source not as single O-type stars but as a more realistic evolving young star cluster. They confirm that young, massive stars can indeed generate optical emission-line spectra that match those of transition objects, and, under some plausible conditions, even those of *bona fide* LINERs. But there is an important caveat: the star cluster must be formed in an instantaneous burst, and its age must coincide with the brief phase ($\sim$3–5 Myr after the burst) during which sufficient Wolf-Rayet stars are present to supply the extreme-UV photons necessary to boost the low-ionization lines. The necessity of a sizable population of Wolf-Rayet stars is also emphasized in the recent study by Gabel & Bruhweiler (2002). As discussed in Ho et al. (2003), the main difficulty with this scenario, and indeed with all models that appeal to young stars (e.g., Taniguchi, Shioya, & Murayama 2000), is that the nuclear stellar population of the host galaxies of the majority of nearby AGNs, irrespective of spectral class, is demonstrably *old*. Stellar absorption indices indicative of young or intermediate-age stars are seldom seen, and the telltale emission features of Wolf-Rayet stars are notably absent. These empirical facts seriously undermine stellar-based photoionization models of AGNs.

19.5.3 *Transition Objects*

The physical origin of transition nuclei continues to be a thorny, largely unresolved problem. Since these objects make up a significant fraction of emission-line nuclei, this complication unfortunately casts some uncertainty into the demography of local AGNs.

In two-dimensional optical line-ratio diagrams (Fig.   19.3), transition nuclei are empirically defined to be those sources that lie sandwiched between the loci of "normal" H II regions and LINERs. This motivated Ho et al. (1993a) to proposed that transition objects may be composite systems consisting of a LINER nucleus plus an H II region component. The latter could arise from neighboring circumnuclear H II regions or from H II regions randomly projected along the line of sight. A similar argument, based on decomposition of line profiles, has been made by Véron, Gonçalves, & Véron-Cetty (1997) and Gonçalves, Véron-Cetty, & Véron (1999).

If transition objects truly are LINERs sprinkled with a frosting of star formation, one would expect that their host galaxies should be largely similar to those of LINERs, namely bulge-dominated systems (§ 19.4.7), modulo minor differences due to the "excess" contaminating star formation. The study of Ho et al. (2003) largely supports this picture. The host galaxies of transition nuclei exhibit systematically higher levels of recent star formation, as indicated by their far-infrared emission and broad-band optical colors, compared to LINERs of matched morphological types. Moreover, the host galaxies of transition nuclei tend to be slightly more inclined than LINERs. Thus, all else being equal, transition-type spectra seem to be found precisely in those galaxies whose nuclei have a higher probability of being contaminated by extra-nuclear emission from star-forming regions.

This story, however, has some holes. If simple spatial blending of circumnuclear H II regions is sufficient to transform a regular LINER into a transition object, the LINER nucleus should reveal itself unambiguously in spectra taken with angular resolution sufficiently high to isolate it. This test was performed by Barth, Ho, & Filippenko (2003), who obtained *HST*/STIS spectra, taken with a 0.''2-wide slit, of a well-defined subsample of 15 transition objects selected from the Palomar catalog. To their surprise, the small-aperture spectra of the nuclei, for the most part, look very similar to the ground-based spectra; they are *not* more LINER-like.

The "masqueraded-LINER" hypothesis can be further tested by searching for compact radio and X-ray cores using high-resolution images. Recall that this is a highly effective alternative method to filter out weak AGNs (§ 19.4.4 and § 19.5.1). Filho et al. (2000, 2002a, 2004) have systematically surveyed the full sample of Palomar transition objects using the VLA. They find that ∼25% of the population contains arcsecond-scale radio cores. These cores appear to be largely nonstellar in nature. The brighter subset of these sources that are amenable to follow-up VLBI observations (Filho et al. 2004) all reveal more compact (milliarcsecond-scale) cores with flat radio spectra and high brightness temperatures ($T_B \gtrsim 10^7$ K). In their preliminary analysis of a *Chandra* survey of Palomar galaxies, Ho et al. (2001) noted that transition objects show a marked deficit of X-ray cores. Although based on small-number statistics, the frequency of X-ray cores for the transition objects in the Ho et al. study also turns out to be 25%.

The above considerations suggest a conservative lower limit of ∼25% for the AGN fraction in transition objects, or a reduction from 13% to ∼3% of the overall galaxy population. In turn, the total AGN fraction (LINERs, Seyferts, and accretion-powered transition objects) for all local galaxies decreases from 43% to 33%. These revised rates are lower limits because of the imperfect correspondence between "genuine" AGNs and the presence of radio and X-ray cores. After all, for reasons that are not yet understood, clearly not all Seyfert nuclei are detected in the radio (e.g., Ho & Ulvestad 2001; Ulvestad & Ho 2001), and X-rays in the 2–10 keV band will be extinguished for gas with sufficiently large column densities ($\gtrsim 10^{24}$ cm^{-2}). Such low-luminosity Compton-thick nuclei, if present, can be uncovered with sensitive, high-resolution observations at harder X-ray energies.

If the majority of transition objects are not AGNs, we are faced with a new conundrum. What are they? For the reasons explained above, the source of their line excitation is unlikely to be shock heating or photoionization by hot, massive stars. Here I suggest two possibilities worth considering. First, the ionizing radiation field might originate from hot, *evolved* stars. This idea has been advocated by Binette et al. (1994), who proposed that post-asymptotic giant branch stars, which can attain effective temperatures as high as ∼ 10^5 K, might be responsible for photoionizing the extended ionized gas often observed in elliptical galaxies. The emission-line spectrum of these nebulae, in fact, tend to be of relatively low ionization (Demoulin-Ulrich, Butcher, & Boksenberg 1984; Phillips et al. 1986; Zeilinger et al. 1996). Invoking evolved stars has the obvious appeal of not conflicting with the dominant old stellar population found in the centers of nearby galaxies. Second, the integrated (off-nuclear) X-ray emission of the central regions of galaxies may contribute nonnegligibly to the ionizing photon budget. Recent *Chandra* and *XMM-Newton* images of the centers of nearby, "ordinary" galaxies have resolved the X-ray emission into two components: discrete sources and diffuse, hot gas. The discrete X-ray source population consists mainly of X-ray binaries, mostly of the low-mass variety (see Fabbiano & White 2004 for a review). While X-ray–emitting plasma has long been known to be pervasive in giant elliptical galaxies, it now appears that it may be a generic constituent even in spheroids of lower mass. For example, diffuse, hot gas has been detected in the central regions of the Milky Way (Baganoff et al. 2004), M31 (Shirey et al. 2001), and M32 (Ho, Terashima, & Ulvestad 2003). Since X-ray binaries and X-ray–emitting gas have "hard" spectra (compared to, say, O-type stars), they would naturally be conducive to producing strong low-ionization optical lines when used as an ionizing source. These unconventional sources of ionization—hot, evolved stars, X-ray binaries, and X-ray plasma—*must* contribute at some level, insofar as we know empirically that they exist. Their ubiquitous presence likely maintains a pervasive, diffuse, ionizing radiation field, which may be sufficient to sustain a "baseline" level of weak optical line emission. It would be fruitful to further explore these issues quantitatively with photoionization models.

19.6   Summary

This review argues that the demographics of AGN activity in nearby galaxies can inform us much about the demographics of massive BHs. While ultimately there is no substitute for direct dynamical mass measurements, such an approach is often neither practical nor feasible. AGN statistics provide important complementary information. The following points are the most germane to BH demography.

(1) Nuclear activity is extremely common in the nearby Universe. Over 40% of all nearby galaxies qualify as AGNs or AGN candidates according to their emission-line spectral properties.

(2) LINERs are the most common variety of local AGN candidates.

(3) The majority of LINERs appear to be genuinely accretion-powered systems. Thus, most LINERs should be considered AGNs.

(4) Nuclear activity preferentially occurs in bulge-dominated galaxies. Galaxies with Hubble types later than Sbc become progressively dominated by nuclear star formation.

(5) The physical origin of the so-called transition objects remains largely unknown, although at least 25% of them appear to be AGNs. This uncertainty affects the quantitative conclusions from (1) and (2), but the qualitative picture remains unchanged.

(6) Inasmuch as central BHs are a precondition for AGN activity, the detection rate of AGNs establishes a lower limit on the incidence of massive BHs in nearby galactic nuclei. The above findings support the prevailing belief, based on dynamical studies and energy arguments (see Barth 2004; Kormendy 2004; Richstone 2004), that massive BHs are a ubiquitous feature of massive galaxies.

(7) Central massive BHs, while perhaps uncommon in pure-disk or dwarf galaxies, do not completely shun such environments. Some late-type galaxies definitely harbor lightweight nuclear BHs, which extend the BH mass function down to the regime of $M_\bullet \approx 10^4 - 10^5 \, M_\odot$,

although the frequency of such objects is not yet well established. These intermediate-mass BHs seem to obey the $M_\bullet$-σ relation established by the supermassive ($10^6 - 10^{9.5} M_\odot$) BHs.

(8) Local AGNs are generically weak, characterized by highly sub-Eddington luminosities. Most central BHs in nearby galaxies are either quiescent or only weakly active.

Acknowledgements. My research is supported by the Carnegie Institution of Washington and by NASA grants from the Space Telescope Science Institute (operated by AURA, Inc., under NASA contract NAS5-26555). I would like to recognize the significant contributions of my collaborators, especially A. J. Barth, A. V. Filippenko, D. Maoz, E. C. Moran, C. Y. Peng, A. Ptak, H.-W. Rix, W. L. W. Sargent, J. C. Shields, Y. Terashima, and J. S. Ulvestad. I thank A. J. Barth, A. V. Filippenko, and W. L. W. Sargent for permission to cite material in advance of publication. A. J. Barth, A. V. Filippenko, and J. S. Mulchaey gave thoughful comments on the manuscript.

References

Alonso-Herrero, A., Rieke, M. J., Rieke, G. H., & Shields, J. C. 2000, ApJ, 530, 688

Anderson, J. M., Ulvestad, J. S., & Ho, L. C. 2004, ApJ, in press

Armus, L., Heckman, T. M., & Miley, G. K. 1990, ApJ, 364, 471

Baganoff, F. K., et al. 2004, ApJ, in press

Baldwin, J. A., Phillips, M. M., & Terlevich, R. 1981, PASP, 93, 5

Balzano, V. A. 1983, ApJ, 268, 602

Barth, A. J. 2002, in Issues in Unification of AGNs, ed. R. Maiolino, A. Marconi, & N. Nagar (San Francisco: ASP), 147

——. 2004, in Carnegie Observatories Astrophysics Series, Vol. 1: Coevolution of Black Holes and Galaxies, ed. L. C. Ho (Cambridge: Cambridge Univ. Press), in press

Barth, A. J., Filippenko, A. V., & Moran, E. C. 1999a, ApJ, 515, L61

——. 1999b, ApJ, 525, 673

Barth, A. J., Ho, L. C., & Filippenko, A. V. 2003, in Active Galactic Nuclei: from Central Engine to Host Galaxy, ed. S. Collin, F. Combes, & I. Shlosman (San Francisco: ASP), 387

Barth, A. J., Ho, L. C., Filippenko, A. V., Rix, H.-W., & Sargent, W. L. W. 2001, ApJ, 546, 205

Barth, A. J., Ho, L. C., Filippenko, A. V., & Sargent, W. L. W. 1998, ApJ, 496, 133

Barth, A. J., Ho, L. C., Rutledge, R. E., & Sargent, W. L. W. 2004, ApJ, in press

Barth, A. J., Reichert, G. A., Filippenko, A. V., Ho, L. C., Shields, J. C., Mushotzky, R. F., & Puchnarewicz, E. M. 1996, AJ, 112, 1829

Barth, A. J., Reichert, G. A., Ho, L. C., Shields, J. C., Filippenko, A. V., & Puchnarewicz, E. M. 1997, AJ, 114, 2313

Barth, A. J., & Shields, J. C. 2000, PASP, 112, 753

Binette, L. 1985, A&A, 143, 334

Binette, L., Magris, C. G., Stasińska, G., & Bruzual A., G. 1994, A&A, 292, 13

Böker, T., van der Marel, R. P., Laine, S., Rix, H.-W., Sarzi, M., Ho, L. C., & Shields, J. C. 2002, AJ, 123, 1389

Bonatto, C., Bica, E., & Alloin, D. 1989, A&A, 226, 23

Boyle, B. J, Shanks, T., Croom, S. M., Smith, R. J., Miller, L., Loaring, N., & Heymans, C. 2000, MNRAS, 317, 1014

Burbidge, E. M., & Burbidge, G. 1962, ApJ, 135, 694

——. 1965, ApJ, 142, 634

Chen, K., & Halpern, J. P. 1989, ApJ, 344, 115

Colbert, E. J. M., & Mushotzky, R. F. 1999, ApJ, 519, 89

Collins, J. A., & Rand, R. J. 2001, ApJ, 551, 57

Combes, F. 2003, in Active Galactic Nuclei: from Central Engine to Host Galaxy, ed. S. Collin, F. Combes, & I. Shlosman (San Francisco: ASP), 411

Costero, R., & Osterbrock, D. E. 1977, ApJ, 211, 675

Danziger, I. J., Fosbury, R. A. E., & Penston, M. V. 1977, MNRAS, 179, 41P

Demoulin-Ulrich, M.-H., Butcher, H. R., & Boksenberg, A. 1984, ApJ, 285, 527

Disney, M. J., & Cromwell, R. H. 1971, ApJ, 164, L35

Dopita, M. A., Koratkar, A. P., Allen, M. G., Tsvetanov, Z. I., Ford, H. C., Bicknell, G. V., & Sutherland, R. S. 1997, ApJ, 490, 202

Dopita, M. A., & Sutherland, R. S. 1995, ApJ, 455, 468

Elvis, M., et al. 1994, ApJS, 95, 1

Eracleous, M., & Halpern, J. P. 1994, ApJS, 90, 1

Fabbiano, G., & Juda, J. Z. 1997, ApJ, 476, 666

Fabbiano, G., & White, N. E. 2004, in Compact Stellar X-ray Sources, ed. W. Lewin & M. van der Klis (Cambridge: Cambridge Univ. Press), in press

Fabian, A. C., Arnaud, K. A., Nulsen, P. E. J., & Mushotzky, R. F. 1986, ApJ, 305, 9

Falcke, H., Nagar, N. M., Wilson, A. S., & Ulvestad, J. S. 2000, ApJ, 542, 197

Ferland, G. J., & Netzer, H. 1983, ApJ, 264, 105

Filho, M. E., Barthel, P. D., & Ho, L. C. 2000, ApJS, 129, 93

——. 2002a, ApJS, 142, 223

——. 2002b, A&A, 385, 425

Filho, M. E., Fraternali, F., Nagar, N. M., Barthel, P. D., Markoff, S., Yuan, F., & Ho, L. C. 2004, A&A, in press

Filippenko, A. V. 2003, in Active Galactic Nuclei: from Central Engine to Host Galaxy, ed. S. Collin, F. Combes, & I. Shlosman (San Francisco: ASP), 387

Filippenko, A. V., & Halpern, J. P. 1984, ApJ, 285, 458

Filippenko, A. V., & Ho, L. C. 2003, ApJ, 588, L13

Filippenko, A. V., Ho, L. C., & Sargent, W. L. W. 1993, ApJ, 410, L75

Filippenko, A. V., & Sargent, W. L. W. 1985, ApJS, 57, 503

——. 1986, in Structure and Evolution of Active Galactic Nuclei, ed. G. Giuricin, et al. (Dordrecht: Reidel), 21

——. 1988, ApJ, 324, 134

——. 1989, ApJ, 342, L11

Filippenko, A. V., & Terlevich, R. 1992, ApJ, 397, L79

Fosbury, R. A. E., Melbold, U., Goss, W. M., & Dopita, M. A. 1978, MNRAS, 183, 549

Fosbury, R. A. E., Melbold, U., Goss, W. M., & van Woerden, H. 1977, MNRAS, 179, 89

Gabel, J. R., & Bruhweiler, F. C. 2002, AJ, 124, 737

Gabel, J. R., Bruhweiler, F. C., Crenshaw, D. M., Kraemer, S. B., & Miskey, C. L. 2000, ApJ, 532, 883

Gebhardt, K., Rich, R. M., & Ho, L. C. 2002, ApJ, 578, L41

Georgantopoulos, I., Panessa, F., Akylas, A., Zezas, A., Cappi, M., & Comastri, A. 2002, A&A, 386, 60

Gerssen, J., van der Marel, R. P., Gebhardt, K., Guhathakurta, P., Peterson, R. C., & Pryor, C. 2002, AJ, 124, 3270 (addendum: 2003, 125, 376)

Gliozzi, M., Sambruna, R. M., Brandt, W. N., Mushotzky, R. F., & Eracleous, M. 2004, A&A, in press

Gonçalves, A. C., Véron-Cetty, M.-P., & Véron, P. 1999, A&AS, 135, 437

Grandi, S. A., & Osterbrock, D. E. 1978, ApJ, 220, 783

Greene, J. E., & Ho, L. C. 2004, ApJ, submitted

Halderson, E. L., Moran, E. C., Filippenko, A. V., & Ho, L. C. 2001, AJ, 122, 637

Halpern, J. P., & Steiner, J. E. 1983, ApJ, 269, L37

Hao, L., & Strauss, M. A. 2004, Carnegie Observatories Astrophysics Series, Vol. 1: Coevolution of Black Holes and Galaxies, ed. L. C. Ho (Pasadena: Carnegie Observatories, http://www.ociw.edu/ociw/symposia/series/symposium1/proceedings.html)

Heckman, T. M. 1980a, A&A, 87, 142

——. 1980b, A&A, 87, 152

——. 2004, in Carnegie Observatories Astrophysics Series, Vol. 1: Coevolution of Black Holes and Galaxies, ed. L. C. Ho (Cambridge: Cambridge Univ. Press), in press

Heckman, T. M., Balick, B., & Crane, P. C. 1980, A&AS, 40, 295

Heckman, T. M., Baum, S. A., van Breugel, W. J. M., & McCarthy, P. 1989, ApJ, 338, 48

Heller, C. H., & Shlosman, I. 1994, ApJ, 424, 84

Hernquist, L. 1989, Nature, 340, 687

Ho, L. C. 1996, in The Physics of LINERs in View of Recent Observations, ed. M. Eracleous et al. (San Francisco: ASP), 103

——. 1999a, ApJ, 510, 631

——. 1999b, ApJ, 516, 672

——. 2002a, ApJ, 564, 120

——. 2002b, in Issues in Unification of AGNs, ed. R. Maiolino, A. Marconi, & N. Nagar (San Francisco: ASP), 165

——. 2003, in Active Galactic Nuclei: from Central Engine to Host Galaxy, ed. S. Collin, F. Combes, & I. Shlosman (San Francisco: ASP), 379

——. 2004, in preparation

Ho, L. C., et al. 2001, ApJ, 549, L51

Ho, L. C., & Filippenko, A. V. 1993, Ap&SS, 205, 19

Ho, L. C., Filippenko, A. V., & Sargent, W. L. W. 1993a, ApJ, 417, 63

——. 1995, ApJS, 98, 477

——. 1996, ApJ, 462, 183

——. 1997a, ApJS, 112, 315

——. 1997b, ApJ, 487, 568

——. 1997c, ApJ, 487, 579

——. 1997d, ApJ, 487, 591

——. 2003, ApJ, 583, 159

——. 2004, in preparation

Ho, L. C., Filippenko, A. V., Sargent, W. L. W., & Peng, C. Y. 1997e, ApJS, 112, 391

Ho, L. C., & Peng, C. Y. 2001, ApJ, 555, 650

Ho, L. C., Rudnick, G., Rix, H.-W., Shields, J. C., McIntosh, D. H., Filippenko, A. V., Sargent, W. L. W., & Eracleous, M. 2000, ApJ, 541, 120

Ho, L. C., Shields, J. C., & Filippenko, A. V. 1993b, ApJ, 410, 567

Ho, L. C., Terashima, Y., & Ulvestad, J. S. 2003, ApJ, 589, 783

Ho, L. C., & Ulvestad, J. S. 2001, ApJS, 133, 77

Huchra, J. P., & Burg, R. 1992, ApJ, 393, 90

Humason, M. L., Mayall, N. U., & Sandage, A. R. 1956, AJ, 61, 97

Kauffmann, G., et al. 2003, MNRAS, in press

Keel, W. C. 1983a, ApJS, 52, 229

——. 1983b, ApJ, 269, 466

Kewley, L. J., Heisler, C. A., Dopita, M. A., & Lumsden, S. 2001, ApJS, 132, 37

Khachikian, E. Ye., & Weedman, D. W. 1974, ApJ, 192, 581

Kim, D.-C., Sanders, D. B., Veilleux, S., Mazzarella, J. M., & Soifer, B. T. 1995, ApJS, 98, 129

Köhler. T., Groote, D., Reimers, D., & Wisotzki, L. 1997, A&A, 325, 502

Komossa, S., Böhringer, H., & Huchra, J. P. 1999, A&A, 349, 88

Koratkar, A. P., Deustua, S., Heckman, T. M., Filippenko, A. V., Ho, L. C., & Rao, M. 1995, ApJ, 440, 132

Kormendy, J. 2004, in Carnegie Observatories Astrophysics Series, Vol. 1: Coevolution of Black Holes and Galaxies, ed. L. C. Ho (Cambridge: Cambridge Univ. Press), in press

Koski, A. T., & Osterbrock, D. E. 1976, ApJ, 203, L49

Krolik, J. H. 1998, Active Galactic Nuclei (Princeton: Princeton Univ. Press)

Kukula, M. J., Pedlar, A., Baum, S. A., O'Dea, C. P. 1995, MNRAS, 276, 1262

Kunth, D., Sargent, W. L. W., & Bothun, G. D. 1987, AJ, 92, 29

Lawrence, A. 1991, MNRAS, 252, 586

Lehnert, M. D., & Heckman, T. M. 1994, ApJ, 426, L27

Lira, P., Lawrence, A., & Johnson, R. A. 2000, MNRAS, 319, 17

Malkan, M. A., & Sargent, W. L. W. 1982, ApJ, 254, 22

Maoz, D., Filippenko, A. V., Ho, L. C., Rix, H.-W., Bahcall, J. N., Schneider, D. P., & Macchetto, F. D. 1995, ApJ, 440, 91

Maoz, D., Koratkar, A. P., Shields, J. C., Ho, L. C., Filippenko, A. V., & Sternberg, A. 1998, AJ, 116, 55

Moran, E. C., Eracleous, M., Leighly, K. M., Chartas, G., Filippenko, A. V., Ho, L. C., & Blanco, P. R. 2004, ApJ, submitted

Moran, E. C., Filippenko, A. V., Ho, L. C., Shields, J. C., Belloni, T., Comastri, A., Snowden, S. L., & Sramek, R. A. 1999, PASP, 111, 801

Murray, N., & Chiang, J. 1995, ApJ, 454, L105

Nagar, N. M., Falcke, H., Wilson, A. S., & Ho, L. C. 2000, ApJ, 542, 186

Nagar, N. M., Falcke, H., Wilson, A. S., & Ulvestad, J. S. 2002, A&A, 392, 53

Nicholson, K. L., Reichert, G. A., Mason, K. O., Puchnarewicz, E. M., Ho, L. C., Shields, J. C., & Filippenko, A. V. 1998, MNRAS, 300, 893

Osmer, P. S. 2004, in Carnegie Observatories Astrophysics Series, Vol. 1: Coevolution of Black Holes and
    Galaxies, ed. L. C. Ho (Cambridge: Cambridge Univ. Press), in press
Osterbrock, D. E., & Martel, A. 1993, ApJ, 414, 552
Osterbrock, D. E., & Miller, J. S. 1975, ApJ, 197, 535
Osterbrock, D. E., & Shaw, R. A. 1988, ApJ, 327, 89
Panessa, F., & Bassani, L. 2002, A&A, 394, 435
Peimbert, M., & Torres-Peimbert, S. 1981, ApJ, 245, 845
Pellegrini, S., Baldi, A., Fabbiano, G., & Kim, D.-W. 2003, ApJ, in press
Pellegrini, S., Fabbiano, G., Fiore, F., Trinchieri, G., & Antonelli, A. 2002, A&A, 383, 1
Peng, C. Y., Ho, L. C., Impey, C. D., & Rix, H.-W. 2002, AJ, 124, 266
Penston, M. V., & Fosbury, R. A. E. 1978, MNRAS, 183, 479
Péquignot, D. 1984, A&A, 131, 159
Phillips, M. M. 1979, ApJ, 227, L121
Phillips, M. M., Charles, P. A., & Baldwin, J. A. 1983, ApJ, 266, 485
Phillips, M. M., Jenkins, C. R., Dopita, M. A., Sadler, E. M., & Binette, L. 1986, AJ, 91, 1062
Ptak, A., Serlemitsos, P. J., Yaqoob, T., & Mushotzky, R. 1999, ApJS, 120, 179
Ptak, A., Terashima, Y., Ho, L. C., & Quataert, E. 2004, ApJ, in press
Quataert, E., Di Matteo, T., Narayan, R., & Ho, L. C. 1999, ApJ, 525, L89
Raimann, D., Storchi-Bergmann, T., Bica, E., & Alloin, D. 2001, MNRAS, 324, 1087
Ravindranath, S., Ho, L. C., Peng, C. Y., Filippenko, A. V., & Sargent, W. L. W. 2001, AJ, 122, 653

Richstone, D. O. 2004, in Carnegie Observatories Astrophysics Series, Vol. 1: Coevolution of Black Holes and
    Galaxies, ed. L. C. Ho (Cambridge: Cambridge Univ. Press), in press
Roberts, T. P., Schurch, N. J., & Warwick, R. S. 2001, MNRAS, 324, 737
Roberts, T. P., & Warwick, R. S. 2000, MNRAS, 315, 98
Rubin, V. C., & Ford, W. K., Jr. 1971, ApJ, 170, 25
Sabra, B. M., Shields, J. C., Ho, L. C., Barth, A. J., & Filippenko, A. V. 2003, ApJ, 584, 164
Sadler, E. M., Jenkins, C. R., & Kotanyi, C. G. 1989, MNRAS, 240, 591
Sandage, A., & Bedke, J. 1994, The Carnegie Atlas of Galaxies (Washington, DC: Carnegie Inst. of Washington)
Schmidt, M. 1968, ApJ, 151, 393
Schmitt, H. R. 2001, AJ, 122, 2243
Schmitt, H. R., Antonucci, R. R. J., Ulvestad, J. S., Kinney, A. L., Clarke, C. J., & Pringle, J. E. 2001, ApJ, 555,
    663
Schmitt, H. R., Kinney, A. L., Calzetti, D., & Storchi-Bergmann, T. 1997, AJ, 114, 592
Searle, L., & Sargent, W. L. W. 1968, ApJ, 153, 1003
Shields, G. A. 1978, Nature, 272, 706
Shields, J. C. 1992, ApJ, 399, L27
Shields, J. C., Rix, H.-W., McIntosh, D. H., Ho, L. C., Rudnick, G., Filippenko, A. V., Sargent, W. L. W., & Sarzi,
    M. 2000, ApJ, 534, L27
Shih, D. C., Iwasawa, K., & Fabian, A. C. 2003, MNRAS, 341, 973
Shirey, R., et al. 2001, A&A, 365, L195
Shuder, J. M. 1981, ApJ, 244, 12
Slee, O. B., Sadler, E. M., Reynolds, J. E., & Ekers, R. D. 1994, MNRAS, 269, 928
Stasińska, G. 1984, A&A, 135, 341
Stauffer, J. R. 1982a, ApJS, 50, 517
——. 1982b, ApJ, 262, 66
Stauffer, J. R., & Spinrad, H. 1979, ApJ, 231, L51
Storchi-Bergmann, T., Baldwin, J. A., & Wilson, A. S. 1993, ApJ, 410, L11
Sugai, H., & Malkan, M. A. 2000, ApJ, 529, 219
Sulentic, J. W., Marziani, P., & Dultzin-Hacyan, D. 2000, ARA&A, 38, 521
Taniguchi, Y., Shioya, Y., & Murayama, T. 2000, AJ, 120, 1265
Terashima, Y., Ho, L. C., & Ptak, A. F. 2000, ApJ, 539, 161
Terashima, Y., Ho, L. C., Ptak, A. F., Mushotzky, R. F., Serlemitsos, P. J., Yaqoob, T., & Kunieda, H. 2000, ApJ,
    533, 729
Terashima, Y., Iyomoto, N., Ho, L. C., & Ptak, A. F. 2002, ApJS, 139, 1
Terashima, Y., & Wilson, A. S. 2003, ApJ, 583, 145
Terlevich, R., & Melnick, J. 1985, MNRAS, 213, 841

Terlevich, R., Melnick, J., & Moles, M. 1987, in Observational Evidence of Activity in Galaxies, ed. E. Ye. Khachikian, K. J. Fricke, & J. Melnick (Dordrecht: Reidel), 499

Thean, A., Pedlar, A., Kukula, M. J., Baum, S. A., & O'Dea, C. P. 2000, MNRAS, 314, 573

Tremaine, S., et al. 2002, ApJ, 574, 740

Ulvestad, J. S., & Ho, L. C. 2001a, ApJ, 558, 561

——. 2001b, ApJ, 562, L133

——. 2002, ApJ, 581, 925

Ulvestad, J. S., & Wilson, A. S. 1989, ApJ, 343, 659

van der Marel, R. P. 2004, in Carnegie Observatories Astrophysics Series, Vol. 1: Coevolution of Black Holes and Galaxies, ed. L. C. Ho (Cambridge: Cambridge Univ. Press), in press

Van Dyk, S. D., & Ho, L. C. 1997, in IAU Colloq. 164, Radio Emission from Galactic and Extragalactic Compact Sources, ed. A. Zensus, G. Taylor, & J. Wrobel (San Francisco: ASP), 205

Veilleux, S., & Osterbrock, D. E. 1987, ApJS, 63, 295

Véron, P., Gonçalves, A. C, & Véron-Cetty, M.-P. 1997, A&A, 319, 52

Véron, P., & Véron-Cetty, M.-P. 1986, A&A, 161, 145

Véron-Cetty, M.-P., & Véron, P. 1986, A&AS, 66, 335

——. 2000, A&A Rev., 10, 81

Walcher, C. J., Häring, N., Böker, T., Rix, H.-W., van der Marel, R. P., Gerssen, J., Ho, L. C., & Shields, J. C. 2004, in Carnegie Observatories Astrophysics Series, Vol. 1: Coevolution of Black Holes and Galaxies, ed. L. C. Ho (Pasadena: Carnegie Observatories, http://www.ociw.edu/ociw/symposia/series/symposium1/proceedings.html)

Weedman, D. W. 1976, ApJ, 208, 30

Weedman, D. W., Feldman, F. R., Balzano, V. A., Ramsey, L. W., Sramek, R. A., & Wu, C.-C. 1981, ApJ, 248, 105

Wrobel, J. M., Fassnacht, C. D., & Ho, L. C. 2001, ApJ, 553, L23

Wrobel, J. M., & Heeschen, D. S. 1991, AJ, 101, 148

Yee, H. K. C. 1980, ApJ, 241, 894

Zeilinger, W. W., et al. 1996, A&AS, 120, 257

20

The evolution of quasars

PATRICK S. OSMER

Department of Astronomy, The Ohio State University

Abstract

This article reviews and discusses (1) the discovery and early work on the evolution of quasars and AGNs, (2) the different techniques used to find quasars and their suitability for evolutionary studies, (3) the current status of our knowledge of AGN evolution for $0 < z < 6$, (4) the new results and questions that deep radio and X-ray surveys are producing for the subject, (5) the relation of AGNs to the massive black holes being found in local galaxies and what they tell us about both galaxy and AGN evolution, and (6) current research problems and future directions in quasar and AGN evolution.

20.1   Introduction and Background

The subject of the evolution of the active galactic nucleus (AGN) population began in the late 1960s with Schmidt's (1968, 1970) discoveries that the space densities of both radio and optically selected quasars increased significantly with redshift. The effect was so strong that it was detectable in samples as small as 20 objects. He developed and applied the V/V_m test for analyzing the space distribution in his samples and showed that there was a strong evolution in the space density of quasars toward higher redshift, increasing by more than a factor of 100 from redshift 0 to 2. This was a striking and unexpected result that posed a question that is still crucial today—What causes the sharp decline since $z = 2$?

In this article I will review and discuss the following subjects:

- The techniques used to discover quasars and AGNs, their selection effects, and the surveys used to study the evolution of the AGN population.
- The general picture of evolution up to 1995, when the first well-defined, quantitative surveys of the evolution of high-redshift quasars were published.
- The current status of major optical surveys such as 2dF and SDSS.
- Radio and X-ray surveys and how they are critical to understanding AGN evolution.
- The relation of AGNs to their host galaxies and how studies of massive black holes in spheroids provide constraints on AGN evolution.
- Current research problems, such as measuring the quasar luminosity function at high redshift and faint magnitudes; relating observed to physical evolution; the framework for connecting observations, accretion processes, and the growth of black hole masses; and how to estimate black hole masses.

Before proceeding, let us define and discuss terms used in this article to aid the clarity of the presentation.

© The Observatories of the Carnegie Institution of Washington 2004.

An AGN is one not powered by normal stellar processes, although active star formation may be occurring in the vicinity. The working hypothesis is that AGNs contain massive black holes and are powered by accretion processes. Their luminosities range from as low as $M_B = -9$ mag to as high as $M_B = -30$ mag ($L_X = 10^{38}$ to 10^{48} erg s^{-1}). Quasars are the high-luminosity ($M_B < -23$ mag, $L_X > 10^{44}$ erg s^{-1}) members of the AGN family.

Traditionally, evolution of AGNs or quasars has meant the evolution with redshift of their luminosity function or space density (which is the integral of the luminosity function over some range of luminosities). However, evolution can also refer to changes with redshift of the spectral energy distribution (SED) or the emission-line spectra of AGNs. In general, *observed* evolution will refer to changes with redshift of any observed property of AGNs.

Ultimately, we wish to map and understand the *physical* evolution of AGNs, by which we mean how their central black holes form and grow with cosmic epoch and how their accretion processes and rates, which determine the luminosities and SEDs we observe from AGNs, evolve with cosmic epoch. The discovery of the ubiquity of black holes in the spheroids of nearby galaxies makes us realize that the physical evolution of AGNs is closely connected with and is an important part of the larger subject of how galaxies in general form and evolve. It appears that virtually every spheroidal system went through an AGN phase at some time in its history—thus the subject of our meeting: "The Coevolution of Black Holes and Galaxies."

However, the persistent question of how many AGNs are hidden because of weak emission lines, obscuration by dust, or absorption in X-rays has continued to impede progress in the mapping of the observational evolution of AGNs and must be addressed in any attempt to determine the properties of the overall AGN population. Fortunately, the advent of powerful new space observatories such as *Chandra* and *XMM-Newton*, in conjunction with sensitive radio and infrared surveys, provides new tools for attacking this problem, as will be addressed below.

At the same time, the formulation and application of the appropriate observational definitions of AGNs continue to be critical issues in current research, especially for low-luminosity objects, which can be hard to find within the glare of their host galaxy or to separate from normal stars. For example, the work of Ho (2003) suggests that some AGNs may have X-ray luminosities down to 10^{36} erg s^{-1}, or less than stellar X-ray sources.

If we are really to understand the global population of AGNs and their relation to galaxies, these problems must be solved. This will be one of the themes to be developed in this article.

20.2 Observational Techniques, Selection Effects, and Surveys

The general principle for discovering AGNs is to make use of one or more of the ways in which they are not like stars or galaxies, for example, how they differ in the SEDs or emission-line spectra. The pointlike, i.e., spatially unresolved, nature of the nuclei is another distinguishing factor. It is also possible to make use of their great distances and correspondingly undetectable proper motions in quasar and AGN searches or their variability in brightness. In this article we will concentrate on techniques that make use of their SEDs and spectral-line properties.

The SEDs of AGNs are remarkable for their broad extent in frequency, from radio to γ-rays, which is much greater than for normal, thermal sources of astronomical radiation. The UV/optical emission-line spectra stand out for the strength and breadth of the principal emission lines and for the wide range of ionization. Typical line widths of permitted lines are 5000 km s^{-1} or more. The strongest individual lines are those of hydrogen (Lyα, Hα, and Hβ), C IV, C III], Mg II, and N V, while broad emission complexes of Fe II are visible. In addition, forbidden lines of [O I], [O II], [O III], and [S II] are prominent.

Some of the observable properties of AGNs provide diagnostic probes of the physical nature of the central engine. For example, the X-ray emission originates in regions as close as a few Schwarzschild radii of the central black hole and yields information about the inner part of the accretion disk and coronal region. The broad UV/optical emission lines are produced within a few light days of the central engine. One main goal of AGN research is to combine multiwavelength and spectral observations of AGNs with theoretical models of the accretion processes so that physical properties such as accretion rates and efficiencies can be inferred from observable data. When success is achieved in this subject, it will yield a significant advance in our understanding of AGN evolution.

20.2.1   Techniques for Finding Quasars and AGNs

20.2.1.1   Quasars

Historically, quasars were first discovered (Hazard, Mackey, & Shimmins 1963; Schmidt 1963) via the optical identification of radio sources, a technique that was both effective and efficient because normal stars and galaxies are much weaker sources of radio emission. It was soon realized (Sandage 1965) that the bulk of the quasar population was radio quiet and could be identified through the excess UV (UVX) radiation that quasars demonstrated relative to normal stars*. We now know that the UVX technique is effective for redshifts up to about 2.2, the point at which Lyα emission shifts into the observed B band and quasars begin to lose their characteristic UV excess. We also know now that only about 10% of quasars and AGNs in the early samples were strong radio emitters, or radio-loud objects.

At higher redshifts, different techniques must be used to find quasars. The problem becomes difficult for two reasons: (1) at redshifts around 3, the optical/UV SEDs of quasars are hard to distinguish from stars, and (2) the space density of quasars at $z > 3$ declines rapidly with increasing redshift.

The slitless-spectrum technique pioneered by Smith (1975) and developed by Osmer & Smith (1976) provided a color-independent method of finding high-redshift quasars through the direct detection of their strong, broad emission lines, in particular Lyα, on low-dispersion objective-prism photographs. The technique was then applied to large telescopes through the use of a transmission grating/prism combination (grism) as the dispersing device (Hoag 1976). However, it was also realized that the slitless-spectrum technique was subject to an important selection effect in that it favored the detection of quasars with strong emission lines†.

Schmidt, Schneider, & Gunn (1986) made an important advance on this problem by using a digital detector with a grism at the Hale 5-m telescope and by developing and applying a numerical selection algorithm for identifying emission-line objects whose properties and efficiencies could be quantified. The effectiveness of their approach is well demonstrated in

---

* For the record, most stars are UV faint and quasars have relatively flat optical/UV SEDs in νf_ν space.

† Of course, all observational techniques for discovering quasars and AGNs are subject to selection effects. This has been a long-standing problem in the determination of their luminosity function and its evolution. Nonetheless, it appears that the slitless-spectrum technique indeed discovers the bulk of the high-z population, although it obviously misses objects with weak or no emission lines.

Figure 2 of their paper, which shows the grism spectra for a variety of high-redshift quasars from their survey.

The advent of rapid plate-scanning machines such as COSMOS and APM enabled the extension to higher redshift of color-based techniques for discovering quasars. Warren et al. (1987) found the first quasar with $z = 4$ in this way. The machines and multi-color techniques made it possible to use more sophisticated combinations of colors to separate quasars from stars and to provide quantitative estimates of the selection efficiency as a function of redshift and apparent magnitude, which were crucial for determining the luminosity function. Subsequently, the Sloan Digital Sky Survey (SDSS; York et al. 2000) combined the multicolor technique with a dedicated survey telescope and the largest digital camera built until that time to open a new frontier in extragalactic research by undertaking a digital survey of 10,000 deg^2 in five filters; the initial results are described in more detail below.

20.2.1.2 AGNs

We now know that the discovery of AGNs preceded quasars by 20 years (Seyfert 1943), although the connection and understanding was not achieved until the mid-1970s. Seyfert's classic paper described the properties of nearby galaxies with unusually bright nuclei, which also had unusual emission-line spectra, in particular, broad lines and a wide range of ionization. Seyfert galaxies and related AGNs such as LINERs (low-ionization nuclear emission-line regions), which are less luminous than $M_B = -23$ mag, constitute the bulk of the AGN population. Their most prominent members can be discovered through imaging and spectroscopic surveys following in the footsteps of Seyfert. However, the discovery of lower-luminosity, more elusive members of the class requires much more care, as the work of Ho, Filippenko, & Sargent (1995) has shown. They examined carefully *all* galaxies within a magnitude-limited survey with high- quality, narrow-slit spectra for evidence of an active nucleus.

Most recently, Heckman (2004) and Hao & Strauss (2004) have demonstrated that careful application of stellar population synthesis modeling to SDSS galaxy spectra can pull out otherwise unrecognizable emission-line and AGN signatures through the careful subtraction of the young stellar and nebular emission population. Their work indicates that the presence of weak AGN activity is much more common than originally thought and is found in the majority of early- and middle-type galaxies.

20.2.1.3 Radio and X-Ray Techniques

The discovery and identification of quasars and AGNs by radio and X-ray techniques is perhaps the most straightforward of all, because normal stars and galaxies are weak emitters in these wavelengths. One requires sufficient sensitivity to compact sources and positional accuracies of $\sim 1''$ on the sky. Objects in radio and/or X-ray catalogs are then matched to optical catalogs for identifications and follow-up optical spectra with a large telescope are used to confirm the identification and establish the redshift of the object. Radio and hard X-ray sources offer the important advantage that they are not affected by dust obscuration that may occur along the line of sight to the AGN. If spectral information is available in the 1 keV range, then estimates of the column density of any absorbing gas along the line of sight may be made.

Until recently, radio surveys were hampered by the fact that, as mentioned previously, only about 10% of AGNs are radio loud and thus radio surveys included only a small fraction of the total population. However, with the advent of deep, wide-area surveys such as FIRST (Becker, White, & Helfand 1995), important new opportunities have arisen. FIRST, which reaches to milli-Jansky flux limits, is sufficiently sensitive to detect *radio-quiet* quasars. When used in conjunction with the multi-color imaging data of the SDSS, it has enabled the discovery of new classes of AGNs (e.g., reddened broad absorption-line quasars that are radio sources) and added a new perspective on the issue of dust obscuration.

The combination of FIRST and SDSS data overcomes another problem with earlier radio surveys, namely, the difficulty of achieving effective redshift preselection for candidate objects. The difficulty was that follow-up spectroscopy of a large number of candidates had to be carried out to find high-redshift or rare types of quasars and AGNs. However, the multicolor SDSS data now can be used to pre-sort candidate objects into the desired groups for follow-up work.

X-ray data have been important to the study of quasars and AGNs since their first detections in X-rays because it was realized that the emission likely originated from very close to the central black hole. Indeed, it can be argued that X-ray emission is the defining characteristic of AGNs (e.g., Elvis et al. 1978). However, the point was somewhat moot at the time because of the lack of sensitivity of the original X-ray observatories. Now, following the work with *ROSAT* and the initial results from *Chandra* and *XMM-Newton*, the tables are turned—the deepest X-ray surveys are picking up objects not previously noted in optical surveys.

20.2.1.4 Summary

Discovery and survey techniques for quasars and AGNs at X-ray, UV/optical, and radio wavelengths are now sufficiently well developed, quantified, and sensitive that we have the main tools in hand to settle many of the most fundamental observational questions about the evolution of the AGN population. The combinations of multiwavelength data that are now possible add even more opportunities for research on the nature of AGNs.

20.3   Evolution of the AGN Population

20.3.1   Results through 1995

Schmidt's discovery of the evolution of the quasar luminosity function immediately stimulated work on the nature of the evolution. While powerful for showing the existence of evolution, the V/V_m test by itself was not capable of delineating the nature of the evolution. Furthermore, the available quasar samples were too small to permit analyses in much detail. Schmidt explored different forms of density evolution, i.e., evolution of the number density with cosmic epoch. He found that both a power-law evolution of the form $(1+z)^k$ and an exponential function of look-back time could fit the data up to redshift 2. Mathez (1976, 1978), building on the work of Lynds & Petrosian (1972), demonstrated that luminosity evolution, in which the characteristic luminosity of quasars increased with redshift also provided a satisfactory fit to the data. Schmidt & Green (1983) presented results from the 92 quasars in the Palomar Bright Quasar Survey that showed the increase of space density with redshift to depend on the luminosity of the objects. This indicated that a simple parameterization of either pure density or pure luminosity evolution did not fit the data well.

Subsequently, the work of Boyle, Shanks, & Peterson (1988, hereafter BSP), using the UVX technique, marked a significant advance in sample size and limiting magnitude. They compiled a sample of 420 quasars to $B < 20.9$ mag from UK Schmidt plates scanned with

COSMOS. They found that a two-power law luminosity function and luminosity evolution adequately describe the data for objects with $M_B < -23$ mag and $z < 2.2$, a result that has been widely used and is consistent with recent 2dF results, as described below.

Thus, the situation by the late 1980s was that either the space density of quasars increased by more than a factor of 100 between redshift 0 and 2 (Schmidt & Green 1983), or their characteristic luminosity increased by a factor of 30 (BSP).

It was also understood that the density and luminosity evolution pictures led to significantly different estimates of the lifetimes of quasars, about 10^7 years in the density evolution picture and 10^9 years or more in the luminosity evolution picture. Another consequence was that most galaxies would pass through a quasar phase in the density evolution model; for luminosity evolution, only a few percent of galaxies would be active.

A different and also important question was raised early on in studies of quasar evolution: What happened at high redshift, $z > 2$? The redshift histograms of quasar catalogs showed a marked decline in numbers at $z > 2$ (e.g., Hewitt & Burbidge 1980), with the implication that the evolution also declined. However, it was also realized that the traditional UVX method was not suitable for finding high-redshift quasars, and the lack of suitably defined samples blocked progress. As mentioned above, the slitless-spectrum technique provided an efficient means of discovering high-redshift quasars, and Osmer (1982) showed from a differential study with the CTIO 4-m telescope and grism that there was strong evidence for a decline in the space density of quasars at $z > 3$. Nonetheless, he could only provide an upper limit on the decline because no quasars with $z > 3$ were found in his survey, and it was clear that more work was needed. Also, it was pointed out by Heisler & Ostriker (1988) that dust absorption by intervening galaxies along the line of sight could produce a decline in the observed space density determined from flux-limited samples.

Significant advances occurred in the 1990s, when the first large, digital surveys for high-redshift quasars were carried out. Warren, Hewett, & Osmer (1991a,b, 1994, hereafter WHO) made use of APM scans of UK Schmidt plates in six colors, u, b_j, v, or, r, i to cover an effective area of 43 deg^2. Their sample contained 86 objects with $16 < m_{or} < 20$ mag and $2.2 < z < 4.5$. They developed a numerical modeling technique to determine the selection probabilities for their objects as a function of redshift and magnitude, allowing for different spectral slopes and emission-line strengths. They used the selection probabilities to make two different estimates of the quasar luminosity function and its evolution. They found strong evidence for a decline in the space density beyond $z = 3.3$ by a factor of 6 for the interval $3.5 \leq z < 4.5$ for luminous quasars with $M_C < -25.6$ mag*.

Schmidt, Schneider, & Gunn (1995, hereafter SSG, and references therein) used the Palomar Transit Grism Survey (Schneider, Schmidt, & Gunn 1994) to establish a sample of 90 objects with Lyα emission and redshifts $2.75 < z < 4.75$ to $AB_{1450} < 21.7$ mag in an area of 61.5 deg^2. Their digital survey used CCD detectors, and they determined the completeness and selection effects for their sample based on the line fluxes and signal-to-noise ratio of the data. Their sample contained 8 objects with $z > 4$, and they found a decline in the space density of a factor of 2.7 per unit redshift for quasars with $M_B < -26$ mag and $z > 2.7$. This result was very important because it used a survey technique different from WHO and had many more quasars with $z > 4$.

Kennefick, Djorgovski, & de Carvalho (1995) made use of three colors, J, F, N, in

---

* M_C is the absolute magnitude on the AB system for the continuum level at Lyα (1216Å).

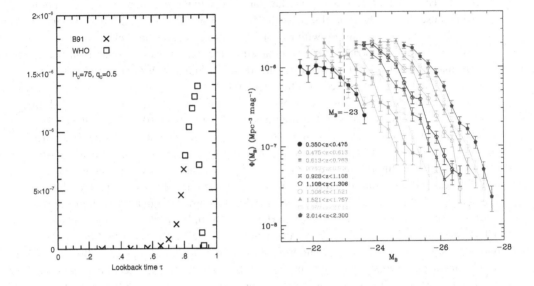

Fig. 20.1. *Left:* A linear plot of the space density of luminous quasars versus look-back time for the BSP and WHO samples. *Right:* The observed luminosity functions for the quasars with $0.35 < z < 2.3$ in the 2dF sample compiled by Boyle et al. (2000).

the second Palomar Sky Survey in a program covering 681 deg^2 in the magnitude range $16.5 < r < 19.6$. They had 10 quasars with $z > 4$ in their sample and found a decline in space density of a factor of 7 at $z = 4.35$ relative to $z = 2.0$.

Taken together, the three surveys agreed well within their respective estimated errors and provided convincing evidence for a steep decline at $z > 3$ in the observed space density of luminous, optically selected quasars. When the results are combined with those of BSP for lower redshifts and plotted on linear scales of space density versus look-back time (Fig. 20.1, *left*), the behavior is dramatic and indicates a remarkable spike of quasar activity when the Universe was 15%–20% of its current age.

At the same time, a number of important questions remained about the nature of the evolution of the quasar luminosity function: (1) At what redshift does the peak of the space density occur? This is a result of the optical SEDs of $z = 3$ quasars being similar to those of stars. (2) How do lower-luminosity quasars and AGNs, which constitute the bulk of the population, evolve? This requires deeper surveys. (3) What is the form of the evolution and its possible dependence on redshift? Hewett, Chaffee, & Foltz (1993) showed from a study of the 1049 quasars and AGNs from the Large Bright Quasar Survey, which cover $0.2 < z < 3$ and $16.5 < m_{B_J} < 18.85$, that the data are not fit well by a pure luminosity evolution model with a two-power law luminosity function. They found that the slope of the luminosity function became steeper at higher redshifts, the rate of evolution was slower for $0.2 < z < 2$ than the Boyle et al. results, and the evolution continued, more slowly, until $z \approx 3$. Thus, more work needs to be done.

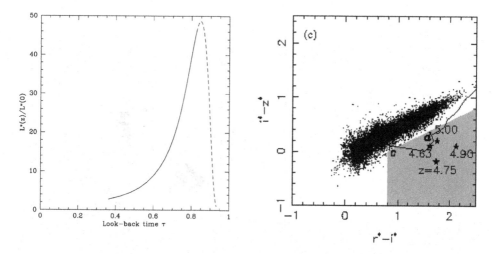

Fig. 20.2. *Left:* A linear plot of the characteristic luminosity of quasars from the 2dF (Boyle et al. 2000) and SSG samples versus look-back time, for the luminosity evolution model. *Right:* A $i^* - z^*$ versus $r^* - i^*$ plot of SDSS data, showing the stellar locus (black area and points) and how quasars with $4.6 < z < 5.0$ separate from the stellar locus because of the presence of Lyα emission in the i^* band and of Lyα forest absorption in the r^* band (Fan et al. 1999).

20.3.2 Recent Large Optical Surveys

The above-mentioned questions on the evolution of quasars provided a significant part of the motivation for two surveys significantly larger than anything previously attempted, the 2dF survey (Boyle et al. 2000, and references therein) and the SDSS (York et al. 2000). The goal of the 2dF was to cover 750 deg^2 of sky to $B < 21$ mag and find $> 25,000$ quasars with $z < 2.3$ via the UVX technique. The SDSS objective was to survey 10,000 deg^2 of sky in five colors, $u'g'r'i'z'$, to find 100,000 quasars covering all redshifts up to 5.8.

Boyle et al. (2000) have estimated the luminosity function from the first 6684 quasars in the 2dF quasar survey. The survey was based on u, b_J, r UK Schmidt plates and is primarily a UVX technique. It was estimated to be 90% complete for $z < 2$. Their final sample included 5057 objects from 196 deg^2 of sky with $M_B < -23$ mag, $18.25 < b_J < 20.85$, $0.35 < z < 2.3$. They combined their sample with 867 objects from the LBQS (Hewett, Foltz, & Chaffee 1995) and fitted a two-power law form to the luminosity function data. The data are shown in Figure 20.1 (*right*). They found that a polynomial evolution of $L^*_{B(z)}$ fits the data well (Fig. 20.2, *left*). Thus, their main conclusion is that for 2dF+LBQS, $-26 < M_B < -23$ mag ($q_0 = 0.5$) and $0.35 < z < 2.3$, pure luminosity evolution works fine, with the characteristic luminosity of quasars increasing by a factor of 40 at the peak of activity. However, the EQS (Edinburgh Quasar Survey, Miller et al., unpublished) and HEQS (Hamburg/ESO Quasar Survey, Köhler et al. 1997) do not fit as well, particularly for $z < 0.5$, where the LBQS has few or no data.

The SDSS opened an important window for the search for high-redshift quasars through its large areal coverage of the sky and its use of the z^* filter with $\lambda_{\text{eff}} \approx 9100$Å, which extended the discovery space in redshift to $z = 5$ and beyond. Figure 20.2 (*right*) (Fan et al.

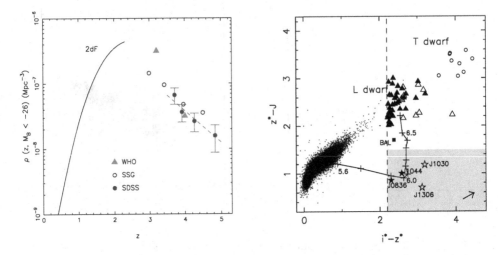

Fig. 20.3. *Left:* The space density of quasars with $M_B < -26$ mag as a function of redshift for the 2dF, SDSS, SSG, and WHO surveys (Fan et al. 2001b). *Right:* The $z^* - J$ versus $i^* - z^*$ diagram for the SDSS, showing how the infrared J band enables the separation of $z > 5.8$ quasars from the numerous L and T dwarfs (Fan et al. 2001c).

1999) illustrates how the r^*, i^*, z^* filters were used to discover the first quasars with $z \approx 5$ by their clear separation from the stellar locus.

Fan et al. (2001a,b) then used the technique to compile a well-defined, color-selected sample of 39 quasars with $3.6 < z < 5.0$ and $i^* \leq 20$ mag in 182 deg^2 of sky. They estimated the luminosity function for objects with $27.5 < M_{1450} < -25.5$ and its evolution with redshift. Their results, which are shown in Figure 20.3 (*left*) agree within the errors at $z \approx 4$ with the previous results of WHO, SSG, and KDC and give a value of the decline in space density of a factor of 3 per unit redshift for $z > 3.6$. They find a flatter slope for the luminosity function at $z \approx 4$ relative to earlier surveys for $z < 3$. These results confirm that pure luminosity evolution does not match the data between redshifts 2 and 5.

Then, Fan et al. (2001c) extended their work to discover four quasars with $z > 5.8$ by using observation in the infrared J band to help eliminate nearby and numerous L and T dwarf stars (Fig. 20.3, *right*). They showed that the space density at $z = 6$ is about a factor of 2 below that at $z = 5$ and follows the decline with redshift just described. These objects provide the strongest evidence yet for a detection of the Gunn-Peterson absorption and thus evidence for reionization at redshifts in the vicinity of $z \approx 6^*$.

To summarize, the 2dF and SDSS results now cover the range $0.3 < z < 6.3$ for high-luminosity, optically selected quasars and reach close to the epoch of reionization. They provide by far the best data on the evolution of such quasars that exist to date.

20.3.3   *Spectral Evolution*

We should also comment on the spectral properties of high-redshift quasars, which look surprisingly like their low-redshift counterparts. The emission-line spectra of the first

---

* Note added in proof. Fan et al. (2003) announced the discovery of three additional quasars with $z > 6$.

quasars discovered at $z \approx 5$ (e.g., Schneider, Schmidt, & Gunn 1991; Fan et al. 1999) show C, N, O, and Si lines in the strengths normally seen at low redshifts. The Dietrich et al. (2002) compilation of spectra covering $0.5 < z < 5$ shows remarkably little evolution with redshift. While it is dangerous to jump to conclusions about abundances based on the appearance of strong emission lines, their lack of evolution is consistent with more detailed analyses. Put another way, we have no evidence for chemical evolution in quasar spectra, except that, if anything, some abundances were *higher* at high redshifts (Hamann & Ferland 1999).

20.3.4 Evolution of Radio Sources

At radio wavelengths, Hook, Shaver, & McMahon (1998), building on the work of Shaver et al. (1996), have carried out an important survey that bears on the question of possible obscuration by dust at high redshifts. Their sample contains 442 radio sources with $S_{2.7GHz} \geq 0.25$ Jy and stellar identifications. The highest redshift object has $z = 4.46$. For objects with radio power $P_{lim} > 7.2 \times 10^{26}$ W Hz^{-1} sr^{-1}, they find an evolution of the space density very similar to WHO, SSG, and the SDSS. This is strong evidence against dust reddening being the main cause of the decline at high redshifts.

At the same time, Webster et al. (1995) and Gregg et al. (2002) have argued that the finding of significant numbers of radio-selected quasars with very red values of $B - K$ in the Parkes and FIRST surveys indicates that up to 80% of the population is being missed in traditional optical surveys because of dust obscuration. This interpretation has been challenged by Benn et al. (1998) and Whiting, Webster, & Francis (2001) on the grounds that the brightness in K can arise from the emission of the host galaxy and/or synchrotron radiation, not from obscuration in the B band by dust. Until this issue can be resolved, the question of dust-obscured quasars remains important to our understanding of quasar evolution, as indicated by the new X-ray results described below. The finding of reddened quasars in the 2MASS survey (Marble et al. 2003, and references therein) is also contributing important information on this subject.

20.4 Estimating Black Hole Masses

Until this point we have discussed the *observed* aspects of the evolution of quasars and AGNs, primarily the evolution of their luminosity functions. Now let us begin to consider their *physical* evolution, for which estimates of the masses of the central black holes are crucial. Such estimates will enable us to map the growth of the black holes with cosmic epoch.

In this meeting we have heard about three ways to estimate black hole masses: (1) from their gravitational influence on the stellar velocity distributions or gas kinematics in the centers of galaxies, (2) reverberation mapping of the broad-line emission region (Barth 2004), and (3) the use of emission-line widths and continuum luminosities (e.g., C IV, Vestergaard 2002, 2004; Mg II, McClure & Jarvis 2004). The first two provide the underpinnings for the mass estimates but are limited to nearby galaxies and AGNs. The third method, while indirect and subject to more uncertainties, has great potential value because it provides the only practical way we have at the moment of estimating the masses of quasars and AGNs at high redshift.

Vestergaard & Osmer (in preparation) are using methods 2 and 3 to make estimates of the mass functions of quasar samples at low (the BQS, Schmidt & Green 1983) and high

(SDSS, Fan et al. 2001a) redshift. Their preliminary results indicate that the SDSS quasars have already achieved masses of $> 10^9 M_\odot$ at $z > 3.6$, and their cumulative mass density is more than a order of magnitude above the BQS sample. This indicates that luminous quasars at high redshift built up their masses early (see also Vestergaard 2004). The BQS cumulative mass density, on the other hand, is an indicator of how the luminous activity has declined rapidly by the present time, when luminous quasars are quite rare. Interestingly, the SDSS cumulative mass densities appear to fit on the extension of the results from the Padovani, Burg, & Edelson (1990) sample of Seyfert galaxies at low redshift. This is consistent with the idea that both the low-redshift Seyferts and low-luminosity AGNs and the high-redshift SDSS quasars have achieved a substantial fraction of their final black hole mass growth.

20.5   Theoretical Considerations: How the Masses Grow

All the recent observational data on quasars and AGNs, in combination with theoretical studies of their evolution and the accretion processes that produce both their luminosity and growth in mass, are now enabling new global studies of their history (e.g., Yu & Tremaine 2002; Yu 2004; Steed, Weinberg, & Miralda-Escudé, in preparation). The goal is to determine how an initial black hole mass function evolves into the one observed today in the local Universe by considering the continuity equation and how the the masses grow with accretion processes. The simple equation $L = \epsilon(\dot{m}/m_{\text{Edd}})Mc^2$, where L is the luminosity produced by an accretion rate $\dot{m}$ in Eddington units with efficiency ϵ for a black hole of mass M, tells us that if we could observationally determine L and ϵ along with black hole masses, for example, we would have enough information to model the evolution of the black holes in galaxies. Put another way, the general goal is to combine the black hole mass function, the time history of accretion, and the distribution of accretion rates and efficiencies to see if we can match the observed luminosity and mass functions for AGNs and black holes. One immediate problem at present is that we do not have a way of separately estimating ϵ and $\dot{m}/m_{\text{Edd}}$; typically people assume that ϵ is 0.1 or some range of values depending on the accretion models they adopt. Another problem is accounting properly for the number of obscured sources in flux-limited samples.

Nonetheless, there are enough existing data to permit interesting progress on the problem. For example, the combination of the black hole mass function for local galaxies and the X-ray background provide integral constraints that must be satisfied by any model. The mass function represents the end point of the accretion processes, while the X-ray background provides a measure of the integrated luminosity produced by accretion over the history of the Universe. The improved optical data on the quasar luminosity functions provide additional constraints on how and when this all occurred, because they map out the evolution of the emitted light with cosmic time. At the same time, the deep X-ray and radio surveys and related optical observations provide crucial information on the contribution of obscured sources to the accretion history of the Universe.

Yu & Tremaine (2002) find that the quasar luminosity functions and local black hole mass functions are consistent if $\epsilon \approx 0.1$ and the black hole mass growth occurred during the optically bright phase. The lifetime of luminous quasars would be of order 10^8 years. At the same time, there remain important questions about the accretion efficiency of lower luminosity quasars and AGNs and its dependence on accretion rate, for example.

20.6   Current Research Programs

20.6.1   Optical/Infrared Surveys

Building on the success of 2dF and SDSS in delineating the evolution of optically selected quasars at high luminosity, a next logical and important observational step is to map the evolution of lower luminosity objects. They constitute the bulk of the AGN population, and they are also crucial for understanding the nature of the extragalactic ionizing background radiation. At high redshift, there is already evidence that the numbers of quasars are too few to account for the observed level of ionization of the intergalactic medium (McDonald & Miralda-Escudé 2001; Schirber & Bullock 2003).

On the observational front, the slope of the luminosity function of high-redshift quasars is quite uncertain. Although Fan et al. (2001b) have made estimates, and the upper limits on the number of AGNs in the HDF also sets constraints (e.g., Conti et al. 1999), direct observations are needed, because it is the slope that determines the number of faint AGNs.

Among the current surveys for fainter quasars at high redshift are the BTC40 (Monier et al. 2002), COMBO-17 (Wolf et al. 2001), and the NOAO DEEP survey (Januzzi & Dey 1999). All are multi-color imaging surveys. In addition, Steidel et al. (2002) are investigating the AGN population found in a deep spectroscopic survey of Lyman-break galaxies. The main properties of these surveys are:

(1) BTC40 covered 40 deg^2 in B, V, I and 36 deg^2 in z and was designed to find quasars with $4.8 < z < 6$. It reached 3σ limiting magnitudes of $V = 24.5$, $I = 22.9$, and $z = 22.9$. To date it has yielded two quasars with redshifts of 4.6 and 4.8 and produced candidates down to $I = 22$ for future spectroscopy on 8–10 m telescopes.

(2) COMBO-17 uses 17 filters covering the $0.37 - 2.2\mu$ wavelength range to achieve in effect low-resolution spectroscopy for a 1 deg^2 area down to $R = 26$ mag. This survey, which will include 50,000 galaxies as well as quasars and AGNs, is very ambitious and promises to yield very significant results when completed.

(3) The NOAO DEEP survey is covering 18 deg^2 of sky in B_W, R, I, J, H, K to optical magnitudes of 26 and near-infrared magnitudes of 21. Because of its broad wavelength coverage and faint limiting magnitudes, it will provide very important data on the evolution of AGNs.

(4) The Steidel et al. survey complements the multi-color ones in that it investigates the spectroscopic properties of galaxies at $z \approx 3$ down to $R_{AB} \approx 25.5$ mag. It has found that about 3% of the galaxies are AGNs, many of which would not have been detected in deep X-ray surveys, and thus is sampling a part of parameter space not covered to date in other work.

An additional value of all these surveys will be to combine their results with those of the deep radio and X-ray surveys mentioned above. This will help define better the luminosity functions and statistics of quasars, AGNs of different types, and normal galaxies and thereby help improve our knowledge of the evolution of all these objects.

20.6.2   Evolution of X-ray Sources

The *Chandra* and *XMM-Newton* observatories are enabling X-ray surveys with sensitivity limits 2 orders of magnitude fainter than was previously possible because of their large collecting areas and exposure times of a million seconds. In the *Chandra* Deep Field South (Rosati et al. 2002), Figure 20.4 shows that the surface density of sources is greater than 3000 deg^{-2}. Thus, for the first time, the surface density of the deepest X-ray selected

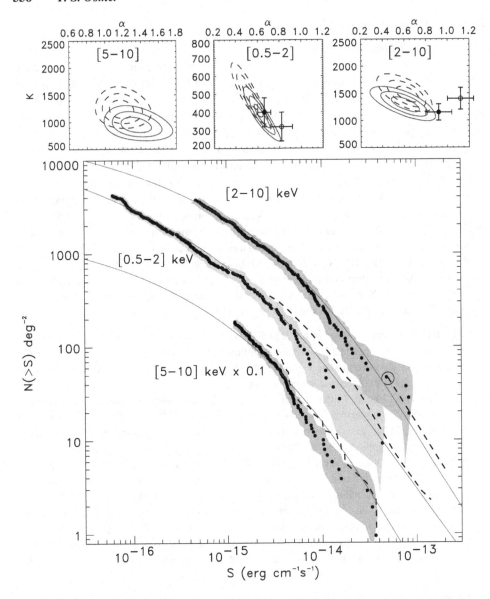

Fig. 20.4. The log *N*–log *S* data for the *Chandra* Deep Field South survey for three different energy bands. The upper boxes show the fits for the normalization constant, κ, and slope, α, of the faint end of the data (Rosati et al. 2002).

AGNs exceeds the values of a few hundred deg^{-2} that were achieved in early deep optical surveys. However, it is also now possible to carry out optical imaging and spectroscopic observations for sources that are 2 orders of magnitude fainter than the nominal limit of SDSS, for example. These capabilities have led both to important discoveries and to the opening up of important areas of research.

For example, the deep source counts show that most of the X-ray background (XRB) can

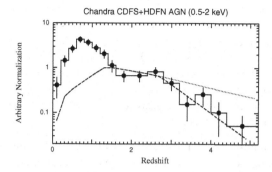
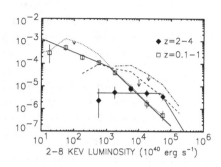

Fig. 20.5. *Left:* The redshift distribution for 243 AGNs in the *Chandra* Deep Field South and Hubble Deep Field North surveys, from Hasinger (2004), compared to population synthesis models by Gilli et al. (2001), where the dashed line is for the redshift decline of SSG for $z > 2.7$ and the dotted line is for a constant space density for $z > 1.5$. Note the observed excess of objects at $z < 1$. *Right:* Luminosity functions derived from *Chandra*, *ROSAT*, and *ASCA* surveys by Cowie et al. (2003), based on their redshift measurements and estimates. Note that the values for sources with $z < 1$ and $L_X < 10^{43}$ erg s^{-1} are well above those for the $2 < z < 4$ objects.

be resolved and accounted for by faint discrete sources. However, to match the SED of the XRB requires the existence of a substantial number of absorbed AGNs at relatively low redshifts (e.g., Gilli, Salvati, & Hasinger 2001, building on much previous work, such as that of Setti & Woltjer 1989).

Interestingly, optical identifications and follow-up spectroscopy are now demonstrating the presence of these sources (e.g., Barger et al. 2001; Hasinger private communication), which have generally escaped notice in previous surveys either because of their faintness or their unremarkable optical appearance and spectra. Cowie et al. (2003) and Hasinger (2004) have assembled large enough samples of objects to show clearly the excess of low-redshift AGNs compared to the expectations of the evolutionary fits found for the optical samples described above (Fig. 20.5). Martini et al. (2002) find an unexpectedly high fraction of X-ray selected AGNs in the cluster Abell 2104, only one of which has the characteristic emission lines of an AGN in its optical spectrum. These studies demonstrate the important power that deep X-ray observations bring to the studies of low-luminosity AGNs.

Thus, we are making good progress on mapping both the total contribution of AGNs to the X-ray background and the evolution with time of their X-ray emission, which in turn will lead to a measure of the accretion history of discrete sources in the Universe. We may look forward to substantial advances in the next decade as the observational data continue to improve and the physics of accretion is better understood.

20.7 Next Steps

In looking ahead, we can see that important next steps in this field include:

- Completing the mapping of the X-ray and optical luminosity functions for quasars and AGNs down to luminosities that include the bulk of the integrated radiation.
- Refining methods for mass determinations and applying them to the full observed range of redshifts and luminosities.

- Finding an observable spectral signature for accretion modes and efficiencies that will allow us to make reliable estimates of accretion rates. Are low-luminosity AGNs a result of low accretion rate, low efficiency, or low mass?
- Determining the numbers of obscured sources and establishing the correlation between, for example, absorption in X-rays and UV/optical obscuration by dust. See if the results are consistent with deep radio and sub-mm observations.
- Achieving a self-consistent fit of the population of observed, discrete X-ray sources with the overall intensity level and SED of the X-ray background.

If, in the end, we can match the observational data for AGNs over their entire redshift range to the local mass function of black holes in galaxies, we will have made a significant leap in our understanding of the coevolution of black holes and galaxies.

Acknowledgements. I am grateful to Eric Monier for assistance with the preparation of this article, especially the figures, and to David Weinberg, Marianne Vestergaard, Brad Peterson, and the anonymous referee for valuable comments on the first drafts. I thank the organizers for the opportunity to speak at the meeting.

References

Barger, A. J., Cowie, L. L., Bautz, M. W., Brandt, W. N., Garmire, G. P., Hornschemeier, A. E., Ivison, R. J., & Owen, F. N. 2001, AJ, 122, 2177

Barth, A. J. 2004, in Carnegie Observatories Astrophysics Series, Vol. 1: Coevolution of Black Holes and Galaxies, ed. L. C. Ho (Cambridge: Cambridge Univ. Press), in press

Becker, R. H, White, R. L., & Helfand, D. J. 1995, ApJ, 450, 559

Benn, C. R., Vigotti, M., Carballo, R., Gonzalez-Serrano, J. I., & Sánchez, S. F. 1998, MNRAS, 495, 451

Boyle, B. J., Shanks, T., Croom, S. M., Smith, R. J., Miller, L., Loaring, N., & Heymans, C. 2000, MNRAS, 317, 1014

Boyle, B. J., Shanks, T., & Peterson, B. A. 1988, MNRAS, 235, 935

Conti, A., Kennefick, J. D., Martini, P., & Osmer, P. S. 1999, AJ, 117, 645

Cowie, L. L., Barger, A. J., Bautz, M. W., Brandt, W. N., & Garmire, G. P. 2003, ApJ, 584, L57

Dietrich, M., Hamann, F., Shields, J. C., Constantin, A., Vestergaard, M., Chaffee, F., Foltz, C. B., & Junkkarinen, V. T. 2002, ApJ, 581, 912

Elvis, M., Maccacaro, T., Wilson, A. S., Ward, M. J., Penston, M. V., & Fosbury, R. A. E. 1978, MNRAS, 183, 129

Fan, X. et al. 1999, AJ, 118, 1

——. 2001a, AJ, 121, 31

——. 2001b, AJ, 121, 54

——. 2001c, AJ, 122, 2833

——. 2003, AJ, 125, 1649

Gregg, M. D., Lacy, M., White, R. L., Glikman, E., Helfand, D. J., Becker, R. H., & Brotherton, M. S. 2002, ApJ, 564, 133

Gilli, R., Salvati, M., & Hasinger, G. 2001, A&A, 366, 407

Hamann, F., & Ferland, G. 1999, ARA&A, 37, 487

Hao, L., & Strauss, M. A. 2004, in Carnegie Observatories Astrophysics Series, Vol. 1: Coevolution of Black Holes and Galaxies, ed. L. C. Ho (Pasadena: Carnegie Observatories, http://www.ociw.edu/ociw/symposia/series/symposium1/proceedings.html)

Hasinger, G. 2004, in IAU Symp. 214, High Energy Processes and Phenomena in Astrophysics, ed. X. Li, Z. Wang, & V. Trimble (San Francisco: ASP), in press (astro-ph/0301040)

Hazard, C., Mackey, M. B., & Shimmins, A. J. 1963, Nature, 197, 1037

Heckman, T. 2004, in Carnegie Observatories Astrophysics Series, Vol. 1: Coevolution of Black Holes and Galaxies, ed. L. C. Ho (Cambridge: Cambridge Univ. Press), in press

Heisler, J., & Ostriker, J. P. 1988, ApJ, 332, 543

Hewett, P. C., Foltz, C. B., & Chaffee, F. H. 1993, ApJ, 406, L43

——. 1995, AJ, 109, 1498

Hewitt, A., & Burbidge, G. 1980, ApJS, 43, 57

Ho, L. C. 2003, in IAU Colloq. 184, AGN Surveys, ed. R. F. Green, E. Ye. Khachikian, & D. B. Sanders (San Francisco: ASP), 13

Ho, L. C., Filippenko, A. V., & Sargent, W. L. W. 1995, ApJS, 98, 477

Hoag, A. A. 1976, PASP, 88, 860

Hook, I. M., Shaver, P., & McMahon, R. G. 1998, in The Young Universe: Galaxy Formation and Evolution at Intermediate and High Redshift, ed. S. D'Odorico, A. Fontana, & E. Giallongo (San Francisco: ASP), 17

Jannuzi, B. & Dey, A. 1999, in The Hy-Redshift Universe: Galaxy Formation and Evolution at High Redshift, ed. A. J. Bunker & W. J. M. van Breugel (San Francisco: ASP), 258

Kennefick, J. D., Djorgovski, S. G., & de Carvalho, R. R. 1995, AJ, 110, 2553

Köhler, T., Groote, D., Reimers, D., & Wisotzki, L. 1997, A&A, 325, 502

Lynds, R., & Petrosian, V. 1972, ApJ, 175, 591

Marble, A. R., Hines, D. C., Schmidt, G. D., Smith, P. S., Surace, J. A., Armus, L., Cutri, R. C., & Nelson, B. O. 2003, ApJ, 590, 707

Martini, P., Kelson, D. D., Mulchaey, J. S., & Trager, S. C. 2002, ApJ, 576, L109

Mathez, G. 1976, A&A, 53, 15

——. 1978, A&A, 68, 17

McClure, R. J., & Jarvis, M. J. 2004, in Carnegie Observatories Astrophysics Series, Vol. 1: Coevolution of Black Holes and Galaxies, ed. L. C. Ho (Pasadena: Carnegie Observatories, http://www.ociw.edu/symposia/series/symposuium1/proceedings.html)

McDonald, P., & Miralda-Escudé, J. 2001, ApJ, 549, L11

Monier, E. M., Kennefick, J. D., Hall, P. B., Osmer, P. S., Smith, M. G., Dalton, G. B., & Green, R. F. 2002, AJ, 124, 2971

Osmer, P. S. 1982, ApJ, 253, 28

Osmer, P. S., & Smith, M. G. 1976, ApJ, 210, 267

Padovani, P., Burg. R., & Edelson, R. A. 1990, ApJ, 353, 438

Rosati, P., et al. 2002, ApJ, 566, 667

Sandage, A. 1965, ApJ, 141, 328

Schirber, M., & Bullock, J. S. 2003, ApJ, 584, 110

Schmidt, M. 1963, Nature, 197, 1040

——. 1968, ApJ, 151, 393

——. 1970, ApJ, 162, 371

Schmidt, M., & Green, R. F. 1983, ApJ, 269, 352

Schmidt, M., Schneider, D. P., & Gunn, J. E. 1986, ApJ, 306, 411

——. 1995. AJ, 110, 68

Schneider, D. P., Schmidt, M., & Gunn, J. E. 1991, AJ, 102, 837

——. 1994, AJ, 107, 1245

Setti, G., & Woltjer, L. 1989, A&A, 224, L21

Seyfert, C. K. 1943, ApJ, 97, 28

Shaver, P. A., Wall, J. V., Kellermann, K. I., Jackson, C. A., & Hawkins, M. R. S. 1996, Nature, 384, 439

Smith, M. 1975, ApJ, 202, 591

Steidel, C. C., Hunt, M. P., Shapley, A. E., Adelberger, K. L., Pettini, M., Dickinson, M., & Giavalisco, M. 2002, ApJ, 576, 653

Vestergaard M., 2002, ApJ, 571, 733

——. 2004, in Carnegie Observatories Astrophysics Series, Vol. 1: Coevolution of Black Holes and Galaxies, ed. L. C. Ho (Pasadena: Carnegie Observatories, http://www.ociw.edu/symposia/series/symposuium1/proceedings.html)

Warren, S. J., Hewett, P. C., Irwin, M. J., McMahon, R. G., & Bridgeland, M. T. 1987, Nature, 325, 131

Warren, S. J., Hewett, P. C., & Osmer, P. S. 1991a, ApJS, 76, 1

——. 1991b, ApJS, 76, 23

——. 1994, ApJ, 421, 412

Webster, R. L., Francis, P. J., Peterson, B. A., Drinkwater, M. J., & Masci, F. J. 1995, Nature, 375, 469

Whiting, M. T., Webster, R. L., & Francis, P. J. 2001, MNRAS, 323, 718

Wolf, C., Borch, A., Meisenheimer, K., Rix, H.-W., Kleinheinrich, M., & Dye, S. 2001, Astronomische Gesellschaft Abstract Series, Vol. 18., abstract MS 05 39

York, D., et al. 2000, AJ, 120, 1579

Yu, Q. 2004, in Carnegie Observatories Astrophysics Series, Vol. 1: Coevolution of Black Holes and Galaxies, ed.

L. C. Ho (Pasadena: Carnegie Observatories,
http://www.ociw.edu/symposia/series/symposuium1/proceedings.html)
Yu, Q., & Tremaine, S. 2002, MNRAS, 335, 965

21

Quasar hosts and the black hole-spheroid connection

JAMES S. DUNLOP

Institute for Astronomy, University of Edinburgh, Royal Observatory

Abstract

I review our current understanding of the structures and ages of the host galaxies of quasars, and the masses of their central black holes. At low redshift, due largely to the impact of the *Hubble Space Telescope*, there is now compelling evidence that the hosts of quasars with $M_V < -24$ mag are virtually all massive ellipticals, with basic properties indistinguishable from those displayed by their quiescent counterparts. The masses of these spheroids are as expected given the relationship between black hole and spheroid mass now established for nearby galaxies, as is the growing prevalence of significant disk components in the hosts of progressively fainter active nuclei. In fact, from spectroscopic measurements of the velocity of the broad-line region in quasars, it has now proved possible to obtain an independent dynamical estimate of the masses of the black holes that power quasars. I summarize recent results from this work, which can be used to demonstrate that the black hole-spheroid mass ratio in quasars is the same as that found for quiescent galaxies, namely $M_\bullet = 0.0012 M_{sph}$. These results offer the exciting prospect of using observations of quasars and their hosts to extend the study of the black hole-spheroid mass ratio out to very high redshifts ($z > 2$). Moreover, there is now good evidence that certain ultraviolet quasar emission lines can provide robust estimates of black hole masses from the observed optical spectra of quasars out to $z > 2$, and perhaps even at $z > 4$. By combining such information with deep, high-resolution infrared imaging of high-redshift quasar hosts on 8-m class telescopes, there is now a real prospect of clarifying the evolution of the black hole spheroid connection over cosmological time scales.

21.1   Introduction

With the discovery that all spheroids (i.e., elliptical galaxies and disk galaxy bulges) appear to house a massive black hole of proportionate mass (Magorrian et al. 1998; Gebhardt et al. 2000a; Merritt & Ferrarese 2001), the nature and evolution of quasar host galaxies has grown from a subject explored primarily by AGN researchers, into an area of interest for all astronomers concerned with the formation and evolution of galaxies and of compact objects.

Indeed, black hole and spheroid formation/growth are now recognized as potentially intimately related processes (Silk & Rees 1998; Fabian 1999; Granato et al. 2001; Archibald et al. 2002), with the evolution of quasar host galaxies as a function of redshift now seen as a key measurement in observational cosmology (e.g., Kauffmann & Haehnelt 2000). In this review I have chosen to focus on what can be learned about the nature of the black hole spheroid connection from observations of quasars and their hosts. I have

© The Observatories of the Carnegie Institution of Washington 2004.

therefore deliberately avoided detailed discussion of many other topics of interest related to quasar host galaxy research, such as the triggering of quasar activity, the origin of radio loudness, and the nature of possible links between quasars and ultraluminous infrared galaxies (ULIRGs).

This Chapter is divided into three sections as follows. First I summarize what has been learned about the host galaxies of low-redshift quasars from deep imaging/spectroscopy over the last decade. Second I discuss the latest dynamical estimates of the masses of the black holes that power quasars. Third I consider the immediate future prospects for extending these two prongs of measurement to higher redshift, to explore the nature of the black hole spheroid connection as a function of cosmological time.

Unless otherwise stated, an Einstein-de Sitter Universe with $H_0 = 50\,\mathrm{km\,s^{-1}Mpc^{-1}}$ has been assumed for the calculation of physical quantities.

21.2   The Host Galaxies of Low-redshift Quasars

Many imaging studies and several spectroscopic studies of "nearby" ($z < 0.3$) quasar host galaxies have been attempted over the last quarter of a century, but it is only in the last decade that a clear picture of the nature of quasar hosts has emerged from this work. This progress can be attributed first to the advent of deep near-infrared imaging (Dunlop et al. 1993; McLeod & Rieke 1994; Taylor et al. 1996), and second to the high angular resolution provided by the refurbished *Hubble Space Telescope (HST)* (e.g., Disney et al. 1995; Bahcall et al. 1997; Hooper, Impey, & Foltz 1997; McLure et al. 1999).

Some workers have chosen to focus on some of the morphological peculiarities and evidence of "action" revealed by this deep imaging, such as tidal tails, and nearby companions (perhaps responsible for triggering the nuclear activity). However, the clearest results, and most meaningful insights have emerged from studies that have focused on determining the properties of the mass-dominant stellar populations in quasar hosts, and exploring how these compare with those of quiescent galaxies.

Figure 21.1 provides an example of how clearly the basic structure of low-redshift quasar host galaxies can be discerned with an exposure of ~ 1 hour on the *HST*. This image (taken from Dunlop et al. 2003) demonstrates not only that the host galaxy is well resolved, but also the extent to which the vast majority of the optical light from the host can generally be attributed to a simple, symmetric, "normal" galaxy (in this case an elliptical, with an $r^{1/4}$ de Vaucouleurs luminosity profile, and a half-light radius of 7.5 kpc).

For simplicity I have chosen to center the following summary of what has been learned from such images around the main results from our own, recently completed, *HST* imaging study of the hosts of radio-loud quasars (RLQs), radio-quiet quasars (RQQs) and radio galaxies (RGs) at $z \simeq 0.2$ (Dunlop et al. 2003). However, wherever appropriate, I have also endeavored to discuss (and if possible explain) the extent to which other authors do or do not agree with our findings.

21.2.1   *Host Galaxy Luminosity, Morphology, and Size*

After some initial confusion (e.g., Bahcall, Kirhakos, & Schneider 1994), recent *HST*-based studies have now reached agreement that the hosts of all luminous quasars ($M_V < -23.5$ mag) are bright galaxies with $L > L^*$ (McLure et al. 1999; McLeod & McLeod 2001; Dunlop et al. 2003). This result is illustrated by Figure 21.2 (left panel), taken from

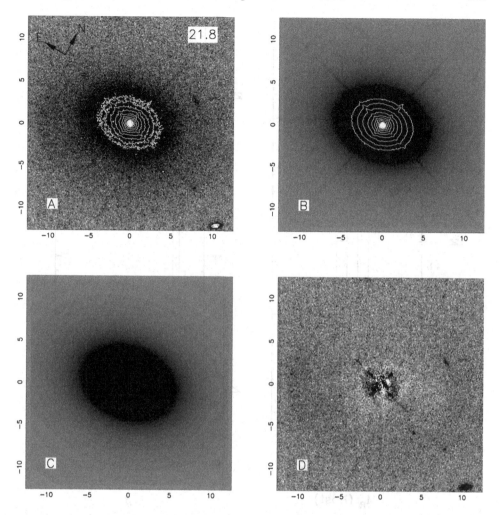

Fig. 21.1. An example of deep *HST* imaging of the host galaxy of a low-redshift quasar. A greyscale/contour representation of an *R*-band image of the $z = 0.1$ RQQ 0204+292 obtained with the WFPC2 is shown in the upper-left panel (the image is $25'' \times 25''$ in size). The upper-right panel shows the best-fitting model of this image (after convolution with the *HST* point-spread function), which comprises a de Vaucouleurs law elliptical galaxy (of half-light radius $r_{1/2} = 7.5$ kpc) along with an unresolved nuclear component. The lower-left image shows the best-fitting host galaxy as it would appear if the nucleus were inactive, while the lower-right panel is the residual image that results from subtraction of the complete model from the image. Further details of the modeling procedure used can be found in McLure, Dunlop, & Kukula (2000) and Dunlop et al. (2003).

Dunlop et al. (2003). However, it can be argued, with justification, that this much had already been established from earlier ground-based studies (e.g., Taylor et al. 1996).

In fact the major advance offered by the *HST* for the study of quasar hosts is that it has enabled host luminosity profiles to be measured over sufficient angular and dynamic range

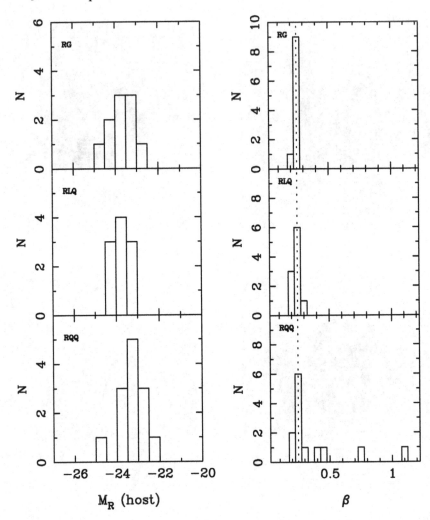

Fig. 21.2. *Left:* Histograms of host galaxy integrated *R*-band absolute magnitudes for the RG, RLQ, and RQQ subsamples imaged with the *HST* by Dunlop et al. (2003). For comparison, the integrated *R*-band absolute magnitude of an L^* galaxy is $M_R = -22.2$ mag. *Right:* Histograms of the best-fit values of β, where host galaxy surface brightness is proportional to $\exp(-r^\beta)$, for the same three subsamples. The dotted line at $\beta = 0.25$ indicates a perfect de Vaucouleurs law, and all of the radio-loud hosts and all but three of the radio-quiet hosts are consistent with this to within the errors. Two of the three RQQs with hosts for which $\beta > 0.4$ transpire to be the two least luminous nuclei in the sample, and should be reclassified as Seyferts.

to allow a de Vaucouleurs $r^{1/4}$-law spheroidal component to be clearly distinguished from an exponential disk, at least for redshifts $z < 0.5$.

In our own study this is the reason that we have been able to establish unambiguously that, at low z, the hosts of both RLQs *and* RQQs are undoubtedly massive ellipticals with (except for one RQQ in our sample) negligible disk components (McLure et al. 1999; Dunlop et al. 2003). This result is illustrated in the right panel of Figure 21.2. This figure confirms that the hosts of RQQs and RGs all follow essentially perfect de Vaucouleurs profiles, in good agreement with the results of other studies. Perhaps the more surprising aspect is the extent to which the RQQ sample is also dominated by spheroidal hosts. At first sight this might seem at odds with the results of some other recent studies, such as those of Bahcall et al. (1997) and Hamilton, Casertano, & Turnshek (2002), who report that approximately one-third to one-half of RQQ lie in disk-dominated hosts. However, on closer examination, it becomes clear that there is no real contradiction provided one compares quasars of similar power. Specifically, if attention is confined to quasars with nuclear magnitudes $M_V < -23.5$ mag we find that 10 out of the 11 RQQs in our sample lie in ellipticals, 6 out of 7 similarly luminous quasars in the sample of Bahcall et al. lie in ellipticals, and at least 17 out of the 20 comparably luminous RQQs in Hamilton et al.'s archival sample also appear to lie in spheroidal hosts.

It is thus now clear that above a given luminosity threshold we enter a regime in which AGNs can only be hosted by massive spheroids, regardless of radio power (a result confirmed by the recent *HST* study of the most luminous low-redshift quasars by Floyd et al. 2004). It is also clear that, within the radio-quiet population, significant disk components become more common at lower nuclear luminosities. This dependence of host galaxy morphology on nuclear luminosity is nicely demonstrated by combining our own results with those of Schade, Boyle, & Letawsky (2000) who have studied the host galaxies of lower-luminosity X-ray selected AGNs. This is shown in Figure 21.3 where the ratio of bulge to total host luminosity is plotted as a function of nuclear optical power. Figure 21.3 is at least qualitatively as expected if black hole mass is proportional to spheroid mass, and black hole masses $> 2 \times 10^8 M_\odot$ are required to produce quasars with $M_R < -23.5$ mag.

In our *HST* study we have also been able to break the well-known degeneracy between host galaxy surface brightness and size. This point is illustrated by the fact that we have, for the first time, been able to demonstrate that the hosts of RLQs and RQQs follow Kormendy's (1977) relation (i.e., the photometric projection of the fundamental plane; Fig. 21.4). Moreover the slope (2.90 ± 0.2) and normalization of this relation are identical to that displayed by normal quiescent massive ellipticals.

21.2.2 Host Galaxy Ages

It is well known from simulations that the merger of two disk galaxies can produce a remnant that displays a luminosity profile not dissimilar to a de Vaucouleurs $r^{1/4}$ law. This raises the possibility that the apparently spheroidal nature of the quasar hosts discussed above might be the result of a recent major merger that could also be responsible for stimulating the onset of nuclear activity. This would also be the natural prediction of suggested evolutionary schemes in which ULIRGs are presumed to be the precursors of RQQs. Could a recent merger of two massive, gas-rich disks be simultaneously responsible for the triggering of nuclear activity and the production of an apparently spheroidal host?

The answer, at least at low redshift, appears to be no. First, as mentioned above, the Kormendy relation displayed by quasar hosts appears to be indistinguishable from that of quiescent, well-evolved massive ellipticals. Moreover, as discussed by Genzel et al. (2001) and Dunlop et al. (2003), ULIRGs generally lie in a different region of the fundamental plane to quasar hosts, with the former apparently destined to evolve into lower or intermediate-mass spheroidal galaxies.

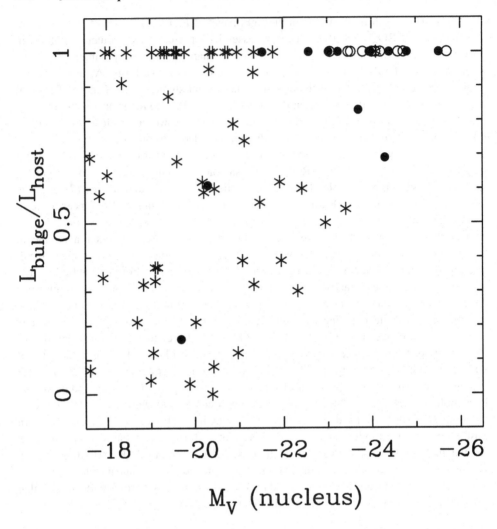

Fig. 21.3. The relative contribution of the spheroidal component to the total luminosity of the host galaxy plotted against the absolute V-band magnitude of the nuclear component. The plot shows the results from Dunlop et al. (2003) (RLQs as open circles, RQQs as filled circles) along with the results from Schade et al. (2000) for a larger sample of X-ray selected AGNs spanning a wider but lower range of optical luminosities (asterisks). This plot illustrates very clearly how disk-dominated host galaxies become increasingly rare with increasing nuclear power, as is expected if more luminous AGNs are powered by more massive black holes, which, in turn, are housed in more massive spheroids.

Secondly, direct attempts to determine the ages of the dominant stellar populations in the quasar hosts provide little evidence of recent, widespread star formation activity. Within the Dunlop et al. sample we have attempted to estimate the ages of the host galaxies both from optical-infrared colors and from deep optical off-nuclear spectroscopy (Nolan et al. 2001). The results of this investigation are that the hosts of both radio-loud and radio-quiet quasars are dominated by old, well-evolved stellar populations (with typically less than 1% of stellar

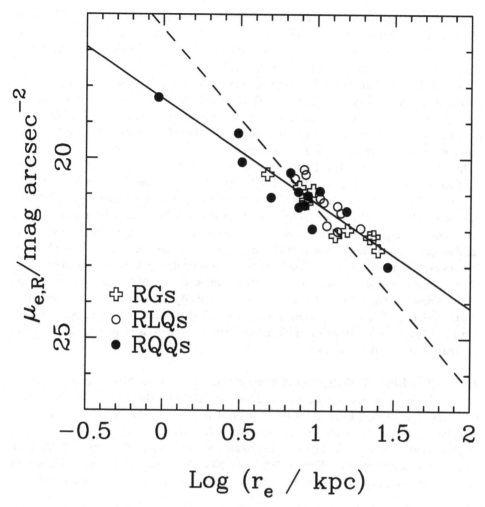

Fig. 21.4. The Kormendy (1977) relation followed by the hosts of all 33 powerful AGNs studied by Dunlop et al. (2003) with the *HST*. The solid line is the least-squares fit to the data that has a slope of 2.90 ± 0.2, in excellent agreement with the slope of 2.95 found by Kormendy for inactive ellipticals. For the few RQQs that have a disk component the best-fitting bulge component has been plotted. The dashed line has a slope of 5, indicative of the "pseudo Kormendy relation" expected if the scale lengths of the host galaxies had not been properly constrained (see Dunlop et al. 2003). (Adapted from Dunlop et al. 2003.)

mass involved in recent star formation activity). There are currently no comparably extensive studies of host galaxy stellar populations with which this result can be compared. However, Canalizo & Stockton (2000) have published results from a more detailed spectroscopic study of three objects, one of which, Mrk 1014, is also in the Dunlop et al. RQQ sample. This is in fact the only quasar host in the Dunlop et al. sample for which Nolan et al. (2001) found clear spectroscopic evidence of A-star features and a significant (albeit still only $\simeq 2\%$ by mass) young stellar population. It is presumably no coincidence that this is also the only quasar in the Dunlop et al. sample that was detected by *IRAS*, and the only host that displays spectacular tidal tail features comparable to those commonly found in images of ULIRGs. However, even for this apparently star-forming quasar host, Canalizo & Stockton agree that $\simeq 95\%$ of the host is dominated by an old, well-evolved stellar population (although they argue that 5%–8% of the galaxy has been involved in recent star formation).

Finally, despite claims to the contrary, recent measurements of molecular gas in AGN host galaxies reported by Scoville et al. (2003) are completely consistent with this picture. Scoville et al. detected substantial molecular gas masses in the hosts of lower luminosity quasars with known substantial disk components, and failed to detect molecular gas in the hosts of the three most luminous quasars in their sample.

In summary, the available evidence indicates that the hosts of quasars with $M_V < -23.5$ mag are virtually all massive elliptical galaxies. Moreover, quasar hosts appear to be "normal" ellipticals in the sense that their basic structural properties, and the ages of their dominant stellar populations are, at least to first order, indistinguishable from those of their quiescent counterparts. Both the universality of elliptical hosts for the most luminous low-redshift quasars and the growing prevalence of significant disk components in the hosts of progressively fainter active nuclei can be viewed as a natural reflection of the proportionality of black hole and spheroid mass now established for nearby quiescent galaxies. In the next section I describe the results of recent attempts to obtain dynamical estimates of the masses of the black holes that power quasars. Such studies allow a direct test of whether or not the constant of proportionality between black hole and spheroid mass is the same in the active and inactive galaxy populations.

21.3   The Black Hole-Spheroid Mass Ratio in Low-redshift Quasars

If one assumes that quasars emit at the Eddington limit, it is straightforward to obtain a very rough estimate of the masses of their central black holes. However, a potentially much more reliable estimate of black hole mass can be obtained via an analysis of the velocity widths of the Hβ lines in quasar nuclear spectra. This has been a growth industry in recent years (e.g., Wandel 1999; Laor 2000), bolstered by estimates of the size of the broad-line region (BLR) from reverberation mapping of low-redshift, broad-line AGNs.

21.3.1   *The Virial Black Hole Mass Estimator*

The underlying assumption behind the virial black hole mass estimator is that the motion of the broad-line emitting material in AGNs is virialized. Under this assumption the width of the broad lines can be used to trace the Keplerian velocity of the broad-line gas, and thereby allow an estimate of the central black hole mass via the formula $M_\bullet = G^{-1} R_{\mathrm{BLR}} V_{\mathrm{BLR}}^2$, where R_{BLR} is the BLR radius and V_{BLR} is the Keplerian velocity of the BLR gas. Currently, the most direct measurements of the central black hole masses of powerful AGNs are for 17 Seyferts and 17 PG quasars for which reverberation mapping has provided a direct measurement of R_{BLR} (Wandel, Peterson & Malkan 1999; Kaspi et al. 2000).

An important outcome from these studies is the discovery of a correlation between R_{BLR} and the monochromatic AGN continuum luminosity at 5100 Å (e.g., $R_{\mathrm{BLR}} \propto \lambda L_{5100}^{0.7}$; Kaspi et al. 2000). By combining this luminosity-based R_{BLR} estimate with a measure of the BLR velocity based on the FWHM of the Hβ emission line, it is now possible to produce a virial black hole mass estimate from a single spectrum covering Hβ. This technique has recently been widely employed to investigate how the masses of quasar black holes relate to the

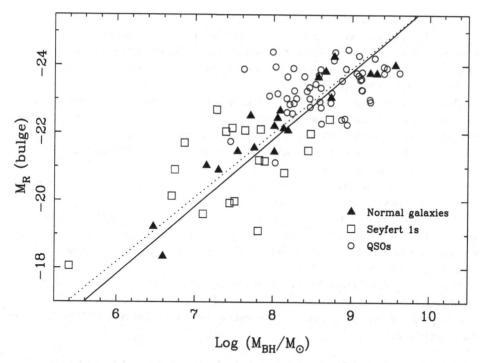

Fig. 21.5. Absolute *R*-band bulge magnitude versus black hole mass plotted for 72 AGNs and 18 inactive elliptical galaxies. The black hole masses for the 72 AGNs are derived from their Hβ line widths using a disklike BLR model (see McLure & Dunlop 2002). The black hole masses of the inactive galaxies (triangles) are dynamical estimates as compiled by Kormendy & Gebhardt (2001). Also shown is the formal best fit (solid line) and the best-fitting linear relation between spheroid and black hole mass (dotted line). (Adapted from McLure & Dunlop 2002.)

properties of the surrounding host galaxies (e.g., Laor 2001; McLure & Dunlop 2001,2002) and the radio luminosity of the central engine (Lacy et al. 2001; Dunlop et al. 2003).

Of most direct relevance to the topic of interest in this review is the result shown in Figure 21.5. This shows how the relationship between host galaxy luminosity and black hole mass derived for quasars and Seyferts compares with that derived for normal galaxies (McLure & Dunlop 2002).

Under the assumption that $M_\bullet \propto M_{sph}$ the best-fitting constant of proportionality derived from the fit to the quasar and Seyfert data points in Figure 21.5 is 0.0012. This is essentially identical to the value (0.0013) for nearby inactive galaxies derived by Kormendy & Gebhardt (2001), and to the value (0.0012) derived by Merritt & Ferrarese (2001). While the virtually exact agreement between these numbers may be fortuitous, the similarity of the mass relationships derived for the active and inactive samples can be fairly viewed as providing confirmation both that the $M_\bullet - M_{sph}$ relation is the same in active and inactive galaxies, and that the assumption of gravitational equilibrium made in applying the Hβ virial mass estimator is valid (see also Gebhardt et al. 2000b).

21.3.2   *Eddington Ratios*

Having confirmed the constant of proportionality between host spheroid and black hole mass for quasars one can then re-address the issue of how the actual nuclear luminosities of quasars compare with their predicted Eddington-limited values (as inferred from the luminosities of their host galaxies).

This is illustrated in Figure 21.6, in which host galaxy absolute *V*-band magnitude is plotted against quasar nuclear absolute magnitude for an expanded sample of quasars assembled from five recent studies [see figure caption and Floyd et al. (2004) for details]. Also shown in this plot are the predicted relations for black holes emitting at 100%, 10% and 1% of the Eddington limit.

This plot provides (perhaps surprisingly good) evidence that, at any given host luminosity, the most luminous quasar nuclei are emitting at the predicted Eddington limit (as calculated on the basis of a black hole mass inferred from host luminosity using the relation shown in Fig. 21.5). It also shows that the majority of low-redshift quasars studied to date are emitting at between 10% and 100% of the Eddington limit, and that their host galaxies range in luminosity from L^* to $10\,L^*$.

In concluding this section, I note that on the basis of this plot it would be predicted that the most luminous quasars found in the high-redshift Universe, with $M_V < -27$ mag, can only be produced by the black holes at the centers of the most massive ($10L^*$) ellipticals, or their progenitors.

21.4   Cosmological Evolution of the Black Hole-Spheroid Mass Ratio

The two main results presented in the last two sections can be summarized as follows. First, it is clear that the host galaxies of low-redshift quasars are normal massive ellipticals. Second, it appears that by combining deep host galaxy imaging with the spectroscopic Hβ virial black hole mass estimator, low-redshift quasars can be used to provide an unbiased estimate of the black hole-spheroid mass ratio in the present-day inactive elliptical galaxy population.

These two results provide confidence that, through the study of quasars at higher redshifts, we can establish the cosmological evolution of the black hole-spheroid mass ratio in the general elliptical galaxy population. This is important for two reasons. First, from a purely practical point of view, we are forced to study quasars to explore the redshift evolution of this mass relationship. This is simply because a virial mass estimator based on bright, observable, emission lines offers the only realistic method by which to measure black hole masses in high-redshift objects. Second, it can certainly be argued that, to all intents and purposes, the high-redshift elliptical galaxy population *is* the high-redshift quasar population. Whereas only 1 in $10^4 - 10^5$ present-day ellipticals is active, Figure 21.6 coupled with a comparison of the present-day elliptical and high-redshift quasar luminosity functions leads to the conclusion that at least 10% of the progenitors of present-day massive ellipticals were active quasars at $z \simeq 2.5$.

So, to explore the cosmological evolution of the black hole-spheroid mass ratio in massive galaxies, we require a version of the Hβ virial mass estimator that can be applied to high-redshift quasars, coupled with a means to estimate the masses of high-redshift quasar hosts. Below I consider the current status of these two observational challenges, starting with the problem of black hole mass estimation at high redshift.

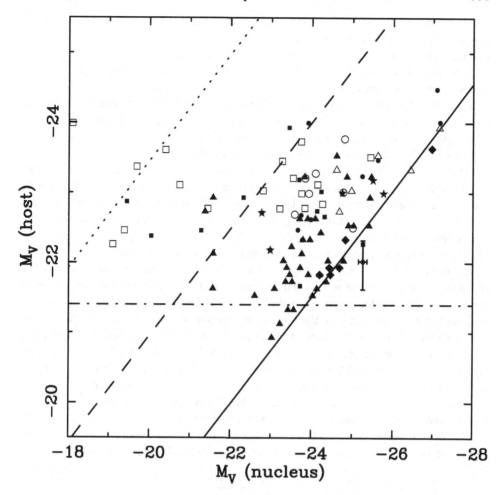

Fig. 21.6. Host absolute magnitude plotted versus nuclear absolute magnitude for the quasars studied by Floyd et al. (2004; circles), Dunlop et al. (2003; squares), McLeod & Rieke (1994; triangles), McLeod & McLeod (2001; diamonds), and also for five objects reimaged with *HST* by Percival et al. (2001; stars). The solid line illustrates the predicted limiting relation on the assumption of Eddington-limited accretion, with the dashed and dotted lines denoting 10% and 1% of the Eddington limit, respectively. The one object in this combined sample that appears to be more luminous than the Eddington limit is the luminous quasar 1252+020. However, as indicated by the large error bars, this is also the object for which Floyd et al. (2004) have least confidence in the robustness with which host and nuclear luminosity have been separated (see Floyd et al. 2004 for further details). (Adapted from Floyd et al. 2004).

21.4.1   *Black Hole Mass Measurement in High-redshift Quasars*

The well-studied Hβ emission line is observable from the ground out to $z \simeq$ 3. However, because it is redshifted into the near-infrared at a redshift of $z \sim 1$, it is observationally expensive to use Hβ to estimate the black hole masses of $z > 1$ quasars. Consequently, a concerted effort has recently been invested to establish whether or not any of the ultraviolet (UV) emission lines, so prominent in the observed optical spectra of high-redshift quasars, can be exploited and trusted to yield a comparably accurate and unbiased estimate of black hole mass.

Two studies have recently been published that provide evidence that this can indeed be achieved. First, Vestergaard (2002) has proposed and calibrated a UV black hole mass estimator based on the FWHM of the C IV emission line ($\lambda = 1549$Å) and the continuum luminosity at 1350Å. Second, McLure & Jarvis (2002) have proposed and confirmed the robustness of a UV black hole mass estimator based on the FWHM of the Mg II emission line ($\lambda = 2799$Å) and the continuum luminosity at 3000Å.

In terms of accessible redshift range, these two proposed mass estimators are reasonably complementary, and in the near future it will be interesting to see how well they can be bootstrapped together to explore the black hole-spheroid mass ratio over a broad baseline in redshift. However, at present it is probably fair to say that while the C IV-based estimator in principle allows black hole mass estimation from optical spectroscopy out to $z \simeq 5$, the Mg II-based estimator appears to be more robust and is better understood.

The main reason for adopting Mg II as the UV tracer of BLR velocity is that, like Hβ, Mg II is a low-ionization line. Furthermore, due to the similarity of their ionization potentials, it is reasonable to expect that the Mg II and Hβ emission lines are produced by gas at virtually the same radius from the central ionizing source. Although care has to be taken in dealing with Fe II contamination in the vicinity of the Mg II line, this presumption has now been directly tested and confirmed by McLure & Jarvis (2002) through a comparison of Mg II FWHM and Hβ FWHM for a sample of 22 objects with reverberation mapping results for which it also proved possible to obtain Mg II FWHM measurements.

Building on this result, McLure & Jarvis (2002) have produced a calibrated, reliable, Mg II virial black hole mass estimator that can be applied over the redshift range $0.3 < z < 2.5$ from straightforward optical spectroscopy. In terms of a useful formula the final calibration of this UV black hole mass estimator is given by McLure & Jarvis as:

$$\frac{M_\bullet}{M_\odot} = 3.37 \left(\frac{\lambda L_{3000}}{10^{37}\,\mathrm{W}} \right)^{0.47} \left(\frac{\mathrm{FWHM(Mg\ II)}}{\mathrm{km\,s^{-1}}} \right)^2$$

The robustness of this new black hole mass estimator is illustrated in Figure 21.7. The left-hand panel shows a comparison of the results derived from the established optical (Hβ) black hole mass estimator plotted against the results from this new UV (Mg II) black hole mass estimator for a combined sample of 150 objects [see McLure & Jarvis (2002) for sample details]. Also shown is the BCES bisector fit to the data, which has the form

$$\log M_\bullet(\mathrm{H}\beta) = 1.00(\pm 0.08)\log M_\bullet(\mathrm{Mg\ II}) + 0.06(\pm 0.67),$$

perfectly consistent with a linear relation. The right-hand panel shows a histogram of $\log M_\bullet(\mathrm{Mg\ II}) - \log M_\bullet(\mathrm{H}\beta)$ for this quasar sample. The solid line shows the best-fitting Gaussian, which has $\sigma = 0.41$. These results lead McLure & Jarvis to conclude that, compared to the traditional optical black hole mass estimator, the new UV estimator provides results which are unbiased and of equal accuracy.

In concluding this subsection I note that an exciting demonstration of how, through near-infrared spectroscopy, this estimator can be applied to estimate the black hole masses of the most distant known quasars at $z > 6$ has recently been provided by Willott, McLure, & Jarvis (2003).

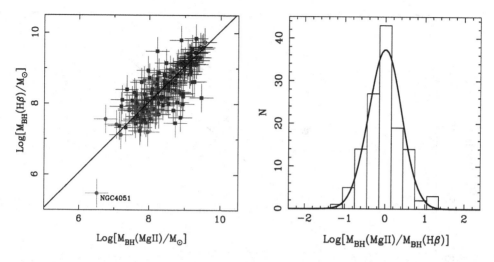

Fig. 21.7. *Left:* The optical (Hβ) versus UV (Mg II) virial black hole estimators for 150 objects from the combined RM (filled circles), LBQS (filled squares), and MQS (open circles) samples described by McLure & Jarvis (2002). The solid line is the BCES bisector fit to the 128 objects from the MQS and LBQS samples and has a slope of 1.00 ± 0.08. The outlying narrow-lined Seyfert NGC 4051 has been highlighted. *Right:* Histogram of $\log M_\bullet(\text{Mg II}) - \log M_\bullet(\text{H}\beta)$ for the 128 objects from the LBQS and MQS samples. Also shown is the best-fitting Gaussian, which has $\sigma = 0.41$. (Adapted from McLure & Jarvis 2002.)

21.4.2 *Host Galaxy Mass Measurement at High Redshift*

The price to be paid for having a bright quasar nucleus from which to make emission-line based black hole mass estimates at high redshift is, of course, that the measurement of the mass of the host galaxy becomes a challenge. The combination of generally unfavorable K corrections and strong surface brightness dimming means that the effective study of quasar hosts beyond $z \simeq 1$ is much harder than the study of low-redshift hosts discussed above.

It is thus natural and sensible to consider whether there is any alternative to host galaxy imaging that might be utilized to estimate host galaxy mass at high redshift. This is the motivation behind the recent suggestion by Shields et al. (2003) that the FWHM of the [O III] emission line in high-redshift quasars provides a measure of the velocity dispersion of the stars in the central regions of the host galaxy. If true, then the black hole-spheroid mass ratio in high-redshift quasars could be estimated simply from high-quality optical and near-infrared spectroscopy.

Unfortunately, this claim seems to be optimistic. While there is reasonable evidence that [O III] FWHM can be used as a proxy for central stellar velocity dispersion in low-luminosity, low-redshift AGNs (Nelson 2000), there are many reasons why this is unlikely to be the case in more luminous objects, especially at high-redshift (Boroson 2003). Moreover, as demonstrated in Figure 21.8, there is little evidence of a statistically useful correlation between the proposed [O III] estimator of spheroid mass and spheroid luminosity for the quasars and AGNs considered earlier in Figure 21.5. Unfortunately, therefore, to obtain

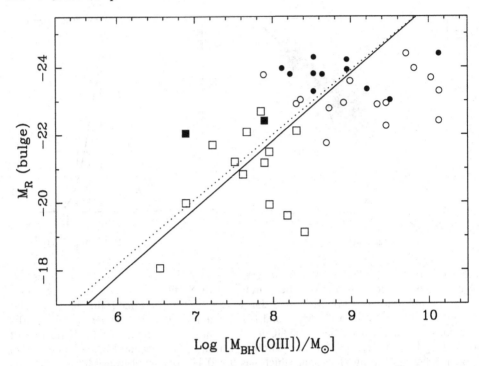

Fig. 21.8. Host spheroid absolute magnitude plotted against the [O III]-based mass estimator proposed by Shields et al. (2003) for the subset of quasars and Seyferts shown in Figure 21.5 for which a reliable FWHM for [O III] could be measured. Comparison of this "relationship" with the tight correlation shown in Figure 21.5 provides little confidence that [O III] can be used as a reliable estimator of stellar velocity dispersion in the hosts of high-redshift quasars. Removal of the radio-loud objects (black data points) does not improve the significance of the correlation.

meaningful estimates of host galaxy masses for high-redshift quasars there currently appears to be no alternative but to attempt to measure host galaxy luminosities.

What, then, are the prospects for determining the masses of high-redshift quasar host galaxies from deep imaging data? Obviously high angular resolution and a sound knowledge of the detailed form of the point-spread function remain a necessity. Also, to minimize the uncertainty in galaxy mass estimation introduced by the evolution of the host stellar population, it is desirable to undertake observations of high-redshift quasar hosts at near-infrared wavelengths.

The advent of the NICMOS camera on *HST* allowed its unique ability to provide images with robust and repeatable point-spread functions to be extended into the near-infrared. Although NICMOS is only effective out to the *H* band at 1.6 μm, this was sufficient to allow Kukula et al. (2001) to extend the restframe *V*-band study of the hosts of moderate luminosity quasars from the $z = 0.2$ regime probed by Dunlop et al. (2003) out to $z \simeq 2$.

Kukula et al. (2001) defined two new quasar samples at $z \simeq 1$ and 2, confined to the luminosity range $-24 > M_V > -25$ mag, and by observing these with NICMOS through the *J* and *H* band, respectively, obtained line-free images of the quasar hosts at both redshifts in the restframe *V* band.

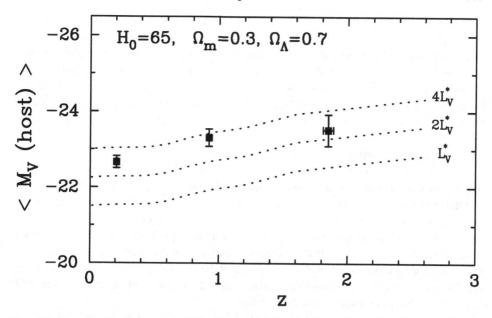

Fig. 21.9. Mean absolute magnitude versus redshift for the quasar host galaxies from the *HST* imaging programs of McLure et al. (1999), Kukula et al. (2001) and Dunlop et al. (2003). The dotted lines show the passive evolutionary tracks for present-day L^*, $2L^*$, and $4L^*$ galaxies assuming that they formed in a single starburst event at $z \approx 5$. At all redshifts the average mass of the quasar host galaxies is consistent with that of a present-day $3L^*$ elliptical, under the assumption of passive evolution.

At $z \simeq 1$ Kukula et al. found it was still possible, on the basis of the NICMOS data, to estimate the scale lengths of the host galaxies with sufficient accuracy to demonstrate that they were at least consistent with the (passively evolved) Kormendy relation derived by Dunlop et al. at $z \simeq 0.2$ (see Fig. 21.4). Therefore, just as at low redshift, the host galaxies of quasars at $z \simeq 1$ appear to be large, luminous systems. However, while for three of the $z \simeq 1$ quasars Kukula et al. found strong evidence that the hosts follow a de Vaucouleurs surface brightness profile, in the majority of cases the data did not allow an unambiguous fit.

By $z \simeq 2$, despite deliberately deeper imaging, the increased size of the H-band point-spread function, coupled presumably with the impact of additional surface brightness dimming, meant that Kukula et al. were unable to determine unambiguous morphologies for any host galaxy, and only highly uncertain scale lengths in most cases. However, extended starlight was still detected in every object and it proved possible to still obtain meaningful measurements of host luminosity.

The most robust result from this study that can be extracted at all redshifts is the average luminosity of the quasar host galaxies in the restframe V band. This is plotted against redshift in Figure 21.9, which shows that, under the assumption of passive evolution, the hosts of comparably luminous quasars are basically unchanged in mass out to $z \simeq 2$.

Although Figure 21.9 represents an interesting first attempt to determine the mass evolution of quasar hosts, this result cannot yet be regarded as anything like as secure as the results on low-redshift quasar hosts presented in § 21.2. First, the samples are small. Second, at present the mass estimates remain vulnerable to the validity of assuming passive evolution. Third, while the complete quasar host sample appears, on average, to have the same stellar mass from $z \simeq 0.2$ to $z \simeq 2$, there is evidence within the data that the RQQ hosts are less massive at high redshift by a factor of $2-3$. This is consistent with, although less extreme than, the decrease in radio-quiet host galaxy mass with increasing redshift reported by Ridgway et al. (2001) from an analysis of NICMOS data. Recently, Lacy et al. (2002) have also found evidence for a slight decline in host galaxy mass by $z \simeq 1$, using ground-based near-infrared imaging coupled with active/adaptive optics.

21.5   Future Prospects

To obtain improved estimates of the masses of high-redshift quasar hosts will require color information (to test the assumption of passive evolution; Kukula et al. 2004) and the extension of high-resolution imaging observations into the K-band with the largest available telescopes. This work is already underway (e.g., Hutchings 2003) and over the next few years, with careful study design (e.g., selection of quasars within a few arcseconds of an appropriate star for reliable PSF determination) and the necessary major investment of telescope time, it is not unreasonable to expect that the Gemini telescopes and the VLT can revolutionize the effective study of high-redshift quasar hosts in much the same way as *HST* has revolutionized the study of low-redshift hosts.

In the very near future the new UV viral black hole mass estimators described above will be applied to the extensive databases of quasar optical spectra now being released by, for example, the Sloan Digital Sky Survey.

Therefore, within the next 2 years or so, it is not unreasonable to anticipate the construction of the first robust measurements of the redshift dependence of black hole-spheroid mass ratio within the bright quasar population out to $z \simeq 5$. Such measurements promise to provide fundamental new insights into our understanding of the relationship between black hole and galaxy formation.

References

Archibald, E. N., Dunlop, J. S., Jimenez, R., Friaça, A. C. S., McLure, R. J., & Hughes, D. H. 2002, MNRAS, 336, 353
Bahcall, J. N., Kirhakos, S., Saxe, D. H., & Schneider, D. P. 1997, ApJ, 479, 642
Bahcall, J. N., Kirhakos, S., & Schneider, D. P. 1994, ApJ, 435, L11
Boroson, T. A. 2003, ApJ, 585, 647
Canalizo, G., & Stockton, A. 2000, AJ, 120, 1750
Disney, M. J., et al. 1995, Nature, 376, 150
Dunlop, J. S., Taylor, G. L., Hughes, D. H., & Robson, E. I. 1993, MNRAS, 264, 455
Dunlop, J. S., McLure, R. J., Kukula, M. J., Baum, S. A., O'Dea, C. P., & Hughes, D. H. 2003, MNRAS, 340, 1095
Fabian, A. C. 1999, MNRAS, 308, 39
Floyd, D. J. E., et al. 2004, MNRAS, submitted
Gebhardt, K., et al. 2000a, ApJ, 539, L13
——. 2000b, ApJ, 543, L5
Genzel, R., Tacconi, L. J., Rigopoulou, D., Lutz, D., & Tecza, M. 2001, ApJ, 563, 527
Granato, G. L., Silva, L., Monaco, P., Panuzzo, P., Salucci, P., De Zotti, G., & Danese, L. 2001, MNRAS, 324, 757
Hamilton, T. S., Casertano, S., & Turnshek, D. A. 2002, ApJ, 576, 61
Hooper, E. J., Impey, C. D., & Foltz, C. B. 1997, ApJ, 480, L95
Hutchings, J. B. 2003, AJ, 125, 1053
Kaspi, S., Smith, P. S., Netzer, H., Maoz, D., Jannuzi, B. T., & Giveon, U. 2000, ApJ, 533, 631
Kauffmann, G., & Haehnelt, M. 2000, MNRAS, 311, 576
Kormendy, J. 1977, ApJ, 217, 406

Kormendy, J., & Gebhardt, K. 2001, in The 20th Texas Symposium on Relativistic Astrophysics, ed. H. Martel & J. C. Wheeler (New York: AIP), 363

Kukula, M. J., et al. 2004, in preparation

Kukula, M. J., Dunlop, J. S., McClure, R. J., Miller, L., Percival, W. J., Baum, S. A., & O'Dea, C. P. 2001, MNRAS, 326, 1533

Lacy, M., Gates, E. L., Ridgway, S. E., de Vries, W., Canalizo, G., Lloyd, J., & Graham, J. R. 2002, AJ, 124, 3023

Lacy, M., Laurent-Meuleisen, S. A., Ridgway, S. E., Becker, R. H., & White, R. L. 2001, ApJ, 551, L17

Laor, A. 2000, ApJ, 543, L111

——. 2001, ApJ, 553, 677

Magorrian, J., et al. 1998, AJ, 115, 2285

McLeod, K. K., & McLeod, B. A. 2001, ApJ, 546, 782

McLeod, K. K., & Rieke, G. H. 1994, ApJ, 431, 137

McLure R. J., & Dunlop J. S. 2001, MNRAS, 327, 199

——. 2002, MNRAS, 331, 795

McLure, R. J., Dunlop, J. S., & Kukula, M. J. 2000, MNRAS, 318, 693

McLure, R. J., Dunlop, J. S., Kukula, M. J., Baum, S. A., O'Dea, C. P., & Hughes, D. H. 1999, MNRAS, 308, 377

McLure R. J., & Jarvis M. J. 2002, MNRAS, 337, 109

Merritt, D., & Ferrarese, L. 2001, MNRAS, 320, L30

Nelson, C. H. 2000, ApJ, 544, L91

Nolan, L. A., Dunlop, J. S., Kukula, M. J., Hughes, D. H., Boroson, T., & Jimenez, R. 2001, MNRAS, 323, 308

Percival, W., Miller, L., McLure, R. J., & Dunlop, J. S. 2001, MNRAS, 322, 843

Ridgway, S. E., Heckman, T. M., Calzetti, D., & Lehnert, M. 2001, ApJ, 550, 122

Schade, D., Boyle, B. J., & Letawsky, M. 2000, MNRAS, 315, 498

Scoville, N. Z., Frayer, D. T., Schinnerer, E., & Christopher, M. 2003, ApJ, 585, L105

Shields, G. A., Gebhardt, K., Salviander, S., Wills, B. J., Xie, B., Brotherton, M. S., Yuan, J., & Dietrich, M. 2003, ApJ, 583, 124

Silk, J., & Rees, M. J. 1998, A&A, 331, L1

Taylor, G. L., Dunlop, J. S., Hughes, D. H., & Robson, E. I. 1996, MNRAS, 283, 930

Vestergaard, M. 2002, ApJ, 571, 733

Wandel, A. 1999, ApJ, 519, L39

Wandel, A., Peterson, B. M., & Malkan, M. A. 1999, ApJ, 526, 579

Willott, C. J., McLure, R. J., & Jarvis, M. J. 2003, ApJ, 587, L15

22

Star formation in active galaxies: a spectroscopic perspective

TIMOTHY M. HECKMAN

Center for Astrophysical Sciences, Department of Physics & Astronomy,
The Johns Hopkins University

Abstract

I review the relationship between star formation and black hole building, based on spectroscopic observations of the stellar population in active galactic nuclei (AGNs) and their host galaxies. My emphasis is on large, well-defined local samples of AGNs, whose optical continua are dominated by starlight. I summarize the spectroscopic tools used to characterize their stellar content. Surveys of the nuclei of the nearest galaxies show that the stellar population in most low-power AGNs is predominantly old. In contrast, young (< 1 Gyr) stars are detected in about half of the powerful type 2 Seyfert nuclei. I summarize Sloan Digital Sky Survey spectroscopy of the host galaxies of 22,000 AGNs. The AGN phenomenon is commonplace only in galaxies with high mass, high velocity dispersion, and high stellar surface mass density. Most normal galaxies with these properties have old stellar populations. However, the hosts of powerful AGNs (Seyfert 2s) have young stellar populations. The fraction of AGN hosts that have recently undergone a major starburst also increases with AGN luminosity. A powerful AGN presumably requires a massive black hole (host with a substantial bulge) plus an abundant fuel supply (a star-forming ISM). This combination is rare today, but would have been far more common at early epochs.

22.1   Introduction

22.1.1   *Motivation*

There is no doubt that that there is a profound physical connection between the creation of supermassive black holes and the formation of bulges and elliptical galaxies. The remarkably tight correlation between the stellar velocity dispersion and black hole mass in these systems (Ferrarese & Merritt 2000; Gebhardt .et al. 2000) provides powerful evidence for this connection. The similarities and differences between the strong cosmic evolution of active galactic nuclei (AGNs) and star formation (e.g., Steidel et al. 1999; Fan et al. 2001) then lead naturally to the speculation that AGN evolution traces the build up of the spheroidal component of galaxies (e.g., Haehnelt & Kauffman 2000; Kauffmann & Haehnelt 2000).

To put the connection between black hole and galaxy building in context, it is interesting to note that the "universal" ratio of $\sim 10^{-3}$ between the black hole and stellar mass in spheroids implies that a moderately powerful AGN with a bolometric luminosity of $10^{11} L_\odot$ fueled by accretion with radiative efficiency of 10% would require an associated average star formation rate of 65 $M_\odot$ yr^{-1} if the spheroid and black hole were to be built on the same time scale. The young stars would outshine the AGN by nearly an order of magnitude in this case!

Several interrelated questions immediately arise, and have indeed formed the basis for most of the discussion at this conference. What are the relevant astrophysical processes that connect the formation of bulges and black holes? What conditions foster the fueling of a powerful AGN? Given our current understanding of galaxy evolution, can we account for the cosmic evolution of the AGN population? Can we see a scaled-down version of bulge building surrounding powerful AGNs in the local Universe, or, instead, are powerful AGNs today just the temporary rejuvenation of a pre-existing black hole with little associated star formation? Can we make any sense of the complex gas/star/black hole ecosystem with all its messy gastrophysics?

22.1.2   My Perspective

My goal is to review what is presently known about the possible connection between star formation and the AGN phenomenon, both in active galaxies and in their nuclei. This is sometimes called the "starburst-AGN connection" (e.g., Terlevich 1989; Heckman 1991), a vast, sprawling topic that cannot be adequately reviewed in a single short paper. Excellent overviews may be found in the volumes edited by Filippenko (1992) and Aretxaga, Kunth, & Mujica (2001a). Other recent reviews that complement the present paper include those by Veilleux (2001) and Cid Fernandes, Schmitt, & Storchi-Bergmann (2001a).

This paper will be focused in scope and methodology. I will just review observations, and only discuss spectroscopic measures of the stellar populations (ignoring indirect evidence for star formation like molecular gas, and far-IR, UV, and radio continuum emission). I will strongly emphasize systematic investigations of large well-defined samples. These restrictions will allow me to make rather robust statements about the empirical basis for a connection between AGNs and star formation. However, these restrictions effectively limit me to the local Universe. While the black hole/bulge connection was largely established at high redshift, I hope to show that we can gain key insights from the fossil record and from local analogs.

22.1.3   Spectral Diagnostics of Young Stars

For young massive stars the strongest spectral features with the greatest diagnostic power lie in the vacuum UV regime between the Lyman break and $\sim$2000 Å (e.g., Leitherer et al. 1999; de Mello, Leitherer, & Heckman 2000). These include the strong stellar wind lines of the O VI, N V, Si IV, and C IV resonance transitions and a host of weaker stellar photospheric lines. Most of the photospheric lines arise from highly excited states and their stellar origin is unambiguous. While resonance absorption lines may have an interstellar origin, the characteristic widths of the stellar wind profiles make them robust indicators of the presence of massive stars. Unfortunately, observations in this spectral regime are difficult. Only a handful of local type 2 Seyferts and LINERs have nuclear UV fluxes that are high enough to enable a spectroscopic investigation. While this small sample may not be representative, the available data firmly establish the presence of a dominant population of young stars (Heckman et al. 1997; González Delgado et al. 1998; Maoz et al. 1998; Colina & Arribas 1999).

While the optical spectral window is far more accessible, the available diagnostic features of massive stars are weaker and less easy to interpret. Old stars are cool and have many strong spectral features in the optical due to molecules and low-ionization metallic species. Hot young stars have relatively featureless optical spectra*. Thus, the spectroscopic impact of the presence of young stars is mostly an indirect one: as they contribute an increasing fraction of the light (as the luminosity-weighted mean stellar age decreases) most of the strongest spectral features in the optical weaken. This effect is easy to measure. Unfortunately, the effect of adding "featureless" nonstellar continuum from an AGN and young starlight will be similar in this regard.

The strongest optical absorption lines from young stars are the Balmer lines. These reach peak strength in early A-type stars, and so they are most sensitive to a stellar population with an age of $\sim$100 Myr to 1 Gyr (e.g., González Delgado et al. 1998). Thus, the Balmer lines do not uniquely trace the youngest stellar population*. On the plus side, they can be used to characterize past bursts of star formation (e.g., Dressler & Gunn 1983; Kauffmann et al. 2003a).

The situation is summarized in Figure 22.1, which plots the strength of the Balmer Hδ absorption line (the Hδ_A index) vs. the 4000 Å break strength [the $D_n(4000)$ index] for a set of 32,000 model galaxies with different star formation histories (see Kauffmann et al. 2003a). For "well-behaved" (continuous) histories of past star formation there is a tight inverse relation between these two parameters. The addition of a starburst moves the galaxy below and to the left of the main locus at early times ($\leq$ 100 Myr) times and above it at intermediate times (100 Myr to 1 Gyr).

Note that the effect of adding featureless AGN continuum will carry an old stellar population located at roughly (2, $-$2) in Figure 22.1 to (1, 0). This "mixing line" lies well under the loci of the stellar populations [the latter having much stronger Hδ_A at a given $D_n(4000)$]. This underscores the importance of the Balmer absorption lines, and of the need to properly account for the contaminating effects of nebular Balmer emission lines. The relative nebular contamination is minimized for the high-order Balmer lines in the blue and near-UV.

So far, I have equated young stars with hot stars. This is certainly true for main-sequence stars. However, a population of red supergiants will contribute significantly to the near-IR light in young stellar populations†. The spectral features produced by red supergiants are qualitatively similar to those produced by red giants that dominate the near-IR light in an old stellar population. A robust method to determine whether old giants or young supergiants dominate is to measure the M/L ratio in the near-IR using the stellar velocity dispersion. So far this technique has been applied to only a small sample, but the results are tantalizing (Oliva et al. 1999; Schinnerer et al. 2004).

In an old stellar population, cool stars provide most of the light in the optical and near-IR and so the associated metallic and molecular spectral features are strong in both bands. In

---

* The most direct optical signatures of young massive stars are the photospheric lines of He I and the broad emission features due to Wolf-Rayet stars. The former reach peak strength in early B stars and hence in stellar populations with ages of tens of Myr (González Delgado, Leitherer, & Heckman 1999). The latter trace the most massive stars and reach peak strengths for ages of several Myr (e.g., Leitherer et al. 1999). Both the He I and Wolf-Rayet features are weak and contaminated by nebular emission lines. They have been detected in only a handful of AGNs (Heckman et al. 1997; González Delgado et al. 1998; Storchi-Bergmann, Cid Fernandes, & Schmitt 1998).

* In view of this, and in view of the importance of the Balmer lines, throughout this review I will use the term "young" to mean stellar populations with ages less than a Gyr.

† For an instantaneous burst of star formation, red supergiants do not appear until $\sim$5 to 6 Myr have elapsed. Thus, they are absent only in very young bursts of very short duration.

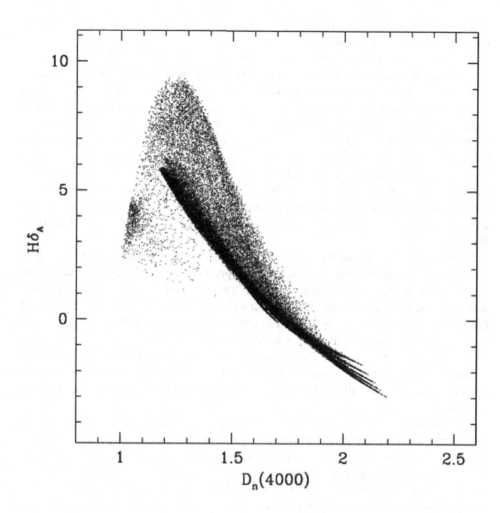

Fig. 22.1. Indices measuring the 4000 Å break and Hδ absorption line are plotted for a grid of 32,000 model galaxies with different star formation histories. The dark diagonal line is an age sequence of galaxies with continuous star formation histories (young at upper left and old at lower right). The other points represent models in which a strong burst of star formation has occurred within the last ∼ 1 Gyr. (See Kauffmann et al. 2003a.)

contrast, the optical (near-IR) continuum in a young stellar population will be dominated by hot main-sequence stars (cool supergiants). The metallic/molecular lines are therefore weak in the optical and strong in the near-IR. This combination of properties provided some of the first direct evidence for a young stellar population in AGNs (Terlevich, Díaz, & Terlevich 1990; Nelson & Whittle 1999).

22.1.4   *Emission-Line Diagnostics and AGN Classification*

As noted above, I will be emphasizing optical spectroscopy in this review of the stellar populations in AGNs. Historically, this has also been the most widely used technique to detect and classify AGNs themselves.

AGNs can be broadly classified depending upon whether the central black hole and its associated continuum and broad emission-line region is viewed directly (a "type 1" AGN) or is obscured by a surrounding dusty medium (a "type 2" AGN). Since this obscuring medium does not fully cover 4π steradians as seen from the central AGN. some of the AGN radiation escapes the central region and photoionizes surrounding circumnuclear ($\sim 10^2$ to 10^3 pc-scale) gas, leading to strong, relatively narrow permitted and forbidden emission lines from the "narrow-line region" (NLR). In type 1 AGNs, the optical continuum is dominated by nonstellar emission, making it difficult to study the stellar population. In what follows, I will therefore ignore this type of AGN. In type 2 AGNs the observed optical continuum is predominantly starlight, with some contribution by light from the obscured AGN scattered into our line of sight and from nebular continuum associated with the NLR (e.g., Tran 1995; Wills et al. 2002).

In the simplest version of the "unified" model for AGNs (Antonucci 1993) the type 1 and type 2 AGNs are drawn from the same parent population and differ only in our viewing angle. In this case the stellar content of the two types will be the same. On the other hand, if the solid angle covered by the dusty obscuring medium varies substantially, then type 1 (2) AGNs will be preferentially drawn from those objects with smaller (larger) covering fractions for this medium. In such a case systematic differences might exist in stellar content between type 1 and type 2 AGNs (especially if the dusty obscurer is related to star formation). There have been recurring suggestions to this effect (e.g., Maiolino et al. 1995; Malkan, Gorjian, & Tam 1998; Oliva et al. 1999). It is important to keep this possible bias in mind.

Narrow emission lines can also be produced via photoionization by hot young stars. The most unambiguous discrimination between excitation of the NLR by young stars vs. an AGN is provided by lines due to species with ionization potentials greater than 54 eV (above the He II edge), since a young stellar population produces a negligible supply of such high-energy photons. Unfortunately, most such high-ionization lines are weak in the optical spectra of AGNs. In practice, classification is therefore usually based on the flux ratios of the strongest lines (Heckman 1980a,b; Baldwin, Phillips, & Terlevich 1981; Veilleux & Osterbrock 1987).

An example of such a classification diagram is shown in Figure 22.2, based on nearly 56,000 emission-line galaxies from the Sloan Digital Sky Survey (SDSS) sample discussed below. In this plot of the [O III] $\lambda5007$/Hβ vs. the [N II] $\lambda6583$/Hα NLR line ratios, two distinct sequences are present. Star-forming galaxies define a narrow locus in which the metallicity increases from the upper left to the lower center. The plume that ascends from the high-metallicity end of the starformer sequence toward higher values of both [O III]/Hβ and [N II]/Hα is the AGN population. In simple physical terms, the high-energy continuum of an AGN results in much greater photoelectric heating per ionization, which raises the temperature in the ionized gas, and therefore strengthens the collisionally excited forbidden lines that cool the gas.

It is important to realize that a contribution to the emission-line spectrum by regions of star formation is almost inevitable in many of the AGNs in Figure 22.2: at the median

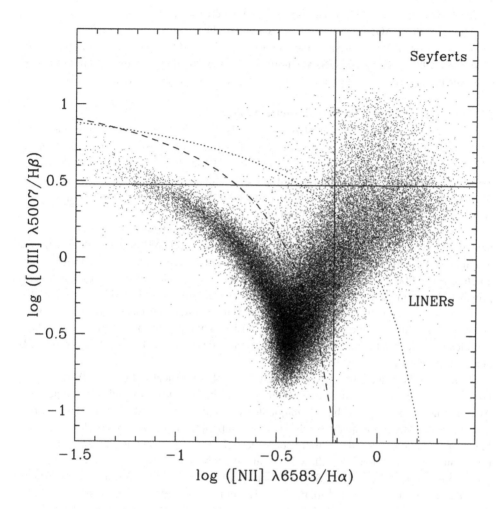

Fig. 22.2. Diagnostic flux-ratio diagram for a sample of nearly 56,000 emission-line galaxies from the SDSS. According to Kewley et al. (2001) galaxies dominated by an AGN will lie above and to the right of the dotted line. Galaxies lying below and to the left of the dashed line are dominated by star-forming regions. Galaxies lying between the dashed and dotted curves are transition objects (AGN and star formation are both important). The locations of classic AGN-dominated Seyfert nuclei and LINERs are indicated. (From Kauffmann et al. 2003c.)

redshift of the SDSS main galaxy sample ($z \approx 0.1$) the 3″ diameter SDSS fiber corresponds to a projected size of nearly 6 kpc. Thus, a galaxy's position along the AGN plume will be determined by the relative contribution of the AGN and star formation to the emission-line spectrum. This is particularly important in the context of this review because it means that the signature of young stars is present in the nebular emission lines as well as in the stellar absorption lines.

The AGN plume is quite broad at its upper end, where the AGN contribution is dominant. Traditionally, these AGN-dominated objects are classified as Seyferts if [O III]/Hβ > 3 and as low-ionization nuclear emission-line regions (LINERs) if [O III]/Hβ < 3 (e.g., Veilleux & Osterbrock 1987). The latter AGNs are more common, but do not attain the high luminosities of powerful Seyfert nuclei (e.g., Heckman 1980b; Ho, Filippenko, & Sargent 2003).

22.2   Young Stars in Active Galactic Nuclei

Spectroscopy of the nearest AGNs affords the opportunity to study the starburst-AGN connection on small physical scales (> a few pc). The drawback is that the nearest AGNs have low luminosities, and we might expect that the amount of star formation associated with black hole fueling would scale in some way with AGN power. To get a complete picture it is thus important to examine both the nearest AGNs and more powerful AGNs. I will review these two regimes in turn.

The earliest investigation of the stellar population for a moderately large sample of the nearest AGNs was by Heckman (1980a,b), who discussed 30 LINERs found in a survey of a sample of 90 optically bright galaxies. The typical projected radius of the spectroscopic aperture was ~200 pc. The spectra covered the range from 3500 to 5300 Å. LINERs were primarily found in galaxies of early Hubble type (E through Sb). Based on the strengths of the stellar metallic lines and the high-order Balmer lines, the nuclear continuum was dominated by old stars in about 3/4 of the LINERs, while a contribution of younger stars was clearly present in the remainder. Typical luminosities of the [O III] λ5007 and Hα NLR emission lines were ~ 10^5 to $10^6 L_\odot$.

Ho et al. (2003) have recently examined the nuclear (typical radius ~ 100 pc) stellar population in a complete sample of ~500 bright, nearby galaxies (of which 43% contain an AGN). In nearly all respects this is a major step forward from my old analysis. The sample is considerably larger, the quality of the spectra is superior, and the analysis of the emission-line properties more careful and more sophisticated. The larger sample size and improved treatment of the emission lines allow Ho et al. to study statistically significant samples of low-luminosity LINERs, type 2 Seyferts, and transition nuclei, and to span a larger range in AGN luminosity ($L_{H\alpha} \approx 10^4$ to $10^7 L_\odot$). The only disadvantage of these spectra is that they do not extend shortward of 4230 Å and so they miss the Hδ and higher-order Balmer lines that most effectively probe young stars. Nevertheless, their results are qualitatively consistent with those of Heckman (1980a,b). The AGNs are hosted by early-type galaxies (E through Sbc). With a few exceptions, the AGNs have predominantly old stellar populations. The exceptions are primarily transition nuclei whose emission-line spectra suggest excitation by a mix of an AGN and massive stars.

My colleagues (Cid Fernandes, González Delgado, Schmitt, Storchi-Bergmann) and I have recently undertaken a program of near-UV spectroscopy of 43 LINER and transition nuclei taken from the Ho, Filippenko, & Sargent (1997) survey. Our specific goal was to access the high-order stellar Balmer absorption lines. We detect these lines in about half of the transition nuclei, but in very few LINERs. The cases in which hot stars are present are primarily Sb, Sbc, or Sc galaxies. These results are consistent with the idea that transition nuclei are composite AGN/starformers.

Let me now discuss significantly more powerful AGNs (classical type 2 Seyferts). It was recognized very early-on (e.g., Koski 1978) that the optical spectra of powerful type 2 Seyfert nuclei could not be explained purely by an old stellar population. An additional "featureless continuum" that typically produced 10% to 50% of the optical continuum was present. Until relatively recently, it was tacitly assumed that this component was light from the AGN (plausibly light from the hidden type 1 Seyfert nucleus that had been reflected into our line of sight by free electrons and/or dust). However, a detailed spectropolarimetric investigation by Tran (1995) and related arguments by Cid Fernandes & Terlevich (1995) and Heckman et al. (1995) showed that only a small fraction of this continuum could be attributed to scattered AGN light. While nebular continuum emission must be present (e.g., Tran 1995; Wills et al. 2002), Cid Fernandes & Terlevich (1995) and Heckman et al. (1995) argued that young stars were the main contributors.

This has been confirmed by several major optical spectroscopic investigations of moderately large samples of type 2 Seyfert nuclei (Schmitt, Storchi-Bergmann, & Cid Fernandes 1999; González Delgado, Heckman, & Leitherer 2001; Cid Fernandes et al. 2001b; Joguet et al. 2001). The latter two papers examined the nuclear stellar population in 35 and in 79 type 2 Seyferts, respectively. The projected aperture sizes range from a few hundred pc to about a kpc. On average, these are considerably more powerful AGNs than those in the Heckman (1980b) or Ho et al. (2003) samples, with [O III] luminosities of 10^6 to $10^9 L_\odot$.

The principal conclusion is that a young (< 1 Gyr) stellar population is clearly present in about half of the Seyfert 2 nuclei. Cid Fernandes et al. (2001b) find that the fraction of nuclei with young stars is $\sim 60\%$ when $L_{[O\ III]} > 10^7 L_\odot$ but is only $\sim 10\%$ for the less powerful nuclei. They also found that the "young" Seyfert 2 nuclei were hosted by galaxies with much larger far-IR luminosities ($\sim 10^{10}$ to $10^{12} L_\odot$) than the "old" nuclei ($\sim 10^9$ to $10^{10} L_\odot$), suggesting that the global star formation rate was correspondingly higher (a topic considered in some detail below). Interestingly, Joguet et al. (2001) found that the Hubble type distribution for the host galaxies was roughly the same for the "old" and "young" nuclei (S0 to Sc). Using morphological classifications based on the *Hubble Space Telescope* imaging survey of (Malkan et al. 1998), Storchi-Bergmann et al. (2001) find a reasonably good correspondence between the presence of a young nuclear population and a late "inner Hubble type" (the presence of dust lanes and spiral features in the inner few kpc).

To date, spectroscopic investigations of the nuclear stellar populations in powerful radio-loud type 2 AGNs (the so-called "narrow-line radio galaxies") have been restricted to relatively small samples (Schmitt, Storchi-Bergmann & Cid Fernandes 1999; Aretxaga et al. 2001b; Wills et al. 2002; Tadhunter et al. 2002). A young nuclear stellar population has been detected in about 1/3 of the cases.

22.3 Young Stars in AGN Host Galaxies

In the previous section I have reviewed spectroscopy of the stellar population in the nuclear region (the centralmost $\sim 10^2$ to 10^3 pc). I now turn my attention to more global properties of AGN hosts.

Raimann et al. (2003) have recently used long-slit optical/near-UV spectroscopy to measure the radial variation in the stellar population for the sample of type 2 Seyfert galaxies whose nuclear properties were investigated by González Delgado et al. (2001). They characterize the stellar content by fitting each spectrum with a set of spectra of stellar clusters spanning a range in age and metal abundance.

They find that the Seyfert 2's have younger mean ages than their comparison sample of normal galaxies at both nuclear and off-nuclear locations. The sample-averaged "vector" of the fractional contribution to the continuum at 4000 Å by stellar populations with ages of 10 Gyr, 1 Gyr, 30 Myr, and 3 Myr is (0.36, 0.25, 0.24, 0.15) at the nucleus, (0.36, 0.31, 0.20, 0.13) at 1 kpc, and (0.32, 0.30, 0.27, 0.11) at 3 kpc. Thus, the stellar population shows no strong systematic radial gradient in the hosts of powerful Seyfert 2 nuclei (at least out to radii of several kpc).

22.3.1   The Sloan Digital Sky Survey

So far, I have reported on spectroscopy of modest-sized samples of AGNs. The ongoing SDSS (York et al. 2000; Stoughton et al. 2002; Pier et al. 2003; Blanton et al. 2004) provides us with the opportunity to investigate the structure and stellar content of the hosts of tens of thousands of AGNs! In the remainder of the paper I will give a progress report on an on-going program in this vein that is a close collaboration between groups at JHU and MPA-Garching (Kauffmann et al., in preparation). The program was made possible by the inspiration and perspiration of the dedicated team that over the past 13 years has created and operated the SDSS.

The SDSS is using a dedicated 2.5-meter wide-field telescope at the Apache Point Observatory to conduct an imaging and spectroscopic survey of about a quarter of the sky. The imaging is conducted in the u, g, r, i, and z bands (Fukugita et al. 1996; Gunn et al. 1998; Hogg et al. 2001; Smith et al. 2002), and spectra are obtained with a pair of multi-fiber spectrographs built by Alan Uomoto and his team at JHU. When the survey is complete, spectra will have been obtained for nearly 10^6 galaxies and 10^5 QSOs selected from the imaging data. Details on the spectroscopic target selection for the "main" galaxy sample and QSO sample can be found in Strauss et al. (2002) and Richards et al. (2002), respectively. We will be summarizing results based on spectra of $\sim$123,000 galaxies contained in the the the SDSS Data Release One (DR1). These data are to be made publically available early in 2003.

Since I will primarily be discussing results derived from the spectra, it is useful to summarize their salient features. Spectra are obtained through $3''$ diameter fibers. At the median redshift of the main galaxy sample ($z \approx 0.1$) this corresponds to a projected aperture size of $\sim$6 kpc which typically contains 20% to 40% of the total galaxy light. Thus, the SDSS spectra are closer to global than to nuclear spectra. At the median redshift the spectra cover the rest-frame wavelength range from $\sim$3500 to 8500 Å with a spectral resolution $R \approx 2000$ ($\sigma_{instr} \approx 65$ km s^{-1}). The spectra are spectrophotometrically calibrated through observations of subdwarf F stars in each 3-degree field. By design, the spectra are well-suited to the determinations of the principal properties of the stars and ionized gas in galaxies.

22.3.2   Our Methodology

For the convenience of the reader we give a brief summary of our methodology below. For a much more complete description see Kauffmann et al. (2003a).

The rich stellar absorption-line spectrum of a typical SDSS galaxy is both a blessing and a curse. While the lines provide unique information about the stellar content and galaxy dynamics, they make the measurement of weak nebular emission lines quite difficult. To deal with this, we have performed a careful subtraction of the stellar absorption-line spectrum before measuring the nebular emission lines. This is accomplished by fitting the emission-line-free regions of the spectrum with a model galaxy spectrum. The model spectra are based on the new population synthesis code of Bruzual & Charlot (2003), which incorporates high-resolution stellar libraries. A set of 39 model template spectra were used that span the relevant range in age and metallicity. After convolving the template spectra to the measured stellar velocity dispersion (σ_*) of an individual SDSS galaxy, the best fit to the galaxy spectrum is constructed from a linear combination of the template spectra.

As diagnostics of the stellar population we have used the amplitude of the 4000 Å break [the $D_n(4000)$ index of Balogh et al. 1999] and the strength of the Hδ absorption line (the Lick Hδ_A index of Worthey & Ottaviani 1997). The diagnostic power of these indices is shown in Figure 22.1 above. In both cases the indices measured in the SDSS galaxy spectra are corrected for the flux of the emission lines in their bandpasses.

Using a library of 32,000 star formation histories that span the relevant range in metallicity, we have used the measured $D_n(4000)$ and Hδ_A indices to estimate the SDSS z-band mass-to-light ratio for each galaxy. By comparing the colors of our best-fitting model to those of each galaxy, we have estimated the dust extinction of starlight in the z-band (e.g., Calzetti, Kinney, & Storchi-Bergmann 1994; Charlot & Fall 2000).

The SDSS imaging data are then used to provide the basic structural parameters. The z-band absolute magnitude and the derived values of M/L and A_z yield the stellar mass (M_*). The half-light radius in the z-band and the stellar mass yield the effective stellar surface mass density ($\mu_* = M_*/2\pi r_{50,z}^2$). As a proxy for Hubble type we use the SDSS "concentration" parameter C, which is defined as the ratio of the radii enclosing 90% and 50% of the galaxy light in the r band (see Stoughton et al. 2002). Strateva et al. (2001) find that galaxies with $C > 2.6$ are mostly early-type galaxies, whereas spirals and irregulars have $2.0 < C < 2.6$.

22.3.3 Results

As discussed in Kauffmann et al. (2003b), the overall SDSS galaxy population is remarkably bimodal in nature. There is a rather abrupt transition in properties at a critical stellar mass $M_* \approx 3 \times 10^{10} M_\odot$. Below this mass, galaxies are young [small $D_n(4000)$ and large Hδ_A], are disk dominated (low concentration, $C < 2.6$), and show a strong increase in surface mass density (μ_*) with increasing mass. Above this mass, galaxies are old [large $D_n(4000)$ and small Hδ_A], are bulge dominated ($C > 2.6$), and have a uniform μ_*. There is also a strong transition in mean stellar age at a characteristic surface mass density of $3 \times 10^8 M_\odot$ kpc^{-2} and a concentration index of 2.6.

Where do AGN hosts fit into this landscape? We find that AGNs are commonly present only in galaxies with large masses ($> 10^{10} M_\odot$), stellar velocity dispersions (> 100 km s^{-1}), and surface mass densities ($> 10^8 M_\odot$kpc^{-2})*. Their stellar content and structure varies significantly as a function of AGN luminosity.

Before proceeding, a brief digression is necessary. Any attempt to characterize the stellar population in the AGN hosts with these spectra must, by necessity, include the transition objects that comprise the majority of the AGNs in Figure 22.2. Excluding these would bias the sample against host galaxies with significant amounts of on-going star formation. The only disadvantage of including the transition class is that it is then difficult (on the basis of

---

* Tremonti et al. (in preparation) find the SDSS galaxies obey a strong mass-metallicity relation. Since AGNs are hosted by massive galaxies, they have high metallicity. This explains why the "AGN plume" ascends from the region of metal-rich starformers in Figure 22.2.

the emission-line ratios alone) to classify them as Seyferts with star formation or LINERs with star formation.

It has long been known that LINERs do not attain the high luminosities of powerful Seyfert nuclei (e.g., Heckman 1980b; Ho et al. 2003)*. Thus, rather than classifying AGNs according to their line ratios, it makes more sense to examine how the properties of AGN hosts vary as a function of AGN luminosity (using the [O III] $\lambda5007$ line as a measure). This allows us to include the transition class AGNs and thus mitigate the selection effects described above. For reference, the change-over from a LINER-dominated to Seyfert-dominated population occurs at an extinction-corrected value of $L_{[O\ III]} \approx 10^7 L_\odot$ in the SDSS sample.

We find that the stellar content of the AGN hosts is a strong function of $L_{[O\ III]}$ (Fig. 22.3). This figure shows that the galaxy population as a whole is strongly bimodal (young, low mass, and low density, *or* old, high mass, and high density). What is very striking is that while the weak AGNs roughly follow the same trend as the normal massive galaxies, the powerful AGNs do not. The powerful AGNs are hosted by galaxies that are massive and dense, but relatively young.

Could the relatively small value for the $D_n(4000)$ index be caused by featureless AGN continuum rather than young stars? Decisive evidence against this is provided by the Balmer absorption lines, which are strong in the hosts of powerful AGNs. This is illustrated in Figure 22.4 which compares a high signal-to-noise ratio composite spectrum of several hundred of the most powerful AGNs ($L_{[O\ III]} > 10^8 L_\odot$) to a similar composite spectrum of normal high-mass, star-forming galaxies. More generally, nearly all the powerful AGNs in our sample lie on or above the locus of the continuous models of star formation in Figure 22.1. They have strong Balmer absorption lines as well as a relatively small $D_n(4000)$ index, so the latter is not due to dilution by AGN light.

Kauffmann et al. (2003b) found that the fraction of galaxies that have undergone a major burst of star formation within the past Gyr declines strongly with increasing galaxy mass. The fraction of all SDSS galaxies that have experienced a major burst (at better than the 97.5% confidence level) is only about 0.1% to 0.3% over the range in M_* appropriate for typical AGN hosts. In contrast, the fraction of "high-confidence" bursts in the AGN hosts rises with increasing AGN luminosity, from ∼1% at the lowest luminosities to nearly 10% at the highest.

Thus, it appears that powerful type 2 AGNs (Seyferts) reside in galaxies that have a combination of properties that are rare in the galaxy population as a whole: they are massive (and dense) but have relatively young stellar populations. This may have a very simple explanation. The two necessary ingredients for a powerful AGN are a massive black hole and an abundant fuel supply. Only massive galaxies contain massive black holes, and only galaxies with significant amounts of recent/ongoing star formation have the requisite fuel supply (ISM). This combination is rare today. The most massive black holes live mostly in the barren environment of a massive early-type galaxy. This was evidently not the case during the AGN era.

---

* In the SDSS sample, the luminosity of the [O III] $\lambda5007$ emission line (the strong line from the NLR that is least contaminated by a contribution from galactic H II regions) is larger in type 2 Seyferts than in LINERs by an average factor of about 30 to 100 for galaxies with the same stellar mass or velocity dispersion. If we assume that AGNs hosts obey the (in)famous relation between black hole mass and stellar velocity dispersion, this would imply that on average LINERs are AGNs operating in a mode with a much lower L/L_{Edd} than Seyfert nuclei. The same trend is seen by Ho (2002, 2004) in nearby galaxies

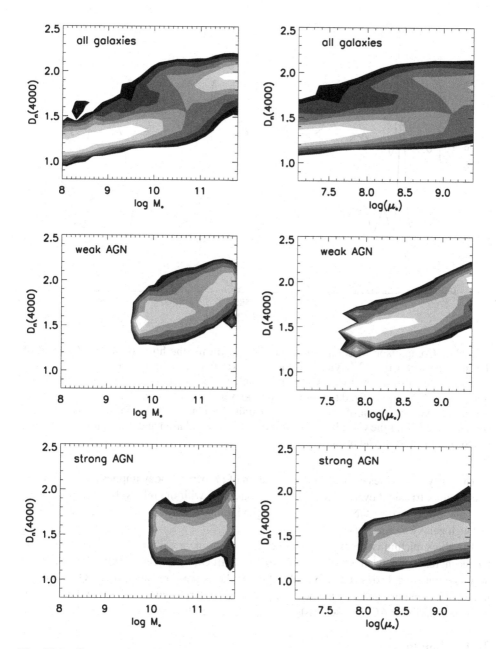

Fig. 22.3. Contour plots of conditional density distributions (see Kauffmann et al. 2003b) showing trends in the age-sensitive index $D_n(4000)$ as a function of stellar mass M_* and surface mass density μ_* for all galaxies (*top*), weak AGNs with $L_{[O\ III]} < 10^7 L_\odot$ (*middle*), and powerful AGNs with $L_{[O\ III]} > 10^7 L_\odot$ (*bottom*). Galaxies have been weighted by $1/V_{max}$, and the bivariate distribution function has been normalized to a fixed number of galaxies in each bin of $\log M_*$ (*left*) and $\log \mu_*$ (*right*). The plots are only made over ranges in M_* and μ_* where there are at least 100 objects per bin (so the plots for the AGN hosts do not extend to low M_* or μ_*). (From Kauffmann et al. 2003c.)

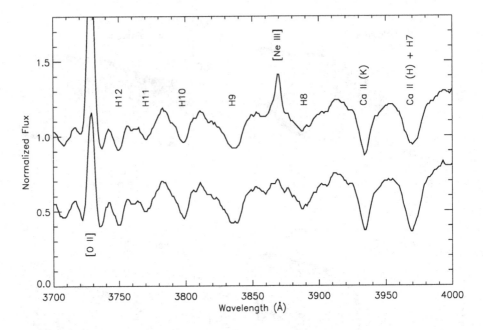

Fig. 22.4. Comparison of composite near-UV spectra for the hosts of several hundred of the most powerful type 2 Seyferts in the SDSS (*top*) with a similar composite spectrum of normal star-forming galaxies with similar stellar masses as the Seyferts (*bottom*). Both spectra have been normalized to unit flux at 4200 Å and then the composite spectrum for the normal galaxies has been offset by −0.5 flux units for clarity. Note that both spectra have similar strengths for the Ca II lines from an old stellar population and the high-order Balmer absorption lines from young stars.

The rarity of galaxies having this combination of properties today implies that we may be witnessing a transient event. This would be consistent with the relatively large fraction of the hosts of powerful AGNs that have undergone a major starburst within the last ∼ 1 Gyr (a rough galaxy merger time scale).

The above picture is appealingly simple. However, one possible "fly in the ointment" is the recent result that the hosts of powerful low-redshift QSOs are very massive but otherwise normal elliptical galaxies (Dunlop et al. 2003). If the stellar population of the QSO hosts is in fact old, then our results on the hosts of powerful Seyfert 2s could mean that the standard unified scenario for AGNs is seriously incomplete at high luminosities.

22.4   Summary

I have reviewed our understanding of the connection between star formation and the AGN phenomenon, emphasizing large surveys that have used direct spectroscopic diagnostics to search for young (< 1 Gyr) stars in AGNs and in their host galaxies in the local Universe. I have considered only those objects in which the blinding glare of the central AGN is obscured from our direct view, so that the optical continuum is dominated by starlight. For these "type 2" AGNs I have used the luminosity of the narrow-line region as an indicator of AGN power.

The major conclusions are as follows:

Galactic Nuclei ($\sim 10^2$ to 10^3 pc): The majority of galaxies of early/middle Hubble type contain AGNs, usually of low luminosity. In most cases, the nuclear stellar population in low-power AGNs is predominantly old. In contrast, a young stellar population is detected in about half of the powerful type 2 Seyfert nuclei.

Host Galaxies ($\sim 10^4$ pc): The host galaxies of AGNs have large stellar masses and surface mass densities. Normal galaxies like this have predominantly old stellar populations. While the stellar content of the hosts of low-power AGNs appears normal, the hosts of powerful AGNs have young stellar populations. A significant fraction of these young hosts have undergone major starbursts in the last ~ 1 Gyr.

The observational evidence for a connection between star formation and black hole fueling is thus clearer in powerful AGNs. This is true both in the nucleus itself and in its host galaxy. Such a dependence is astrophysically plausible, and is qualitatively consistent with the processes that established the correlation between black hole and stellar mass in galaxy bulges.

The most natural explanation for why powerful AGNs are hosted by massive galaxies with young stellar populations is that a powerful AGN requires both a high mass black hole (massive host galaxy) and an abundant fuel supply (a star-forming ISM). While this combination is rare today, it would have been much more common during the AGN era at high redshift.

Acknowledgements. I would like to thank Carnegie Observatories, in particular Luis Ho, for organizing and hosting such a successful meeting. I would also like to thank all my collaborators on the work reported in this paper, with a special thanks to Guinevere Kauffmann who is leading the SDSS project on AGN hosts.

Funding for the creation and distribution of the SDSS Archive has been provided by the Alfred P. Sloan Foundation, the Participating Institutions, the National Aeronautics and Space Administration, the National Science Foundation, the U.S. Department of Energy, the Japanese Monbukagakusho, and the Max Planck Society. The SDSS Web site is http://www.sdss.org/.

The SDSS is managed by the Astrophysical Research Consortium (ARC) for the Participating Institutions. The Participating Institutions are The University of Chicago, Fermilab, the Institute for Advanced Study, the Japan Participation Group, The Johns Hopkins University, Los Alamos National Laboratory, the Max-Planck-Institute for Astronomy (MPIA), the Max-Planck-Institute for Astrophysics (MPA), New Mexico State University, University of Pittsburgh, Princeton University, the United States Naval Observatory, and the University of Washington.

References

Antonucci, R. R. J. 1993, ARA&A, 31, 473

Aretxaga, I., Kunth, D., & Mujica, R. 2001a, Advanced Lectures on the Starburst-AGN Connection (Singapore: World Scientific)

Aretxaga, I., Terlevich, E., Terlevich, R. J., Cotter, G., & Díaz, A. I. 2001b, MNRAS, 325, 636

Baldwin, J. A., Phillips, M. M., & Terlevich, R. 1981, PASP, 93, 5

Balogh, M. L., Morris, S. L., Yee, H. K. C., Carlberg, R. G., & Ellingson, E. 1999, ApJ, 527, 54

Blanton, M. R., Lupton, R. H., Maley, F. M., Young, N., Zehavi, I., & Loveday, J. 2004, AJ, in press

Bruzual A., G., & Charlot, S. 2003, MNRAS, 344, 1000

Calzetti, D., Kinney, A. L., & Storchi-Bergmann, T. 1994, ApJ, 429, 582

Charlot, S., & Fall, S. M. 2000, ApJ, 539, 718

Cid Fernandes, R., Jr., Heckman, T. M., Schmitt, H. R., Golzález Delgado, R. M., & Storchi-Bergmann, T. 2001b, ApJ, 558, 81

Cid Fernandes, R., Jr., Schmitt, H. R., & Storchi-Bergmann, T. 2001a, RMxAC, 11, 133

Cid Fernandes, R., Jr., & Terlevich, R. 1995, MNRAS, 272, 423

Colina, L., & Arribas, S. 1999, ApJ, 514, 637

de Mello, D., Leitherer, C., & Heckman, T. M. 2000, ApJ, 530, 251

Dressler, A., & Gunn, J. E. 1983, ApJ, 270, 7

Dunlop, J. S., McLure, R. J., Kukula, M. J., Baum, S. A., O'Dea, C. P., & Hughes, D. H. 2003, MNRAS, 340, 1095

Fan, X., et al. 2001, AJ, 121, 54

Ferrarese, L., & Merritt, D. 2000, ApJ, 539, L9

Filippenko, A. V. 1992, ed., Relationships between Active Galactic Nuclei and Starburst Galaxies (San Francisco: ASP)

Fukugita, M., Ichikawa, T., Gunn, J. E., Doi, M., Shimasaku, K., & Schneider, D. P. 1996, AJ, 111, 1748

Gebhardt, K., et al. 2000, ApJ, 539, L13

González Delgado, R. M., Heckman, T., & Leitherer, C. 2001, ApJ, 546, 845

González Delgado, R. M., Heckman, T., Leitherer, C., Meurer, C., Krolik, J. H., Wilson, A. S., Kinney, A. L., & Koratkar, A. P. 1998, ApJ, 505, 174

González Delgado, R. M., Leitherer, C., & Heckman, T. 1999, ApJS, 125, 489

Gunn, J. E., et al. 1998, AJ, 116, 3040

Haehnelt, M., & Kauffmann, G. 2000, MNRAS, 318, L35

Heckman, T. M. 1980a, A&A, 87, 142

——. 1980b, A&A, 87, 152

——. 1991, in Massive Stars in Starbursts, ed. C. Leitherer et al. (Cambridge: Cambridge Univ. Press), 289

Heckman, T. M., et al. 1995, ApJ, 452, 549

Heckman, T. M., González Delgado, R. M., Leitherer, C., Meurer, G. R., Krolik, J., Wilson, A. S., Koratkar, A>, & Kinney, A. 1997, ApJ, 482, 114

Ho, L. C. 2002, in Issues in Unification of AGNs, ed. R. Maiolino, A. Marconi, & N. Nagar (San Francisco: ASP), 165

——. 2004, in Carnegie Observatories Astrophysics Series, Vol. 1: Coevolution of Black Holes and Galaxies, ed. L. C. Ho (Cambridge: Cambridge Univ. Press), in press

Ho, L. C., Filippenko, A. V., & Sargent, W. L. W. 1997, ApJS, 112, 315

——. 2003, ApJ, 583, 159

Hogg, D. W., Finkbeiner, D. P., Schlegel, D. J., & Gunn, J. E. 2001, AJ, 122, 2129

Joguet, B., Kunth, D., Melnick, J., Terlevich, R., & Terlevich, E. 2001, A&A, 380, 19

Kauffmann, G., et al. 2003a, MNRAS, 341, 33

——. 2003b, MNRAS, 341, 54

——. 2003c, MNRAS, 346, 1055

Kauffmann, G., & Haehnelt, M. 2000, 311, 576

Kewley, L. J., Dopita, M. A., Sutherland, R. S., Heisler, C. A., & Trevena, J. 2001, ApJ, 556, 121

Koski, A. T. 1978, ApJ, 223, 56

Leitherer, C., et al. 1999, ApJS, 123, 3

Maiolino, R., Ruiz, M., Rieke, G. H., & Keller, L. D. 1995, ApJ, 446, 561

Malkan, M. A., Gorjian, V., & Tam, R. 1998, ApJS, 117, 25

Maoz, D., Koratkar, A. P., Shields, J. C., Ho, L. C., Filippenko, A. V., & Sternberg, A. 1998, AJ, 116, 55

Nelson, C. H., & Whittle, M. 1999, AdSpR, 23, 891

Oliva, E., Origlia, L., Maiolino, R., & Moorwood, A. F. M. 1999, A&A, 350, 9

Pier, J. R., Munn, J. A., Hindsley, R. B., Hennessy, G. S., Kent, S. M., Lupton, R. H., & Ivezić, Z. 2003, AJ, 125, 1559

Raimann, D., Storchi-Bergmann, T., Gonzalez Delgado, R. M., Cid Fernandes, R., Heckman, T., Leitherer, C., & Schmitt, H. 2003, MNRAS, 339, 772

Richards, G. T., et al. 2002, AJ, 123, 2945

Schinnerer, E., Colbert, E., Armus, L., Scoville, N. Z., & Heckman, T. M. 2004, in Coevolution of Black Holes and Galaxies, ed. L. C. Ho (Pasadena: Carnegie Observatories, http://www.ociw.edu/ociw/symposia/series/symposium1/proceedings.html)

Schmitt, H. R., Storchi-Bergmann, T., & Cid Fernandes, R. 1999, MNRAS, 303, 173

Smith, J. A., et al. 2002, AJ, 123, 2121

Steidel, C. C., Adelberger, K. L., Giavalisco, M., Dickinson, M., & Pettini, M. 1999, ApJ, 519, 1

Storchi-Bergmann, T., Cid Fernandes, R., & Schmitt, H. R. 1998, ApJ, 501, 94

Storchi-Bergmann, T., González Delgado, R. M., Schmitt, H. R., Cid Fernandes, R., & Heckman, T. 2001, ApJ, 559, 147

Stoughton, C., et al. 2002, AJ, 123, 485 (erratum: 123, 3487)

Strateva, I., et al. 2001, AJ, 122, 1104

Strauss, M., et al. 2002, AJ, 124, 1810

Tadhunter, C. N., Dickinson, R., Morganti, R., Robinson, T. G., Villar-Martin, M., & Hughes, M. 2002, MNRAS, 330, 977

Terlevich, E., Díaz, A. I., & Terlevich, R. 1990, MNRAS, 242, 271

Terlevich, R. 1989, in Evolutionary Phenomena in Galaxies, ed. J. Beckman, & B. Pagel (Cambridge: Cambridge Univ. Press), 149

Terlevich, R. & Melnick, J. 1985, MNRAS, 213, 841

Tran, H. 1995, ApJ, 440, 597

Veilleux, S. 2001, in Starburst Galaxies: Near and Far, ed. L. Tacconi & D. Lutz (Heidelberg: Springer-Verlag), 88

Veilleux, S., & Osterbrock, D. E. 1987, ApJS, 63, 295

Wills, K. A., Tadhunter, C. N., Robinson, T. G., & Morganti, R. 2002, MNRAS, 333, 211

Worthey, G., & Ottaviani, D. L. 1997, ApJS, 111, 377

York, D. G., et al. 2000, AJ, 120, 1579

23

AGN feedback mechanisms

MITCHELL C. BEGELMAN

JILA, University of Colorado at Boulder

Abstract

Accreting black holes can release enormous amounts of energy to their surroundings, in various forms. Such feedback may profoundly influence a black hole's environment. After briefly reviewing the possible types of feedback, I focus on the injection of kinetic energy through jets and powerful winds. The effects of these outflows may be especially apparent in the heating of the X-ray–emitting atmospheres that pervade clusters of galaxies. Analogous heating effects, during the epoch of galaxy formation, could regulate the growth of supermassive black holes.

23.1   Introduction

Active galactic nuclei (AGNs) release large amounts of energy to their environments, in several forms. In luminous AGNs such as Seyfert nuclei and quasars, the most obvious output is radiative; indeed, radiation can affect the environment through both radiation pressure and radiative heating. Although jets and winds are usually associated with radio galaxies, recent theoretical and observational developments suggest that the kinetic energy output may be as important as (or more important than) the radiative output for most accreting black holes. Finally, significant outputs of energetic particles, whether charged ("cosmic rays") or neutral (relativistic neutrons, neutrinos), cannot be ruled out.

In this review, I focus on the effects of kinetic energy feedback, since these are probably most relevant to the coevolution of black holes and galaxies. Recent observations of galaxy clusters suggest that AGN feedback plays a crucial role in regulating the thermodynamics of the intracluster medium (ICM). I discuss the nature of this interaction, then extrapolate to similar effects that may have operated in protogalaxies, during the era when supermassive black holes were growing toward their present masses.

23.2   Forms of Feedback

Before specializing to the case of kinetic energy injection in cluster atmospheres, I briefly review the various forms of energy injection and summarize their likely effects. This section is an updated version of the discussion given in Begelman (1993).

23.2.1   *Radiation Pressure*

Radiation pressure can exert a force on the gas via electron scattering, scattering and absorption on dust, photoionization, or scattering in atomic resonance lines. Electron scattering is the simplest mechanism to treat, with a cross section of $\langle \sigma/H \rangle \sim 7 \times 10^{-25}x$

© The Observatories of the Carnegie Institution of Washington 2004.

cm^2 per hydrogen atom, where x is the ionized fraction. The maximum column density over which the force can be exerted is given by $N_{H,max} \sim \langle \sigma/H \rangle^{-1} \sim 2 \times 10^{24} x^{-1}$ cm^{-2}.

If the radiation flux from the nucleus does not greatly exceed the Eddington limit of the central black hole, the radiation force exerted through electron scattering will have a relatively minor dynamical effect on the gas in the host galaxy, compared to gravitational and thermal pressure forces. In contrast, radiation pressure acting on dust can exert a much larger force per H atom, although over a correspondingly smaller column density. If we assume a dust-to-gas ratio (by mass) of 0.01, a typical grain size a, and a total cross section per grain of the same order as the geometric cross section (a reasonable assumption for UV and soft X-ray photons hitting ~ 0.1 μm grains), then the cross section per H atom exceeds that for electron scattering in fully ionized gas by a factor of order $10^5 (a/0.1$ μm$)^{-1}$. The force exerted per particle is higher by the same factor, but the column density affected is only $2 \times 10^{19}(a/0.1$ μm$)$ cm^{-2}. Dopita and collaborators have argued that this form of pressure could dominate the dynamics of narrow emission-line regions in AGNs, under certain conditions, and could be crucial for regulating the ionization state (Dopita et al. 2002; Dopita 2003).

The force exerted on $\sim 10^4 - 10^5$ K gas as a result of steady-state photoionization and recombination is characterized by the mean cross section $\langle \sigma/H \rangle \sim 10^{-18}(1-x)$ cm^2. Photoionization equilibrium predicts that $(1-x) \sim 10^{-4} p_{gas}/p_{rad}$, where p_{gas} and p_{rad} are the gas pressure and pressure of ionizing radiation, respectively. (Strictly speaking, one should use $4\pi J/c$ instead of p_{rad}, where J is the mean intensity. However, use of p_{rad} is quantitatively correct and promotes a more physically intuitive discussion.) Thus, we can write $\langle \sigma/H \rangle \sim 10^{-22} p_{gas}/p_{rad}$ cm^2. Note that p_{gas}/p_{rad} is just the reciprocal of the "ionization parameter" Ξ defined by Krolik, McKee, & Tarter (1981). The force per H atom exerted through photoionization is $200 p_{gas}/p_{rad}$ times greater than that exerted through electron scattering.

The largest forces per H atom are possible through scattering in UV resonance lines. For species i, the effective cross section is

$$\left\langle \frac{\sigma}{H} \right\rangle \sim \frac{1}{\Delta \nu_D} \frac{\pi e^2}{m_e c}(A_i X_i f_i), \tag{23.1}$$

where $\Delta \nu_D$ is the Doppler width of the line and A_i, X_i, and f_i are, respectively, the abundance of element i, the fraction of the element in the relevant ionization state, and the oscillator strength of the transition. For important resonance lines such as those of C IV, Si IV, and N V, $A_i X_i f_i$ can attain values of order 10^{-4} for cosmic abundances. The fractional Doppler width, at $T \sim 10^4$ K, is $\Delta \nu_D / \nu \sim 10^{-4}$; hence, we find that the mean cross section can be as large as 10^{-17} cm^2 and the corresponding force as much as *seven orders of magnitude* larger than electron scattering. The drawback is that the bandwidth over which resonance-line scattering is effective is extremely small, fractionally of order $\sim 10^{-4}$ of the ionizing spectrum for each strong resonance line. The amount of momentum available from such a small bandwidth is very small. Therefore, in order for resonance-line scattering to be dynamically important, (1) the gas must accelerate, so that new portions of the spectrum are continuously Doppler shifted into the line (the basis for the Sobolev approximation), and (2) there must be a significant number of lines contributing at different wavelengths. Models of UV resonance-line acceleration for O-star winds (Castor, Abbott, & Klein 1975a) have been adapted to explain the fast $(v \rightarrow 0.1c)$ outflows in broad absorption-line (BAL) QSOs (Arav

& Li 1994; Arav, Li, & Begelman 1994; Murray et al. 1995), where there is circumstantial evidence for acceleration by radiation pressure (Arav 1996; Arav et al. 1999).

For AGN radiation acting on general interstellar matter, the radiation pressure force will be exerted mainly through dust and photoionization. Dynamical effects can be significant if $p_{rad} > p_{ISM}$, where p_{ISM} is the pressure in the undisturbed gas and

$$p_{rad} = \frac{L}{4\pi R^2 c} = 3 \times 10^{-9} L_{46} R_{kpc}^{-2} \text{ dyne cm}^{-2} \tag{23.2}$$

for gas situated R_{kpc} kpc from an isotropic source of ionizing radiation with luminosity $10^{46} L_{46}$ erg s^{-1}.

The maximum column density in a slab of gas that can be fully ionized by AGN continuum is given by

$$N_{H,ion} \sim 10^{22} \ln\left(1 + \frac{p_{rad}}{p_{ISM}}\right) \text{ cm}^{-2} \tag{23.3}$$

if dust absorption is neglected, and only slightly smaller if a cosmic dust abundance is taken into account. If $p_{ISM} < p_{rad}$, then $N_{H,ion}$ gives the depth to which an irradiated cloud will be "pressurized" by the radiation. Differential pressure forces between the front and back of a cloud can lead to a "pancake effect" (Mathews 1982), which tends to squash irradiated clouds down to a column density of order $N_{H,ion}$.

23.2.2   *Radiative Heating*

Gas exposed to ionizing radiation from an AGN tends to undergo an abrupt transition from the typical H II region temperature, $\sim 10^4$ K, to a higher temperature and ionization state when p_{gas}/p_{rad} falls below some critical value, which lies in the range $\sim 0.03 - 0.1$ (McCray 1979; Krolik et al. 1981). The "hot phase" equilibrium temperature is close to the temperature at which Compton cooling balances inverse Compton heating, which can be $T_{IC} \sim 10^6 - 10^7$ K for AGN spectra.

Clouds with column densities greater than $N_{H,ion}$ will not heat up all at once. Only the surface layers will be ablated, and the back-pressure of the ablated gas will keep the cloud interior at a high enough pressure to avoid immediate heating (Begelman, McKee, & Shields 1983; Begelman 1985). Depending on the size of the cloud and the timescales involved, the heated gas may reach a temperature $T_s \sim 10^5 - 10^6$ K at the point at which it becomes supersonic with respect to the cloud surface. A critical condition at the sonic point determines the mass flux per unit area, which is proportional to $p_{rad}/T_s^{1/2}$:

$$\dot{N} \sim (6 \times 10^6 - 6 \times 10^7) \frac{L_{46}}{R_{kpc}^2} \text{ cm}^{-2} \text{ s}^{-1}. \tag{23.4}$$

The lifetime of a cloud against ablation by X-ray heating is given by

$$\frac{N}{\dot{N}} \sim (5 \times 10^6 - 5 \times 10^7) \frac{R_{kpc}^2}{L_{46}} \frac{N_H}{10^{22} \text{ cm}^{-2}} \text{ yr}, \tag{23.5}$$

which corresponds to a global ablation rate of

$$\dot{M}_{abl} \sim (20 - 200) C L_{46} \, M_\odot \text{ yr}^{-1} \tag{23.6}$$

for gas with a covering factor C. The effects of X-ray heating on the evolution of the ISM in a spiral galaxy were studied by Begelman (1985) and by Shanbhag & Kembhavi (1988).

Possible observable consequences of X-ray heating include: (1) elimination of cool ISM phases from the inner parts of the galaxy, which would allow UV radiation to penetrate to much larger distances than would otherwise be possible; (2) modification of ISM phase structure by preferential destruction of small clouds; only clouds (e.g., giant molecular clouds) with a sufficiently large column density would survive long enough to be observed; and (3) generation of peculiar cloud velocities > 100 km s^{-1}, via the "rocket effect," with a random component introduced through cloud-cloud shadowing. The importance of these effects is highly uncertain as they depend critically on the geometric distribution of the dense phases of the ISM, and on the replenishment of ISM through stellar evolutionary and other processes.

23.2.3 *Energetic Particles*

In addition to charged relativistic particles that can diffuse through the surrounding medium ("cosmic rays"), AGNs may also emit "exotic" particle outflows, consisting, for example, of relativistic neutrons and neutrinos (Begelman, Rudak, & Sikora 1990). These would tend to deposit their energy in a volume-distributed fashion, rather than in impulsive fashion at a shock front. Even neutrinos, if they are sufficiently energetic—TeV or above—could be absorbed by nearby stars and heat their interiors (Czerny, Sikora, & Begelman 1991).

Ultrarelativistic neutrons could have particularly interesting effects since relativistic time dilation would allow them to travel large distances unimpeded before they decay and couple to the ambient plasma (Sikora, Begelman, & Rudak 1989). This form of energy injection could drive powerful, fast winds that start far from the central engine (e.g., Begelman, de Kool, & Sikora 1991). We note, however, that to date there is no compelling evidence that most AGNs emit a large fraction of their energy in this form. Dynamical effects of neutron winds have been invoked recently to explain certain features of gamma-ray burst afterglows (Beloborodov 2003; Bulik, Sikora, & Moderski 2004).

23.2.4 *Kinetic Energy*

In addition to the obvious example of radio galaxies, which often release the bulk of their power in the form of jets (Rees et al. 1982; Begelman, Blandford, & Rees 1984), most if not all accreting black holes could produce substantial outflows. Numerical simulations suggest that accretion disks, which transfer angular momentum and dissipate binding energy via magnetorotational instability, may inevitably produce magnetically active coronae (Miller & Stone 2000). These likely generate outflows that are further boosted by centrifugal force (Blandford & Payne 1982). We have already mentioned the possible role of radiation pressure in accelerating winds in BAL QSOs—such outflows could also be boosted hydromagnetically. Whatever the acceleration mechanism(s), these winds are probably accelerated close to the central engine, and therefore should be regarded as part of the kinetic energy output. New spectral analyses of BAL QSOs, made possible by observations with the *Hubble Space Telescope*, imply that the absorption can be highly saturated (Arav et al. 2001) and may originate far from the nucleus (de Kool et al. 2001). This indicates that the kinetic energy in the BAL outflow is larger than previously thought and can approach the radiation output. Moreover, new evidence suggests that relativistic jets are common or ubiquitous in X-ray binaries containing black hole candidates. While the energetics of these outflows are not yet fully established, their environmental impacts may be substantial (Heinz 2002).

Unless radiation removes at least 2/3 of the liberated binding energy, very general theoretical arguments indicate that rotating accretion flows *must* lose mass. The physical reason is that viscous stresses transport energy outward, in addition to angular momentum. If radiation does not remove most of this energy, then a substantial portion of the gas in the flow will gain enough energy to become unbound (Narayan & Yi 1995; Blandford & Begelman 1999). Blandford (this volume) discusses this effect and its consequences in more detail. While it may sometimes be possible to tune the system so that the gas circulates without escaping, any excess dissipation (i.e., increase of entropy) near a free surface of the flow will lead to outflow. There are several possible sources of such dissipation, including magnetic reconnection, shocks, radiative transport, and the magnetocentrifugal coupling mentioned above. If radiative losses are very inefficient, outflows can remove all but a small fraction of the matter supplied at large radii.

23.3   Energy Budget

At an energy conversion efficiency of ϵc^2 per unit of accreted mass, an accreting black hole liberates $10^{19}(\epsilon/0.01)$ erg per gram. In principle the efficiency could be ~ 6 to more than 40 times larger than this (depending on the black hole spin and boundary conditions near the event horizon: Krolik 1999; Agol & Krolik 2000), but we have chosen deliberately to be conservative. ϵ might be viewed as the efficiency of kinetic energy production, since this is probably the most effective means by which black holes affect their surroundings. In a galactic bulge with a velocity dispersion of $200\sigma_{200}$ km s^{-1}, the accretion of one gram liberates enough energy to accelerate $2 \times 10^4(\epsilon/0.01)\sigma_{200}^2$ gm to escape speed— provided that most of the energy goes into acceleration. Given a typical ratio of black hole mass to galactic bulge mass of $\sim 10^{-3}$, feedback from a supermassive black hole growing toward its final mass could easily exceed the binding energy of its host galaxy's bulge.

Under many circumstances, feedback via kinetic energy injection can be quite efficient. Both radiative heating and acceleration by radiation pressure, on the other hand, have built-in inefficiencies. In both photoionization heating and Compton heating, only a fraction of the photon energy goes into heat, the majority being reradiated. For example, the Compton heating efficiency per scattering is $\sim kT_{IC}/m_e c^2 < 10^{-2}$ for typical AGN Compton temperatures. Acceleration by radiation pressure extracts only a fraction $\sim v/c$ of the available energy per scattering, where v is the speed of the accelerated gas. For outflows in BAL QSOs, with velocities of up to $\sim 0.2c$, this can represent an efficient energy injection mechanism, even for single scattering. The energetic efficiency of radiative acceleration is increased if the photons scatter multiple times, to $\tau v/c$, where τ is the optical depth.

Even the efficacy of kinetic energy injection depends on the structure of the medium in which it is deposited. The speed of a shock or sound wave propagating through a medium with a "cloudy" phase structure will be highest in the phase with the lowest density (the intercloud medium). Dense regions will be overrun and left behind by the front, as first pointed out by McKee & Ostriker (1977) in connection with supernova blast waves propagating into the interstellar medium. Consequently, most of the energy goes into the gas which has the lowest density (and is the hottest) to begin with. The global geometric structure of the ambient gas is important as well. Since a wind or hot bubble emanating from an AGN will tend to follow the "path of least resistance," a disklike structure can lead to a "blowout" of the hot gas along the axis. A more spherically symmetric gas distribution will tend to keep the AGN energy confined in a bubble.

23.4 AGN Feedback in Clusters

AGN feedback due to kinetic energy injection is perhaps most evident in clusters of galaxies. Recent observations of intracluster gas by *Chandra* and *XMM-Newton* indicate that some energy source is quenching so-called "cooling flows" in clusters of galaxies (Allen et al. 2001; Fabian et al. 2001; Peterson et al. 2001). Energy injected by intermittent radio galaxy activity at the cluster center is the most likely culprit. The same form of energy input, spread over larger scales, could be responsible for an inferred "entropy floor" in the gas bound to clusters of galaxies (Valageas & Silk 1999; Nath & Roychowdury 2002).

23.4.1 Evolution of Radio Galaxies

Radio galaxies evolve through three stages, only the first of which is dominated by the jet momentum. Although the radio morphologies in powerful (Fanaroff-Riley class II) sources are dominated by the elongated lobes and compact hotspots, most of the energy accumulates in a faint "cocoon" that has a thick cigar shape (Blandford & Rees 1974; Scheuer 1974). The same is probably true of the weaker FR I sources, which appear to be dominated by emission from turbulent regions along the jet. The cocoon is overpressured with respect to the ambient medium, and drives a shock "sideways" at the same time as the jets are lengthening their channels by depositing momentum. The sideways expansion quickly becomes competitive with the lengthening. The archetypal FR II source Cygnus A, for example, which appears to be long and narrow in the radio, displays an aspect ratio < 3 in X-rays (although the hotspots remain prominent in the X-ray image: Wilson, Young, & Shopbell 2000).

Dynamically, active radio galaxies with overpressured cocoons resemble spherical, supersonic stellar wind bubbles (Castor, McCray, & Weaver 1975b; Begelman & Cioffi 1989). To zeroth order, the evolution of the bubble can be described by a self-similar model in which the internal and kinetic energy are comparable, and share the integrated energy output of the wind. The speed of expansion is then

$$v \sim \left(\frac{L_j}{\rho}\right)^{1/5} R^{-2/3},$$
(23.7)

where L_j is the power of the jets, ρ is the ambient density, and R is the radius of the shock. The supersonic expansion phase ends when the expansion speed drops below the sound speed in the ambient medium. This occurs at a radius

$$R_{\text{sonic}} \sim 5 \left(\frac{\langle L_{43}\rangle}{n}\right)^{1/2} T_{\text{keV}}^{-1/4} \text{ kpc},$$
(23.8)

where $\langle L_{43}\rangle$ is the time-averaged jet power in units of 10^{43} erg s^{-1}, n is the ambient particle density in units of cm^{-3}, and T_{keV} is the ambient temperature in units of keV. Thereafter the evolution is dominated by buoyancy (Gull & Northover 1973). We have chosen fiducial parameters that are fairly typical of conditions in cD galaxies at the centers of rich clusters— note how small R_{sonic} is, compared to a typical cluster core radius, or even the core radius of the host galaxy. Cygnus A, which has been expanding for several million years, is hundreds of kpc across, and is still overpressured by a factor $\sim 2-3$ with respect to the ambient medium, is the exception rather than the rule. It is a very powerful source expanding into a relatively tenuous ambient medium (Smith et al. 2002). In the X-ray emitting clusters that we discuss below the central radio galaxies seem to have evolved into the buoyancy-driven stage fairly early.

At least two additional caveats must be taken into account in considering the effects of radio galaxies on their surroundings. First, the active production of jets is probably intermittent—there is indirect statistical evidence that the duty cycle may be as short as 10^5 yr (Reynolds & Begelman 1997). During "off" periods, the overpressured bubble continues to expand as a blast wave with fixed total energy, but the radio emissivity may rapidly fade. Second, the direct influence of the radio galaxy on its surroundings does not end with the onset of buoyancy-driven evolution. As I describe below, the buoyant bubbles of very hot (possibly relativistic) plasma seem fairly immiscible with their surroundings. They can "rise" for considerable distances, spreading the AGN's energy output widely. Both of these points will figure prominently in the next section.

23.4.2   *Quenching of Cooling Flows*

Radiative cooling of gas in the central regions of galaxy clusters often occurs on a timescale much shorter than the Hubble time. In the absence of any heat sources, this implies that the ICM must settle subsonically toward the center in order to maintain hydrostatic equilibrium with the gas at larger radii. The mass deposition rates predicted by this "cooling flow" model are very high and range typically from 10 to 1000 solar masses per year. X-ray observations made prior to the launch of *Chandra* seemed to be broadly consistent with this picture (Fabian 1994). However, the picture has *not* held up well in the era of *Chandra* and *XMM-Newton*. Although both gas temperatures and cooling times are observed to decline toward cluster cores, new observations show a remarkable lack of emission lines from gas at temperatures below ~ 1 keV in the central regions of clusters (Allen et al. 2001; Fabian et al. 2001; Peterson et al. 2001), suggesting a temperature "floor" (Peterson et al. 2003) at about $1/2 - 1/3$ of the temperature at the "cooling radius" (where the cooling timescale equals the Hubble time). Moreover, the mass deposition rates obtained with *Chandra* and *XMM-Newton* using spectroscopic methods are many times smaller than earlier estimates based on *ROSAT* and *Einstein* observations, as well as more recent morphological estimates based on cooling rates alone (David et al. 2001; McNamara et al. 2001; Peterson et al. 2001, 2003). The strong discrepancies between these results indicate that the gas is prevented from cooling by some heating process.

The most plausible candidates for heating the ICM are thermal conduction from the outer parts of the cluster (Bertschinger & Meiksin 1986; Narayan & Medvedev 2001; Voigt et al. 2002; Zakamska & Narayan 2003) and kinetic energy injected by a central AGN. Given the metallicities and colors of the galaxies hosting cooling flows, heating by supernovae and hot stellar winds seems marginally adequate at best (Wu, Fabian, & Nulsen 2000). AGN heating is especially attractive because $\sim 70\%$ of cD galaxies in the centers of cooling flow clusters are radio galaxies (Burns 1990). The *ensemble-averaged* power from radio galaxies is more than sufficient to offset the mean level of cooling (Peres et al. 1998; Böhringer et al. 2002), although not every cluster shows strong radio galaxy activity at the present time. Moreover, AGN heating is naturally concentrated toward the center of the cluster, where the risk of runaway cooling ("cooling catastrophe") is greatest.

It is one thing to argue, on energetic grounds, that AGN feedback is capable of replenishing the heat lost to radiation in cooling flows. It is quite another to determine how this happens in detail. Because of the steep dependence of the radiative cooling function on density, it has proven notoriously difficult to "stabilize" cooling flows, so that heating approximately balances cooling at all radii. For example, Meiksin (1988) found that conduction could not stop cooling catastrophes in the central regions of clusters, although it could offset cooling in the outer parts if the temperature gradient were not too large. But even this requires fine tuning of the conductivity (via a "magnetic suppression factor" relative to the Spitzer value) and boundary conditions, as too large a conduction rate will lead to a nearly isothermal temperature distribution, contrary to observations. Indeed, as Loeb (2002) points out, a large enough conductivity to suppress cooling flows, if extrapolated to cluster envelopes, would cause them to evaporate. Early attempts to offset cooling using a central heat source (e.g., Loewenstein, Zweibel, & Begelman 1991, using cosmic rays) ran into similar fine-tuning problems.

The episodic model for cluster heating (Binney & Tabor 1995; Ciotti & Ostriker 1997, 2001) attempts to avoid these generic difficulties. No steady state is sought. Instead, the cluster atmosphere goes through repeated cycles of cooling and infall—which fuel the central AGN—followed by heating and outflow. The rapid heating and expansion of the ICM turns off the fuel supply to the AGN, the initial conditions of the cluster atmosphere are "reset," and the process repeats. The energy injection process, whether due to jets (as in Binney & Tabor) or inverse Compton heating (as in Ciotti & Ostriker), is violent and heats the ICM from the inside out. This creates an observational challenge for this class of models, since they generally predict that the temperature should decrease outward during the heating phase—the opposite to what is seen. Nor are the expected strong shocks observed. If one concludes from this that the heating episodes somehow elude observation, the same can be said for the cooling catastrophes—the inevitable conclusion of each cooling phase. These are also not seen, although Kaiser & Binney (2003) point out that they may be sufficiently short-lived to have evaded detection in existing datasets. (We also note that the Compton temperatures assumed by Ciotti & Ostriker are based on extreme—and highly beamed—spectra of blazars, and are probably far too high to be realistic. Using more realistic AGN Compton temperatures will weaken this mechanism to the point where it is probably not effective. Thus, if episodic mechanisms work at all, it is probably through kinetic energy injection.)

The absence of strong shocks bounding radio galaxy lobes is a major observational surprise. For example, the X-ray–bright rims surrounding the radio lobes of 3C 84 (NGC 1275) in the Perseus cluster are cooler than their surroundings (Fabian et al. 2000; Fig. 23.1), contrary to predictions (Heinz et al. 1998). This probably results from the entrainment and lifting of low-entropy gas, from the cluster center, into regions where the ambient entropy is higher (Reynolds, Heinz, & Begelman 2001, 2002; Quilis, Bower, & Balogh 2001; Brighenti & Mathews 2002; Nulsen et al. 2002). It is probably not due to the *in situ* radiative cooling of shock-compressed gas bounding the radio lobes. This shows that 3C 84 has already evolved to the buoyancy-driven stage; similar conclusions can be drawn for other cluster cores with prominent radio sources.

Another surprise is the apparent "immiscibility" of the hot (possibly relativistic) plasma injected by the jets and the thermal ICM. It has been known since the time of *ROSAT* (Böhringer et al. 1993; McNamara, O'Connell, & Sarazin 1996) that the plasma in radio lobes can displace cooler thermal gas, creating "holes" in the X-ray emission. More sensitive *Chandra* imaging has shown not only how common such holes are, but also how long they can persist. In particular, numerous examples of "ghost cavities" have been found

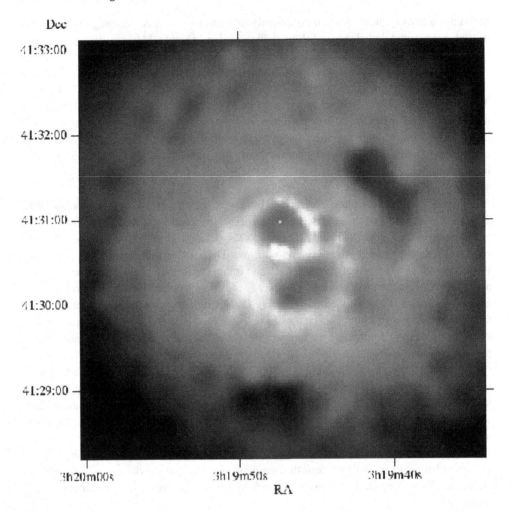

Fig. 23.1. Adaptively smoothed, 0.5–7 keV *Chandra* image of the core of the Perseus cluster (from Fabian et al. 2000). The radio lobes of 3C 84 coincide with the central X-ray holes, which are bounded by bright (cool) rims. The more distant "ghost cavities" are thought to be buoyant bubbles created by earlier episodes of activity.

(e.g., McNamara et al. 2001; Johnstone et al. 2002; Mazzotta et al. 2002). These are presumably buoyant bubbles left over from earlier epochs of activity; several examples are seen in Fig. 23.1.

The persistence of highly buoyant bubbles may be key to understanding how AGNs heat cluster atmospheres. Strongly positive entropy gradients observed in cluster atmospheres (David et al. 2001; Böhringer et al. 2002) appear to rule out standard convection. But the Schwarzschild criterion refers to heat transport by marginally buoyant fluid elements, not the highly buoyant bubbles that appear to be present. Numerical simulations are beginning to address how buoyant plumes of plasma injected by jets can increase the potential and thermal energy of the ICM (Quilis et al. 2001; Reynolds et al. 2001, 2002; Churazov et al. 2002; Brüggen & Kaiser 2002); spread out laterally (into "mushroom clouds": Churazov et al. 2001), yielding more even distribution of the injected energy; and persist long after the observable radio lobes have faded (Brüggen et al. 2002; Reynolds et al. 2002; Basson & Alexander 2003). The latter point is especially important given statistical (Reynolds & Begelman 1997) and morphological (e.g., Virgo cluster: Young, Wilson, & Mundell 2002; Forman et al. 2004; Perseus cluster: Fabian et al. 2000, 2002) evidence that typical radio galaxy activity is intermittent, possibly with a short duty cycle. Moreover, both radio and X-ray observations suggest that the energy ultimately gets distributed remarkably evenly (e.g., Owen, Eilek, & Kassim 2000), despite the apparent immiscibility noted above. Whether this mixing is due to the propagation of (magneto-)acoustic waves, buoyancy, Kelvin-Helmholtz instabilities, unsteadiness in the jets, mixing by "cluster weather" (due, e.g., to galaxy motion or cluster mergers), or other effects remains unclear.

Bubbles rising subsonically do *pdV* work on their surroundings as they traverse the pressure gradient. Since the timescale for the bubbles to cross the cluster (of order the free-fall time) is much shorter than the cooling timescale, the flux of bubble energy through the ICM approaches a steady state, implying that details of the energy injection process—such as the number flux of bubbles (e.g., one big one or many small ones), the bubble size, filling factor, and rate of rise—do not affect the mean heating rate. If we assume that the acoustic energy generated by the *pdV* work is dissipated within a pressure scale height of where it is generated, we can devise an average volume heating rate for the ICM, as a function of radius (Begelman 2001b):

$$\mathcal{H} \sim \frac{\langle L \rangle}{4\pi r^3} \left(\frac{p}{p_0} \right)^{1/4} \left| \frac{d \ln p}{d \ln r} \right|. \tag{23.9}$$

In eq. (23.9), $\langle L \rangle$ is the time-averaged power output of the AGN, $p(r)$ is the pressure inside the bubbles (and p_0 is the pressure where the bubbles are formed), and the exponent $1/4$ equals $(\gamma - 1)/\gamma$ for a relativistic plasma (the exponent would be $2/5$ for a nonrelativistic gas). A major assumption of the model, that the *pdV* work is absorbed and converted to heat within a pressure scale height, will have to be assessed using 2-D and 3-D numerical simulations and studies of the microphysics of cluster gas. We have also assumed that the energy is spread evenly over 4π sr, a likely consequence of buoyancy.

The most important property of the above "effervescent heating" rate is its proportionality to the pressure gradient (among other factors), since this determines the rate at which *pdV* work is done as the bubbles rise. Since thermal gas that suffers excess cooling will develop a slightly higher pressure gradient, the effervescent heating mechanism targets exactly those regions where cooling is strongest. Therefore, it has the potential to stabilize radiative cooling (Begelman 2001b). This potential is borne out in 1-D, time-dependent numerical simulations (Ruszkowski & Begelman 2002), which show that the flow settles down to a steady state that resembles observed clusters, for a wide range of parameters and without fine-tuning the initial or boundary conditions (Fig. 23.2). Even though these models include conduction (at 23% of the Spitzer rate), which may be necessary for global stability (Zakamska & Narayan 2003), the heating is overwhelmingly dominated by the AGN ($\geq 4 : 1$, at all radii; see Fig. 23.3). The mass inflow rate through the inner boundary, which determines the AGN feedback in these simulations, stabilizes to a reasonable value far below that predicted by cooling flow models.

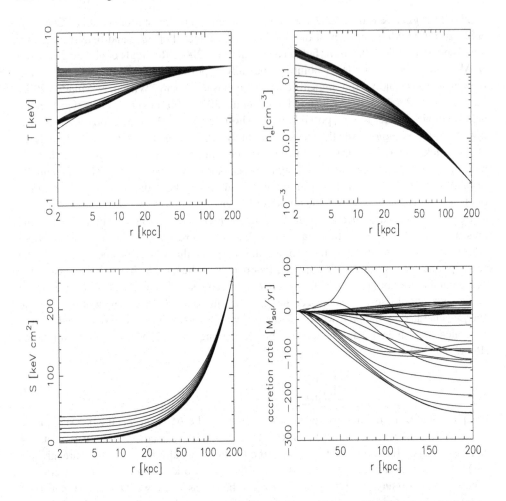

Fig. 23.2. ZEUS 1-D simulation of cluster evolution over one Hubble time, including effervescent heating, thermal conduction, and AGN feedback. The figure shows time sequences of (clockwise from upper left) temperature, density, accretion rate, and specific entropy profiles. The model settles down to a stable, steady state, which is visible via the dense concentration of curves. See Ruszkowski & Begelman (2002) for further details.

23.4.3   The "Entropy Floor"

AGN feedback may do more than offset radiative losses in the cores of certain clusters. Cluster X-ray luminosities and gas masses increase with temperature more steeply than predicted by hierarchical merging models (Markevitch 1998; Nevalainen, Markevitch, & Forman 2000). In other words, the atmospheres in less massive clusters and groups are hotter than they should be, given the gravitational interactions that assembled them. These correlations apply to regions of clusters well outside the cooling radius, as well as to clusters without cooling cores. They can be interpreted as evidence for an entropy "floor" (Lloyd-Davies, Ponman, & Cannon 2000), indicating that low-entropy material is removed either by cooling and mass dropout (presumably to form stars: Bryan 2000; Voit et al. 2002) or

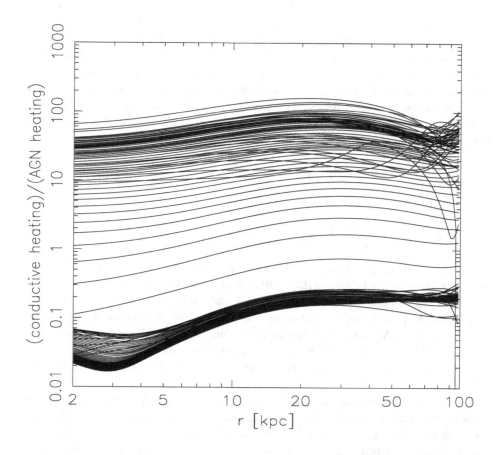

Fig. 23.3. Time sequence showing ratio of conductive heating rate to local AGN heating rate, as a function of radius, for the model displayed in Fig. 23.2. Conduction coefficient is set to 0.23 of Spitzer value. Although conduction dominates to begin with, it offsets only $\sim 10-15\%$ of the radiative cooling in the final state; the rest is due to the AGN.

by substantial AGN heating before or during cluster assembly (e.g., Valageas & Silk 1999; Nath & Roychowdury 2002; McCarthy et al. 2002, and references therein).

Mechanisms like those described above should also operate on these larger scales. For example, if the pressure gradient inside the cooling radius is not too steep, the effervescent heating model predicts that a substantial fraction of the injected energy will escape to radii where the cooling time is longer than the Hubble time. The calculations have not been done yet.

23.5   Feedback and the Growth of Supermassive Black Holes

The correlations measured between black hole masses and the velocity dispersions and/or masses of their host galaxies' bulges (Magorrian et al. 1998; Ferrarese & Merritt 2000; Gebhardt et al. 2000) suggest a direct relationship between supermassive black hole (SMBH) formation and galaxy formation. Inventories of quasar light (Sołtan 1982; Yu &

Tremaine 2002, and references therein) and the hard X-ray background (Fabian & Iwasawa 1999) suggest that much of the black hole growth occurred through (radiatively efficient) accretion, rather than through hierarchical mergers of smaller black holes. If even a few percent of the liberated energy emerged in kinetic form, as seems very likely, then there would have been more than enough energy to unbind the gas in the protogalactic host. Thus, several authors have suggested that SMBHs limited their own growth, or even the growth of the host galaxy, by depositing this energy in their surroundings (e.g., Silk & Rees 1998; Blandford 1999; Fabian 1999). If the energy is deposited adiabatically, the feedback luminosity L_f can unbind the gas provided that

$$L_f > \frac{\sigma^5}{G} = 5 \times 10^{43} \sigma_{200}^5 \text{ erg s}^{-1}. \tag{23.10}$$

This implies a limiting black hole mass

$$M_\bullet \sim 10^8 \left(\frac{L_f}{0.004 L_E} \right)^{-1} \sigma_{200}^5 \, M_\odot, \tag{23.11}$$

where L_E is the Eddington limit. SMBHs must have accreted a significant fraction of their masses at close to the Eddington limit (Blandford 1999; Fabian 1999), since statistical arguments suggest that black hole growth took only a few Eddington e-folding times ($\sim$ few $\times$ 40 Myr), and since at least some supermassive black holes existed only 1–2 Gyr after the Big Bang. Arguments based on the quasar luminosity function suggest that the most massive black holes might even have grown at several times the Eddington rate (Begelman 2001a, 2002) or with a higher radiative efficiency than is normally assumed (Yu & Tremaine 2001).

The above estimate assumes weak radiative losses in the protogalactic medium. However, the protogalactic environment is likely to be even more cooling-dominated than cluster cores (White & Rees 1978; Fabian 1999). If cooling is important it would require more energy to unbind the gas, by a factor that could be as large as $c/\sigma = 1500 \sigma_{200}^{-1}$. The limiting mass is then

$$M_\bullet \sim 6 \times 10^8 \left(\frac{L_f}{L_E} \right)^{-1} \sigma_{200}^4 \, M_\odot. \tag{23.12}$$

The difference between these two limiting cases is analogous to the difference between the expansion of a supernova remnant in the energy-conserving Sedov (blast wave) phase, and its rapid deceleration during the momentum-conserving (radiative) "snowplow" phase. Reality is likely to be somewhere in between, with the flow of AGN energy partially trapped by density inhomogeneities that result from rapid cooling.

23.6   Conclusions

AGN feedback effects, potentially enormous on the basis of energetic arguments, depend sensitively on both the form of feedback and the detailed structure of the environment. The efficiency of feedback due to radiation is often small, except under particular circumstances. Kinetic energy injected by the AGN tends to be trapped by the ambient medium, leading to a higher efficiency. Recent theoretical advances suggest that accreting black holes often return a large fraction of the liberated energy to the environment in the form of winds and jets.

The X-ray emitting atmospheres in clusters of galaxies provides an excellent testbed for the effects of AGN feedback. Recent X-ray observations show that radio galaxies can blow long-lasting "holes" in the ICM, and may offset the effects of radiative losses well enough to hamper large-scale inflows. Further observations, and sophisticated numerical simulations, will be needed to fully understand these interactions.

Acknowledgements. This work was supported in part by NSF grant AST–9876887. I thank Mateusz Ruszkowski for many useful discussions, and for supplying Fig. 23.3.

References

Agol, E., & Krolik, J. H. 1999, ApJ, 528, 161

Allen, S. W., et al. 2001, MNRAS, 324, 842

Arav, N. 1996, ApJ, 465, 617

Arav, N., et al. 2001, ApJ, 561, 118

Arav, N., Korista, K. T., de Kool, M., Junkkarinen, V. T., & Begelman, M. C. 1999, ApJ, 516, 27

Arav, N., & Li, Z. Y. 1994, ApJ, 427, 700

Arav, N., Li, Z. Y., & Begelman, M. C. 1994, ApJ, 432, 62

Basson, J. F., & Alexander, P. 2003, MNRAS, 339, 353

Begelman, M. C. 1985, ApJ, 297, 492

——. 1993, in The Environment and Evolution of Galaxies, ed. J. M. Shull & H. A. Thronson Jr. (Dordrecht: Kluwer), 369

——. 2001a, ApJ, 551, 897

——. 2001b, in Gas and Galaxy Evolution, ed. J. E. Hibbard, M. P. Rupen, & J. H. van Gorkum (San Francisco: ASP), 363

——. 2002, ApJ, 568, L97

Begelman, M. C., Blandford, R. D., & Rees, M. J. 1984, Rev. Mod. Phys., 56, 255

Begelman, M. C., & Cioffi, D. F. 1989, ApJ, 345, L21

Begelman, M. C., de Kool, M., & Sikora, M. 1991, ApJ, 382, 416

Begelman, M. C., McKee, C. F., & Shields, G. A. 1983, ApJ, 271, 70

Begelman, M. C., Rudak, B., & Sikora, M. 1990, ApJ, 362, 38 (Erratum: 370, 791 [1991])

Beloborodov, A. M. 2003, ApJ, 585, L19

Bertschinger, E., & Meiksin, A. 1986, ApJ, 306, L1

Binney, J., & Tabor, G. 1995, MNRAS, 276, 663

Blandford, R. D. 1999, in Galaxy Dynamics, ed. D. R. Merritt, M. Valluri, & J. A. Sellwood (San Francisco: ASP), 87

Blandford, R. D., & Begelman, M. C. 1999, MNRAS, 303, L1

Blandford, R. D., & Payne, D. G. 1982, MNRAS, 199, 883

Blandford, R.D., & Rees, M. J. 1974, MNRAS, 169, 395

Böhringer, H., Matsushita, K., Churazov, E., Ikebe, Y., & Chen, Y. 2002, A&A, 382, 804

Böhringer, H., Voges, W., Fabian, A. C., Edge, A. C., & Neumann, D. M. 1993, MNRAS, 318, L25

Brighenti, F., & Mathews, W. G. 2002, ApJ, 574, L11

Brüggen, M., & Kaiser, C. R. 2002, Nature, 418, 301

Brüggen, M., Kaiser, C. R., Churazov, E., & Enßlin, T. A. 2002, MNRAS, 331, 545

Bryan, G. L. 2000, ApJ, 544, L1

Bulik, T., Sikora, M., & Moderski, R. 2004, in Proc. XXXVIIth Rencontres de Moriond, Les Arcs, France, 4-9 March 2002, in press (astro-ph/0209339)

Burns, J. O. 1990, AJ, 99, 14

Castor, J. I., Abbott, D. C., & Klein, R. I. 1975a, ApJ, 195, 157

Castor, J., McCray, R., & Weaver, R. 1975b, ApJ, 200, L107

Churazov, E., Brüggen, M., Kaiser, C. R., Böhringer, H., & Forman, W. 2001, ApJ, 554, 261

Churazov, E., Sunyaev, R., Forman, W., & Böhringer, H. 2002, MNRAS, 332, 729

Ciotti, L., & Ostriker, J. P. 1997, ApJ, 487, L105

——. 2001, ApJ, 551, 131

Czerny, M., Sikora, M., & Begelman, M. C. 1991, in Relativistic Hadrons in Cosmic Compact Objects, Lecture Notes in Physics, Vol. 391 (Berlin: Springer-Verlag), 23

David, L. P., Nulsen, P. E. J., McNamara, B. R., Forman, W., Jones, C., Ponman, T., Robertson, B., & Wise, M. 2001, ApJ, 557, 546

de Kool, M., Arav, N., Becker, R. H., Gregg, M. D., White, R. L., Laurent-Muehleisen, S. A., Price, T., & Korista, K. T. 2001, ApJ, 548, 609

Dopita, M. A. 2003, Ap&SS, 284, 569

Dopita, M. A., Groves, B. A., Sutherland, R. S., Binette, L., & Cecil, G. 2002, ApJ, 572, 753

Fabian, A. C. 1994, ARA&A, 32, 277

——. 1999, MNRAS, 308, L39

Fabian, A. C., et al. 2000, MNRAS, 318, L65

Fabian, A. C., Celotti, A., Blundell, K. M., Kassim, N. E., & Perley, R. A. 2002, MNRAS, 331, 369

Fabian, A. C., Mushotzky, R. F., Nulsen, P. E. J., & Peterson, J. R. 2001, MNRAS, 321, L20

Ferrarese, L., & Merritt, D. 2000, ApJ, 539, L9

Forman, W., Jones, C., Markevitch, M., Vikhlinin, A., & Churazov, E. 2004, in XIII Rencontres de Blois 2001, ed. L. M. Celnikier, in press (astro-ph/0207165)

Gebhardt, K., et al. 2000, ApJ, 539, L13

Gull, S. F., & Northover, K. J. E. 1973, Nature, 224, 80

Heinz, S. 2002, A&A, 388, L40

Heinz, S., Reynolds, C. S., & Begelman, M. C. 1998, ApJ, 501, 126

Johnstone, R. M., Allen, S. W., Fabian, A. C., & Sanders, J. S. 2002, MNRAS, 336, 299

Kaiser, C. R., & Binney, J. J. 2003, MNRAS, 338, 837

Krolik, J. H. 1999, ApJ, 515, L73

Krolik, J. H., McKee, C. F., & Tarter, C. B. 1981, ApJ, 249, 422

Lloyd-Davies, E. J., Ponman, T. J., & Cannon, D. B. 2000, MNRAS, 315, 689

Loeb, A. 2002, NewA, 7, 279

Loewenstein, M., Zweibel, E. G., & Begelman, M. C. 1991, ApJ, 377, 392

Magorrian, J., et al. 1998, AJ, 115, 2285

Markevitch, M. 1998, ApJ, 504, 27

Mathews, W. G. 1982, ApJ, 252, 39

Mazzotta, P., Kaastra, J. S., Paerels, F. B., Ferrigno, C., Colafrancesco, S., Mewe, R., & Forman, W. R. 2002, ApJ, 567, L37

McCarthy, I. G., Babul, A., & Balogh, M. L. 2002, ApJ, 573, 515

McCray, R. 1979, in Active Galactic Nuclei, ed. C. Hazard & S. Mitton (Cambridge: Cambridge Univ. Press), 227

McKee, C. F., & Ostriker, J. P. 1977, ApJ, 218, 148

McNamara, B. R., et al. 2001, ApJ, 562, L149

McNamara, B. R., O'Connell, R. W., & Sarazin, C. L., 1996, AJ, 318, 91

Meiksin, A. 1988, ApJ, 334, 59

Miller, K. A., & Stone, J. M. 2000, ApJ, 534, 398

Murray, N., Chiang, J., Grossman, S. A., & Voit, G. M. 1995, ApJ, 451, 498

Narayan, R., & Medvedev, M. V. 2001, ApJ, 562, L129

Narayan, R., & Yi, I. 1995, ApJ, 444, 231

Nath, B. B., & Roychowdhury, S. 2002, MNRAS, 333, 145

Nevalainen, J., Markevitch, M., & Forman, W. 2000, ApJ, 532, 694

Nulsen, P. E. J., David, L. P., McNamara, B. R., Jones, C., Forman, W. R., & Wise, M. 2002, ApJ, 568, 163

Owen, F. N., Eilek, J. A., & Kassim, N. E. 2000, ApJ, 543, 611

Peres, C. B., Fabian, A. C., Edge, A. C., Allen, S. W., Johnstone, R. M., & White, D. A. 1998, MNRAS, 298, 416

Peterson, J. R., et al. 2001, A&A, 365, L104

Peterson, J. R., Kahn, S. M., Paerels, F. B. S., Kaastra, J. S., Tamura, T., Bleeker, J. A. M., Ferrigno, C., & Jernigan, J. G. 2003, ApJ, 590, 207

Quilis, V., Bower, R. G., & Balogh, M. L. 2001, MNRAS, 328, 1091

Rees, M. J., Begelman, M. C., Blandford, R. D., & Phinney, E. S. 1982, Nature, 295, 17

Reynolds, C. S., & Begelman, M. C. 1997, ApJ, 487, L135

Reynolds, C. S., Heinz, S., & Begelman, M. C. 2001, ApJ, 549, L179

——. 2002, MNRAS, 332, 271

Ruszkowski, M., & Begelman, M. C. 2002, ApJ, 581, 223

Scheuer, P. A. G. 1974, MNRAS, 166, 513

Shanbhag, S., & Kembhavi, A. 1988, ApJ, 334, 34

Sikora, M., Begelman, M. C., & Rudak, B. 1989, ApJ, 341, L33

Silk, J., & Rees, M. J. 1998, A&A, 331, L1

Smith, D. A., Wilson,A. S., Arnaud, K. A., Terashima, Y., & Young, A. J. 2002, ApJ, 565, 195

Sołtan, A. 1982, MNRAS, 200, 115

Valageas, P., & Silk, J. 1999, A&A, 350, 725

Voigt, L. M., Schmidt, R. W., Fabian, A. C., Allen, S. W., & Johnstone, R. M. 2002, MNRAS, 335, L7

Voit, G. M., Bryan, G. L., Balogh, M. L., & Bower, R. G. 2000, ApJ, 576, 601

White, S. D. M., & Rees, M. J. 1978, MNRAS, 183, 341

Wilson, A. S., Young, A. J., & Shopbell, P. L. 2000, ApJ, 544, L27

Wu, K. K. S., Fabian, A. C., & Nulsen, P. E. J. 2000, MNRAS, 318, 889

Young, A. J., Wilson, A. S., & Mundell, C. G. 2002, ApJ, 579, 560

Yu, Q., & Tremaine, S. 2002, MNRAS, 335, 965

Zakamska, N. L., & Narayan, R. 2003, ApJ, 582, 162

24

Pieces of the galaxy formation puzzle: where do black holes fit in?

RACHEL S. SOMERVILLE
Space Telescope Science Institute

Abstract

Over the past twenty years we have developed a fairly credible story of how galaxies form in the context of the hierarchical structure formation paradigm. A major challenge for the coming years is to understand the coevolution of galaxies and the supermassive black holes that we now believe inhabit most of them. I discuss the status of our understanding of the main processes that shape galaxies, comparing where possible the results of the complementary techniques of numerical hydrodynamics and semi-analytic modeling. I suggest that recent progress in understanding the connection of the structural properties (matter density and angular momentum profiles) of dark matter halos will help us to understand the inner structure of galaxies and hence to better model black hole formation and fueling. I also discuss some of the outstanding problems in galaxy formation, and how including the formation of supermassive black holes might help alleviate them.

24.1 Introduction

In the modern hierarchical paradigm of structure formation, galaxies are molded by a complex web of interconnected processes. They are harbored within dark matter halos, which in turn are shaped by their larger scale environment and their history. They receive fuel for star formation via radiative cooling of hot gas that is trapped within the potential wells of these halos. By some poorly understood process, this cold gas is eventually transformed into stars, and the most massive of these stars return heavy elements and energy to the interstellar medium (ISM) and perhaps to the hot gas halo. In some cases, supernovae may even drive large-scale winds, which spew material far beyond the potential well of the galaxy, into the intergalactic medium, polluting the material that will collapse to form future generations of galaxies. While this basic picture forms a plausible story of how galaxies form and what determines their properties, and detailed calculations based upon it have yielded some nontrivial successes in predicting or interpreting various galaxy observations, there are some indications that there are some important pieces missing in our understanding. It is the growing realization that perhaps we need to understand the *coevolution* of supermassive black holes (SMBHs) and galaxies in order to fill in some of these missing pieces that has inspired this meeting, and that will no doubt be the subject of much thought in the near future.

In this article, I review the status of recent attempts to model these key processes, making wherever possible explicit comparisons between the complementary methods of "semi-analytic" modeling and numerical (or *N*-body) hydrodynamic simulations. While

© The Observatories of the Carnegie Institution of Washington 2004.

to date there have been very few attempts to explicitly model the coevolution of SMBHs and galaxies, I will attempt to highlight aspects of our understanding of galaxy formation that might help us to understand the formation of SMBHs, and aspects of SMBH formation and the associated AGN phase that might impact our understanding of galaxy formation.

24.2   The Internal Structure of Dark Matter Halos

24.2.1   *Halo Density Profiles and Mass Accretion History*

Over the past five to ten years, the vast improvement in resolution and dynamic range accessible to N-body techniques with modern supercomputers has led to a characterization of the internal structure of dark matter (DM) halos that form in the cold dark matter (CDM) paradigm. It is a now familiar result that the radial density profiles of DM halos may be well described by a universal function, which has come to be known as the "Navarro-Frenk-White" (NFW) profile (Navarro, Frenk, & White 1997). The NFW profile allows halos (assumed to be spherically symmetric) to be characterized by two parameters, for example, a total mass and a characteristic radius or density. The characteristic radius or density may also be expressed as a *concentration*, which represents how much of the halo's mass lies near the center versus the outskirts. While early work (Navarro et al. 1997) emphasized the correlation between these two parameters (lower mass halos are more concentrated), studies of larger, statistically complete samples of halos have revealed that there is also a significant scatter in halo concentration at fixed mass (Bullock et al. 2001a).

The origin of the correlation between concentration and mass, and the scatter at fixed mass, is now understood as being connected to the *mass accretion history* of the halo. By studying *structural merger trees*, in which the merger histories of dark matter halos were traced through a high-resolution N-body simulation in conjunction with the halo structural properties, Wechsler et al. (2002) showed that early-forming halos have high concentrations, while late-forming halos have lower concentrations. This accounts for the mass-concentration correlation, as larger mass halos form later on average than smaller mass halos. Moreover, Wechsler et al. (2002) presented a recipe for modeling halo concentration based on semi-analytic merger trees, and showed that the scatter in concentration at fixed mass seen in the N-body simulations can be accounted for by the spread in formation histories for halos of a given mass. This recipe will also allow future semi-analytic models to properly include the correlations between merger history and halo structure, which will probably have important implications for predicted galaxy properties.

24.2.2   *Halo Spin and Merger History*

Another fundamental property of a dark matter halo is its angular momentum, often characterized by the dimensionless spin parameter, λ (Peebles 1969). It is well established that the distribution function of spins for DM halos formed in CDM simulations has the form of a log-normal (Barnes & Efstathiou 1987), and is insensitive to cosmic epoch, halo mass or environment, and cosmology or power spectrum (Lemson & Kauffmann 1999). The origin of this angular momentum is generally attributed to tidal torques from the large-scale density field, experienced before turnaround, while the proto-halo perturbations are in the linear regime (Peebles 1969; Doroshkevich 1970; White 1984). This picture, combined with the invariance of the *distribution* of spins with cosmic epoch, may have led to the impression that spin is a property that remains invariant over the lifetime of a halo. However, it appears that halos acquire angular momentum in *mergers*, via transfer of orbital angular

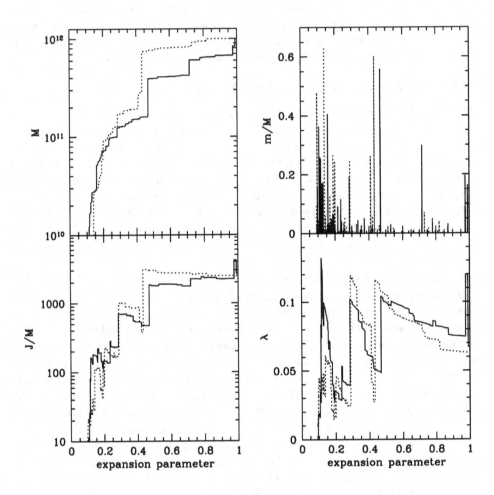

Fig. 24.1. *Left panels:* The build-up of mass (top) and specific angular momentum (bottom) for two different halos (solid and dotted) as a function of expansion factor $a = (1+z)^{-1}$. *Right panels:* Mass ratios of merging satellites (top) and spin parameter (bottom) as a function of a. (Reproduced from Vitvitska et al. 2002.)

momentum to internal spin. Because minor mergers (i.e., events in which one member of the merging pair is much smaller in mass than the other) are frequent and more or less isotropic, the angular momentum from these events mostly cancels out. However, major mergers (in which the lumps have comparable masses) are rare and therefore tend to result in a significant net gain of angular momentum (depending, of course, on the impact parameter of the merger). This is illustrated, based on the same structural merger trees discussed above, in Figure 24.1, and has obvious significance for our understanding of the structure of galactic disks (discussed in more detail below). Based on these ideas, Vitvitska et al. (2002) and Maller, Dekel, & Somerville (2002) proposed analytic recipes that could be included in semi-analytic merger trees to model halo spin, and showed that such a scheme reproduces the log-normal distribution obtained in the *N*-body simulations. Maller et al. (2002) further showed that this approach could reproduce the known insensitivity of halo spin to mass, redshift, and cosmology, as well as the correlation with merger history noted above.

24.3   Formation of Galactic Disks

The basic picture of disk formation in the context of CDM was outlined by White & Rees (1978) and Fall & Efstathiou (1980), which incorporated the earlier work of Mestel (1963), Crampin & Hoyle (1964), and others. The theory has been further developed in many other works (e.g., Blumenthal et al. 1986; Flores et al. 1993; Dalcanton, Spergel, & Summers 1997; Mo, Mao, & White 1998; van den Bosch 2000; Avila-Reese & Firmani 2000; Klypin, Zhao, & Somerville 2002). In this picture, halos (both dark matter and diffuse gas) obtain angular momentum from tidal torques as discussed above, and this gas then conserves its angular momentum as it collapses to form a disk. This leads to a direct relationship between the angular momentum (or spin) of the halo and the size (scale length) of the disk. The analytic work showed that as long as most of the initial angular momentum is conserved during the process of disk formation, then disk galaxies are predicted to have sizes consistent with observed galaxies at $z = 0$. Put another way, if this picture is correct, then somehow cosmological tidal torques seem to impart just enough spin to the proto-galactic gas to make galaxies with about the right amount of specific angular momentum—a rather profound result.

24.3.1   The Angular Momentum Catastrophe

Unfortunately, when we try to simulate disk formation in a CDM Universe directly using numerical techniques the results are not so successful. When cooling is included in hydrodynamic *N*-body codes, coherently rotating, disklike structures are formed—but in the first generation of simulations to attempt this calculation, the disks were deficient in specific angular momentum by about an order of magnitude compared to observed disk galaxies (Navarro & Benz 1991; Navarro & White 1994; Sommer-Larsen, Gelato, & Vedel 1999; Navarro & Steinmetz 2000). Evidently, the idealized picture of disk formation outlined above does not describe what actually happens in the hydro simulations, where gas efficiently loses most of its angular momentum during cooling and collapse, resulting in disks that are too compact. This result has come to be known as the "angular momentum catastrophe."

The angular momentum catastrophe has been linked to the "bottom-up" nature of structure formation in CDM, and the very efficient early cooling in small proto-galactic clumps. Once gas is locked up in these small clumps, it tends to lose angular momentum very efficiently via dynamical friction. Indeed, it seems that if one can somehow delay cooling, the angular momentum catastrophe is much alleviated. This has been demonstrated in a variety of ways: by simply switching cooling on by hand only at a relatively late epoch (Weil, Eke, & Efstathiou 1998; Eke, Efstathiou, & Wright 2000), by implementing an inefficient star formation algorithm (Sáiz et al. 2001), by adopting an alternative cosmology with late structure formation such as warm dark matter (Sommer-Larsen & Dolgov 2001), or by including strong supernova feedback (Thacker & Couchman 2001). However, it is known that our Galaxy and M31 have very old stars even in the outer parts of their disks (e.g., Ferguson & Johnson 2001)—delaying cooling and star formation may lead to conflict with this observational fact.

It also seems that recent higher resolution simulations may suffer a less severe catastrophe than their predecessors, suggesting that insufficient resolution may remain a concern (Governato et al. 2004). A recent study by Abadi et al. (2003) suggests that perhaps the problem is not with the properties of the disks themselves, but that galaxies are too spheroid-dominated. They argue that, *taken in isolation*, i.e. with the contribution of the spheroid removed, their simulated disk is not significantly deficient in angular momentum for its circular velocity or luminosity. (They present results for a single galaxy only, which was selected from a larger, lower resolution cosmological box and re-simulated at high resolution.) However, the galaxy they analyzed has half its stellar mass in a bulge, and resembles a very early-type spiral or lenticular galaxy. Because these simulations are computationally very expensive, it is difficult to produce statistically complete samples of galaxies, and thus to make general statements about populations. These results suggest, however, that the problem may be one of the morphological mix in the simulations: too much stellar mass ends up in low-angular momentum spheroids.

To summarize, there are at least three rather different ideas prevalent in the community regarding the root cause of the angular momentum catastrophe:

- resolution or other numerical effects
- too much early merging, perhaps reflecting an excess of small-scale power in the current "standard Λ" cosmological model, resulting in a too-spheroid-dominated population
- lack of understanding of astrophysical processes such as star formation and feedback

Note that the approach needed to solve the problem will be quite different depending on which of these is the most important effect. In the first case, we just need to wait for faster computers, while in the second case, we need to modify the assumed cosmology or power spectrum, and in the third case, we need to modify the treatment of star formation and/or feedback. Of course, it is entirely possible that a combination of more than one, or even of all three, of these effects is contributing to the problem. More detailed analytic and semi-analytic studies may help to disentangle these effects (e.g., Maller & Dekel 2002).

24.3.2 *The Angular Momentum Profile Mismatch*

The more detailed analytic models run into problems of their own, however. In order to make a connection between the structural properties of the halo and those of the galaxy that forms within it, we must know either the functional form of the initial (pre-collapse) *radial distribution* of angular momentum in the halo, *or* the form of the final (post-collapse) radial profile of the disk. In the absence of detailed knowledge of the angular momentum profiles of halos (which require very high-resolution simulations to determine numerically), it became a common practice to assume that the post-collapse gas surface density profile has an exponential form (e.g., Blumenthal et al. 1986; Flores et al. 1993; Mo et al. 1998; Klypin et al. 2002). This is generally justified by arguing that the radial surface brightness profiles of observed stellar disks are typically well approximated by exponentials over most of their extent.

Recent studies have revealed that the radial angular momentum profile of material within dark matter halos, like the matter density profiles discussed earlier, have a universal form when scaled appropriately. The specific angular momentum as a function of radius $j(r)$ is found to be well fit by a power-law form: $j(r) \propto r^\alpha$ with $\alpha = 1.1 \pm 0.3$ (Bullock et al. 2001b; see also van den Bosch et al. 2002). The value of the profile slope α has a significant scatter, and obeys a Gaussian distribution. With the usual assumption that the angular momentum is well aligned, that mass shells do not cross and angular momentum is conserved during collapse, we can now compute the resulting radial profile of a disk that forms in such a halo—and it is not an exponential (Bullock et al. 2001b; van den Bosch 2001, 2002). There is a "cusp" of excess low angular momentum material in the inner part, and a "tail" of high angular momentum material in the outer part of the disk. This is what I refer to as the "angular momentum profile mismatch."

It has been proposed that the low-angular material may form a bulge, while the high-angular material may be inefficient at forming stars and remain gaseous. This does not work, of course, to explain bulgeless galaxies with pure exponential profiles, which, while rare, do exist. An alternate explanation is that disks are not born with exponential profiles, but evolve into them via secular processes (Lin & Pringle 1987; Slyz, Devriendt, & Silk 2002), but it is probably difficult for secular evolution to move low-angular momentum material *outward*, to larger radii in the disk—perhaps this material has to be removed from the galaxy entirely.

What does any of this have to do with black holes, since black hole properties seem to correlate with spheroid, not disk, properties? First, mergers efficiently strip angular momentum from the gas and stars in a galaxy, and angular momentum loss on kpc scales is the first step in the "feeding chain" that funnels material to the accretion disk around the black hole, as discussed by Wada (this volume). If the angular momentum catastrophe is a symptom of a deeper malady, that the predominant CDM paradigm predicts too much merging, then calculations of black hole growth via mergers (e.g., Volonteri, Haardt, & Madau 2003), quasar merger models such as those discussed by Haehnelt (this volume), and all related quantities will be in error. Second, if the solution to the catastrophe is strong feedback, which delays cooling until the merger rate has decreased, then this process, which regulates spheroid growth, must also regulate black hole growth in order to maintain the observed link between the two. This *may* be a key to understanding the black hole-spheroid scaling relations. Alternatively, perhaps the direction of the causality is reversed: suppose that after a merger, gas is efficiently funneled into the center of the galaxy, not only producing a spheroid but also feeding the central black hole and producing AGN activity. In turn, when the accretion-powered jets turn on, they drive the remaining gas out to large distances, where it may take a very long time to cool and fall back in. In this case, the key to solving the angular momentum catastrophe for galactic disks may lie in the observed connection between black holes and spheroids. Third, if we want to partially solve the angular momentum profile mismatch by converting the low-j material in disks into spheroids, then we have to come to grips with two separate mechanisms for forming spheroids (merging and secular evolution). Observational evidence that there are two different populations of spheroids is already mounting (see the paper by Carollo, this volume, and references therein). If this is the case, and if it is also the case that the "pseudo-bulges" produced via secular evolution have the same scaling relations as the *bona fide* ones produced by mergers (see papers by Kormendy and Richstone, this volume), then we need to understand *why*.

24.4   Radiative Cooling

Gas in the low-density IGM is cold and, after the epoch of reionization, photoionized. This gas may be heated to higher temperature (a few $\times 10^6$ K) by shocks during gravitational collapse. Thus the gas in collapsed structures such as halos and even filaments is hot enough and dense enough to cool efficiently by atomic line cooling. This is the dominant process in producing the fuel for star formation in galaxies.

24.4.1 *Semi-analytic Models versus Hydro*

Numerical techniques such as smoothed particle hydrodynamics (SPH) or grid-based hydrodynamic codes treat the processes of gravitational collapse in an expanding Universe with a CDM power spectrum and the shock heating and radiative cooling of gas described above in order to track the rate at which gas cools and becomes available for star formation. In the absence of complications such as heavy element pollution and non-gravitational heating, the physics of these processes are well understood and the main uncertainties arise from numerical issues. An alternative approach is to use a simple spherical model (as proposed by White & Frenk 1991) applied within a semi-analytic halo merger tree to estimate the bulk rate of cooling, as is the common practice in semi-analytic models of galaxy formation (e.g., Kauffmann, White, & Guiderdoni 1993; Cole et al. 1994, 2000; Kauffmann et al. 1999; Somerville & Primack 1999). This model relies on several assumptions:

- halos are spherical, have no substructure, and have a prescribed density profile (typically isothermal or NFW)
- all gas is shock heated to the virial temperature of the halo upon collapse
- the hot halo gas is in collisional equilibrium
- the cooling time is equal to the time needed for the gas to radiate all its thermal energy at the rate given by the radiative cooling function, for gas of the specified density and temperature

Because the density decreases with increasing radius, this model predicts that the halo cools from the inside out. In practice, for a given density profile one can calculate a "cooling radius" (see, e.g., Somerville & Primack 1999), within which the gas is assumed to be "cold," and outside of which it is assumed to be "hot." While this picture is obviously rather artificial, and several if not all of the assumptions listed above are known to be violated in the numerical simulations (for example, most of the cooling occurs in filaments, and gas is never shock heated to the virial temperature of the halo), there is apparently quite good agreement between the results of the hydrodynamic simulations and the semi-analytic models, at least for bulk properties such as the cold gas fraction as a function of halo mass (Benson et al. 2001; Yoshida et al. 2002; Helly et al. 2003).

24.4.2 *Problems: Overcooling, Cooling Flows, and Entropy Floors*

Unfortunately, though not surprisingly, there are several indications that reality is more complicated than the simple case described above, which we can understand and simulate readily. There are three separate but interconnected problems related to cooling, all of which hint at the need for an additional source of non-gravitational heating.

It was pointed out in early explorations of galaxy formation within the CDM paradigm (White & Rees 1978; White & Frenk 1991) that in the absence of some braking process, far too much gas is able to cool, leading to an overly large fraction of baryons in the form of stars or cold gas today. For a recent update on both theoretical and observational aspects of this problem, which is often referred to as the "overcooling problem," see Balogh et al. (2001). The overcooling problem is one compelling indication of the need for a process that either prevents gas from cooling as efficiently as expected, or that heats it up again.

Supernova (SN) feedback is generally invoked to do the job, although it is not clear that the energetics of SN-driven winds are up to the task (more discussion of this question follows). Another candidate, which has received less attention, is heating by AGNs.

A second problem, long suspected but only recently made compelling, is the cooling flow conundrum (see the recent review by Fabian 2004). From arguments like those presented above, cooling times in cluster cores are expected to be short, and so one expects a cooling flow to form. However, *XMM-Newton* spectra show no evidence that cooling below 1–2 keV is taking place (e.g., Kaastra et al. 2001; Peterson et al. 2001). *Chandra* images of these clusters show holes coincident with radio lobes and cold fronts, indicating that these systems are complex (Fabian 2004, and references therein). The most popular solutions may be divided into two broad categories: an additional heating source, such as AGN or substructure, or heating by thermal conduction. In some (but not all) of the systems, there is direct evidence for AGN heating (e.g., observed radio jets), and the energetics are plausible, but the heating must occur over very large scales, which is a bit of a puzzle. Thermal conduction works by using the large reservoir of hotter gas in the outskirts of the cluster to heat the gas in the core. It has been considered reasonable to neglect conduction in clusters because of the conventional wisdom that magnetic fields strongly suppress conduction perpendicular to the field. Recently, however, this idea has been re-examined, and it has been pointed out that in a turbulent magnetized plasma, the field lines become tangled, allowing for more efficient conduction (Narayan & Medvedev 2001). The current situation seems to be that for some clusters, AGN heating seems to be the most plausible mechanism, while in others, conduction appears likely to be able to do the job (Fabian, Voigt, & Morris 2002; Voigt et al. 2002), suggesting that both may play a role.

The third problem is in understanding the observed scaling relations between X-ray properties of clusters, such as the X-ray luminosity-temperature (L_X–T_X) relation. The observed relation deviates from the self-similar behavior expected from gravitational heating and a constant baryon fraction, becoming steeper on the low-T end (Allen & Fabian 1998; Markevitch 1998; Arnaud & Evrard 1999). This may be expressed as the presence of an entropy "floor," a minimum entropy that is negligible in larger systems but dominates in smaller, group-sized systems (Ponman, Cannon, & Navarro 1998; Lloyd-Davies, Ponman, & Cannon 2000). The question is: what produces this entropy floor? Again, feedback from SN and/or AGNs are obvious candidates. It has been argued (e.g., Kravtsov & Yepes 2000; Wu, Fabian, & Nulsen 2000) that the energy produced by SN is insufficient, but this is difficult to show conclusively as it depends on the density and temperature of the ICM or proto-ICM at the time at which the energy is injected. Nath & Roychowdhury (2002) show, based on a semi-analytic calculation, that the energy injected by radio jets and broad absorption-line quasars brings the predicted L_X–T_X relation into agreement with observations. See the paper by Begelman (this volume) for discussion of much more detailed models of the impact of AGNs on the cluster entropy problem.

I have discussed these three problems in some detail because all of them point out holes in our understanding of the processes that shape galaxy formation. Overcooling in clusters remains a problem in all existing semi-analytic models and hydrodynamic simulations. It is generally "cured" in semi-analytic models by invoking an *ad hoc* means of suppressing cooling in large halos. It is possible, even likely, that the energy released during the fueling of supermassive black holes in galaxies plays an important role in solving one or even all of these problems.

24.5   Star Formation and Feedback

24.5.1   *Star Formation*

Star formation in cosmological simulations remains a bit like that Far Side cartoon where, in the midst of the calculation on the blackboard it is written "and then, a miracle happens...." Once gas has cooled down to about 10^4 K, it is turned into stars with an efficiency that is determined by some sort of empirical/phenomenological recipe. In the approach most commonly used in cosmological hydrodynamic simulations, star formation is assumed to occur when some combination of the following conditions is satisfied:

- overdensity greater than threshold value
- temperature less than threshold value
- physical density greater than threshold value
- convergent flow ($\nabla \cdot v < 0$)
- Jeans unstable

and the star formation time scale is then generally assumed to be proportional to the dynamical time or the minimum of the cooling and dynamical times. An alternative method is to impose a relationship between star formation rate and local gas density, i.e. a Schmidt law $\dot{\rho}_* \propto \rho_{gas}^N$ with $N \approx 1.5$, based on the empirical relations described by, for example, Kennicutt (1989, 1998). A very nice comparison of different star formation recipes in SPH simulations has recently been presented by Kay et al. (2002). This work showed that the fraction of baryons in stars at $z = 0$ may vary by up to a factor of 2 for the range of recipes and parameter values appearing in the literature.

It is clear that to make real progress, we must grapple with the multi-phase nature of the ISM and with exchange between different phases; for instance, molecular hydrogen clouds form out of cold, dense, gas, while subsequent star formation may evaporate the surrounding clouds and return material to the warm, low-density ISM. There have been a few recent attempts to include a multi-phase star formation algorithm in a hydrodynamical code (e.g., Yepes et al. 1997; Springel & Hernquist 2003a). However, at some level, the results still rely on unknown free parameters.

In semi-analytic codes, only gross information at best is available about the structure of the galaxy and the density of the cold gas, and star formation must therefore be treated even more schematically than in the numerical simulations. Typically, the star formation rate is modeled using an equation of the general form:

$$\dot{\rho}_* = \frac{m_{\text{cold}}}{\tau_*}, \tag{24.1}$$

where m_{cold} is the mass of cold gas in the galaxy (Kauffmann et al. 1993, 1999; Cole et al. 1994, 2000; Somerville & Primack 1999). A common choice for the star formation time scale $\tau_* \propto t_{\text{dyn}}$, with t_{dyn} the dynamical time of the disk. This then resembles the Kennicutt law and is also similar to the recipes often used in the hydro codes. However, the typical t_{dyn} (as well as t_{dyn} for a halo of a given mass) decreases rapidly with increasing redshift, because of the higher density of objects that collapsed at high redshift, when the background density was also much higher than today. This implies that the characteristic time scale for star formation is much shorter at high redshift than today. On the face of it, this appears to be in agreement with observations, but it turns out that if the Kennicutt-style scaling remains unchanging with cosmic time, star formation is *so efficient* in the very dense galaxies at high

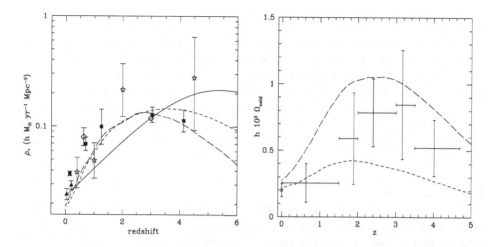

Fig. 24.2. *Top:* The global star formation rate density of the Universe as a function of redshift. Symbols show observational estimates from the compilation in SPF. The solid line shows the fit to the results of Springel & Hernquist (2003b) from hydrodynamical simulations with a multi-phase treatment of star formation and feedback. The dashed lines are results from semi-analytic models similar to those of SPF. The short dashed line shows a model using the "Kennicutt" scaling for star formation in normal galaxies, and has starbursts in major mergers. The long dashed line is a model with constant star formation efficiency ($\tau_* = $ const; see text) and bursts in both major and minor mergers, as in the "collisional starburst" model of SPF. *Bottom:* The global matter density in the form of cold neutral gas (in units of the critical density). Symbols show observational estimates (see SPF), while the dashed lines show the same models shown in the top panel.

redshift that, despite higher cooling rates, nearly all available gas is consumed very rapidly. This has a number of bad effects. It leaves insufficient cold gas lying around at $z \approx 3$ to account for the observed number of damped Lyα systems (Somerville, Primack, & Faber 2001, hereafter SPF). The brightest Lyman-break galaxies at $z \gtrsim 2$ have star formation rates of several hundred solar masses per year, when dust extinction is accounted for (Adelberger & Steidel 2000), while the objects detected in the sub-mm by SCUBA, believed to be at similar redshifts, have implied star formation rates of thousands of solar masses per year (Blain et al. 2002, and references therein). It is very difficult to understand how to produce these extreme objects unless large reservoirs of cold gas are available at high redshift. With the Kennicutt-style star formation recipe, galaxy mergers tend to be fairly gas poor at all redshifts, making it difficult to account for the steep decline of the number density of bright quasars at $z \lesssim 2$ in a model in which black hole fueling is triggered by mergers (Kauffmann & Haehnelt 2000). A modified star formation recipe, in which the star formation time scale in "normal" (i.e., non-starburst) galaxies remains *constant* with redshift, while a more efficient starburst mode is triggered by galaxy mergers, is much more successful in reproducing all of these observations (Kauffmann & Haehnelt 2000; SPF). However, this implies that the empirical scaling relations that guide our understanding of star formation at low redshift are not invariant in cosmic time—if correct, quite an important conclusion.

Given the different levels of detail in the implementation of star formation algorithms in the numerical hydrodynamic simulations and in the semi-analytic models, one might expect the results to be quite different even though the recipes are based on similar ideas. Although no detailed comparisons have been done to date, it seems that the results are actually surprisingly similar. For example, the masses and clustering properties of galaxies at $z \approx 3$ in the hydrodynamic simulations of Weinberg, Hernquist, & Katz (2002) are quite similar to those predicted by semi-analytic models (SPF; Wechsler et al. 2001). The star formation history produced by the recent SPH simulations with multi-phase star formation of Springel & Hernquist (2003a,b) also looks reasonably similar to the one produced by semi-analytic models (Fig. 24.2).

24.5.2 Supernova Feedback

Massive stars and supernova heat the ISM and can in some cases drive large-scale winds that lead to mass loss. I have already discussed the compelling need for some form of efficient feedback in CDM-based models of galaxy formation in order to avert the overcooling crisis and perhaps to solve the cooling flow conundrum, the entropy floor problem, and the angular momentum catastrophe. In addition, supernova feedback might be the best way to expel heavy elements from the dense inner parts of galaxies where they are created and strew them throughout the low-density IGM. By examining the steep slope of the CDM mass function at the low-mass end ($dN/dM \propto -2$), compared with the much flatter slope of the galaxy luminosity function, we can also quickly conclude that this feedback needs to be strongly *differential*, i.e., it needs to be more efficient in low-mass halos than in high mass halos (White & Frenk 1991).

Although there is a long list of problems that we hope supernova feedback can solve, we do not know how to simulate it and there are few direct observational constraints. Some recent observational work (Martin 1999; Heckman et al. 2000) suggests that in nearby starburst galaxies, the rate at which cold gas is heated and returned to the warm diffuse ISM is approximately equal to the star formation rate. The reheating rate seems to be independent of the galaxy circular velocity, but gas may only be able to escape the potential well of the galaxy entirely (blow-away) if the circular velocity of the galaxy is less than ~ 100 km s^{-1} (Martin 1999). At high redshift, there is evidence for powerful winds and high mass loss rates in the $z \approx 3$ Lyman-break galaxies (Pettini et al. 2001), and, intriguingly, the "Martin-Heckman" law based on low-redshift observations (reheating rate $\simeq$ star formation rate) seems to hold.

As with star formation, current cosmological N-body simulations simply do not have the dynamic range to directly model the effect of supernova feedback on a galaxy, and therefore again various phenomenological recipes are used. In hydrodynamical simulations, a typical approach is to return some fraction of the energy of each SN (really, of each "star particle," each of which usually weighs at least several million solar masses) to the ISM in the form of thermal energy (e.g., Katz 1992). In practice, this tends to have little effect on the galaxy, as the star formation takes place where the gas is dense and so the energy is returned to an environment where cooling times are short. The gas quickly radiates away the excess energy and very little (if any) mass is actually ejected from the galaxy, so the net effect is small. Various tricks have been used to attempt to get around this problem, including imposing a time lag by turning off cooling for some period of time following the reheating (Thacker & Couchman 2001), and keeping the energy in a "turbulent reservoir" for some time period before letting it leak out (Springel 2000). These kinds of tricks do help to reduce the efficiency of star formation, but still do not lead to large amounts of mass loss from the galaxy.

An alternative approach involves imparting kinetic energy to the gas by giving small "kicks" to the neighboring particles (e.g., Navarro & White 1993). This is more effective in removing gas from the dense parts of the galaxy, but can lead to numerical problems, particularly in particle-based codes such as SPH—because the local density is found by smoothing over the N nearest neighbors (typically $N \simeq 32$), when particles escape into very underdense regions with very few neighboring particles it becomes impossible to calculate the density and temperature accurately. In addition, when applied within simulations of individual disk galaxies, the "kinetic" method was found to vertically thicken the disk too much unless very low fractions ($\sim 1.0 \times 10^{-5}$) of the SN energy were assumed to be manifested as kinetic energy (e.g., Mihos & Hernquist 1994), in which case the feedback again had very little effect. Kay et al. (2002) have presented a useful comparison of several of these commonly used feedback techniques implemented within an SPH code.

Mac Low & Ferrara (1999) simulated the effect of supernova explosions on individual galaxies at very high resolution and found that significant mass loss from SN-driven winds ("blow-away") occurs only in very low-mass galaxies ($M_{gas} \approx 10^8 M_\odot$).

As with star formation, to make progress, probably one must consider the multi-phase nature of the ISM. Several recent analytic and semi-analytic works have approached this problem, with promising results (Efstathiou 2000; Silk 2001). Springel & Hernquist (2003a) have presented a feedback scheme based on the multi-phase picture of the ISM outlined by McKee & Ostriker (1977), implemented within SPH. Their results are stable to numerical resolution, unlike many previous implementations, but they still have to introduce large-scale winds essentially by hand.

To summarize, while there are many suggestions that some sort of feedback is needed that removes a significant fraction of the gas and metals from galaxies, there has been little success in simulating this process in a physical way under the assumption that the putative winds are driven by massive stars and/or supernovae. Once again, the powerful jets that are known to be associated with accretion onto black holes, which have so far been entirely neglected in detailed cosmological simulations, may have an important role to play in shaping the lives of galaxies.

24.6 Conclusions

The discovery of the tight correlation between the masses of supermassive black holes and the properties of the galactic spheroids they inhabit points to the inevitable conclusion that either galaxies know about their black holes and respond accordingly, that black hole growth is regulated by the galaxy, or that some external force shapes both the black hole and the galaxy. In this article, I have outlined ways in which things that we think we understand about galaxy formation could help us understand the formation of SMBHs, and ways in which things that we *do not* understand about galaxy formation might be informed by considering the influence of the SMBH that we now believe to reside in nearly all galaxies, especially the energy released during the feeding of these beasts.

Here is a summary of the main points:

(1) We are in the process of gaining a better understanding of the internal structure of dark matter halos (their matter density and angular momentum) and the connection with the mass

accretion and merger history. This will help improve modeling of the inner structure of galaxies, which will be crucial to understanding the formation and fueling of central black holes.

(2) There is a series of possibly interconnected problems in CDM-based models of galaxy formation:

- the angular momentum catastrophe and the angular momentum profile mismatch
- the overcooling crisis
- the cooling flow conundrum
- the entropy floor problem

All suggest the need for efficient non-gravitational heating and mass loss from galaxies. Supernovae and AGNs are both good candidates for driving these processes, and it remains to be seen what the balance of importance will be between the two. Heavy elements may be able to provide important clues here.

(3) Modeling of star formation and supernova feedback in cosmological simulations remains extremely primitive, partly due to our lack of understanding of these processes and partly due to the numerical challenge of the very large dynamical range involved. Progress in both areas will probably only be made by explicitly treating the multi-phase nature of the ISM.

A recurring theme in this paper has been the need for some process or processes that make cooling and star formation less efficient by heating or physically removing cold gas from galaxies. So far, it seems difficult to accomplish this at the needed level with supernovae. This may reflect limitations in simulation techniques; however, it may hint that the powerful winds associated with accretion onto SMBHs (e.g., manifested as radio galaxies and broad absorption-line quasars) have an important role to play in filling in some of the missing pieces in our understanding of galaxy formation. There is no doubt that pursuing the interconnection between galaxies, quasars, and the supermassive black holes they harbor will be an important topic of investigation in the years to come.

References

Abadi, M. G., Navarro, J. F., Steinmetz, M., & Eke, V. R. 2003, ApJ, 591, 499
Adelberger, K. L., & Steidel, C. C. 2000, ApJ, 544, 218
Allen, S. W., & Fabian, A. C. 1998, MNRAS, 297, L57
Arnaud, M., & Evrard, A. E. 1999, MNRAS, 305, 631
Avila-Reese, V., & Firmani, C. 2000, RevMexAA, 36, 23
Balogh, M. L., Pearce, F. R., Bower, R. G., & Kay, S. T. 2001, MNRAS, 326, 1228
Barnes, J. E., & Efstathiou, G. 1987, ApJ, 319, 575
Benson, A. J., Pearce, F. R., Frenk, C. S., Baugh, C. M., & Jenkins, A. 2001, MNRAS, 320, 261
Blain, A. W., Smail, I., Ivison, R. J., Kneib, J.-P., & Frayer, D. T. 2002, Phys. Rep., 369, 111
Blumenthal, G. R., Faber, S. M., Flores, R., & Primack, J. R. 1986, ApJ, 301, 27
Bullock, J. S., Dekel, A., Kolatt, T. S., Kravtsov, A. V., Klypin, A. A., Porciani, C., & Primack, J. R. 2001, ApJ, 555, 240
Bullock, J. S., Kolatt, T. S., Sigad, Y., Somerville, R. S., Kravtsov, A. V., Klypin, A. A., Primack, J. R., & Dekel, A. 2001, MNRAS, 321, 559
Cole, S., Aragón-Salamanca, A., Frenk, C. S., Navarro, J. F., & Zepf, S. E. 1994, MNRAS, 271, 781
Cole, S., Lacey, C., Baugh, C., & Frenk, C. S. 2000, MNRAS, 319, 168
Crampin, D. J., & Hoyle, F. 1964, ApJ, 140, 99
Dalcanton, J. J., Spergel, D. N., & Summers, F. J. 1997, ApJ, 482, 659
Doroshkevich, A. G. 1970, Astrofizika, 6, 581
Efstathiou, G. 2000, MNRAS, 317, 697
Eke, V., Efstathiou, G., & Wright, L. 2000, MNRAS, 315, L18

Fabian, A. C. 2004, in Galaxy Evolution: Theory and Observations, ed. V. Avila-Reese et al., RevMexAA SC, in press (astro-ph/0210150)

Fabian, A. C., Voigt, L. M., & Morris, R. G. 2002, MNRAS, 335, L71

Fall, S. M., & Efstathiou, G. 1980, MNRAS, 193, 189

Ferguson, A. M. N., & Johnson, R. A. 2001, ApJ, 559, L13

Flores, R., Primack, J. R., Blumenthal, G. R., & Faber, S. M. 1993, ApJ, 412, 443

Governato, F., et al. 2004, ApJ, submitted (astro-ph/0207044)

Heckman, T. M., Lehnert, M. D., Strickland, D. K., & Armus, L. 2000, ApJS, 129, 493

Helly, J. C., Cole, S., Frenk, C. S., Baugh, C. M., Benson, A., Lacey, C., & Pearce, F. R. 2003, MNRAS, 338, 913

Kaastra, J., Ferrigno, C., Tamura, T., Paerels, F. B. S., Peterson, J. R., & Mittaz, J. P. D. 2001, A&A, 365, L99

Katz, N. 1992, ApJ, 391, 502

Kauffmann, G., Colberg, J. M., Diaferio, A., & White, S. D. M. 1999, MNRAS, 303, 188

Kauffmann, G., & Haehnelt, M. 2000, MNRAS, 311, 576

Kauffmann, G., White, S. D. M., & Guiderdoni, B. 1993, MNRAS, 264, 201

Kay, S. T., Pearce, F. R., Frenk, C. S., & Jenkins, A. 2002, MNRAS, 330, 113

Kennicutt, R. C. 1989, ApJ, 344, 685

——. 1998, ApJ, 498, 181

Klypin, A. A., Zhao, H. S., & Somerville, R. S. 2002, ApJ, 573, 597

Kravtsov, A. V., & Yepes, G. 2000, MNRAS, 318, 227

Lemson, G., & Kauffmann, G. 1999, MNRAS, 302, 111

Lin, D. N. C., & Pringle, J. E. 1987, ApJ, 320, L87

Lloyd-Davies, E. J., Ponman, T. J., & Cannon, D. B. 2000, MNRAS, 315, 689

Mac Low, M.-M., & Ferrara, A. 1999, ApJ, 513, 142

Maller, A. H., & Dekel, A. 2002, MNRAS, 335, 487

Maller, A. H., Dekel, A., & Somerville, R. S. 2002, MNRAS, 329, 423

Markevitch, M. 1998, ApJ, 504, 27

Martin, C. L. 1999, ApJ, 513, 156

McKee, C. F., & Ostriker, J. P. 1977, ApJ, 218, 148

Mestel, L. 1963, MNRAS, 126, 553

Mihos, J. C., & Hernquist, L. 1994, ApJ, 437, 611

Mo, H. J., Mao, S., & White, S. D. M. 1998, MNRAS, 295, 319

Narayan, R., & Medvedev, M. V. 2001, ApJ, 562, L129

Nath, B., & Roychowdhury, S. 2002, MNRAS, 333, 145

Navarro, J. F., & Benz, W. 1991, ApJ, 380, 320

Navarro, J. F., Frenk, C. S., & White, S. D. M. 1997, ApJ, 490, 493

Navarro, J. F., & Steinmetz, M. 2000, ApJ, 538, 477

Navarro, J. F. & White, S. D. M. 1993, MNRAS, 265, 271

——. 1994, MNRAS, 267, 401

Peebles, P. J. E. 1969, ApJ, 155, 393

Peterson, J. R., et al. 2001, A&A, 365, L104

Pettini, M., Shapley, A. E., Steidel, C. C., Cuby, J.-G., Dickinson, M., Moorwood, A. F. M., Adelberger, K. L., & Giavalisco, M. 2001, ApJ, 554, 981

Ponman, T. J., Cannon, D. B., & Navarro, J. F. 1999, Nature, 397, 135

Sáiz, A., Domínguez-Tenreiro, R., Tissera, P. B., & Courteau, S. 2001, MNRAS, 325, 119

Silk, J. 2001, MNRAS, 324, 313

Slyz, A., Devriendt, J., Silk, J., & Burkert, A. 2002, MNRAS, 333, 894

Springel, V. 2000, MNRAS, 312, 859

Springel, V., & Hernquist, L. 2003a, MNRAS, 339, 289

——. 2003b, MNRAS, 339, 312

Somerville, R. S., & Primack, J. R. 1999, MNRAS, 310, 1087

Somerville, R. S., Primack, J. R., & Faber, S. M. 2001, MNRAS, 320, 504 (SPF)

Sommer-Larsen, J., & Dolgov, A. 2001, ApJ, 551, 608

Sommer-Larsen, J., Gelato, S., & Vedel, H. 1999, ApJ, 519, 501

Tamura, T., et al. 2001, A&A, 365, L87

Thacker, R. J., & Couchman, H. M. P. 2001, ApJ, 555, L17

van den Bosch, F. C. 2000, ApJ, 530, 177

——. 2001, MNRAS, 327, 1334

———. 2002, MNRAS, 332, 456

van den Bosch, F. C., Abel, T., Croft, R. A. C., Hernquist, L., & White, S. D. M. 2002, ApJ, 576, 21

Vitvitska, M., Klypin, A. A., Kravtsov, A. V., Wechsler, R. H., Primack, J. R., & Bullock, J. S. 2002, ApJ, 581, 799

Voigt, L. M., Schmidt, R. W., Fabian, A. C., Allen, S. W., & Johnstone, R. M. 2002, MNRAS, 335, L7

Volonteri, M., Haardt, F., & Madau, P. 2003, ApJ, 582, 559

Wechsler, R. H., Bullock, J. S., Primack, J. R., Kravtsov, A. V., & Dekel, A. 2002, ApJ, 568, 52

Wechsler, R. H., Somerville, R. S., Bullock, J. S., Kolatt, T. S., Primack, J. R., Blumenthal, G. R., & Dekel, A. 2001, ApJ, 554, 85

Weil, M. L., Eke, V. R., & Efstathiou, G. 1998, MNRAS, 300, 773

Weinberg, D. H., Hernquist, L., & Katz, N. 2002, ApJ, 571, 15

White, S. D. M. 1984, MNRAS, 286, 38

White, S. D. M., & Frenk, C. S. 1991, ApJ, 379, 52

White, S. D. M., & Rees, M. J. 1978, MNRAS, 183, 341

Wu, K. K. S., Fabian, A. C., & Nulsen, P. E. J. 2000, MNRAS, 381, 889

Yepes, G., Kates, R., Khokhlov, A., & Klypin, A. 1997, MNRAS, 284, 235

Yoshida, N., Stoehr, F., Springel, V., & White, S. D. M. 2002, MNRAS, 335, 762

25

Joint formation of supermassive black holes and galaxies

MARTIN G. HAEHNELT
Institute of Astronomy, Cambridge

Abstract

The tight correlation between black hole mass and velocity dispersion of galactic bulges is strong evidence that the formation of galaxies and supermassive black holes are closely linked. I review the modeling of the joint formation of galaxies and their central supermassive black holes in the context of the hierarchical structure formation paradigm.

25.1   Supermassive Black Holes in Galactic Bulges

25.1.1   *SMBHs — Commonplace Rather Than Rare Disease*

The suggestion that galaxies may harbor supermassive black holes (SMBHs) at their centers dates back to the 1960s (Lynden-Bell 1969; see Rees 1984 for a review of the older literature). Initially black holes were suggested to explain the large efficiencies of transforming matter into radiation necessary to sustain the large luminosities of bright active galactic nuclei (AGNs). For a long time the evidence for the presence of SMBHs was mainly due to observations of such "active" SMBHs. However, the space density of bright AGNs is about a factor of 100 to 1000 smaller than that of normal galaxies. For several decades it was thus a lively debated question whether all normal galaxies (including our own) contain "dormant" SMBHs, or whether only a small fraction of all galaxies and which galaxies may exhibit such a "disease" (see Rees 1990 for a discussion).

The discovery of the tight correlation between black hole mass and the velocity dispersion of the bulge component of galaxies (Gebhardt et al. 2000; Ferrarese & Merritt 2000), enabled by the increased resolution of *HST*, has decided this question beyond any doubt in favor of most galactic *bulges* containing black holes. Note that this also finally rules out the possibility that bright QSOs are long-lived (e.g., Small & Blandford 1992; Choi, Yang, & Yi 1999). In light of the faint-level AGN activity that has been detected in many galaxies for some time (see Ho, this volume, for a review), this may not come too much as a surprise. It even appears now that not only most galaxies with a bulge contain SMBHs holes, but also that most of such galaxies show activity (see Heckman and Ho, this volume).

The fact that the mass of the black hole appears to scale with the mass of the galactic bulge rather than the mass of the galaxy or its dark matter (DM) halo is likely to be a major new clue in understanding the formation of galactic bulges and central SMBHs.

© The Observatories of the Carnegie Institution of Washington 2004.

25.1.2   The $M_\bullet$–σ_* Relation: An Undeciphered Clue?

The tight correlation of the black hole mass with global structural properties of the galaxy, such as the velocity dispersion of its bulge, came as a big surprise. The $M_\bullet$–σ_* relation is generally taken as strong evidence that the growth of SMBHs and the formation of galaxies go hand in hand. This is reflected in the fact that this idea has become the theme of this conference.

Many suggestions have been made as to what may determine the mass of SMBHs in galaxies. The models can be broadly grouped as follows.

- Simple scaling models: The mass of the black holes is assumed to scale with certain properties of the galaxy or its DM halo (e.g., Haehnelt & Rees 1993; Haiman & Loeb 1998).
- Supply-driven models: The mass of accreted gas scales with the amount of available (low-angular momentum) fuel (e.g., Kauffmann & Haehnelt 2000; Adams, Graff, & Richstone 2001; Volonteri, Haardt, & Madau 2002).
- Self-regulating models: The energy output during accretion or the influence of the black hole on the galactic gravitational potential limits the fuel supply (e.g., Norman & Silk 1983; Small & Blandford 1992; Silk & Rees 1998; Haehnelt, Natarajan, & Rees 1998; Sellwood & Moore 1999; Wyithe & Loeb 2002; El-Zant et al. 2003).
- "Exotic" models: Accretion of collisional DM or accretion of stars (e.g., Hennawi & Ostriker 2002; Zhao, Haehnelt, & Rees 2002).

Most of these models succeed in explaining the observed scaling relations. Nevertheless, none appears likely to be the final answer. Some of them fall short of specifying the physical mechanism at work. The rest invoke fine-tuning or only moderately plausible assumptions of some kind or another. Most puzzling in that respect is the remarkable tightness of the observed correlation of the black hole mass with the stellar velocity dispersion on large radii. This correlation is significantly tighter than, for example, the correlation of stellar velocity dispersion and bulge luminosity (Haehnelt & Kauffmann 2000). It appears that either the tightness of the relation for the current sample of nearby galactic bulges is a statistical fluke or that we have not yet fully unravelled the clue it may give us as to how SMBHs grow and form.

25.2   The Assembly of SMBHs

25.2.1   The Observed Accretion History of SMBHs

The next important piece of observational evidence on how SMBHs have assembled comes from the demography of the active population. The observed energy density U_X emitted by active black holes at redshift z in band X can be used to estimate the overall black hole mass density in remnant black holes using Soltan's (1982) argument

$$\rho_\bullet = \epsilon_{\rm acc}^{-1} f_X^{-1} \int \frac{dU_X(z)}{dz}(1+z)dz. \tag{25.1}$$

This elegant argument only requires assumptions about the efficiency of turning accreted rest-mass energy into radiated energy $\epsilon_{\rm acc}$ and the factor f_X, which relates the total emissivity to the emissivity in the chosen band. For optically bright QSOs these estimates have been done for some time and the black hole mass density associated with accretion in optically bright QSOs is well determined. The remnant black hole mass density attributable to

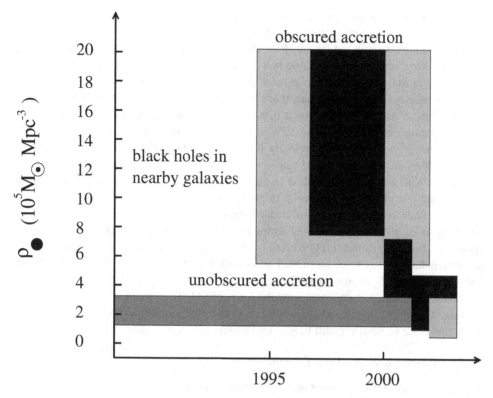

Fig. 25.1. Comparison of estimates of black hole mass density for nearby galactic bulges (black) with that inferred for accretion in optically bright QSOs (unobscured accretion, dark grey) and in hard X-ray selected active SMBHs (obscured accretion, light grey). Estimates are converging to within a factor of 2 to the value inferred from optically bright QSOs.

optically "obscured" accretion, as traced by hard X-ray sources, is much less well known. In the last couple of years it has been claimed that the total mass accreted in optically obscured sources may exceed that accreted in optically bright QSOs by a large factor (Fabian & Iwasawa 1999; Elvis, Risaliti, & Zamorani 2002). However, the bolometric correction factors were large ($\gtrsim$ 30) and uncertain, and the redshift distribution was not yet determined. As discussed by Fabian (this volume), with the data on hard X-ray selected active SMBHs rapidly accumulating, a more moderate contribution to the integrated black hole mass density appears more plausible. With this more moderate contribution and the reduced estimates of the black hole mass density in nearby galactic bulges that accompanied the discovery of the $M_\bullet$–σ_* relation, it appears that optically bright QSOs are a reasonably faithful tracer of the integrated mass accretion history of SMBHs. Figure 25.1 shows schematically how in the last few years estimates of the black hole mass density in nearby galactic bulges and in active SMBHs have converged to within a factor of 2 to the value estimated for accretion in optically bright QSOs (Haehnelt & Kauffmann 2001; Merritt & Ferrarese 2001; Aller & Richstone 2002; Yu & Tremaine 2002; Comastri et al. 2003; Fabian, this volume).

25.2.2   *Theoretical Models for the Assembly of SMBHs*

A wide variety of formation mechanisms for SMBHs has been suggested, which are nicely summarized in the famous flow chart of Martin Rees (Rees 1977, 1978, 1984). The suggestions vary from the collapse of a (rotating) supermassive stars, to the evolution of a dense star cluster, to the direct collapse of a gas cloud, and various combinations thereof. The main message of the flow chart, that there is no lack of plausible routes to a SMBH, is as valid today as it was when it was first compiled. However, for most of the processes the large formation efficiency of $\sim 0.2\%$ of the baryonic mass in a galactic bulge and the tight correlation with the stellar velocity dispersion of the galactic bulge are surprising. *Have we, then, made progress and can we exclude large parts of the diagram?* Probably not much. What has been established so far is that the majority of galactic bulges contain SMBHs and that emission of optically bright QSOs trace the mass accretion history well. We further know from our theoretical understanding of the hierarchical build-up of galaxies that mergers of SMBHs have to play an important role in their build-up. However, at the same time, bright, high-redshift QSOs tell us that about 10% of the black hole mass density in SMBHs had already been assembled at redshift 4, and some black black holes already had a mass of $10^9 M_\odot$ even at redshift 6.

Thus, a consistent picture for the growth history of SMBHs at redshift $z \lesssim 5$ appears to be in place. The more difficult and also more interesting problem of how SMBHs have formed in the first place is, however, still largely unconstrained by observations.

25.3   Hierarchical Galaxy Formation

25.3.1   *The Standard Paradigm of Structure Formation*

The joint theoretical and observational effort of the last three decades has led to a determination or, maybe better, reconstruction of the spatial power spectrum of the initial density fluctuations, which is reliable to better than 20% on scales from a few Mpc to the Hubble radius (see Spergel et al. 2003 for a recent summary). This power spectrum of density fluctuations is hierarchical in nature: fluctuations on large scales are smaller than those on small scales. The power spectrum is tantalizingly close to the ΛCDM variant of the cold dark matter (CDM) scenario, which postulates DM particles with random velocities too small to affect structure formation. Such a power spectrum results in a hierarchical build-up of structures with small objects forming first and larger objects building up by merging. In this model the first pregalactic structures form at redshift 20 to 5, while the first DM halos capable of hosting big galaxies assemble between redshift 5 and 2, and the formation of galaxy clusters occurs at redshift 1 to 0.

25.3.2   *Hierarchical Build-up of Galaxies and Merger Tree Models*

The hierarchical structure formation paradigm has profound implications for the way galaxies and active nuclei form and evolve (White & Rees 1978; Blumenthal et al. 1984; White & Frenk 1991; Haehnelt & Rees 1993). Galaxies typically double their mass in about 20% of the Hubble time. A key ingredient of hierarchical galaxy formation models is thus the frequent merging of DM halos and galaxies. Kauffmann and collaborators were the first to model in detail the basic processes governing galaxy formation following the merging history of DM halos (Kauffmann, White, & Guiderdoni 1993). Gas cooling, star formation, and stellar feedback were described with simple recipes, and the merger history was modeled with extensive Monte Carlo simulations. This model soon became known as

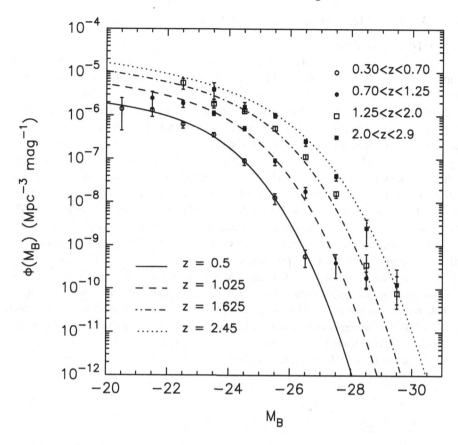

Fig. 25.2. The luminosity function of QSOs in a model where the space density of AGNs is related to the formation rate of DM halos . (Reproduced from Haehnelt & Rees 1993.)

the "semi-analytical model" of galaxy formation. It was shown to be consistent with a wealth of observations at low redshift if the parameterization of the physical processes are chosen in an appropriate way (e.g., Cole et al. 1994; Kauffmann et al. 1999a; Somerville & Primack 1999; Somerville, this volume). High-redshift observations have removed some, but not all, of the parameter degeneracies that exist (Kauffmann et al. 1999b; Somerville, Primack, & Faber 2001). The detailed comparison with the luminosity function and clustering properties of galaxies in different bands nevertheless gives confidence that the models are a fair representation of the hierarchical build-up of the *mass* of observed galaxies, their bulges, and their DM halos. Details of the predicted star formation history, luminosities, and colors at high redshift are, however, much less certain.

Despite their success in reproducing many observed properties of galaxies, the fact that stellar populations in early-type galaxies are generally very old regardless of their location in galaxy clusters or in the field (e.g., van Dokkum et al. 2001) has not yet been fully understood in the context of hierarchical galaxy formation models. I will discuss later how this may be related to a similar problem of understanding the rapid decline of the QSO emissivity at low redshift.

25.4   SMBHs in Hierarchically Merging Galaxies

25.4.1   *SMBHs in DM Halos*

The merging histories and dynamical evolution of DM halos in CDM-like structure formation models has been well established by analytical calculations and numerical simulations (e.g., Jenkins et al. 1998). As mentioned in the last section, the evolution of the baryonic component of these DM halos is more difficult to predict. It is thus reasonable to try to link the evolution of the population of (active) SMBHs directly to the evolution of the population of DM halos, especially at high redshift where our knowledge about the host galaxies of SMBHs has only recently started to build up. When CDM-like models were first seriously discussed, it was not obvious that the rather late emergence of galaxy-size structures in these models was not in conflict with the existence of bright, high-redshift QSOs, which were found in large numbers at the end of the 1980s. Efstathiou & Rees (1988) showed that this is not the case. Haehnelt & Rees (1993) pointed out that in CDM models the build up of galactic-size structures coincides with the peak in AGN activity at redshift 2.5 and that the emergence of the QSO population between redshift 5 and 2.5 arises naturally in CDM-like cosmogonies (see also Cavaliere & Szalay 1986). They further showed that the evolution of the QSO emissivity at $0 < z < 3$ can be reproduced if a suitable scaling of black hole mass with the virial velocity of newly formed halos and redshift was assumed (Fig. 25.2). The formation rate of galactic nuclei was modeled as the time derivative of the space density of DM halos times a "lifetime" t_Q of the QSO phase, while the luminosity was assumed to scale with the Eddington luminosity:

$$
\begin{aligned}
N_{\text{QSO}} &= \dot{N}_{\text{halo}} t_Q, \\
L_{\text{QSO}} &= \alpha_{\text{Edd}} L_{\text{Edd}}(M_\bullet), \\
M_\bullet &= g(M_{\text{halo}}, z) = g(v_{\text{circ}}, z)
\end{aligned}
\tag{25.2}
$$

Haehnelt & Rees (1993) had to choose the scaling with redshift such that the black hole formation efficiency dropped very quickly with decreasing redshift. They argued that this is because at redshifts smaller than 2.5 the amount of available fuel in galaxy-size halos decreases with the decreasing ability of the gas to cool at low redshift (see also Cavaliere & Vittorini 2000 and see Monaco, Salucci, & Danese 2000 for an alternative interpretation). They further found that $t_Q \approx 10^8$ yr gives a good match to the QSO luminosity function for their assumed scaling of black hole formation efficiency with virial velocity of the DM halo. In a similar spirit, Haiman & Loeb (1998) suggested that the black hole mass scales linearly with halo mass and showed that the QSO luminosity function can then be reproduced with a very short lifetime $t_Q \approx 10^6$ yr. Haehnelt et al. (1998) showed that a consistent picture for the high-redshift star-forming ("Lyman-break") galaxies and the AGN population arises within this framework if Lyman-break galaxies host the SMBHs responsible for the observed QSO activity (see Granato et al. 2001 for a discussion of the possible connection between QSOs and sub-mm sources). Haehnelt et al. (1998) showed that the scaling of black hole mass with halo mass or circular velocity is degenerate with the assumed lifetime t_Q. The QSO luminosity function can be matched either with a steep relation between black hole mass and halo mass ($M_\bullet \propto M_{\text{halo}}^{5/3} \propto v_{\text{circ}}^5$), more massive halos for a given black hole mass and a longer lifetime *or* a shallower relation between black hole and halo mass ($M_\bullet \propto M_{\text{halo}} \propto v_{\text{circ}}^3$), less massive halos and a shorter lifetime. Haehnelt et al. (1998) further showed that this degeneracy can be broken with a study of the clustering properties of AGNs, which is

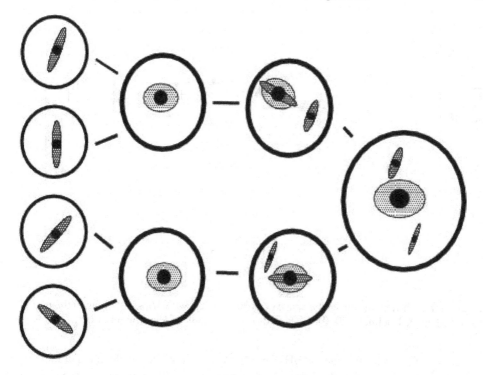

Fig. 25.3. "Merger tree' 'models of galaxies with black holes: Monte Carlo realization of the merging histories of DM halos plus simple recipes for gas cooling, star formation, stellar feedback accretion, and dynamics of galaxies and black holes.

sensitive to the absolute mass scale of the DM halos hosting QSOs. They found that a scaling $M_\bullet \propto v_{\mathrm{circ}}^5$ and $t_Q \approx 10^7$ yr did fit best the QSO luminosity function and the then still rather scarce information on the clustering of high-redshift QSOs. Haiman & Hui (2001) and Martini & Weinberg (2001) investigated in more detail what constraints on the lifetime can be obtained from upcoming QSO surveys (see Martini, this volume, for a summary of this and other constraints on QSO lifetimes). Wyithe & Loeb (2002) showed that a scaling $M_\bullet \propto v_{\mathrm{circ}}^5$ is also consistent with the new high-redshift data from the SDSS QSO survey and the luminosity function of X-ray selected AGNs.

25.4.2   *SMBHs in Merger Tree Models*

When Magorrian et al. (1998) published a large sample of black hole mass estimates for nearby galaxies and confirmed the suggestion that black hole mass and bulge mass are strongly correlated with a nearly linear relation (Kormendy & Richstone 1995), the idea that the formation of galaxies and black holes are closely linked became more widely accepted. Cattaneo, Haehnelt, & Rees (1999) demonstrated that a hierarchical merger tree model could reproduce the observed slope and scatter of the "Magorrian relation," provided that star formation and the growth of black holes are closely linked. They thereby enforced a rapid drop of the gas available in both shallow and very deep potential wells to mimic the effect of feedback on small halos and the inability of the gas to cool in the very massive halos. Kauffmann & Haehnelt (2000) combined the full merger tree models of galaxies

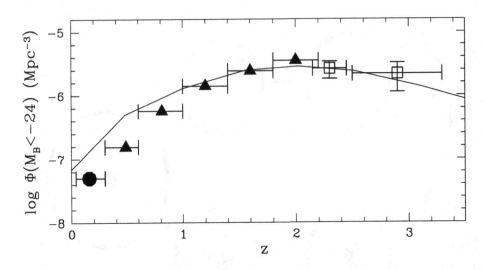

Fig. 25.4. Redshift evolution of the space density of QSOs with $M_B < -24$ mag in the model of Kauffmann & Haehnelt (2000). (Adapted from Haehnelt & Kauffmann 2001.)

(Fig. 25.3) with the idea that the SMBHs form from the cold gas available in major mergers and assumed that the accretion of gas leads to QSO activity of duration $t_Q \approx 10^7$ yr. They demonstrated that such a model is consistent with the evolution of the QSO emissivity (Fig. 25.4), the Magorrian relation, the present-day luminosity function, and global star formation history of galaxies and the evolution of cold gas as probed by damped Lyα systems. It is interesting to note that to match the observations Kauffmann & Haehnelt (2000) had to change the feedback prescription compared to the previous modeling of Kauffmann et al., such that galaxies become progressively more gas rich at high redshift. Kauffmann & Haehnelt (2000) also made some predictions for the luminosities of QSO host galaxies at high redshift. In the model of Kauffmann & Haehnelt the rather steep decline of the QSO emissivity with decreasing redshift can be attributed to a combination of a decrease in the merger rate, a decrease of the amount of cold gas available for fueling, and an increase in the accretion time scale.

Haehnelt & Kauffmann (2000) showed that the Kauffmann & Haehnelt (2000) model is also consistent with the $M_\bullet$–σ_* relation (Fig. 25.5). Somewhat surprisingly, the model also reproduces the tightness of the relation, despite the frequent merging. This is because the black holes move along the $M_\bullet$–σ_* relation during mergers and the cold gas available for accretion during major mergers scales explicitly with the depth of potential well rather than with the mass of the galaxy in their model. One should, however, keep in mind that the dynamical processes were modeled in a simplistic fashion. More detailed modeling of the dynamics of the galaxy and black hole merger most likely will introduce additional scatter. Kauffmann & Haehnelt (2002) investigated the clustering properties of galaxies and QSOs in the model and found that a lifetime of 10^7 yr is also consistent with the 2dF clustering data. They further pointed out that study of the galaxy-QSO correlation function may give further clues on how QSO activity is triggered. The spin distribution of black

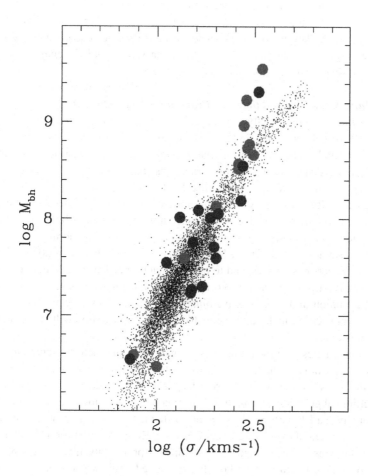

Fig. 25.5. The $M_\bullet$–σ_* relation in the model of Kauffmann & Haehnelt (2000). (Reproduced from Haehnelt 2002.)

holes may give another handle on determining the merger history of black holes (Hughes & Blandford 2003). Cattaneo (2002) investigated the expected evolution of the spin of black holes hosted by hierarchically merging DM halos for a variety of assumptions for the spin-up and spin-down processes. Unfortunately, observationally we know very little about the spin of SMBHs even though it has been suggested that the jets in radio-loud AGNs may be powered by the rotational energy of their central black hole (Blandford & Znajek 1977).

Despite their success in reproducing many observed properties, hierarchical models generally struggle to reproduce the rapid decline of QSO activity toward low redshift. As mentioned earlier, this difficulty is most likely connected to a similar difficulty in explaining the very rapid decrease of star formation activity in galactic bulges. In galactic bulges both the star formation activity and the accretion onto the black holes seem to be "switched off" at higher redshift than star formation in other galaxies. This suggests that the presence of a supermassive black hole is the physical reason (see Granato et al. 2001 for a model along those lines). Feedback effects like radiation pressure, radiative heating, energetic particles, and kinetic energy due to accretion have a large impact on the surrounding (forming) host galaxy and beyond (see Begelman, this volume, for a review). A better understanding of AGN feedback may thus be key to resolving some of the difficulties encountered by the current models (see also Somerville, this volume). Supermassive black holes may so be an essential ingredient in shaping the Hubble sequence of galaxies.

25.4.3   *Binary/Multiple SMBHs and the Core Properties of Galactic Bulges*

If galaxies merge hierarchically and all galactic bulges contain black holes the formation of supermassive binary and multiple black holes is expected to be common. When two galaxies of moderate mass ratio merge, a hard binary will form quickly (e.g., Miloslavjević & Merritt 2001). There are two processes that can shrink the separation of hard binaries: accretion of gas and hardening by three-body interaction with stars (Begelman, Blandford, & Rees 1980). Stellar hardening of supermassive binaries requires the ejection of several times the mass of the binary black hole. Binary black holes may thus play an important role in shaping the core profiles of galactic bulges (Ravindranath, Ho, & Filippenko 2002; Milosavljević et al. 2002; Merritt, this volume). Disk accretion of cold gas following minor mergers has been argued to be an efficient mechanism to merge binary black holes with a small ratio of secondary to primary mass (Armitage & Natarajan 2002). The formation and evolution of a supermassive binary black hole in a major merger of gas rich galaxies has not yet been studied, but it seems likely that gas accretion will also lead to rapid hardening in this case.

Whether binary black holes in typical low-redshift galaxies can reach the separation at which emission of gravitational radiation leads to coalescence within a Hubble time is somewhat uncertain (Yu & Tremaine 2002), but it appears likely that they do in all but the most massive elliptical galaxies (Haehnelt & Kauffmann 2002; Milosavljević & Merritt 2003). Observationally hard binary black holes are difficult to detect, and observational evidence so far is circumstantial (Merritt & Ekers 2002; see Komossa 2003 for a review).

In hierarchically merging galaxies there is a significant probability that a third black hole will fall in before a hard binary black hole has coalesced. This will normally lead to gravitational slingshot ejection of the lightest black hole (Saslaw, Valtonen, & Aarseth 1974; Hut & Rees 1992). The binary will also get a kick velocity. If all three black holes have similar masses, the kick velocity will be sufficient to kick the binary into the outer parts of the galaxy or even to eject it entirely (Hut & Rees 1992; Xu & Ostriker 1994). As discussed by Redmount & Rees (1989), at coalescence supermassive binaries will also get a kick due to the radiation-reaction forces predicted by general relativity. The kick velocity due to the resulting recoil should scale linearly with the mass ratio of the coalescing binary for small mass ratios. The absolute values of the kick velocities are uncertain. They will depend strongly on the radius of the last stable orbit and therefore on the spin and orbital orientation of the binary (Fitchett & Detweiler 1984). While the expected value for a Schwarzschild hole is small for rapidly spinning black holes, the kick velocity may be as large as a few hundred $km\,s^{-1}$ and may remove the merged binary from small and maybe even big galaxies in some cases. Note that in the case of two spinning black holes the asymmetry of the radiation with respect to the plane of the orbit may result in a recoil along the direction of the orbital angular momentum (Redmount & Rees 1989). Three-body interaction and the gravitational-radiation recoil of SMBHs in hierarchically merging galaxies may thus lead to a population of black holes outside galaxies (see also Haehnelt & Kauffmann 2002; Volonteri et al. 2002).

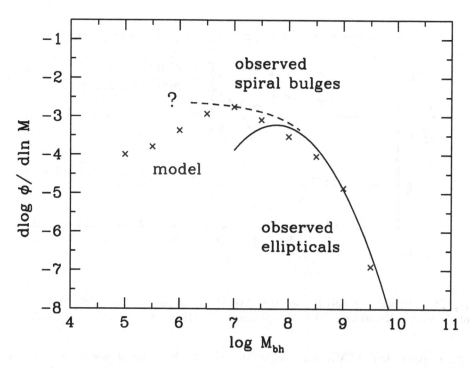

Fig. 25.6. The mass function of the black holes in the model of Haehnelt & Kauffmann (2000) compared to the mass function inferred for early-type galaxies extrapolated to include the contribution from spiral bulges. (Adapted from Haehnelt 2003.)

25.5 Early Evolution

25.5.1 *Do Intermediate-mass Black Holes Form in Shallow Potential Wells?*

While the accretion history of SMBHs between redshift 5 and 0 seems now reasonably well constrained, little is known about the the accretion/formation history at higher redshifts. At redshift 5 the mass of a typical DM halo in the ΛCDM model is about $10^{11} M_\odot$. In the ΛCDM model the matter fluctuation spectrum extends to smaller masses, and hierarchical growth of structure starts at much earlier times. The recent detection of a large polarization signal in the CMB at large scales by the *WMAP* satellite requires that a large volume fraction of the Universe was reionized at $z \approx 20$ (Kogut et al. 2003); however, note that the measurement error is still large. If confirmed, this is strong observational evidence that the matter fluctuation power spectrum extends to scales as small as $10^6 M_\odot$, or even smaller. On these scales nonlinear structures form at $z > 20$. *Should we expect hierarchical growth at similarly early times to contribute significantly to the build-up of SMBHs?* This depends crucially on what happens in shallow potential wells with circular velocities $v_c \lesssim 100\,\mathrm{km\,s^{-1}}$. Modeling of the galaxy luminosity function in a CDM-like hierarchical structure formation model requires that the star formation efficiency drops rapidly in shallow potential wells. Otherwise the faint-end slope of the luminosity function would be much steeper than observed (White & Frenk 1991). The effect of stellar feedback and/or the effect of heating due to photoionization is generally invoked as explanation. It

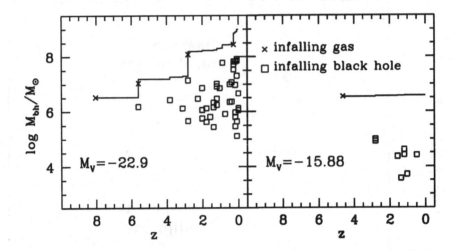

Fig. 25.7. Typical assembly history of a SMBH for a bright and a faint bulge in the model of Haehnelt & Kauffmann (2000). (Reproduced from Haehnelt 2003.)

appears plausible that SMBHs assemble from the same low-angular momentum tail of the cold gas reservoir that also fuels star formation. This was, for instance, assumed in the model of Kauffmann & Haehnelt (2000), in which SMBHs only form efficiently in halos with $v_c \gtrsim 100\,\mathrm{km\,s^{-1}}$. As a result, the mass function of black holes turns over at masses $\lesssim 10^7\,M_\odot$ (Fig. 25.6), and the hierarchical build-up of black holes only starts at redshift $z \lesssim 6{-}8$ in their model (Fig. 25.7). Nevertheless, some star formation has to occur in more shallow potential if reionization starts as early as implied by the large electron optical depth found by *WMAP*. Madau & Rees (2001) suggested that the black hole remnants of massive stars formed in these shallow potential wells will lead to the formation of a population of intermediate-mass black holes. Volonteri et al. (2002, 2003) followed the merging of DM halos containing SMBHs from $z > 20$. They found that hierarchical build-up starting from stellar-mass seed black holes can contribute significantly to the population of SMBHs at redshift five *if* the black holes merge efficiently (see also Islam, Taylor, & Silk 2003). Whether black holes can merge in these shallow potential is, however, very uncertain. It requires rapid sinking of the seed black holes to the center of merged pregalactic structures *plus* either a dense stellar system to provide a sufficient number of stars for binary hardening or the accretion of cold gas. Stars and the cold gas from which they form are, however, expected to be in short supply in shallow potential wells. Furthermore, even if the gravitational radiation recoil were too small to affect the hierarchical build-up of SMBHs in normal galaxies, it may nevertheless efficiently remove a significant fraction of coalescing binary black holes from pregalactic structures with shallow potential wells, should they exist. Currently our best bet to make further progress in this area is to study how the $M_\bullet$–σ_* relation extends to smaller galaxies. There are a number of observational claims for the detection of intermediate-mass black holes (King et al. 2001; Gebhardt, Rich, & Ho 2002; Filippenko & Ho 2003; van der Marel, this volume). Should these detections consolidate, the next step will be to establish whether

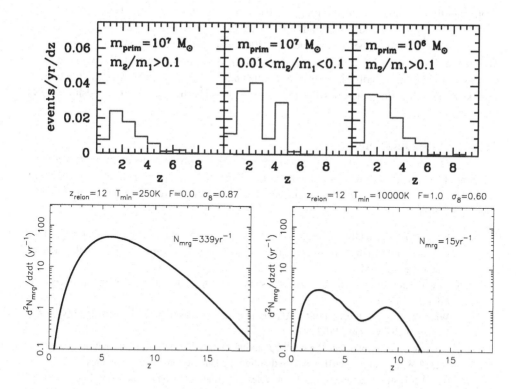

Fig. 25.8. *Top:* Merging rate of galaxies forming supermassive binary black holes for a range of primary black hole mass and mass ratio in the models of Kauffmann & Haehnelt (2000). (Reproduced from Haehnelt 2003.) *Bottom:* Merging rate of DM halos hosting SMBHs predicted by Wyithe & Loeb (2003). (Adapted from Wyithe & Loeb 2003.)

intermediate-mass black holes have formed with similarly large efficiency as the SMBHs in deep potential wells. This would argue for a continuous hierarchical build-up from stellar-mass seeds at very high redshift. With *JWST* it should be possible to probe some of this directly (see Haiman, this volume).

25.5.2 *LISA and the Assembly of SMBHs at $z > 5$*

A somewhat longer shot to establish how the SMBHs observed at redshifts $\gtrsim 5$ did assemble will be observations with the space-based gravitational wave interferometer *LISA*, expected to be launched in the next decade. *LISA* will be sensitive to gravitational waves with frequencies below 1 Hz, which are not accessible from the ground because of seismic noise. The typical dynamical time at the Schwarzschild radius of a SMBHs is $t_{dyn} \sim 3(M/10^5 M_\odot)$ s. The detection of black hole-black hole mergers involving SMBHs is thus a prime objective of *LISA*. The long baseline achievable in space means that for SMBHs *LISA* will not be so much sensitivity limited but rather event-rate limited instead. *LISA* will be sensitive enough to detect equal-mass mergers of $10^4 - 10^7 M_\odot$ black holes at $z < 20$. *LISA* will even be able to determine the luminosity distance and spin for some coalescing binary black holes (Bender & Pollack 2004). However, if the black hole formation efficiency drops rapidly in potential wells with $v_{circ} < 100\,km\,s^{-1}$, event rates are only $0.3 - 1\,yr^{-1}$ (Fig. 25.8). If hierarchical build-up extends to smaller DM halos, event rates could be as large as few tens to a hundred per year (Haehnelt 2003; see also Haehnelt 1994, 1998; Menou, Haiman, & Narayanan 2001; Wyithe & Loeb 2002). Note, however, that this would predict a mass function that rapidly rises toward smaller masses (Fig. 25.6). *LISA* has the exiting prospect to finally settle the question of how the SMBHs that power high-redshift AGNs were assembled.

25.6   Open Questions

The recent observational progress has led to big improvements in our understanding on how black holes were assembled in hierarchically merging galaxies.   There is, nevertheless, a long list of unanswered questions. The following is a certainly incomplete version of such a list:

(1) Is AGN activity really triggered by mergers (beginning, end, multiple)? What is the time scale of QSO activity? What determines it? Why is it apparently shorter than the merging time scale of galaxies?

(2) How much room is there for dark or obscured accretion? Can the accretion rate exceed the Eddington limit?

(3) What is the physical origin of the $M_\bullet - \sigma_*$ relation? Is it as tight as claimed, and if so, why? Does it evolve with redshift?

(4) Does AGN activity affect the cooling/heating budget during galaxy formation in a global sense? What role do SMBHs play in defining the Hubble sequence of galaxies?

(5) Are (hard) supermassive binary black holes common? On which time scale do they merge? Are supermassive binary black holes responsible for the core properties of galactic bulges? Do black holes receive kick velocities that eject them from (small) galaxies?

(6) Do intermediate-mass black holes form in shallow potential wells? Does the $M_\bullet - \sigma_*$ relation extend to smaller black hole masses? Does the hierarchical build-up of SMBHs extend to pregalactic structures at very high redshift?

Many of these questions are already under intense scrutiny. Progress in answering them will hopefully bring us closer to a more complete understanding of the physical processes responsible for the formation of SMBHs.

25.7   Summary

Models where black holes grow by a combination of gas accretion traced by short-lived ($\sim 10^7\,yr$) QSO activity and merging in hierarchically merging galaxies are consistent with a wide range of observations in the redshift range $0 < z < 5$. The rapid decline of both the QSO activity and star formation in galactic bulges suggests that SMBHs may play a bigger role in shaping the Hubble sequence of galaxies than previously anticipated. The frequent merging of galaxies will lead to the ubiquitous formation of supermassive binary black holes. These are generally expected to merge in all but the most massive galaxies by gas infall and stellar hardening. Gravitational slingshot in triple black holes and gravitational radiation recoil of coalescing binaries may lead to a substantial population of SMBHs outside of galaxies. The actual formation of SMBHs is still poorly constrained observationally. Direct collapse of a gas cloud to seed black holes of $10^5 - 10^6\,M_\odot$ and extension of the hierarchical merging all the way to stellar-mass seed black holes from

Population III stars mark two extreme ends of the list of viable possibilities. Further observational and theoretical study of intermediate-mass black holes, and eventually the study of the merging history of black holes with *JWST* and the planned space-based gravitational wave interferometer *LISA* should help to answer the question of how SMBHs formed.

Acknowledgements. I would like to thank Luis Ho with for his patient persistence in urging me to write this review. Thank is also due to Stuart Wyithe for providing material for Figure 25.8.

References

Adams, F. C., Graff, D. S., & Richstone, D. O. 2001, ApJ, 551, L31

Aller, M. C., & Richstone, D. O. 2002, AJ, 124, 3035

Armitage, P. J., & Natarajan, P. 2002, ApJ, 567, L9

Begelman, M. C., Blandford, R. D., & Rees, M. J. 1980, Nature, 287, 307

Bender, P. L., & Pollack, S. 2004, in Carnegie Observatories Astrophysics Series, Vol. 1: Coevolution of Black Holes and Galaxies, ed. L. C. Ho (Pasadena: Carnegie Observatories, http://www.ociw.edu/ociw/symposia/series/symposium1/proceedings.html)

Blandford, R. D., & Znajek, R. L. 1977, MNRAS, 179, 433

Blumenthal G. R., Faber, S. M.,, Primack, J. R., & Rees, M. J. 1984, Nature, 311, 517

Burkert, A., & Silk, J. 2001, ApJ, 554, L151

Cattaneo, A. 2002, MNRAS, 333, 353

Cattaneo, A., Haehnelt, M. J., & Rees, M. J. 1999, MNRAS, 308, 77

Cavaliere, A., & Szalay, A. S. 1986, ApJ, 311, 589

Cavaliere, A., & Vittorini, V. 2000, ApJ, 543, 599

Choi, Y., Yang, J., & Yi, I. 1999, ApJ, 518, L77

Cole, S., Aragón-Salamanca, A., Frenk, C. S., Navarro, J. F., & Zepf, S. E. 1994, MNRAS, 271, 781

Comastri, A., et al. 2003, in The Astrophysics of Gravitational Wave Sources, ed. J. M. Centrella (AIP: New York), 151

Efstathiou, G., & Rees, M. J. 1988, MNRAS, 230, L5

Elvis, M., Risaliti G., & Zamorani G. 2002, ApJ, 565, L75

El-Zant, A., Shlosman I., Begelman, M., & Frank, J. 2003, ApJ, 590, 641

Fabian, A. C., & Iwasawa, K. 1999, ApJ, 303, L34

Ferrarese, L., & Merritt, D. 2000, ApJ, 539, L9

Filippenko, A. V., & Ho, L. C. 2003, ApJ, 588, L13

Fitchett, M. J., & Detweiler 1984, MNRAS, 211, 933

Gebhardt, K., et al. 2000, ApJ, 539, L13

Gebhardt, K., Rich, R. M., & Ho, L. C. 2002, ApJ, 578, L41

Granato, G. L., Silva, L., Monaco, P., Panuzzo, P., Salucci, P., De Zotti, G., Danese, L. 2001, MNRAS, 324, 757

Haehnelt, M. G. 1994, MNRAS, 269, 199

——. 1998, in Second International LISA Symposium, ed. W. M. Folkner (AIP: New York), 45

——. 2002, in A New Era in Cosmology, ed. N. Metcalfe & T. Shanks (San Francisco: ASP), 95

——. 2003, Classical and Quantum Gravity, 20, S31

Haehnelt, M. G., & Kauffmann G. 2000, MNRAS, 318, L35

——. 2001, in Black Holes in Binaries and Galactic Nuclei, ed. L. Kaper, E. P. J. van den Heuvel, & P. A. Woudt (Berlin: Springer), 364

——. 2002, MNRAS, 336, L51

Haehnelt, M. G., Natarajan, P., & Rees, M. J. 1998, MNRAS, 300, 817

Haehnelt, M. G., & Rees, M. J. 1993, MNRAS, 263, 168

Haiman, Z., & Hui, L. 2001, ApJ, 547, 27

Haiman, Z., & Loeb, A. 1998, ApJ, 503, 505

Hennawi J. F., & Ostriker, J. P. 2002, ApJ, 572, 41

Hughes, S. A., & Blandford, R. D. 2003, ApJ, 585, L101

Hut, P., & Rees, M. J. 1992, MNRAS, 259, 27

Islam, R. R., Taylor, J. E., & Silk, J. 2003, MNRAS, 340, 647

Jenkins, A., et al. 1998, ApJ, 499, 20

Kauffmann, G., Colberg, J. M., Diaferio, A., White, S. D. M., 1999a, MNRAS, 303, 188

——. 1999b, MNRAS, 307, 529

Kauffmann, G., & Haehnelt, M. G. 2000, MNRAS, 311, 576

——. 2002, MNRAS, 332, 529

Kauffmann, G., White, S. D. M., & Guiderdoni, B. 1993, MNRAS, 264, 201

King, A. R., Davies, M. B., Ward, M. J., Fabbiano, G., & Elvis, M. 2001, ApJ, 552, L109

Kogut, A., et al. 2003, ApJS, 148, 161

Komossa, S. 2003, in The Astrophysics of Gravitational Wave Sources, ed. J. M. Centrella (AIP: New York), 161

Kormendy, J., & Richstone, D. 1995, ARA&A, 33, 581

Lynden-Bell, D. 1969, Nature, 223, 690

Madau, P., & Rees, M. J. 2001, 511, L27

Magorrian, J., et al. 1998, AJ, 115, 2285

Martini, P., & Weinberg, D. H. 2001, ApJ, 547, 12

Menou, K., Haiman, Z., & Narayanan, V. K. 2001, ApJ, 558, 535

Merritt, D., & Ekers, R. D. 2002, Science, 297, 1310

Merritt, D., & Ferrarese, L. 2001, MNRAS, 320, L30

Milosavljević, M., & Merritt, D. 2001, ApJ, 563, 34

——. 2003, ApJ, submitted (astro-ph/0212459)

Milosavljević, M., Merritt, D., Rest, A., & van den Bosch F. C., 2002, MNRAS, 331, L51

Monaco, P., Salucci, P., & Danese, L. 2000, MNRAS, 317, 488

Norman C. A., & Silk, J. 1983, ApJ, 266, 502

Ravindranath, S., Ho, L. C., & Filippenko, A. V. 2002, ApJ, 566, 801

Redmount, I. H., & Rees, M. J. 1989, Comm. Astrophys., 14, 165

Rees, M. J. 1977, QJRAS, 18, 429

——. 1978, The Observatory, 98, 210

——. 1984, ARA&A, 22, 471

——. 1990, Science, 247, 817

Saslaw, W. C., Valtonen, M. J., & Aarseth, S. J. 1974, ApJ, 190, 253

Sellwood, J. A., & Moore, E. M. 1999, ApJ, 510, 125

Silk, J., & Rees, M. J. 1998, A&A, 331, 1

Small, T. A., & Blandford, R. D. 1992, MNRAS, 259, 725

Sołtan, A. 1982, MNRAS200, 115

Somerville, R. S., & Primack, J. R. 1999, MNRAS, 310, 1087

Somerville, R. S., Primack, J. R., & Faber, S. M. 2001, MNRAS, 320, 504

Spergel, D. N., et al. 2003, ApJS, 149, 175

van Dokkum, P. G., Franx, M., Kelson, D. D., & Illingworth, G. D. 2001, ApJ, 553, L39

Volonteri, M., Haardt, F., & Madau, P. 2002, Ap&SS, 281, 501

——. 2003, ApJ, 582, 559

White, S. D. M., & Frenk, C. S. 1991, MNRAS, 379, 52

White, S. D. M., & Rees, M. J. 1978, MNRAS, 183, 341

Wyithe, J. S. B., & Loeb, A. 2002, ApJ, 581, 886

——. 2003, ApJ, 590, 691

Xu, G., & Ostriker, J. P. 1994, ApJ, 437, 184

Yu, Q. 2002, MNRAS, 331, 935

Yu, Q., & Tremaine, S. 2002, MNRAS, 335, 695

Zhao, H. S., Haehnelt, M., & Rees, M. J. 2002, NewA, 7, 385

26

The formation of spheroidal stellar systems

ANDREAS BURKERT[1] and THORSTEN NAAB[2]

(1) Max-Planck-Institut für Astronomie, Königstuhl 17, D-69117 Heidelberg, Germany
(2) Institute of Astronomy, Madingley Road, Cambridge CB3 0HA, UK

Abstract

We summarize current models of the formation of spheroidal stellar systems. Whereas globular clusters form in an efficient mode of star formation inside turbulent molecular clouds, the origin of galactic spheroids, that is bulges, dwarf ellipticals, and giant ellipticals, is directly coupled with structure formation and merging of structures in the Universe. Disks are the fundamental building blocks of galaxies and the progenitors of galactic spheroids. The origin of the various types of spheroids and their global properties can be understood as a result of disk heating by external perturbations, internal disk instabilities, or minor and major mergers.

26.1   The Realm of Spheroids

Spheroids exist in the Universe with a wide range in masses and length scales. Probably the most simple, classical examples of stellar spheroids are globular star clusters with masses in the range of $10^4 M_\odot$ to $10^6 M_\odot$ and half-mass radii of order 2–10 pc (Harris 1996). These almost spherical systems appear to be stable and very long lived. Although the metallicities of different clusters in the Milky Way vary from [Fe/H] $\approx$ −2.5 to solar or even larger, the strikingly narrow iron abundance spreads of stars within individual clusters (Kraft 1979) and their small age spread indicate that each cluster consists of only one stellar generation that formed on a short time scale from chemically homogenized gas. Peebles & Dicke (1968) proposed that globular clusters are the first objects that formed in the Universe. More recent models assume that globulars formed at the same time as their host galaxies (Fall & Rees 1985; Vietri & Pesce 1995). As giant molecular clouds have similar masses and radii, they are considered to be the primary sites of cluster formation. Unfortunately, the formation of stars and the condensation of molecular clouds into dense, massive star clusters is still not well understood up to now (for a recent review see Lada & Lada 2003). Klessen & Burkert (2000, 2001) and Bate, Bonnell, & Bromm (2002; see also Clarke, this volume) investigated numerically the gravitational collapse of a turbulent cloud. Their models showed that the stabilizing turbulent motion of the molecular gas is dissipated on a short dynamical time scale, resulting in collapse and star formation. These models, however, neglected energetic feedback processes, which are known to play a crucial role in regulating and terminating star formation. In order to form a gravitationally bound, dense stellar cluster, high local star formation efficiencies of order $\eta_{sf} \approx 50\%$ are required (Brown, Burkert, & Truran 1991, 1995; Geyer & Burkert 2001). This is in contradiction with observations, which indicate that the fraction of molecular cloud material that turns into stars

© The Observatories of the Carnegie Institution of Washington 2004.

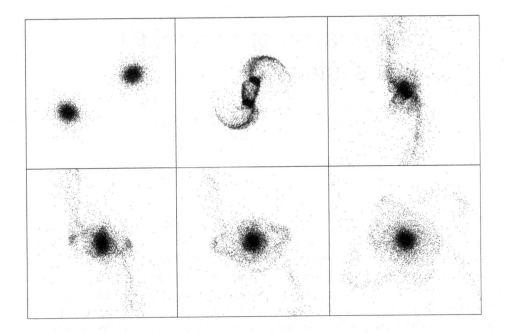

Fig. 26.1. Simulation of an equal-mass spiral galaxy merger. Each box has a size of 210 kpc. The upper-left box shows the initial condition. The other boxes show snapshots of the evolution at 4.6×10^8, 8×10^8, 1.3×10^9, 1.9×10^9, and 2.6×10^9 yr. Dark matter deficient dwarf spheroidals might form through gravitational instabilities inside tidal arms. The merger remnant is surrounded by rings and shells that provide long-term signatures of its merging history.

is typically of order $\eta_{sf} \leq 10\%$ due to gas ionization by the UV field of newly formed high-mass stars (Myers et al. 1986; Williams & McKee 1997; Koo 1999). Ashman & Zepf (1992) argued that globular clusters can form efficiently in interacting galaxies (Schweizer 1999), which indicates that peculiar, galactic non-equilibrium environments might enhance the star formation efficiency in molecular clouds. Under these conditions, supersonic cloud-cloud collisions or cloud implosions induced by an increase of the external gas pressure could destabilize a whole cloud complex, triggering global collapse and efficient star formation.

In contrast to globular clusters, the origin and structure of galactic spheroids (dwarf spheroidals, dwarf ellipticals, giant ellipticals, and bulges) seem to be strongly coupled with the hierarchical merging history of substructures in the Universe. Within the popular cold dark matter (CDM) cosmogony, the visible components of galaxies arise from gas infall into dark matter halos, followed by star formation. Disks are envisioned to form as a result of smooth gas accretion from the intergalactic medium (e.g., Katz & Gunn 1991; Navarro & White 1994; Steinmetz & Müller 1994). Spheroids result from processes that heat and destroy stellar disks. Low-mass dwarf spirals are particularly sensitive to stirring and harassment by the cumulative tidal interactions of high-speed galaxy encounters in galactic clusters (Moore et al. 1996; Moore, Lake, & Katz 1998), leading in the end to dwarf ellipticals. Massive spiral galaxies, on the other hand, can be destroyed by major mergers

(Fig. 26.1) and transform into giant ellipticals and bulges (Toomre 1974; Kauffmann, Charlot, & White 1996).

Recently, numerical simulations have shown that the formation of disks and spheroids might be even more complex. High-resolution cosmological simulations of galaxy evolution including star formation and feedback processes (Steinmetz & Navarro 2002) as well as semi-analytical models (Khochfar & Burkert 2003) indicate that the galaxies change their morphological type frequently. For example, spheroids could form by an early merger of low-mass disks and later on rebuild new disks by smooth gas accretion. These bulge-disk systems could merge again, forming an even larger spheroid. Within the framework of this scenario galactic bulges represent early spheroids that have grown a new, surrounding disk component. It is, however, not clear up to now whether all bulges necessarily formed that way (e.g., Wyse, Gilmore, & Franx 1997). Wyse & Gilmore (1992) argued that the specific angular momentum distribution of the Milky Way's bulge is very similar to that of the stellar halo and very different from that of the disk. This would suggest that the bulge was built up by dissipative inflow (Gnedin, Norman, & Ostriker 2000) of gas that was lost from star-forming regions and substructures in the Galactic halo, suggestive of a monolithic collapse scenario (Eggen, Lynden-Bell, & Sandage 1962). Yet another possibility are disk instabilities (Athanassoula 2002), which lead to barlike structures that later on transform into bulges through a buckling instability (Combes et al. 1990; Pfenniger & Norman 1990; Norman, Sellwood, & Hasan 1996; Noguchi 2000). Balcells et al. (2003) report a lack of bulges with $r^{1/4}$ surface density profiles, expected in the merging scenario, favoring the secondary process. Ellis, Abraham, & Dickinson (2001) find that intermediate-redshift bulges are bluer than their elliptical counterparts, which indicates that bulges are younger than ellipticals, in contradiction with the bottom-up structure formation scenario of the CDM model. It is likely that some bulges formed by disk instabilities and others by early mergers. In this case, two bulge populations should exist, with different kinematic and photometric properties.

Within the cosmological CDM scenario, galactic spheroids are surrounded by dark matter halos. An exception might be tidal tail galaxies. Distinct gaseous and stellar clumps are frequently found in tidal tails of interacting galaxies (Schweizer 1978; Mirabel, Lutz, & Maza 1991). Barnes & Hernquist (1992) used numerical simulations to demonstrate that these self-gravitating systems consist preferentially of gas and stars and form frequently in the thin, expanding tails of merging galaxies. In contrast to structures that form by cosmological merging, the dark matter fraction in tidal tail galaxies is negligibly small. The dwarf spheroidals orbiting the Milky Way might represent such a population of dark matter deficient tidal tail systems (Irwin & Hatzidimitriou 1995; Klessen & Kroupa 1998).

The formation of galactic spheroids as a result of discrete, violent perturbations of galactic disks is supported by the observation that galaxy populations vary strongly with the galaxy density in clusters. It has been recognized early that most early-type systems are found in clusters (e.g., Hubble & Humason 1931). Detailed observations by Dressler (1980) suggested a well-defined relationship between the local density in clusters and galaxy type (see also Whitmore & Gilmore 1991). Postman & Geller (1984) extended the study of this morphology-density relation to poorer groups of galaxies and defined a single morphology-density relation that is valid over 6 orders of magnitude in density. Melnick & Sargent (1977) found a relation between the morphological type of individual galaxies and their distance from the cluster center. It is still a matter of debate whether this morphology-radius relation follows from the morphology-density relation, or vice versa. Whitmore, Gilmore, & Jones (1993) argued on the basis of Dressler's data that the distance from the cluster center is the more fundamental parameter. This conclusion is supported by the study of Sanromá & Salvador-Solé (1990) who showed that the radial variations in cluster properties are preserved independent of substructure.

Hubble Space Telescope images of clusters at intermediate redshifts have confirmed that morphological transformations occur frequently in clusters. Dressler et al. (1997) and Couch et al. (1998) found an abnormally high proportion of spiral and irregular types at redshifts $z \approx 0.5$ and an increase of the fraction of S0 galaxies toward the present time. These observations are in agreement with cosmological models that predict that galaxy mergers lead to ellipticals and S0s, and that in dense, rich clusters no subsequent formation of a new disk component is possible (Kauffmann, White, & Guiderdoni 1993; Baugh, Cole, & Frenk 1996; Kauffmann 1996). Okamoto & Nagashima (2001) combined semi-analytical methods with cosmological N-body simulations to study the formation and evolution of cluster galaxies. Their models can reproduce the morphology-density relation for elliptical galaxies. However they also predict a clear separation between bulge-dominated and disk-dominated galaxy types in clusters. Mixed types like S0 galaxies should be rare, which is not in agreement with the the observations.

26.2   Rotating Spheroids

Stellar equilibrium systems exist in two basic configurations: rotationally supported disks and pressure-supported spheroids. Disks are stabilized by the balance between centrifugal forces and gravity. Their radial surface density distribution is determined primarily by the specific angular momentum distribution of the stellar system and the shape of the gravitational potential well or the total mass distribution. The velocity dispersion σ of stellar disks is, by definition, small compared to their rotation v_{rot}. It therefore does not affect their radial density profiles or rotation curves, while still regulating their vertical thickness.

Disks are called dynamically cold because of their small random velocities $\sigma \ll v_{rot}$. Spheroids are, in contrast, dynamically hot stellar systems with $\sigma \geq v_{rot}$. Even in these systems angular momentum and rotation can still play an important role. The difference between disks and ellipticals is therefore not necessarily a result of differences in the specific angular momentum distribution but rather due to differences in the stellar velocity dispersion. Stellar disks, for example, can easily be converted into spheroids through internal instabilities or external perturbations that increase the particles' vertical velocity dispersion, even if their angular momentum distribution remains unchanged. This scenario is very attractive in explaining the origin of dwarf ellipticals, which have exponential surface brightness profiles, reminiscent of a disk progenitor. A process that could convert exponential disks into spheroids is tidal interaction in clusters (Moore et al. 1998). Galactic harassment, however, would not reduce significantly the rotational velocity, in contrast with recent observations by Geha, Guhathakurta. & van der Marel (2002).

Spheroids could either be flattened by rotation or by an anisotropic velocity distribution. Violent processes that break up disks and lead to ellipticals should in general result in anisotropic systems. However, it has been argued that especially lower-mass, disky ellipticals are rotationally flattened and isotropic systems (Bender 1988a). If the equidensity

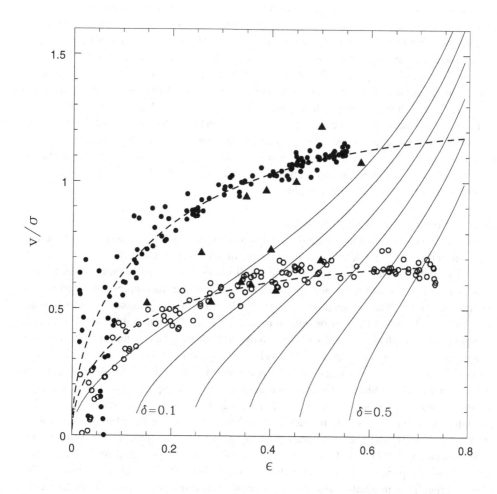

Fig. 26.2. The ratio of line-of-sight rotational velocity to line-of-sight velocity dispersion as a function of ellipticity for disky ellipticals (triangles) and two collisionless merger remnants of disk galaxies (filled and open circles), viewed with different projection angles. The solid lines show theoretical predictions of anisotropic stellar systems with given anisotropy δ. The dashed curves show inclination effects for a system with $(\epsilon, \delta) = (0.78, 0.4)$ and $(\epsilon, \delta) = (0.7, 0.5)$.

surfaces of an oblate spheroid with ellipticity ϵ are all similar, the ratio of line-of-sight rotational velocity v to line-of-sight velocity dispersion σ is (Binney & Tremaine 1987)

$$\frac{v^2}{\sigma^2} = 0.5(1-\delta)\frac{\frac{\arcsin\epsilon}{\epsilon} - \sqrt{1-\epsilon^2}}{\sqrt{1-\epsilon^2} - (1-\epsilon^2)\frac{\arcsin\epsilon}{\epsilon}} - 1 \tag{26.1}$$

where the anisotropy parameter $\delta = 1 - \Pi_{zz}/\Pi_{xx}$ measures the deviation from isotropy and Π_{ii} is the random kinetic energy tensor component in the i'th direction. The solid lines in Figure 26.2 show v/σ versus ϵ for various values of δ. For $\epsilon > 0.2$, inclination (dashed curves) mainly decreases the ellipticity, with no significant change in v/σ. The triangles in Figure

26.2 show observed lower-mass disky ellipticals. They appear to be isotropic, with $\delta < 0.2$. However, some objects, especially those with $v/\sigma \approx 0.5$, could also represent inclined anisotropic ellipticals, with intrinsic anisotropies of $\delta = 0.5$ and high ellipticities ($\epsilon \approx 0.7$). If seen edge-on, these systems would be interpreted as S0 galaxies and therefore would not be classified as disky ellipticals. The open and filled circles show two merger remnants from numerical simulations with mass ratios 3:1 and different initial disk orientations (Burkert, Naab, & Binney 2003, in preparation). Each point represents a different projection angle and follows the theoretical dependence of v/σ and ϵ on the inclination angle. We find that mergers of initially aligned disks result in ellipticals that are indeed intrinsically isotropic (filled circles) and fast rotating, with $v/\sigma = 1$. Misaligned disks, however, form ellipticals that are anisotropic (open circles) with $\delta = 0.5$ and $v/\sigma = 0.5$. These objects could still appear isotropic due to inclination effects.

26.3   Stellar Equilibrium Systems

Any stellar dynamical system is completely specified by its phase space distribution function $f(\vec{x}, \vec{v}, t)$, which determines the number of stars that at time t have positions $\vec{x}$ in a small volume dx^3 and velocities $\vec{v}$ in the small range dv^3. In collisionless systems the flow of points in the 6-dimensional phase space resembles an incompressible fluid and is determined by the Vlasov equation $df/dt = 0$. In equilibrium f must be a steady-state solution ($\partial f/\partial t = 0$) of the Vlasov equation, and the Jeans theorem holds, which says that f depends on the phase space coordinates only through integrals of motion. In the case of spherical symmetry with an isotropic velocity dispersion, f is only a function of the energy: $f = f(E = v^2/2 - \Phi)$, where Φ is the gravitational potential. Obviously there exist an infinite number of equilibrium distribution functions, and stellar spheroids could have a large variety of density distributions. This is not observed, however. Galaxies can be subdivided just into two major groups with respect to their density profiles: giant ellipticals and dwarf ellipticals. Giant ellipticals are characterized by de Vaucouleurs profiles (de Vaucouleurs 1948; Kormendy 1977), dwarfs by exponential profiles. The exponential profiles might be reminiscent of exponential progenitor disks. The origin of the de Vaucouleurs profile and the observed regularity in giant ellipticals is more obscure and still not completely understood.

Internal secular evolution due to two-body relaxation (e.g., Lynden-Bell & Wood 1968) could efficiently erase the information about the initial state, leading to universal structures. This is likely in the case of globular clusters with lifetimes that are large compared to their internal relaxation time scale. The situation, however, is different for galaxies, which have two-body relaxation time scales that by far exceed their age. Hernquist (1990) presented an analytical density distribution $\rho_{\rm H}(r)$ that closely matches the de Vaucouleurs law:

$$\rho_{\rm H}(r) = \frac{M}{2\pi} \frac{a}{r} \frac{1}{(r+a)^3}, \tag{26.2}$$

where M is the total mass and a is a scale length.

The velocity dispersion profile $\sigma(r)$ in the inner region of the Hernquist spheroid is given by

$$\sigma^2 \sim r \ln\left(\frac{a}{r}\right) \tag{26.3}$$

and is characterized by a kinematically cold, power-law density core with a velocity dispersion that decreases toward the center and a density that diverges for $r \to 0$. Numerical simulations of galaxy mergers confirm that kinematically cold cores form as predicted by Equation 26.3. Binney (1982) calculated the fractional energy distribution $N(E)$ that would be required for a stellar systems to follow the $r^{1/4}$ law. He found the interesting result that $N(E)$ is well described by a Boltzmann law

$$N(E) = N_0 \exp(\beta E), \tag{26.4}$$

where $\beta = -2r_e/GM$ represents a negative temperature. Although such an energy distribution is also found in numerical simulations (Spergel & Hernquist 1992) there does not yet exist any analytical theory that could explain its origin.

The origin of universal $r^{1/4}$ profiles might require a phase of strong violent relaxation of the stellar system. Lynden-Bell (1967) noted that strong fluctuations of the gravitational potential during this relaxation phase would change the specific energy distribution of stellar systems and might eventually lead to a universal relaxed state that is independent of the initial conditions. Subsequently, orbital phase mixing will drive the systems toward equilibrium on a time scale of order 2–3 dynamical time scales. Simulations of violently collapsing collisionless stellar systems (van Albada 1982) lead to equilibrium states that were in rough agreement with a de Vaucouleurs profile. A universal state, however, is only achieved if the initial density distribution is very concentrated, as otherwise phase space constraints affect the relaxation and final structure of the inner region (Burkert 1990; Hozumi, Burkert, & Fujiwara 2000). Spergel & Hernquist (1992) adopted a different approach and proposed that violent relaxation can be described by numerous random orbital perturbations that occur preferentially at perigalacticon. In this case, the probability of a particle being scattered into a given state would be proportional to the phase space accessible at perigalacticon, resulting in an exponential energy distribution.

26.4   Fundamental Plane Relations

Stellar systems are characterized by three global physical parameters: central velocity dispersion σ_0, effective radius r_e, and effective surface brightness μ_e, or, in physical units, $\log I_e = -0.4(\mu_e - 27)$. With $L \sim I_e r_e^2$ and assuming virial equilibrium ($M \sim \sigma_0^2 r_e$) Bender, Burstein, & Faber (1992) introduced an orthogonal coordinate system in the 3-space of the observable parameters $\log \sigma_0^2$, $\log r_e$ and $\log I_e$:

$$\kappa_1 \equiv (\log \sigma_0^2 + \log r_e)/\sqrt{2}, \tag{26.5}$$

$$\kappa_2 \equiv (\log \sigma_0^2 + 2\log I_e - \log r_e)/\sqrt{6}, \tag{26.6}$$

$$\kappa_3 \equiv (\log \sigma_0^2 - \log I_e - \log r_e)/\sqrt{3}. \tag{26.7}$$

If we define the luminosity L and the mass M of a galaxy as $L = c_1 I_e r_e^2$ and $M = c_2 \sigma_0^2 r_e$, as given by the virial theorem, with c_1 and c_2 being structure constants, the effective radius can be written as $r_e = (c_1/c_2)(M/L)^{-1}\sigma_0^2 I_e^{-1}$. Then κ_1 is proportional to $\log M$, κ_2 is proportional to $\log(M/L)I_e^3$, and κ_3 is proportional to $\log(M/L)$.

Figure 26.3 shows the distribution of elliptical galaxies and bulges in κ-space. The $\kappa_1 - \kappa_3$ projection shows the plane edge-on. Its tilt is independent of the environment (Jørgensen, Franx, & Kjaergaard 1996) and does in general also exist for S0s and dwarf ellipticals (Nieto et al. 1990). In addition to the optical, a fundamental plane is also found in the infrared, but with a slightly different slope (Mobasher et al. 1999), and probably in the X-ray regime

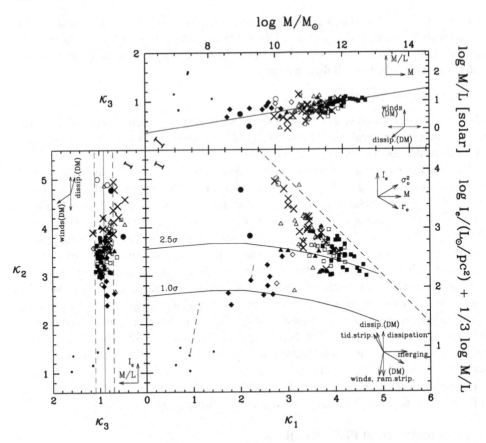

Fig. 26.3. This figure, adopted from Bender, Burstein, & Faber (1997), shows the distribution of all types of dynamically hot galaxies in κ-space. Large squares denote giant ellipticals ($M_T < -20.5$ mag); triangles show ellipticals of intermediate luminosity (-20.5 mag $< M_T < -18.5$ mag). Circles and diamonds denote compact ellipticals and dwarf galaxies, respectively. Open symbols are rotationally flattened galaxies, while filled symbols are anisotropic objects. Bulges are represented by crosses. The five small filled squares at low κ_1 values denote local dwarf spheroidals. The set of arrows indicates how dissipation with and without dark matter, tidal stripping, ram pressure stripping, or merging would move the objects in κ space. The curved lines marked 1.0σ and 2.5σ indicate the range of κ_1 versus κ_2 values expected from a CDM density fluctuation spectrum neglecting dissipation.

(Fukugita & Peebles 1999). The origin of the slope is not well understood up to now. It probably corresponds to variations in the internal structure and to changes in metallicity and age, which seem to correlate well with galaxy mass.

The edge-on view of the fundamental plane can be thought of as a consequence of the virial theorem, independent of initial conditions. The face-on view ($\kappa_1 - \kappa_2$ projection), on the other hand, provides important information about the formation of spheroids. In this plane dwarf ellipticals and giant ellipticals divide into two orthogonal sequences (see also Kormendy 1985; Binggeli & Cameron 1991). Whereas giant ellipticals and bulges with total blue luminosities brighter than $M_{B_T} \approx -18$ mag and stellar masses $M_* > 10^{10} M_\odot$ are characterized by high surface densities that decrease systematically with increasing mass, dwarf ellipticals with $M_{B_T} \geq -18$ mag are diffuse and have surface densities that increase with mass or luminosity. Dissipationless collapse in a CDM Universe would produce structures that lie within the thin solid lines denoted 1.0σ and 2.5σ. Energy dissipation moves galaxies toward larger κ_2 values. Obviously, low-mass giant ellipticals and bulges experienced a large amount of dissipation, leading to high surface densities, compared to the expected dissipationless values. Giant ellipticals, on the other hand, might have formed in gas-poor stellar mergers, which are preferentially dissipationless. The sequence of dwarf ellipticals that runs almost perpendicular to giant ellipticals indicates that these systems might have strongly been affected by wind-driven mass loss (Larson 1974; Arimoto & Yoshii 1986, 1987; Dekel & Silk 1986; Vader 1986; Matteucci & Tornambè 1987; Martinelli, Matteucci, & Colafrancesco 2000), which decreased both κ_2 and κ_1. The galactic wind model can also explain the observed color-magnitude relation (Faber 1973; Bower, Lucey, & Ellis 1992), according to which the integrated colors of dwarf ellipticals become progressively bluer toward fainter luminosities. Gas loss would terminate the epoch of star formation progressively later in more massive ellipticals with deeper potential wells. The stellar populations in brighter galaxies should therefore be more enhanced in heavy elements and would appear redder. Bender et al. (1997) argued, however, that progressively larger amounts of mass loss, starting from a single progenitor galaxy with $\kappa_1 \approx 3.5$ and $\kappa_2 \approx 2.6$ cannot explain the dwarf sequence, which in this case should be much steeper. Dwarf galaxies instead had to form from different progenitors with different initial densities and probably also different amounts of mass loss. It is still not clear up to now why a large range of possible progenitors and the expected strong variations in star formation and galactic mass loss histories should lead to dwarf ellipticals that populate such a narrow one-dimensional sequence in κ-space.

The dichotomy between dwarf and giant ellipticals is clearly visible when investigating their global or central properties. The situation seems to be different when one considers the shape of their light profiles, where the transition appears to be more continuous. Most bright dEs have an inner luminosity excess above the exponential surface brightness profile that is characteristic for low-luminosity dwarfs (Binggeli & Cameron 1991). The profiles of these nucleated dwarfs resemble closely the characteristic $r^{1/4}$ profiles of giant ellipticals. This observed continuity motivated Young & Currie (1994) and subsequently Jerjen & Binggeli (1997) and Binggeli & Jerjen (1998) to fit Sérsic (1968) profiles

$$I(r) = I_0 e^{-(r/r_0)^n} \tag{26.8}$$

to their sample of early-type dwarf and giant galaxies (see also Caon, Capaccioli, & D'Onofrio 1993). They found that the Sérsic index n and the Sérsic parameters I_0 and r_0 vary smoothly with luminosity, indicating that all ellipticals can be reunited into one sequence. The exception are compact ellipticals (Faber 1973; Burkert 1994), which are a rare and special kind of ellipticals with shapes like giants but luminosities like dwarfs. Up to now it is not clear why all ellipticals have surface brightness profiles that vary smoothly with luminosity while, at the same time, their global parameters and also their central parameters (Kormendy 1985) show a clear dichotomy between giant and dwarf ellipticals.

26.5   The Formation of Elliptical Galaxies

Elliptical galaxies have long been thought to be simple spheroidal dynamically relaxed stellar systems that follow a universal de Vaucouleurs $r^{1/4}$ law (de Vaucouleurs 1948) and are classified only by their ellipticity. The traditional formation mechanism for giant ellipticals that would naturally result in a homogeneous family of galaxies is the "monolithic collapse" model. It was motivated by the idea that the oldest stars of the spheroidal halo component of the Galaxy formed during a short period of radial collapse of gas (Eggen et al. 1962). In this case, ellipticals could have formed very early as soon as a finite over-dense region of gas and dark matter decoupled from the expansion of the Universe and collapsed. If during the protogalactic collapse phase star formation was very efficient, a coeval spheroidal stellar system could have formed (Partridge & Peebles 1967; Larson 1969, 1974; Searle, Sargent, & Bagnuolo 1973) before the gas dissipated its kinetic and potential energy and settled into the equatorial plane, forming a disk galaxy. A possible test of this assumption is the redshift evolution of the zero point of the fundamental plane, which is a very sensitive indicator of the age of a stellar population (van Dokkum & Franx 1996). It evolves very slowly, especially for massive elliptical galaxies, indicating a formation redshift of their stars of $z \geq 3$ (Bender et al. 1998; van Dokkum et al. 1998). This scenario would be in agreement with the monolithic collapse picture.

An alternative scenario, proposed by Toomre & Toomre (1972), is that elliptical galaxies formed via a morphological transformation induced by binary mergers of disk galaxies. During the merging phase the stellar disks experienced a phase of violent relaxation due to the strong tidal interactions, resulting in a spheroidal merger remnant. The merging scenario has been tested by observations of rich clusters at intermediate and low redshifts. There is growing evidence that the abundance of spiral galaxies in clusters indeed decreases from a redshift of $z = 0.8$ to $z = 0$ (Dressler et al. 1997; Couch et al. 1998; van Dokkum et al. 2000). A similar trend is observed for the relative numbers of star-forming and post-starburst galaxies (Butcher-Oemler effect) (Butcher & Oemler 1978, 1984; Postman, Lubin, & Oke 1998; Poggianti et al. 1999). At the same time, the early-type fraction increases from 40% to 80% between $z = 1$ and $z = 0$ (van Dokkum & Franx 2001). Semi-analytical models of galaxy formation within the hierarchical merging scenario by Kauffmann (1996) and Kauffmann & Charlot (1998) are also consistent with a low formation redshift for early-type galaxies, which seems to be in contradiction with the ages of their stellar populations. Van Dokkum & Franx (2001) showed that this problem can be solved if the progenitors of present-day ellipticals are not classified as ellipticals at high redshift. In this case, the apparent luminosity and color evolution would look similar to a single age stellar population that formed at very high redshift, independent of the true star formation history.

26.5.1   Boxy and Disky Ellipticals

Further insight into the formation history of ellipticals comes from detailed observations of nearby galaxies, which can be subdivided into two groups with respect to their structural properties (Bender 1988a; Bender, Döbereiner, & Möllenhoff 1988; Kormendy & Bender 1996). Faint giant ellipticals are isotropic rotators with small minor axis rotation and disky deviations of their isophotal contours from perfect ellipses. Their diskiness might be due to a faint secondary disk component that contributes up to 30% to the total light in these galaxies. Disky ellipticals also have power-law inner density profiles (Lauer et al. 1995; Faber et al. 1997) and show little or no radio and X-ray emission (Bender et al. 1989). Bright giant elliptical galaxies with $L_B \geq 10^{11} L_\odot dot$, on the other hand, exhibit nearly elliptical or box-shaped isophotes and show flat cores. Their kinematics are generally more complex than those of disky objects. They rotate slowly, are supported by anisotropic velocity dispersions and have a large amount of minor axis rotation. Boxy galaxies have smaller values of n than disky galaxies. Occasionally, they have kinematically distinct cores (Bender 1988b; Franx & Illingworth 1988; Jedrzejewski & Schechter 1988), which are metal enhanced, indicating that gas infall and subsequent star formation must have played some role during their formation (Bender & Surma 1992; Davies, Sadler, & Peletier 1993). Boxy ellipticals also show stronger radio emission than average and have high X-ray luminosities, consistent with emission from hot gaseous halos (Beuing et al. 1999).

The distinct physical properties of disky and boxy elliptical galaxies demonstrates that the two types of ellipticals could have experienced different formation histories. It has been argued by Kormendy & Bender (1996) and Faber et al. (1997) that the high surface densities (see Fig. 26.3), the secondary disk components, and the central power-law density cusps of disky ellipticals result from substantial gas dissipation during the merging of gas-rich progenitors. Disky ellipticals seem to continue the Hubble sequence from S0s to higher bulge-to-disk ratios. Boxy ellipticals, on the other hand, might have formed by dissipationless mergers between collisionless stellar disks or other ellipticals (Naab & Burkert 2000; Khochfar & Burkert 2003).

26.5.2 *Merger Simulations*

Merger simulations of disk galaxies provide the best access to a direct comparison with observations of individual galaxies. The stellar content of a galaxy is represented by particles that can be analyzed with respect to their photometric and kinematic properties in the same way as an observed galaxy. It has generally been assumed that the progenitors of ellipticals galaxies are disk galaxies. That this assumption is questionable has been demonstrated by Khochfar & Burkert (2003). Their semi-analytical models show that most massive ellipticals actually formed by mixed (elliptical-spiral) or early-type mergers.

Negroponte & White (1983), Barnes (1988), and Hernquist (1992) performed the first fully self-consistent merger models of two equal-mass stellar disks embedded in dark matter halos. The remnants were slowly rotating, pressure-supported, anisotropic, and generally followed an $r^{1/4}$ surface density profile in the outer parts. However, due to phase space limitations (Carlberg 1986) the surface brightness profiles in the inner regions were flatter than observed. To solve this problem a massive central bulge component had to be included in the progenitors (Hernquist 1993a). In this case, the progenitors resembled already early-type galaxies. It seems to be unlikely that all merger progenitors of ellipticals contained a massive central bulge component. On the other hand, these simulations already emphasized that global properties of equal-mass merger remnants resemble those of ordinary, slowly rotating massive elliptical galaxies.

Additional evidence for the merger scenario are tidal tails and shells that are observed in the outer parts of ellipticals and are found to be a natural result of disk mergers (Hernquist & Spergel 1992). In addition, the formation of kinematically decoupled subsystems in merger simulations that include gas strongly support the merger scenario (Hernquist & Barnes 1991). Note, however, that Harsoula & Voglis (1998) proposed an alternative scenario where kinematically distinct subsystems can form directly from an early cosmological collapse without any major mergers thereafter.

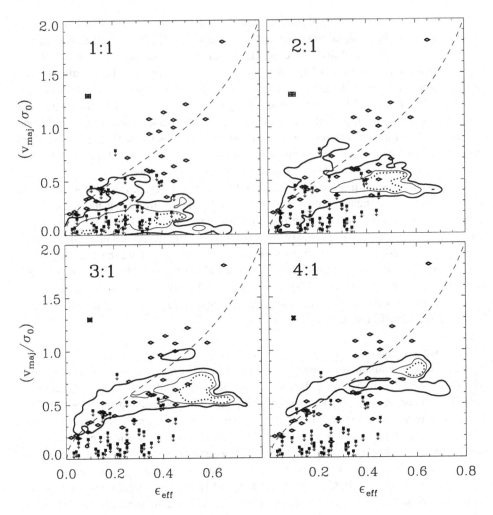

Fig. 26.4. Rotational velocity over velocity dispersion versus characteristic ellipticity for mergers with various mass ratios. Values for observed ellipticals are overplotted. The dashed line shows the theoretically predicted correlation for an oblate isotropic rotator.

More detailed investigations of the isophotal shapes of equal-mass merger remnants have shown that the same remnant can appear either disky or boxy when viewed from different directions (Hernquist 1993b). This result is puzzling since most boxy ellipticals are radio and X-ray luminous, in contrast to disky ellipticals. As radio and X-ray properties should not depend on projection effects, the isophotes should not change with viewing angle.

In contrast to anisotropic, equal-mass mergers, mergers with a mass ratio of 3:1 lead to remnants that are flattened and fast rotating (Bendo & Barnes 2000). Naab, Burkert, & Hernquist (1999) analyzed the photometric and kinematic properties of a typical 3:1 merger remnant and compared the results to observational data of disky elliptical galaxies. They found an excellent agreement and proposed that fast-rotating disky elliptical galaxies can originate from pure collisionless 3:1 mergers, as opposed to slowly rotating, pressure-supported ellipticals, which might form from equal-mass mergers of disk galaxies. Burkert & Naab (2004) and Naab & Burkert (2004) analyzed a large number of high-resolution, statistically unbiased mergers with mass ratios of 1:1, 2:1, 3:1, and 4:1. They concluded that the dichotomy of giant ellipticals can be understood as a sequence of mass ratios of disk-disk mergers (Fig. 26.4). Equal-mass mergers produce anisotropic and slowly rotating remnants with a large amount of minor axis rotation. A subset of initial disk orientations result in purely boxy ellipticals. Only if the initial spins of the disks are aligned will the remnant appear isotropic and disky or boxy depending on the orientation. In contrast, 3:1 and 4:1 mergers form a more homogeneous group of remnants. They have preferentially disky isophotes, and are fast rotating with small minor axis rotation, independent of the assumed projection. 2:1 mergers have intermediate properties, with boxy or disky isophotes depending on the projection and the orbital geometry of the merger.

The influence of gas on the global structure of elliptical galaxies is not well understood. Observations indicate that some giant ellipticals contain a significant amount of gas that is distributed in an extended disklike component (Oosterloo et al. 2002; Young 2002). Such an extended disk naturally forms in gas-rich, fast-rotating, 3:1 merger remnants (Naab & Burkert 2001). Even in 1:1 mergers the remaining gas in the outer parts of the remnant has high enough angular momentum to form extended gas disks as it falls back (Barnes 2002). On the other hand, the simulations of equal-mass mergers also indicate that half of the gas is driven to the center of the remnant, producing a peak in surface density that is not observed (Mihos & Hernquist 1994; Barnes & Hernquist 1996).

The presence of gas in merger simulations influences the stellar structure of the remnants. Even if star formation is neglected, stars in remnants of gas-rich mergers are less likely to be on box orbits than their collisionless counterparts (Barnes & Hernquist 1996), leading to a better agreement with observations of stellar line-of-sight velocity distributions (Bender, Saglia, & Gerhard 1994; Naab & Burkert 2001). The influence of star formation on merger remnants has theoretically been addressed in detail by Bekki & Shioya (1997), Bekki (1998), and more recently by Springel (2000). They found that the rapidity of gas consumption can affect the isophotal shapes. Secular star formation, however, leads to final density profiles that deviate significantly from the observed $r^{1/4}$ profiles in radial regimes where all ellipticals show almost perfect de Vaucouleurs laws (Burkert 1993). As star formation is likely to occur in all disk galaxy mergers this result represents a serious problem for the merger scenario.

26.6   Conclusions

Within the framework of cosmological hierarchical structure formation, galactic disks represent the fundamental building blocks where most of the stars form. Tidal encounters and galaxy mergers heat and destroy these disks, resulting in kinematically hot stellar systems. Galaxy harassment in clusters can preserve the disk structure while increasing the random kinetic energy of the stars perpendicular to the disk. In this case, exponential dwarf ellipticals would form. Galaxy mergers represent more violent processes that lead to strong violent relaxation, erasing the information about the initial state and resulting in a de Vaucouleur's profile as seen in giant elliptical galaxies.

Detailed observations of the kinematic and geometric properties of spheroids, coupled with sophisticated high-resolution simulations, have led to major progress in understanding the origin of these systems. However, many problems still exist and need to be investigated in detail.

Violent relaxation and the origin of the $r^{1/4}$ law is still not understood up to now. Observations of nonrotating, exponential dwarf ellipticals are in contradiction with the harassment scenario. In addition, there exists no theory that can predict why the scale length of dwarf ellipticals is on average in the range of 0.5–1 kpc, independent of luminosity. More observations are required to test the theoretical predictions of two different bulge populations, one with exponential profiles, resulting from disk instabilities, and the other with de Vaucouleur's profiles, resulting from an early, violent merger phase of the protogalaxy. It is also not clear whether the dichotomy of giant ellipticals into disk and boxy objects is preferentially due to variations in the mass ratio of the merger components. Another possibility is an additional gaseous component that settled into an equatorial disk inside the spheroid where it turned into stars. In this case, gas dynamics and dissipation will have affected the structure preferentially in disky ellipticals, which might explain their high surface densities compared to massive, boxy ellipticals that formed preferentially by dissipationless mergers. More simulations including gas dynamics and star formation are required in order to test this scenario. The origin of the most luminous giant ellipticals is currently not understood at all. These objects are much more massive than disk galaxies and therefore could not have formed by major disk mergers. In addition, their metallicities are supersolar and higher than the stellar populations of disk galaxies. Luminous, giant ellipticals probably formed by multiple mergers within dense groups of galaxies followed by an efficient phase of star formation and metal enrichment. Whether this scenario can also explain their large ages and their location in the low-density region of the fundamental plane needs to be explored in greater detail.

Acknowledgements. Andreas Burkert would like to thank Luis Ho for the invitation to a very stimulating and pleasant conference.

References

Arimoto, N., & Yoshii, Y. 1986, A&A, 164, 260
——. 1987, A&A, 173, 23
Ashman, K. M., & Zepf, S. E. 1992, ApJ, 384, 50
Athanassoula, E. 2002, ApJ, 569, L83
Balcells, M., Graham, A. W., Domínguez-Palmero, L., & Peletier, R. F. 2003, ApJ, 582, L79
Barnes, J. E. 1988, ApJ, 331, 699
——. 2002, MNRAS, 333, 481
Barnes, J. E., & Hernquist, L. 1992, Nature, 360, 715
——. 1996, ApJ, 471, 115
Bate, M. R., Bonnell, I. A., & Bromm, V. 2002, MNRAS, 332, L65
Baugh, C. M., Cole, S., & Frenk, C. S. 1996, MNRAS, 283, 1361
Bekki, K. 1998, ApJ, 502, L133
Bekki, K., & Shioya, Y. 1997, ApJ, 478, L17
Bender, R. 1988a, A&A, 193, L7
——. 1988b, A&A, 202, L5
Bender, R., Burstein, D., & Faber, S. M. 1992, ApJ, 399, 462
——. 1997, in Galaxy Scaling Relations: Origins, Evolution and Applications, ed. L. N. da Costa & A. Renzini (Heidelberg: Springer), 95
Bender, R., Döbereiner, S., & Möllenhoff, C. 1988, A&AS, 74, 385
Bender, R., Saglia, R. P., & Gerhard, O. E. 1994, MNRAS, 269, 785
Bender, R., Saglia, R. P., Ziegler, B., Belloni, P., Greggio, L., Hopp, U., & Bruzual A., G. 1998, ApJ, 493, 529
Bender, R., & Surma, P. 1992, A&A, 258, 250
Bender, R., Surma, P., Döbereiner, S., Möllenhoff, C., & Madejsky, R. 1989, A&A, 217, 35
Bendo, G. J., & Barnes, J. E. 2000, MNRAS, 316, 315

Beuing, J., Döbereiner, Böhringer, H., & Bender, R. 1999, MNRAS, 302, 209

Binggeli, B., & Cameron, L. M. 1991, A&A, 252, 27

Binggeli, B., & Jerjen, H. 1998, A&A, 333, 17

Binney, J. 1982, MNRAS, 200, 951

Binney, J., & Tremaine, S. 1987, Galactic Dynamics (Princeton: Princeton Univ. Press)

Bower, R. G., Lucey, J. R., & Ellis, R. S. 1992, MNRAS, 254, 601

Brown, J. H., Burkert, A., & Truran, J. W. 1991, ApJ, 376, 115

——. 1995, ApJ, 440, 666

Burkert, A. 1990, MNRAS, 247, 152

——. 1993, A&A, 278, 23

——. 1994, MNRAS, 266, 877

Burkert, A., & Naab, T. 2004, in Galaxies and Chaos, ed. G. Contopoulos & N. Voglis (Springer), in press

Butcher, H., & Oemler, A., Jr. 1978, ApJ, 219, 18

——. 1984, ApJ, 285, 426

Caon, N., Capaccioli, M., & D'Onofrio, M. 1993, MNRAS, 265, 1013

Carlberg, R. G. 1986, ApJ, 310, 593

Combes, F., Debbasch, F., Friedli, D., & Pfenniger, D. 1990, A&A, 233, 82

Couch, W. J., Barger, A. J., Smail, I., Ellis, R. S., & Sharples, R. M. 1998, ApJ, 497, 188

Davies, R. L., Sadler, E. M., & Peletier, R. F. 1993, MNRAS, 262, 650

Dekel, A., & Silk, J. 1986, ApJ, 303, 39

de Vaucouleurs, G. 1948, Ann. d'Ap., 11, 247

Dressler, A. 1980, ApJ, 236, 351

Dressler, A., et al. 1997, ApJ, 490, 577

Eggen, O. J., Lynden-Bell, D., & Sandage, A. R. 1962, ApJ, 136, 748

Ellis, R. S., Abraham, R. G., & Dickinson, M. 2001, ApJ, 551, 111

Faber, S. M. 1973, ApJ, 179, 423

Faber, S. M., et al. 1997, AJ, 114, 1771

Fall, S. M., & Rees, M. J. 1985, ApJ, 298, 18

Franx, M., & Illingworth, G. D. 1988, ApJ, 327, L55

Fukugita, M., & Peebles, P. J. E. 1999, ApJ, 524, L31

Geha, M., Guhathakurta, R., & van der Marel, R. P. 2002, AJ, 124, 3073

Geyer, M. P., & Burkert, A. 2001, MNRAS, 323, 988

Gnedin, O. Y., Norman, M. L., & Ostriker, J. P. 2000, ApJ, 540, 32

Harris, W. E. 1996, AJ, 112, 1487

Harsoula, M., & Voglis, N. 1998, A&A, 335, 431

Hernquist, L. 1990, ApJ, 356, 359

——. 1992, ApJ, 400, 460

Hernquist, L., & Barnes, J. E. 1991, Nature, 354, 210

Hernquist, L., & Spergel, D. N. 1992, ApJ, 399, L117

Hernquist, L. 1993a, ApJS, 86 389

——. 1993b, ApJ, 409, 548

Hozumi, S., Burkert, A., & Fujiwara, T. 2000, MNRAS, 311, 377

Hubble, E., & Humason, M. L. 1931, ApJ, 74, 43

Irwin, M.. & Hatzidimitriou, D. 1995, MNRAS, 277, 1354

Jedrzejewski, R. I., & Schechter, P. L. 1988, ApJ, 330, L87

Jerjen, H., & Binggeli, B. 1997, The Nature of Elliptical Galaxies, ed. M. Arnaboldi, G. S. Da Costa, & P. Saha (San Francisco: ASP), 239

Jørgensen, I., Franx, M., & Kjaergaard, P. 1996, MNRAS, 280, 167

Katz, N., & Gunn, J. E. 1991, ApJ, 377, 365

Kauffmann, G. 1996, MNRAS, 281, 487

Kauffmann, G., & Charlot, S. 1998, MNRAS, 297, L23

Kauffmann, G., Charlot, S., & White, S. D. M. 1996, MNRAS, 283, L117

Kauffmann, G., White, S. D. M., & Guiderdoni, B. 1993, MNRAS, 264, 201

Khochfar, S., & Burkert, A. 2003, ApJ, submitted (astro-ph/0303529)

Klessen, R. S., & Burkert, A. 2000, ApJS, 128, 287

——. 2001, ApJ, 549, 386

Klessen, R. S., & Kroupa, P. 1998, ApJ, 498, 143

Koo, B. C. 1999, ApJ, 518, 760

Kormendy, J. 1977, ApJ, 218, 333

———. 1985, ApJ, 295, 73

Kormendy, J., & Bender, R. 1996, ApJ, 464, L119

Kraft, R. P. 1979, ARA&A, 17, 309

Lada, C. J., & Lada, E. A. 2003, ARA&A, 41, 57

Larson, R. B. 1969, MNRAS, 145, 405

———. 1974, MNRAS, 169, 229

Lauer, T. R., et al. 1995, AJ, 110, 2622

Lynden-Bell, D. 1967, MNRAS, 136, 101

Lynden-Bell, D., & Wood, R. 1968, MNRAS, 138, 495

Martinelli, A., Matteucci, F., & Colafrancesco, S. 2000, A&A, 354, 387

Matteucci, F., & Tornambè, F. 1987, A&A, 185, 51

Melnick, J., & Sargent, W. L. W. 1977, ApJ, 215, 401

Mihos, J. C., & Hernquist, L. 1994, ApJ, 437, L47

Mirabel, I. F., Lutz, D., & Maza, J. 1991, A&A, 243, 367

Mobasher, B., Guzmán, R., Aragón-Salamanca, A., & Zepf, S. 1999, MNRAS, 304, 225

Moore, B., Katz, N., Lake, G., Dressler, A., & Oemler, A. 1996, Nature, 379, 613

Moore, B., Lake, G., & Katz, N. 1998, ApJ, 495, 139

Myers, P. C., Dame, T. M., Thaddeus, P., Cohen, R. S., Silverberg, R. F., Dwek, E., & Hauser, M. G. 1986, ApJ, 301, 398

Naab, T., & Burkert, A. 2000, in Dynamics of Galaxies: from the Early Universe to the Present, ed. F. Combes, G. Mamon, & V. Charmandaris (San Francisco: ASP), 267

———. 2001, ApJ, 555, L91

———. 2001, in The Central Kpc of Starbursts and AGN: The La Palma Connection, ed. J. H. Knapen et al. (San Francisco: ASP), 735

———. 2004, ApJ, submitted

Naab, T., Burkert, A., & Hernquist, L. 1999, ApJ, 523, L133

Navarro, J. F., & White, S. D. M. 1994, MNRAS, 267, 401

Negroponte, J., & White, S. D. M. 1983, MNRAS, 205, 1009

Nieto, J.-L., Davoust, E., Bender, R., & Prugniel, P. 1990, A&A, 230, L17

Noguchi, M. 2000, MNRAS, 312, 194

Norman, C. A., Sellwood, J. A., & Hasan, H. 1996, ApJ, 462, 114

Okamoto, T., & Nagashima, M. 2001, ApJ, 547, 109

Oosterloo, T. A., Morganti, R., Sadler, E., Vergani, D., & Caldwell, N. 2002, AJ, 123, 729

Partridge, R. B., & Peebles, P. J. E. 1967, ApJ, 147, 868

Peebles, P. J. E., & Dicke, R. H. 1968, ApJ, 154, 891

Pfenniger, D., & Norman, C. 1990, ApJ, 363, 391

Poggianti, B. M., Smail, I., Dressler, A., Couch, W. J., Barger, A. J., Butcher, H., Ellis, R. S., & Oemler, A., Jr. 1999, ApJ, 518, 576

Postman, M., & Geller, M. J. 1984, ApJ, 281, 95

Postman, M., Lubin, L. M., & Oke, J. B. 1998, AJ, 116, 560

Sanromá, M., & Salvador-Solé, E. 1990, ApJ, 360, 16

Schweizer, F. 1978, in Structure and Properties of Nearby Galaxies, ed. E. M. Berkhuijsen & R. Wielebinski (Reidel: Dordrecht), 279

———. 1999, in Spectrophotometric Dating of Stars and Galaxies, ed. I. Hubeny, S. Heap, & R. Cornett (San Francisco: ASP), 135

Sérsic, J. L. 1968, Atlas de Galaxias Australes (Córdoba: Obs. Astron., Univ. Nac. Córdoba)

Searle, L., Sargent, W. L. W., & Bagnuolo, W. G. 1973, ApJ, 179, 427

Spergel, D. N., & Hernquist, L. 1992, ApJ, 397, L75

Springel, V. 2000, MNRAS, 312, 859

Steinmetz, M., & Müller, E. 1994, A&A, 281, L97

Steinmetz, M., & Navarro, J. F. 2002, NewA, 7, 155

Toomre, A. 1974, in IAU Symp. 58, The Formation and Dynamics of Galaxies, ed. J. R. Shakeshaft (Dordrecht: Reidel), 347

Toomre, A., & Toomre, J. 1972, ApJ, 178, 623

Vader, J. P. 1986, ApJ, 305, 390

van Albada, T.S. 1982, MNRAS, 201, 939

van Dokkum, P. G., & Franx, M. 1996, MNRAS, 281, 985

——. 2001, ApJ, 553, 90

van Dokkum, P. G., Franx, M., Fabricant, D., Illingworth, G. D., & Kelson, D. D. 2000, ApJ, 541, 95

van Dokkum, P. G., Franx, M., Kelson, D. D., & Illingworth, G. D. 1998, ApJ, 504, L17

Vietri, M., & Pesce, E. 1995, ApJ, 442, 618

Whitmore, B. C., & Gilmore, D. M. 1991, ApJ, 367, 64

Whitmore, B. C., Gilmore, D. M., & Jones, C. 1993, ApJ, 407, 489

Williams, J. P., & McKee, C. F. 1997, ApJ, 476, 166

Wyse, R. F. G., & Gilmore, G. 1992, AJ, 104, 144

Wyse, R. F. G., Gilmore, G., & Franx, M. 1997, ARA&A, 35, 637

Young, C. K., & Currie, M. J. 1994, MNRAS, 268, L11

Young, L. M. 2002, AJ, 124, 788

27

Massive black holes, gravitational waves, and pulsars

DONALD C. BACKER[1], ANDREW H. JAFFE[2], and ANDREA N. LOMMEN[3]
(1) University of California, Berkeley, CA, USA
(2) Imperial College, London, England
(3) Sterrenkundig Instituut "Anton Pannekoek," Amsterdam, The Netherlands

Abstract

We discuss recent work on two topics. In a recent paper by Jaffe & Backer, we calculate the spectrum of the stochastic background of gravitational radiation from the coalescence of massive black holes throughout the low-redshift Universe. In the recent thesis work of Lommen the observational upper limit on the energy density in the stochastic background of gravitational radiation at nanoHertz frequencies is lowered by an order of magnitude. This limit places an observational bound on the very uncertain merger history of galaxies and subsequent coalescences of massive black holes.

27.1 Introduction

The evidence for massive black holes (MBHs) in the majority of spheroids at zero redshift is increasing, although the origin and mass evolution of MBHs is an outstanding mystery. The evidence for growth of galaxies by mergers is strong, although the merger rate and its evolution with redshift is uncertain. Begelman, Blandford & Rees (1980) wrote a seminal paper on physical processes involved during the merger of galaxies containing MBHs that is followed by the coalescence of the MBHs. Rajagopal & Romani (1995) used the framework in Begelman et al., along with the classic description of gravitational radiation from a point mass binary in Peters & Matthews (1963), to estimate the stochastic background* spectrum of gravitational radiation. In the first section below we provide a summary of Jaffe & Backer (2003) that updates the work of Rajagopal & Romani using new data on both MBH demographics and galaxy merger rate.

The driver for the calculations done in Rajagopal & Romani and in Jaffe & Backer is the potential detection of the gravitational wave background by precision pulsar timing experiments, as had been considered earlier by Sazhin (1978) and Detweiler (1979). Pulsars act as distant clocks, and their electromagnetic emissions to us are perturbed in the same manner as the laser light in gravitational wave interferometers such as LIGO or *LISA*. A mirror is not needed as the source is periodic. Kaspi, Taylor, & Ryba (1994) present precision timing of two millisecond pulsars over a period of seven years. This measurement is used to constrain the energy density in the gravitational wave background to less than 6×10^{-8} in units of the closure density. In the second section we summarize the Berkeley Ph.D. effort of Lommen (2001) that extends the Kaspi et al. experiment to 17 years and lowers the gravitational wave background energy density limit by an order of magnitude.

---

* Some would call this a *foreground* and reserve *background* for primordial waves.

© The Observatories of the Carnegie Institution of Washington 2004.

Prospects for future work are discussed briefly in the final section.

27.2   MBH-MBH Coalescences in the Universe

The goal in Jaffe & Backer is a parameterized model of MBH-MBH coalescences in the Universe that can be adjusted as new data become available. There are five major ingredients, which we list in order of decreasing certainty:

- Binary MBH gravitational wave spectrum
- Cosmological parameters
- Evolution of the MBHs in a merging system
- MBH demographics vs. z
- Galaxy-galaxy merger rate vs. z

As stated in the introduction, the emission of gravitational radiation of a point mass binary in general relativity is a classic result. We use the Peters & Matthews (1963) formulation to derive the wave amplitude h, a dimensionless strain, as a function of MBH masses at a given frequency at the source, which appears redshifted in the frame of the observer. The emitted frequency is twice the orbital frequency for the assumed circular orbits, which, in turn, establishes both the luminosity and the lifetime of the binary system.

Standard cosmological parameters are used. Current uncertainties lead to small changes in the resultant gravitational wave background spectrum.

Begelman et al. (1980) present a framework for the evolution of a pair of galaxies with central MBHs after the galaxies become gravitationally bound. The galaxies merge into each other on an orbital time scale owing to dynamical friction. The process continues with the central cores and their MBHs sinking toward each other. In Begelman et al. and at this conference, the final fate of the MBH binary is questioned. In the simplest of spherical models dynamical friction halts when a loss cone of stars in the orbital plane develops. This may happen before the gravitational radiation time scale is sufficiently short to bring the MBHs to coalescence in a Hubble time. While there is an absence of observational evidence of stalled MBH binaries in subparsec orbits, measurements are very difficult or currently impossible, and so we must say that there is no significant evidence of absence. Figure 27.1 displays the evolution time scale as a function of MBH separation, following a figure in Begelman et al. (1980). In Jaffe & Backer (2003), we assume that nature is efficient in bringing the MBH binary from the galaxy merger separation quickly to the separation where gravitational wave decay of the orbit is rapid. That is, no Green's function is needed to convert between the galaxy-galaxy capture moment and the time of domination of evolution by gravitational wave emission.

We use the results in Merritt & Ferrarese (2001) to provide the MBH demographics in the local Universe. In our current work, we make no adjustment to the population as a function of redshift.

The final ingredient in our model is the galaxy-galaxy merger rate. We use the recent work of Patton et al. (2002). These authors estimate the fraction of bound pairs in a large (4184) field galaxy survey. The pair separation establishes the orbit time and hence the time scale for evolution of the orbit by dynamical friction (see Fig. 27.1). We explore two models of variation of the merger rate with redshift that are within current observational limits. We have approached the model from the viewpoint of observational data at zero redshift rather than the more theoretical approach that uses a merger tree starting at high redshift.

Galaxy Merger -- Massive Black Hole Coalescence

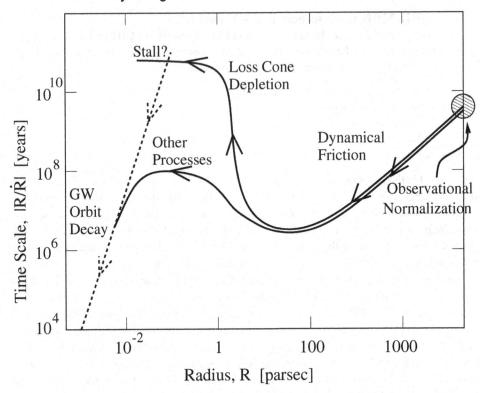

Fig. 27.1. Evolution of the separation of massive black holes during the galaxy-galaxy merger process. Dynamical friction brings them together until a loss cone forms, which slows decrease. Once sufficiently close, the loss of orbital energy and angular momentum to gravitational radiation leads to rapid coalescence.

Our results are summarized in a spectrum (Fig. 27.2) of the characteristic strain, $h_c(f)$, which is derived from the power spectral density spectrum of h by $h_c^2 \equiv f P_h$. This allows for ease of comparison with experimental measurement variances at any frequency, which is the topic of the next section. Both analytical and Monte Carlo results are presented. The most likely redshift for source contribution is ~ 1, and some million MBH binaries are contributing. The significant variance from a smooth spectrum in the Monte Carlo model indicates that a relatively small number dominate.

27.2.1   *Observations of MBH-MBH Coalescence with LISA*

We can extend the formalism from Jaffe & Backer (2003) for other uses. When the *LISA* satellite system is launched in the coming decade, we will be able to monitor the Universe of coalescing MBHs more directly, by observing individual events themselves: they are the brightest objects in the gravitational radiation sky.

As we show in Jaffe & Backer (2003), the nHz–μHz background observable with pulsar timing is sensitive to the most massive, nearby binaries. *LISA*, however, observes the final

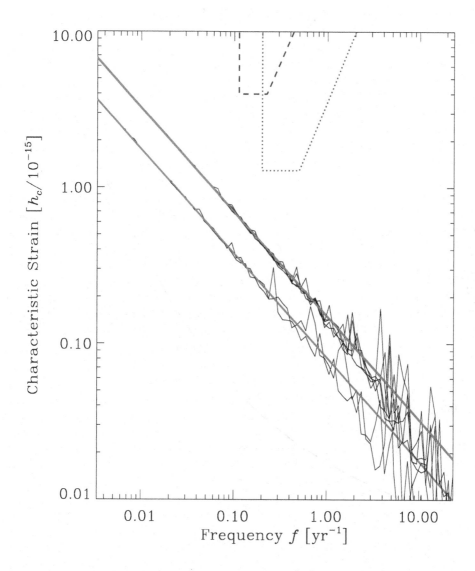

Fig. 27.2. Characteristic strain spectrum $h_c(f)$ for two models discussed in Jaffe & Backer (2003), along with Monte Carlo realizations. The upper thick curve and the associated realizations have strong evolution of merger rate with z, and the lower set of curves has no evolution. The maximum redshift is $z = 3$. The dashed line gives an estimate of the current best limits on the gravitational wave background from pulsar timing observations. The dotted line shows the expected limits from a pulsar timing array, after operation for ~ 8 years.

coalescence events limited not so much by the power of the event, but by whether the frequency of the signal falls within the *LISA* band. Thus, we must determine the total event rate (along the light cone) of MBH-MBH coalescences, and what fraction thereof is observable by *LISA*.

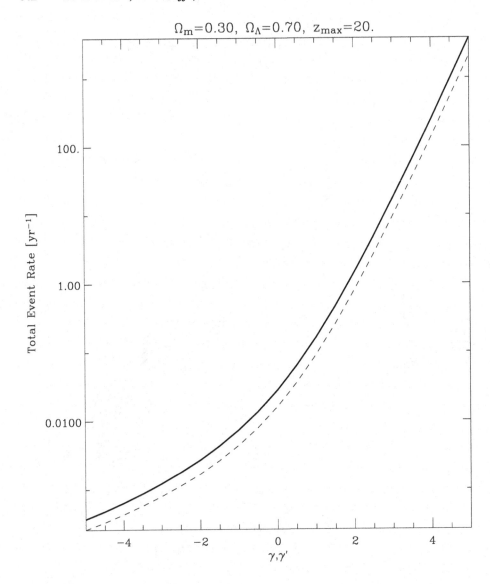

$\Omega_m=0.30, \ \Omega_\Lambda=0.70, \ z_{max}=20.$

Fig. 27.3. The total MBH-MBH coalescence rate for the model of Jaffe & Backer (2003). The index γ gives the power-law index of the merger rate (which is also degenerate with the normalization of the MBH mass function). The solid (dotted) curve is for a merger rate parameterized as a power law in redshift (time).

Obviously, the total MBH-MBH coalescence rate depends on the details of the MBH population and the galaxy merger rate as a function of redshift, as shown in Figure 27.3. We see that for various scenarios the rate can vary by several orders of magnitude, and could be as high as 1000 per year.

Hughes (2001) notes that the gravitational wave signal is dependent only on the combination of parameters $(1+z)M_c$, where z is the redshift of the coalescing pair, and

$M_c^{5/3} = (M_1 + M_2)^{-1/3} M_1 M_2$ gives the "chirp mass" of the system. Hughes also shows that *LISA* can best measure the parameters of those binaries with $(1+z)M_c \sim 10^5 M_\odot$. In the Jaffe & Backer (2003) formalism, the mass of the MBH population does not change with time, and today the mean is considerably higher than this. In reality, of course, the population would likely have had a mean near $10^5 M_\odot/(1+z)$ at a redshift of several. Thus, if we calculate the fraction of pairs around this mass for the present-day population, we can consider this a rough lower bound on the total event rate from binaries, whose properties will be measured well by *LISA*. For the MBH mass function considered in Jaffe & Backer, we find about 10% of MBHs are within a factor of 20 of $10^5 M_\odot$, and so about 1% of pairs would be in that range. However, if the mass function at early times was peaked around this value, a considerably larger fraction of pairs would be detected.

A more realistic calculation (as in Menou et al. 2001) would explicitly take into account the evolution of the MBH mass function. However, since the mechanism for MBH growth is poorly understood (is it dominated by accretion or mergers? how are MBHs originally formed?), it is likely that the theoretical parameter space is as wide as the purely parametric models we have constructed here.

27.3   Gravitational Wave Detection and Pulsar Timing Experiments

Any precision measurement of length or time or energy can be used to detect gravitational radiation. Figure 27.4 presents a large-scale view of the stochastic gravitational wave background spectrum with limits from current and planned experiments. The cosmic microwave background anisotropy $(\delta T/T)$ places a limit on a scale-free primordial spectrum at attoHertz frequencies. Pulsar measurements $(\delta t/t)$ place a limit at nanoHertz frequencies, where the dominant source is likely to be from the MBH binary population in the Universe. The *LISA* and LIGO instruments will operate in the milliHertz and kiloHertz bands, respectively. *LISA* will be sensitive to individual MBH-MBH coalescence events, while pulsar timing will detect only the cacophony of all objects with just a small level of cosmic variance from one frequency to the next (Fig. 27.2).

Gravitational radiation perturbs the spatial part of the spacetime metric from unity by an amount given by the dimensionless strain, h. For electromagnetic radiation propagation through a gravitational radiation field the strain can be treated as a perturbation of the index of refraction from unity in 3-space. Propagation of an electromagnetic wave (and therefore a pulsar pulse) from a distant source in the Galaxy through integer cycles of a gravitational wave leaves no net effect. What *is* embedded in the pulsar arrival time is the result of propagation through fractional cycles of the gravitational wave—at the pulsar after emission and at the earth upon reception. The latter reception effect is correlated among a spatial array of pulsars, and therefore we can use a pulsar timing array as a gravitational wave telescope.

The dimensionless strain that we are sensitive to in pulsar timing is given by $h_c \approx R/(\sqrt{N}T/2\pi)$, where R is the characteristic residual from the timing model, N is the number of degrees of freedom, and T is the duration of the measurement. Given that R is independent of T in the best cases, the strain limit improves with duration. Sensitivity to low-frequency gravitational waves is limited by duration as well as by the need to fit for a neutron star rotation polynomial series in the timing model.

In her recent thesis work at Berkeley, A. Lommen (2001), along with collaborators E. Splaver and D. Nice at Princeton, initiated a precision timing program at the Arecibo Observatory following the upgrade completion in 1996. The *best* star for long-term precision

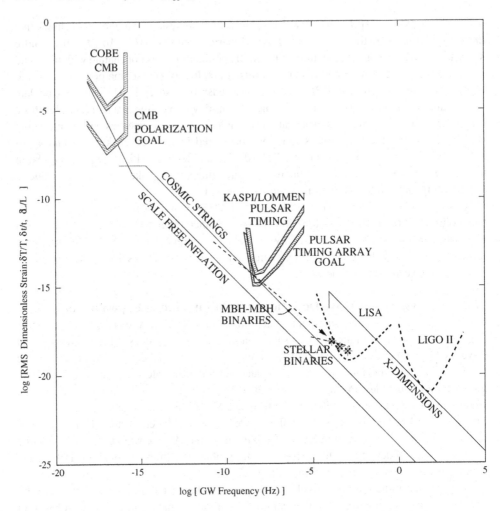

Fig. 27.4. Characteristic strain spectrum covering all observable bands.

is the 5-millisecond pulsar B1855+09. We have tied this new experiment to the earlier Arecibo experiment presented in Kaspi et al. (1994) using lower sensitivity measurements at other telescopes. The resulting limits in Figure 27.2 follow roughly from $R \approx 2$ μs, $N \approx 200$, and $T \approx 17$ yr. Our current limit in h_c is a factor of $\sim$5 above the MBH-MBH model with strong evolution of the merger rate with redshift. We are in the process of defining more quantitatively the parameter space excluded by the measurements, but it seems at this time that the pulsar measurements are not highly restrictive.

27.4   Conclusion

The Jaffe & Backer model of MBH-MBH coalescences provides a revised estimate of the stochastic background of gravitational radiation at nanoHertz frequencies along with a new estimate of the cosmic variance in the spectrum. This model can be improved as new information becomes available. Furthermore, the model can be improved by linking evolutionary components of the model to early-Universe constraints on the growth of structure.

New precision pulsar timing measurements have lowered the bound on the energy density in a stochastic background of gravitational radiation. This constrains extreme models of galaxy-galaxy merger history. Continuing measurements will employ a pulsar timing array that will allow us to detect, rather than simply constrain, the spectrum.

References

Begelman, M. C., Blandford, R. D., & Rees, M. J. 1980, Nature, 287, 307

Detweiler, S. 1979, ApJ, 234, 1100

Hughes, S. 2001, MNRAS, 331, 805

Jaffe, A. H., & Backer, D. C. 2003, ApJ, 583, 616

Kaspi, V. M., Taylor, J. H., & Ryba, M. F. 1994, ApJ, 428, 713

Lommen, A. N. 2001, Ph.D. Thesis, University of California at Berkeley

Merritt, D., & Ferrarese, L. 2001, MNRAS, 320, L30

Patton, D. R., et al. 2002, ApJ, 565, 208

Peters, P., & Matthews, J. 1963, Phys. Rev., 13, 435

Rajagopal, M., & Romani, R. W. 1995, ApJ, 446, 543

Sazhin, M. 1978, Sov. Astron., 22, 36

28

Obscured active galactic nuclei and obscured accretion

ANDREW C. FABIAN

Institute of Astronomy, Cambridge, UK

Abstract

Most of the local active galactic nucleus (AGN) population is obscured and much of the X-ray background originates in obscured AGNs. The contribution of obscured accretion to the growth of massive black holes is discussed here. The recent identification of significant samples of the X-ray sources that dominate the X-ray background intensity has shown a redshift peak at 0.7–0.8, rather than the redshift of 2 found for bright optical quasars. Obscured accretion has a faster evolution than unobscured accretion. The lower redshift and luminosity of most obscured AGNs mean that although they dominate the absorption-corrected intensity of the X-ray background by a factor of about 3 over unobscured objects, they make only an equal contribution to the local mass density in black holes. Obscured and unobscured AGNs together contribute about $4 \times 10^5 \, M_\odot \, \mathrm{Mpc}^{-3}$. Type 2 quasars and Compton-thick objects may give another $10^5 \, M_\odot \, \mathrm{Mpc}^{-3}$, but no more unless direct determinations from the $M_\bullet - \sigma$ relation seriously underestimate the local black hole mass density, or unless most massive black holes are rapidly spinning (so having a higher radiative efficiency than the 10% assumed above). Obscured accretion probably dominates the growth of black holes with masses below a few times $10^8 \, M_\odot$, whereas optically bright quasars dominate at higher masses. The luminosity absorbed by the dusty gas in obscured AGNs is reradiated in the mid-infrared and far-infrared bands. The contribution of AGNs drops from about 20% of the mid-infrared background to just a few percent of the far-infrared background.

28.1   Introduction

The X-ray background (XRB) is dominated by the emission from active galactic nuclei (AGNs). This enables a census to be made of the radiative growth of massive black holes. The infrared and sub-mm backgrounds (hereafter IRB) are dominated by emission from star formation. Together the backgrounds provide measures of the evolution of black holes and galaxies.

The situation is complicated, however, by the fact that the bulk of the XRB is due to highly obscured AGNs. This was first predicted by Setti & Woltjer (1989), elaborated on by Madau, Ghisellini, & Fabian (1994), Comastri et al. (1995), Gilli, Risaliti, & Salvati (1999), and others, and demonstrated by direct resolving of the XRB with *Chandra* by Mushotzky et al. (2000), Brandt et al. (2001), Giacconi et al. (2001), and Rosati et al. (2002), and with *XMM-Newton* by Hasinger et al. (2001). Simple pre-*Chandra/XMM-Newton* estimates (Fabian & Iwasawa 1999), based on a comparison of the intensity of the 2–10 keV XRB, which is dominated by obscured AGNs with that of the soft XRB below 1 keV, which is dominated by unobscured quasars, indicated that most accretion may be obscured (i.e., occurring behind a line-of-sight column density exceeding $N_H = 10^{22}$ cm^{-2}). This assumed that the redshift evolution of obscured and unobscured objects is the same. Recent *Chandra* and *XMM-Newton* data, however, show that this is not the case.

That obscured AGNs are common and need to be included in estimates of accretion power is obvious from the fact that the three nearest AGNs with intrinsic X-ray luminosities above 10^{40} erg s^{-1} (NGC 4945, the Circinus galaxy, and Centaurus A) are all highly obscured with $N_H > 10^{23}$ cm^{-2} (Matt et al. 2000). Two (NGC 4945 and Circinus) are even Compton-thick with $N_H > 1.5 \times 10^{24}$ cm^{-2}. This situation has only been slowly appreciated, perhaps due to NGC 4945 appearing as a starburst galaxy at all non-X-ray wavelengths and to the Circinus galaxy lying close to the Galactic plane.

Nevertheless, it has long been known that the number density of Seyfert 2 galaxies, where the active nucleus is obscured, exceeds that of Seyfert 1 galaxies, although selection effects complicate making a comparison at a fixed bolometric AGN luminosity. Geometrical unification has often been assumed, with Seyfert 2s being Seyfert 1s viewed through a surrounding torus for which the opening angle is about 60°. More recent X-ray studies, particularly with *BeppoSAX*, have been showing that this picture probably applies to only a subset of AGNs.

Here, the evidence for distant obscured AGNs is reviewed and their contribution to the XRB examined. The total energy density due to accretion is then deduced from the spectrum of the XRB, and via Sołtan's (1982) method converted into a local black hole mass density due to radiative growth. Comparison with the locally determined black hole mass density from quiescent galaxies shows that there could be a problem in terms of excessive growth, unless (1) the radiative efficiency of most accretion is higher than that for a standard accretion disk around a Schwarzschild (non-spinning) black hole (e.g., Elvis, Risaliti, & Zamorani 2002), (2) the bolometric correction is lower, or (3) the redshift distribution peaks at $z < 2$.

The problem is illustrated well by the recent estimate for the growth of bright optical quasars by Yu & Tremaine (2002), which allows for little obscured accretion, particularly in massive objects. It is shown here that if their estimate for the local black hole mass density from direct measurements of nearby quiescent galaxies can be revised upward by a factor of 1.5–2, to be in agreement with that of Ferrarese (2002), and factors due to (2) and (3) above are also revised in accord with recent XRB studies, then agreement can be found for a radiative efficiency of 0.1. About equal amounts of the local black hole mass density are then due to obscured and to unobscured accretion.

The X-ray and UV energy absorbed in the obscuring gas is reradiated in the far-infrared and sub-mm bands. A few percent of the energy density in these backgrounds is due to accretion, but most is due to star formation.

28.2 Obscured AGNs

Most Seyfert 2 galaxies contain obscured AGNs, with 2–10 keV X-ray luminosities typically up to about 10^{44} erg s^{-1}. In studies of local, optically selected Seyfert 2s with *BeppoSAX*, Maiolino et al. (1998) have shown that about one-half are Compton thick. In general this half are the classical, optical Seyfert 2 galaxies and the other, Compton-thin, half corresponds to the optical intermediate classes of Seyfert 1.8 and 1.9 (Risaliti, Maiolino, & Salvati 1999).

Distant obscured AGNs (redshift $z > 0.3$) are now being found in large numbers by X-ray observations with *Chandra* and *XMM-Newton* (Mushotzky et al. 2000; Alexander et al. 2001; Barger et al. 2001; Brandt et al. 2001; Crawford et al. 2001; Giacconi et al. 2001; Hasinger et al. 2001; Rosati et al. 2002). Most of the serendipitous sources found in an X-ray image above 1 keV made with these telescopes are obscured. Source variability in many cases makes an AGN identification unambiguous. The determination of column densities requires that the source is identified and its redshift known. Where significant samples are available (e.g., Alexander et al. 2001; Barger et al. 2002; Mainieri et al. 2002), more than two-thirds are Compton thin with $10^{21} < N_H < 10^{23}$ cm^{-2}; most of the remainder are unobscured. The absorption-corrected, 2–10 keV luminosity of the obscured objects is typically in the range of 10^{42}–10^{44} erg s^{-1}. Only a handful of obscured AGNs have yet been found with 2–10 keV luminosities exceeding the level of $\sim 3 \times 10^{44}$ erg s^{-1}, corresponding to a quasar [Crawford et al. 2002; Norman et al. 2002; Stern et al. 2002; Wilman et al. 2003; Mainieri et al. 2002; the last authors define a quasar as $L(0.5-10 \text{ keV}) > 10^{44}$ erg s^{-1}].

Note that there is not complete agreement between optical and X-ray classification of some of the Compton-thin AGNs. Some show X-ray absorption but little optical extinction (Maiolino et al. 2001), and vice versa, and some X-ray obscured objects show no detectable narrow-line region at optical or infrared wavelengths (e.g., Comastri et al. 2002; Gandhi, Crawford, & Fabian 2002). When there is a large covering fraction of the nucleus by dusty gas, there need be little or no optical/UV narrow-line region. The terms Type 2 and Type 1 when applied to X-ray sources are commonly referring to whether there is absorption or not, irrespective of the optical spectrum.

Another class of obscured AGNs are the powerful radio galaxies. These have large column densities (probably in a torus perpendicular to the radio axis) of $\sim 10^{23}$ cm^{-2} or more (e.g., Cygnus A, Ueno et al 1994; 3C 294 at $z = 1.786$, Fabian et al. 2003; B2 0902 at $z = 3.2$, Fabian, Crawford, & Iwasawa 2002; see Fig. 28.1). These are sufficiently rare that their contribution to the XRB intensity is negligible.

Source counts from deep X-ray surveys show that most of the XRB is now resolved (Fig. 28.2), with a major uncertainty being the actual intensity of the XRB measured by wide-beam instruments. The counts flatten below a flux of about 10^{-14} erg cm^{-2} s^{-1} in the 2–10 keV band, above which more than 60% of the XRB intensity originates. At much lower fluxes more and more starburst galaxies are detected (Alexander et al. 2002a; Hornschemeier et al. 2002). They make a negligible contribution to the total XRB intensity (Brandt et al. 2002).

28.3   X-ray Constraints on the Radiative Growth of Massive Black Holes

The basic method for deducing the local density in black holes due to growth by accretion that emitted measurable radiation is that originally due to Sołtan (1982). From

$$E = \eta M c^2,$$

where η is the efficiency with which mass is turned into radiation ($\eta = 0.06$ for a standard thin disk around a non-spinning black hole; a typical assumed value for an accreting black hole is 0.1), we find

$$\varepsilon_{\text{rad}}(1+z) = \eta \rho_\bullet c^2.$$

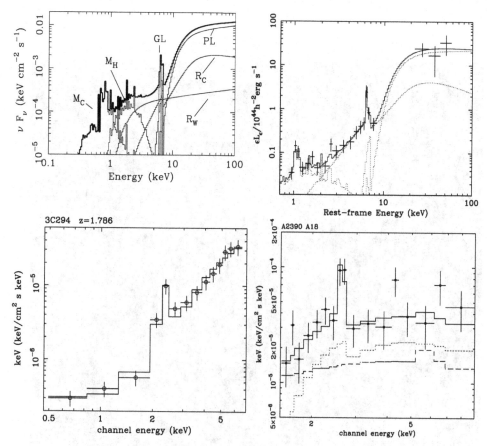

Fig. 28.1. Examples of the νF_ν X-ray spectra of obscured AGNs. *Top left:* Model spectrum that fits the *BeppoSAX* data on NGC 6240 (Vignati et al. 1999). The heavily absorbed power law (PL) and Gaussian line (GL) emission of iron are shown, with hot and cold reflection and emission components. *Top right:* IRAS 09104+4109 has a 2–10 keV luminosity of $\sim 10^{46}$ erg s^{-1} behind a column density of 3×10^{24} cm^{-2} (Franceschini et al. 2000; Iwasawa, Fabian, & Ettori 2001). *Lower left:* is 3C 294, a powerful radio galaxy at $z = 1.786$ with an X-ray luminosity of $\sim 10^{45}$ erg s^{-1} and a column density of 8×10^{23} cm^{-2}. *Lower right:* is an *XMM-Newton* spectrum (Gandhi 2002) of serendipitous source A18 at $z = 1.467$ in the field of the rich galaxy cluster A2390 that has a 2–10 keV luminosity of $\sim 10^{45}$ erg s^{-1} and $N_H \approx 2 \times 10^{23}$ cm^{-2}. The level of contaminating cluster emission to the spectrum is indicated by the dashed line.

$\varepsilon_{\rm rad}$ is the observed energy density in that radiation now, $\rho_\bullet$ is the mean mass density added to the black holes, and z is the mean redshift of the population. Note that the result is independent of the assumed cosmology and requires only that the redshift distribution of the sources be known.

$\varepsilon_{\rm rad}$ is determined from either the images and spectra of the sources themselves or from the background radiation they produce (Fig. 28.3). A bolometric correction κ from the observed band to the total luminosity is required. $\rho_\bullet$ is either determined from the above equation, giving $\rho_\bullet^{\rm AGN}$, or is measured from local galaxies $\rho_\bullet^{\rm direct}$ using the black hole mass to galaxy

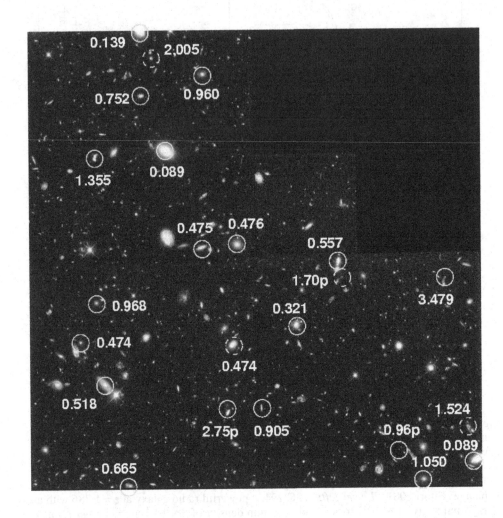

Fig. 28.2. *Chandra* sources from a 2 Ms exposure of the HDF-N superposed on a *Hubble Space Telescope* optical image (kindly provided by W. N. Brandt; see, e.g., Brandt et al. 2002). Source redshifts are indicated ("p" means photometric).

bulge velocity dispersion ($M_\bullet - \sigma$) relation (from Ferrarese & Merritt 2000 or Gebhardt et al. 2000), the less well-known velocity dispersion function for galaxies or a proxy for it, and the Hubble constant (in order to determine volumes). The variables are η, z, H_0, and κ.

Various attempts have recently been made to compare the values of $\rho_\bullet^{AGN}$ and $\rho_\bullet^{direct}$. In units of $10^5 M_\odot \, \mathrm{Mpc}^{-3}$ for $\rho_\bullet$, and adopting $H_0 = 75 \, \mathrm{km\,s}^{-1} \, \mathrm{Mpc}^{-1}$,, Ferrarese (2002) obtained $\rho_\bullet^{direct} = 4-5$, whereas Yu & Tremaine (2002) find $\rho_\bullet^{direct} = 3 \pm 0.5$. Using bright optical quasars only and $\eta = 0.1$, Yu & Tremaine (2002) obtain $\rho_\bullet^{AGN} = 2.2$. They consider

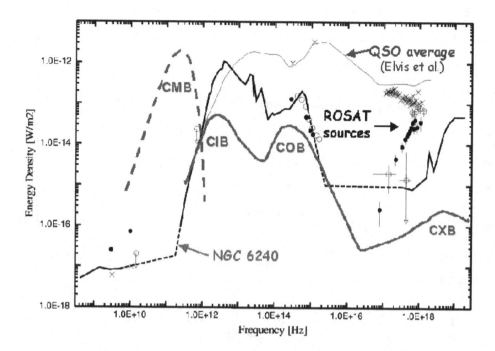

Fig. 28.3. Overview of the intensity of the background radiations (CIB=IRB, COB, and CXB=XRB) together with schematic spectral energy distributions of a nearby obscured AGN, NGC 6240, and of the average unobscured quasar. (Kindly provided by G. Hasinger; see Hasinger 2003.)

that this agrees well enough with their value of $\rho_\bullet^{\text{direct}}$ that there is no room for significant growth by obscured accretion. In other words, they find that the mean density in local black holes can be wholly due to radiatively efficient accretion in an optically bright (and unobscured) quasar phase.

The result of Yu & Tremaine (2002) contrasts with that obtained from the XRB by Fabian & Iwasawa (1999). In order to use the XRB some correction has to be made for absorption. This was accomplished by noting that the mean spectra of unobscured quasars is a power law with an energy index of unity in the 1–20 keV band. Therefore the minimum correction is that required to push the XRB spectrum, which is a power law of index 0.4 in the 2–10 keV band, up to an index of one, matching at the EI_E peak in the XRB (Fig. 28.4). This process emphasizes the importance of obscured accretion since the unobscured objects dominate below 1 keV, which is a level 4 times below that of the resultant absorption-corrected minimum spectrum. In the absorption-corrected sense there is 3 times more energy density in the obscured objects than in the unobscured ones. This could imply that obscured accretion dominates the growth of massive black holes, contrary to the conclusion of Yu & Tremaine (2002).

The value for $\rho_\bullet^{\text{AGN}}$ obtained in this way from the XRB is 6–9 (Fabian & Iwasawa 1999) and 7.5–16.8 (Elvis et al. 2002). The bolometric correction was that relevant for quasars (from the work of Elvis et al. 1994), $\kappa_X = 30-50$, $\eta = 0.1$, and $z = 2$.

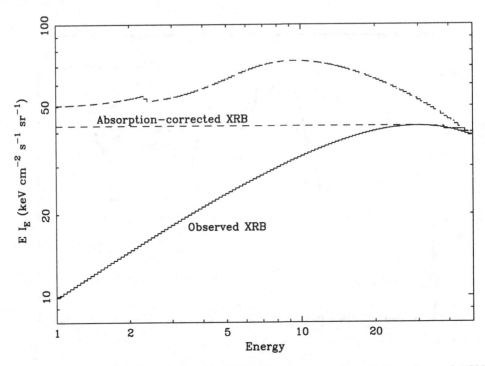

Fig. 28.4. Observed EI_E spectrum of the XRB. The spectrum of a typical unobscured AGN is a horizontal line (energy index of unity). The minimum correction to the XRB is the horizontal line (dashed) that matches the XRB spectral peak. Disk reflection (indicated by top line) could increase this estimate. (Adapted from Fabian & Iwasawa 1999.)

The XRB value for $\rho_{\bullet}^{AGN}$ was below the value of $\rho_{\bullet}^{direct}$ in 1999, when the local quiescent black holes masses (Magorrian et al. 1998) were about 3–5 times higher than are found now. If current values of $\rho_{\bullet}^{direct} = 3 - 5$ are used then it might seem that there is a problem. Elvis et al. (2002) have argued that it implies $\eta > 0.15$ and therefore that all massive black holes are spinning rapidly. This could cause some problems with merger-based galaxy and black hole growth schemes (Hughes & Blandford 2003).

High radiative efficiency also means that the mass-doubling time (assuming that the sources are Eddington limited) exceeds 60 Myr, which could cause problems in growing massive objects from much smaller seed black holes. Perhaps, however, the plunge region within the innermost stable orbit is being tapped by the action of magnetic fields in the disk, thus yielding a higher efficiency without spin (Gammie 1999; Krolik 1999; Agol & Krolik 2000). Another possibility is that many black holes are ejected in mergers, although the required large fraction seems doubtful.

28.4   The Redshift Distribution of Obscured Sources

The above discussion using the XRB assumed that the evolution of the obscured AGNs is the same as that of unobscured quasars, which peaks at $z \approx 2$. Recent results from source identifications (Alexander et al. 2001; Barger et al. 2002; Mainieri et al. 2002; Rosati et al. 2002; Hasinger 2003) have, however, shown that the obscured objects peak at a lower redshift of about $z \approx 0.7$. The identification of a complete sample has not yet been carried out, but, as noted by the above authors, the results from partial samples already show that there are many more sources found below redshift 1 than would be expected from any model based on quasar evolution. This is a very important result.

The immediate effect on the problem in the last section is the drop in z and also of κ. Most, but not all, of the X-ray sources now have absorption-corrected, 2–10 keV luminosities below $10^{44}\,\mathrm{erg\,s^{-1}}$ and are strictly not quasars. Although this luminosity distinction is somewhat arbitrary, the key point is that Seyferts have a 2–10 keV bolometric correction factor κ of 10–20, rather than 30–50 typical of quasars.

It is difficult to determine κ for Seyfert galaxies since the dominant thermal disk emission lies in the rest-frame EUV. Moreover, the low redshift of most of the well-studied ones means that the disk emission, unlike that in quasars, is not shifted into easily observable bands. I have used results from *ASTRO-1, EUVE*, and *FUSE* and find that $\kappa = 12 - 18$, using data on Mrk 335 (Zheng et al. 1995), NGC 3783, (Krolik & Kriss 2001) and NGC 5548 (Magdziarz et al. 1998).

Together, the joint effects of the lower redshift peak and lower bolometric correction reduce the value of $\rho_\bullet^{\mathrm{AGN}}$ for obscured sources by a factor of up to 3. The net result is that optically bright (unobscured) Type 1 quasars give $\rho_\bullet^{\mathrm{AGN}} \approx 2$, and obscured, Type 2 AGNs now give a similar value, totaling about 4, again in units of $10^5\,M_\odot\,\mathrm{Mpc}^{-3}$. This then agrees well with the values from local studies of $\rho_\bullet^{\mathrm{direct}}$.

[Note that the change in κ with luminosity means that a lower limit for the intrinsic 2–10 keV luminosity of quasars is difficult to determine at the present time. The origin of the optical definition is somewhat arbitrary. The properties of objects appear to change around $L(2-10\,\mathrm{keV}) = 3 \times 10^{44}\,\mathrm{erg\,s^{-1}}$, which is the limit used here.]

28.4.1 Some Implications for Obscured Sources

The discovery that obscured AGNs follow a much steeper evolution than optically bright quasars means that they are a different population. The lack of any unobscured counterpart implies that the obscuration covers a large part of the sky as seen by the source itself. There can be no simple torus or geometrical unification picture for these objects. The large covering fraction of absorbing material may explain why there is little in the way of an optical/UV narrow-line region seen for some of the identified X-ray sources (particularly if the absorbing gas is dusty; see, e.g., Gandhi et al. 2002).

Interestingly, there is another population of objects that does evolve in a similar way to the obscured X-ray population, namely dust-enshrouded starburst galaxies. Chary & Elbaz (2001) show that distant luminous and ultraluminous infrared galaxies seen with *ISO* evolve very rapidly to $z \approx 0.8$. There is some overlap between *ISO* and *Chandra/XMM-Newton* X-ray sources in deep images (Wilman, Fabian, & Gandhi 2000; Alexander et al. 2002a; Fadda et al. 2002), and it is plausible that a subset ($\sim 20\%$) of the dusty starburst galaxies are X-ray detectable Type 2 AGNs. The inner parts of the starburst itself may, through winds and supernovae, be responsible for inflating the absorbing gas so that it has a large covering fraction as seen from the center (Fabian et al. 1998; Wada & Norman 2002).

It is not yet clear from identification work quite what fraction of the X-ray sources dominating the XRB are dusty starbursts. Very (and extremely) red objects are reasonably common counterparts of the *Chandra* serendipitous X-ray sources (Alexander et al. 2002b); yet, they are a heterogeneous class (e.g., Smail et al. 2002) that includes both dusty starbursts and old early-type galaxies.

A further point is that Seyfert galaxies typically operate at about 10% of the Eddington limit. If the XRB objects are similar, then from the inferred luminosities, the black hole masses are in the range of $10^6 - 3 \times 10^8 \, M_\odot$, below that generally implied for quasars ($\sim 10^8$ to $> 10^9 \, M_\odot$). Thus, unobscured accretion seen in optically bright quasars may make most of the black holes above $3 \times 10^8 \, M_\odot$ and obscured accretion those at lower masses.

What we do not know is whether quasars passed though an obscured phase early on (see, e.g., Fabian 1999), perhaps when their masses were $10^8 \, M_\odot$ or less. As discussed earlier, there are some Type 2 quasars being found in deep X-ray surveys, but not in large numbers. Also there is little evidence yet for a population of distant Compton-thick objects ($N_H > 1.5 \times 10^{24} \, \mathrm{cm}^{-2}$). Of course, these are likely to be difficult to identify spectroscopically. Wilman (2002) has provided arguments against any significant Compton-thick population. There may still be a Compton-thick population to fill in the νI_ν peak in the XRB, but it could well be at *low* redshift ($z < 0.5$). It requires a instruments like the *Swift* BAT and EXIST to uncover this population in detail.

Unless the estimates for $\rho_\bullet^{\mathrm{direct}}$ are revised upward in the future, there is little room for $\rho_\bullet^{\mathrm{AGN}}$ in terms of Type 2 quasars or Compton-thick objects, with a limit being $\lesssim 1 \times 10^5 \, M_\odot \, \mathrm{Mpc}^{-3}$.

28.5   Models for the Evolution of the XRB

The newly discovered redshift distribution (Fig. 28.5) for the obscured AGNs has prompted some new synthesis models for the XRB in which the obscured objects have a steeper evolution than the unobscured ones (Franceschini, Braito, & Fadda 2002; Gandhi & Fabian 2003).

A difficulty in making such models is the lack of any X-ray luminosity functions for Type 2 AGNs. Gandhi & Fabian (2003) used the 15 μm *ISO/IRAS* infrared luminosity function of Xu et al. (1998). Spectral energy distributions suggest that $L(2-10 \, \mathrm{keV}) \approx L(15 \, \mu m)$, and using that we normalize to the local 2–10 keV X-ray luminosity function of Piccinotti et al. (1982). A power-law distribution in column density is assumed. A reasonable fit to the spectrum of the XRB is obtained if the emissivity due to Type 2 AGNs evolves as $(1+z)^4$ out to $z = 0.7$ and then remains flat to higher redshifts (at least to $z = 1.5$). $\rho_\bullet^{\mathrm{AGN}}$ from this model is in accord with the values in § 28.4. The model can also reproduce well the X-ray source counts, provided that some density evolution is included.

Gandhi & Fabian (2003) tested whether the iron emission line produced by fluorescence in the absorption process is detectable in the spectrum (Matt & Fabian 1994) and found that it should give a small peak in the XRB spectrum over the 3–4 keV band. Interestingly, there is a bump in the *ASCA* spectrum of the XRB at that point (Gendreau 1996), although it is not statistically significant. *XMM-Newton* should do better and could thereby confirm, in an integral manner, the redshift distribution of the component sources of the XRB.

28.6   Some Comments on Fueling and Obscuration

Rapid fueling of a black hole requires a plentiful gas supply, which, if distributed, is likely to coincide with a starburst. Star formation can churn the gas up through winds and supernovae and make the covering fraction of cold absorbing gas large (Fabian et al. 1998; Wada & Norman 2002). The coincidence between the *ISO* dusty starburst population and the obscured Type 2 AGNs should not then be too surprising. What has yet to be explained

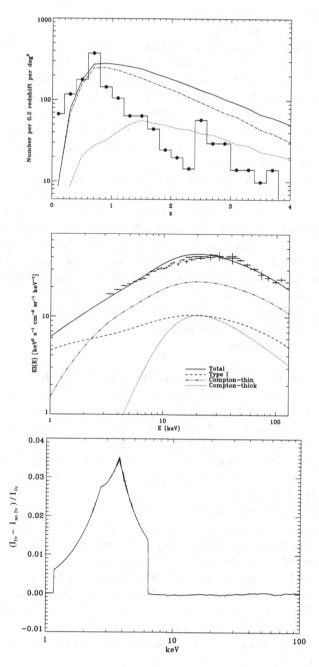

Fig. 28.5. *Top:* The observed serendipitous X-ray source distribution with redshift (Hasinger 2003), with the predictions from the model of Gandhi & Fabian (2003). Type 1 and 2 sources are the dotted and dashed lines, respectively. Source identifications are not complete, so more sources may fill in above $z = 1$. The numbers found below $z = 1$ already far exceed the predictions of the model by Gilli, Salvati, & Hasinger (2001). *Middle:* Matching the model to the XRB spectrum. It is not clear that a significant Compton-thick population is required. *Bottom:* The spectral residuals expected due to the presence of iron fluorescence emission in the sources.

is why there is such a dramatic and rapid decrease in this activity since a redshift of 0.7 [proportional to $(1+z)^4$], over the last 5 Gyr.

The question is then raised of why quasars are unobscured. It could be the high luminosity that drives away all nearby gas. Perhaps, as indicated above, the optically bright quasars are of higher mass (above $10^8 M_\odot$) than the typical Type 2 AGNs, or they are closer to the Eddington limit. Powerful radio galaxies, which have very luminous and presumably massive nuclei do, however, provide counter examples; although the radio outbursts may all be young ($< 10^7$ yr) and due to even higher mass black holes, they could be well sub-Eddington. A significant population of Type 2 quasars may yet emerge from the complete identification of large, deep X-ray samples.

The Eddington limit for a central black hole will always be significantly less than the Eddington limit for the galaxy bulge. That assumes, however, only radiation pressure through electron scattering. A nuclear wind or radiation pressure acting on dusty cold gas (which has a much larger absorption cross section than electron scattering) can make the relevant limiting luminosity for the galaxy bulge more than the Eddington limit for the nucleus itself (Fabian 1999; Fabian, Wilman, & Crawford 2002). Consequently an obscured nucleus can increase in luminosity and then blow away the obscuring gas and its own fuel supply (see also Silk & Rees 1998). This can help relate the final black hole mass to the mass and potential well of its host bulge. The tight $M_\bullet - \sigma$ relation found locally (Ferrarese & Merritt 2000; Gebhardt et al. 2000) does, however, suggest that a single mechanism is acting throughout the entire mass range.

28.7   Contributions to the Far-infrared and Sub-mm Backgrounds

The X-ray and UV luminosity absorbed in dusty Type 2 AGNs is re-emitted in the far-infrared. Much work remains to be done, but it is plausibly reradiated at about $100 \mu m$ in the rest frame. From an *ISO* study Fadda et al. (2002; see also Alexander et al. 2002a) find that AGNs contribute about 17% of the infrared background at $15 \mu m$. In a νF_ν sense this contribution will rise by about a factor of 2 out to a few $100 \mu m$ (see, e.g., Crawford et al. 2002), whereas the IRB rises by about a factor of 10, such that the total contribution of AGNs to the whole IRB will be 3%–4%.

The AGN fractional contribution to the IRB is highest at the shorter wavelengths and drops at the longer wavelengths. Only if there is some as yet unidentified population of Compton-thick objects can this fraction be much larger and important at long wavelengths, in the sub-mm. The margin for such a population, given the agreement between the predicted and observed local black hole densities, is small. The *Chandra* medium-deep detection rate for serendipitous SCUBA sources is low (Bautz et al. 2000; Fabian et al. 2000; Hornschemeier et al. 2000; Barger et al. 2001; Almaini et al. 2003). In the 2 Ms CDF-N, Alexander et al (2003) detect 7 out of 10 bright SCUBA sources and classify 5 as AGNs. They find luminosities that are Seyfert-like and conclude that the sub-mm emission is dominated by starbursts.

Fabian & Iwasawa (1999) determined the IRB contribution to be 2 nW m^{-2} sr^{-1}, while Elvis et al. (2002) found 3.6–8 nW m^{-2} sr^{-1}; these values are to be compared with a total integrated intensity of 40 nW m^{-2} sr^{-1} for the IRB. Revising the bolometric correction factor κ down to 15 then makes these predictions range between $\sim 1-3$ nW m^{-2} sr^{-1}. In other words, AGNs, principally obscured ones, contribute a few percent to the IRB.

Our understanding of the source composition of the IRB will receive an enormous boost from the imminent launch and operation of *SIRTF*.

28.8 Summary

The XRB and X-ray source populations show that there is much obscured accretion in the Universe. This creates a black hole growth crisis if the Yu & Tremaine (2002) analysis stands, unless AGNs are all particularly radiatively efficient ($\eta > 0.15$), perhaps with rapidly spinning black holes (Elvis et al. 2002). However, if the mean local density of black holes is between 4 and $5 \times 10^5 \, M_\odot \, \mathrm{Mpc}^{-3}$, as deduced by Ferrarese (2002), then there is consistency for $\eta \approx 0.1$ between the density predicted for both obscured and unobscured AGNs using the XRB. They contribute roughly equal amounts to the local mass density of black holes, with obscured accretion contributing most for black holes with masses below about $3 \times 10^8 \, M_\odot$ and unobscured quasars contributing most above that value.

The overall picture is that the most massive black holes (above the break in the present mass function at about $3 \times 10^8 \, M_\odot$) are built earlier (by $z \approx 1.5$) than the lower mass ones below the break (which are built by $z \approx 0.7$). This means that the lower mass ones take about twice as long to assemble. As the Universe ages, the black holes that remain active are becoming increasingly obscured.

The fraction of the mass density from obscured accretion is lower than estimated earlier because the sources evolve differently to quasars, peaking at $z = 0.7$ rather than 2. Also their luminosities are in the range of Seyferts which locally have a lower bolometric correction. This makes the contribution of obscured accretion to the mass density in black holes at about 50%. Uncertainties due to the level of spin and efficiency for both obscured and unobscured AGNs remain, and to the exact evolution of complete samples of the X-ray sources. Also there could be populations of Type 2 quasars and distant Compton-thick sources yet to be discovered. Such populations can only make a significant contribution if either optically bright quasars are found to be rapidly spinning or the local mass density in black holes has been seriously underestimated.

The covering fraction of the obscured Type 2 AGNs which dominate the XRB must be very high, or the unobscured fraction would already have been noticed. The obscuring material probably forms part of a compact, inner dusty starburst in the host galaxy. The absorbed luminosity is reradiated in the mid-infrared to far-infrared bands and contributes a few per cent of the IRB.

Complete Type 2 AGNs samples at all redshifts are urgently needed. Obscured accretion, which is best studied at X-ray wavelengths, must be studied in order for the growth and evolution of massive black holes to be understood.

Acknowledgements. I thank Luis Ho for organizing such an interesting meeting, and Niel Brandt, Poshak Gandhi, Günther Hasinger, Kazushi Iwasawa, and Jeremy Sanders for help.

References

Agol, E., & Krolik, J. H. 2000, ApJ, 528, 161

Alexander, D. M., et al. 2003, AJ, 125, 383

Alexander, D. M., Aussel, H., Bauer, F. E., Brandt, W. N., Hornschemeier, A. E., Vignali, C., Garmire, G. P., & Schneider, D. P. 2002a, ApJ, 568, L85

Alexander, D. M., Brandt, W. N., Hornschemeier, A. E., Garmire, G. P., Schneider, D. P., & Bauer, F. E. 2001, AJ, 122, 2156

Alexander, D. M., Vignali, C., Bauer, F. E., Brandt, W. N., Hornschemeier, A. E., Garmire, G. P., & Schneider, D. P. 2002b, AJ, 123, 1149

Almaini, O., et al. 2003, MNRAS, 338, 303

Barger, A. J., Cowie, L. L., Brandt, W. N., Capak, P., Garmire, G. P., Hornschemeier, A. E., Steffen, A. T., & Wehner, E. H. 2002, AJ, 124, 1838

Barger, A. J., Cowie, L. L., Mushotzky, R. F., & Richards, E. A. 2001, AJ, 121, 662

Bautz, M. W., Malm, M. R., Baganoff, F. K., Ricker, G. R., Canizares, C. R., Brandt, W. N., Hornschemeier, A. E., & Garmire, G. P. 2000, ApJ, 543, L119

Brandt, W. N., et al. 2001, AJ, 122, 1

Brandt, W. N., Alexander, D. M., Bauer, F. E., & Hornschemeier, A. E. 2002, Phil. Trans. Roy. Soc., 360, 2057

Chary, R., & Elbaz, D. 2001, ApJ, 556, 562

Comastri, A., et al. 2002, ApJ, 571, 771

Comastri, A., Setti, G., Zamorani, G., & Hasinger, G. 1995, A&A, 296, 1

Crawford, C. S., Fabian, A. C., Gandhi, P., Wilman, R. J., & Johnstone, R. M. 2001, MNRAS, 324, 427

Crawford, C. S., Gandhi, P., Fabian, A. C., Wilman, R. J., Johnstone, R. M., Barger, A. J., & Cowie, L. L. 2002, MNRAS, 333, 809

Elvis, M., et al. 1994, ApJS, 95, 1

Elvis, M., Risaliti, G., & Zamorani, G. 2002, ApJ, 565, L75

Fabian, A. C. 1999, MNRAS, 308, L39

Fabian, A. C., et al. 2000, MNRAS, 315, L8

Fabian, A. C., Barcons, X., Almaini, O., & Iwasawa, K. 1998, MNRAS, 297, L11

Fabian, A. C., Crawford, C. S., & Iwasawa, K. 2002, MNRAS, 331, L57

Fabian, A. C., & Iwasawa, K. 1999, MNRAS, 303, L34

Fabian, A. C., Sanders, J. S., Crawford, C. C., & Ettori, S. 2003, MNRAS, 341, 729

Fabian, A. C., Wilman, R. J., & Crawford, C. S. 2002, MNRAS, 324, 427

Fadda, D., Flores, H., Hasinger, G., Franceschini, A., Altieri, B., Cesarsky, C., Elbaz, D., & Ferrando, P. 2002, A&A, 383, 838

Ferrarese, L. 2002, in Current High-Energy Emission around Black Holes, ed. C.-H. Lee & H.-Y. Chang (Singapore: World Scientific), 3

Ferrarese, L., & Merritt, D. 2000, ApJ, 539, L9

Franceschini, A., Bassani, L., Cappi, L., Granato, G. L., Malaguti, G., Palazzi, E., & Persic, M. 2000, A&A, 353, 910

Franceschini, A., Braito, V., & Fadda, D. 2002, MNRAS, 335, L51

Gammie, C. F. 1999, ApJ, 522, L57

Gandhi, P. 2002, Ph.D. Thesis, Cambridge University

Gandhi, P., Crawford, C. S., & Fabian, A. C. 2002, MNRAS, 337, 781

Gandhi, P., & Fabian, A. C. 2003, MNRAS, 339, 1095

Gebhardt, K., et al. 2000, ApJ, 539, L13

Gendreau, K. 1995, Ph.D. Thesis, Massachusetts Institute of Technology

Giacconi, R., et al. 2001, ApJ, 551, 624

Gilli, R., Risaliti, G., & Salvati, M. 1999, A&A, 347, 424

Gilli, R., Salvati, M., & Hasinger, G. 2001, A&A, 366, 407

Hasinger, G. 2003, in New Visions of the X-ray Universe in the XMM-Newton and Chandra Era, ed. F. Jansen (Nordwijk: ESA), ESA SP-488 (astro-ph/0202430)

Hasinger, G., et al. 2001, A&A, 365, L45

Hornschemeier, A. E., et al. 2000, ApJ, 541, 49

Hornschemeier, A. E., et al. 2001, ApJ, 554, 742

Hornschemeier, A. E., Brandt, W. N., Alexander, D. M., Bauer, F. E., Garmire, G. P., Schneider, D. P., Bautz, M. W., & Chartas, G. 2002, ApJ, 568, 62

Hughes, S. A., & Blandford, R. D. 2003, ApJ, 585, L101

Iwasawa, K., Fabian, A. C., & Ettori, S. 2001, MNRAS, 321, L15

Krolik, J. H. 1999, ApJ, 515, L73

Krolik, J. H., & Kriss, G. A. 2001, ApJ, 561, 684

Madau, P., Ghisellini, G., & Fabian, A. C. 1994, MNRAS, 270, L17

Magdziarz, P., Blaes, O. M., Zdziarski, A. A., Johnson, W. N., & Smith, D. A. 1998, MNRAS, 301, 179

Magorrian, J., et al. 1998, AJ, 115, 2285

Mainieri, V., Bergeron, J., Hasinger, G., Lehmann, I., Rosati, P., Schmidt, M., Szokoly, G., & Della Ceca, R. 2002, A&A, 393, 425

Maiolino, R., Marconi, A., Salvati, M., Risaliti, G., Severgnini, P., La Franca, F., & Vanzi, L. 2001, A&A, 365, 37

Maiolino, R., Salvati, M., Bassani, L., Dadina, M., Della Ceca, R., Matt, G., Risaliti, G., & Zamorani, G. 1998, A&A, 338, 781

Matt, G., & Fabian, A. C. 1994, MNRAS, 267, 187

Matt, G., Fabian, A. C., Guainazzi, M., Iwasawa, K., Bassani, L., & Malaguti, G. 2000, MNRAS, 318, 173

Merritt, D., & Ferrarese, L. 2001, MNRAS, 320, L30

Mushotzky, R. F., Cowie, L. L., Barger, A. J., & Arnaud, K. A. 2000, Nature, 404, 459

Norman, C., et al. 2002, ApJ, 571, 218

Piccinotti, G., Mushotzky, R. F., Boldt, E. A., Holt, S. S., Marshall, F. E., Serlemitsos, P. J., & Shafer, R. A. 1982, ApJ, 253, 485

Risaliti, G., Maiolino, R., & Salvati, M. 1999, ApJ, 522, 157

Rosati, P., et al. 2002, ApJ, 566, 667

Setti, G., & Woltjer, L. 1989, A&A, 224, L21

Silk, J., & Rees, M. J 1998, A&A, 331, L1

Smail, I., Owen, F. N., Morrison, G., Keel, W. C., Ivison, R. J., & Ledlow, M. J. 2002, ApJ, 581, 844

Sołtan, A. 1982, MNRAS, 200, 115

Stern, D., et al. 2002, ApJ, 568, 71

Ueno, S., Koyama, K., Nishida, M., Yamauchi, S., & Ward, M. J. 1994, ApJ, 431, L1

Vignati, P., et al. 1999, A&A, 349, L57

Wada, K., & Norman, C. A. 2002, ApJ, 566, L21

Wilman, R. J. 2002, preprint

Wilman, R. J., Fabian, A. C., Crawford, C. S., & Cutri, R. M. 2003, MNRAS, 338, L19

Wilman, R. J., Fabian, A. C., & Gandhi, P. 2000, MNRAS, 318, L11

Xu, C., et al. 1998, ApJ, 508, 576

Yu, Q., & Tremaine, S. 2002, MNRAS, 335, 965

Zheng, W., et al. 1995, ApJ, 444, 632

29

Conference summary

P. TIM DE ZEEUW
Leiden Observatory

29.1 Introduction

In the past decade, much effort was devoted to measure the masses of supermassive black holes in galactic nuclei, to establish the relation between black hole mass and the global/nuclear properties of the host galaxy, and to understand the role of these objects in the formation and subsequent dynamical evolution of galaxies. This whole area of research is an appropriate and timely topic for the first in this series of symposia. I congratulate the organizers, and in particular Luis Ho, for putting together such an interesting program of talks and posters.

Nearly half the contributions were devoted to the various methods that are being used to estimate masses of central black holes, from nearby globular clusters all the way to quasars at redshifts $z > 6$. This work forms the foundation of our understanding of black hole demography, and provides constraints on scenarios for black hole formation and growth, and on their connection to the formation and properties of the host galaxy. I will address these topics in this same order.

29.2 Black Hole Masses

The basic principle is to deduce the mass distribution in a galactic nucleus from the observed motions of stars or gas, and compare it with the luminous mass associated with the observed surface brightness distribution. If this requires a central mass-to-light ratio M/L larger than anything that can be produced by normal (dynamical) processes, the associated dark mass is considered to be a black hole.

Depending on the nature of the nucleus, and on its distance, one may employ the kinematics of individual stars, the spatially-resolved kinematics of the absorption lines of the integrated stellar light, or of the emission lines of gas (optical/maser), or the unresolved kinematics from reverberation mapping and modeling of line profiles. It is useful to review the contributions on all these approaches by starting with the nearest black holes, and working to larger distances. This takes us from what could be called "primary" black hole mass indicators to "secondary" and "tertiary" estimators.

29.2.1 The Galactic Center

The nearest supermassive black hole is located at the Galactic Center. Despite the large foreground extinction, it is possible to detect individual stars near it in the infrared, and to measure their proper motions and radial velocities. Ghez showed us beautiful results based on adaptive-optics-assisted near-IR imaging with a spatial resolution of $\sim0.''05$. The

© The Observatories of the Carnegie Institution of Washington 2004.

velocities of stars this close to the Center are large: the five-year coverage to date already defines accurate three-dimensional stellar orbits for at least six stars, some of which are very eccentric. The stars appear to have early-type spectra, so may be massive and young. The current record holder recently passed pericenter at only 60 AU from the black hole, with a velocity of $\sim$9000 km s^{-1}!

With data of this quality, measuring the mass at the Galactic Center reduces to the classical astronomy of binary orbits. The current best estimate is $3.6 \times 10^6 M_\odot$. The spatial resolution achieved allows probing well inside the radius of influence of the black hole. The Galactic Center hence provides a unique laboratory for measuring the properties of a star cluster in the regime where the black hole dominates the gravitational potential, and allows studying the formation and evolution of massive stars under atypical conditions (see contributions by Alexander and Scoville).

The generally friendly but nevertheless intense rivalry between the groups at Keck (Ghez) and at ESO (Genzel/Eckart) promises further interesting results in the near future. In particular, the near-IR integral-field spectrographs SINFONI and OSIRIS, and hopefully NIFS on Gemini, will provide proper motions and radial velocities from the same observations, and will push these studies to fainter magnitudes.

29.2.2 Globular Clusters

Measurement of the resolved stellar kinematics is also possible in nearby globular clusters, which might contain "intermediate-mass" black holes, i.e., with masses in the range $10^3 - 10^4 M_\odot$. Ground-based proper motion and radial velocity studies generally lack the spatial resolution to observe the kinematic signature of such black holes, but the *Hubble Space Telescope (HST)* can detect them.

Van der Marel discussed the recent measurement of radial velocities in the central arcsecond of M15 obtained with STIS. The interpretation of the observed radial velocity distribution has led to significant controversy, which centers around the nature of the measured inward increase of M/L. Because dynamical relaxation has been significant in M15, dark remnants must have aggregated in its center. The initial estimate of a black hole of $\sim 3 \times 10^3 M_\odot$ had to be revised because of errors in previously published figures describing a Fokker-Planck model for the cluster. The revised figures suggest that the observed M/L increase could be due entirely to dark remnants, but leaves room for a black hole of $\sim 10^3 M_\odot$. Whatever the final word, this work suggests that it might be profitable to scrutinize other globular clusters, and it pushes the groups that have been producing detailed dynamical models for globular clusters to provide them for specific objects, while fitting all kinematic data.

HST proper motions might also reveal central black holes in globular clusters. The case of ω Centauri is of particular interest. It is a massive cluster that may be the remnant of a galaxy that fell in long ago, shows little evidence of mass segregation, and has ground-based kinematics for $\sim 10^4$ stars. It is the target of an extensive WFPC2 proper motion study (King & Anderson 2002) which should detect a black hole if it has a mass as large as expected from the $M_\bullet - \sigma$ relation for galaxies (see below).

29.2.3   *Stellar Absorption-line Kinematics*

We cannot resolve motions of individual stars in the nuclei of other galaxies, but it is possible to measure the stellar kinematics from the integrated light at high spatial resolution. Ground-based integral-field spectroscopy is crucial for constraining the stellar M/L and the intrinsic shape of the galaxy. The orbital structure in these systems is rich, so a true inward increase of M/L must be distinguished from a possible radial variation of the velocity anisotropy. This requires measuring the shape of the line-of-sight velocity distribution as a function of position on the sky, and then fitting these with dynamical models. Recent work in this area was reviewed by Gebhardt and Richstone, with assists by Kormendy and Lauer.

A number of independent codes have been developed to construct dynamical models with axisymmetric geometry. These are by now all based on Schwarzschild's (1979) numerical orbit superposition method (Rix et al. 1997; van der Marel et al. 1998). The consensus view is that they give consistent black hole masses, as long as one is careful to obtain smooth solutions (by maximum entropy or via regularization, or by including very large numbers of orbits). A case study is provided by M32, which has been modeled by many groups over the past 20 years. The latest analysis is by Verolme et al. (2002), who model major-axis STIS data as well as SAURON integral-field kinematics. This produces an accurate M/L, for the first time fixes the inclination of M32, and sets the mass of its black hole at $2.5 \times 10^6 M_\odot$.

It is useful to keep in mind that two other nearby nuclei differ considerably from that in M32. The nucleus of M31 has long been known to be asymmetric (Light, Danielson & Schwarzschild 1974). TIGER and OASIS integral-field spectroscopy revealed its stellar kinematics (Bacon et al. 2001), and allowed understanding of previous long-slit work, including FOS, FOC and STIS data (Kormendy, Emsellem). Long-slit absorption-line spectroscopy of the nucleus of the Sc galaxy M33 shows that it cannot contain a black hole more massive than $1.5 \times 10^3 M_\odot$ (Gebhardt et al. 2001).

To date, nearly 20 early-type galaxies with distances up to 20 Mpc have been studied in this way. Dynamical modeling of ground-based and STIS spectroscopy provides evidence for black holes in all of them. More than half the determinations come from the long-running *HST* program by the Nuker team, and are based on edge-on axisymmetric models (Gebhardt et al. 2003). Additional cases can be found in the poster contributions (e.g., Cappellari, Cretton). The black hole masses usually lie in the range $5 \times 10^{7-8} M_\odot$, and may be accurate to 30% on average. The inferred internal orbital structure suggests tangential anisotropy near the black hole. This constrains black hole formation scenarios (Sigurdsson, Burkert). However, many of these objects display kinematic signatures of triaxiality, and surely not all of them are edge-on. This casts doubt on the accuracy of the individual masses. Application of the same modeling approach, but now for triaxial geometry, will show how severe these biases are. The machinery for doing this is now available (Verolme et al. 2003).

Detection of stellar kinematic evidence for black holes smaller than $5 \times 10^7 M_\odot$ in galaxies at the typical distance of Virgo requires higher spatial resolution than *HST* offers. Here other methods will be needed. The central surface brightness in the giant ellipticals is too low for the 2.4-m *HST* to provide spectroscopy of sufficient signal-to-noise ratio at high spatial resolution. They are prime targets for near-infrared integral field spectroscopy of the CO bandhead at 2.3 μm, with 8-m class telescopes.

29.2.4   *Optical Emission Lines*

The nuclei of active early-type galaxies, and those of most spirals, contain optical emission-line gas. Its kinematics can be used to constrain the central mass distribution, as long as the spatial resolution is sufficient to resolve the region where the gravity field of the black hole dominates the motions. It is important to account for the proper observational set-up. Maciejewski showed that not doing so can change an inferred black hole mass by a factor as large as 4! It is also important to confirm that the gas is in regular motion, by using an integral-field spectrograph, or by taking multiple parallel slits.

To date, the emission-line gas in about 100 nuclei has been observed with *HST*, for galaxies to distances of up to 100 Mpc (see contributions by Axon, Barth, Marconi, Sarzi, Verdoes Kleijn). Only $\sim$20% of these seem to have circular rotation, and so far only a few black hole masses derived in this way have been published, all of them larger than $\sim10^7 M_\odot$. NGC 3245 is a textbook example (Barth et al. 2001).

The accuracy of these black hole masses is not easy to establish. To date, there are very few cases for which a black hole mass has been determined by more than one independent method. While good results are claimed for NGC 4258 (stellar kinematics/maser emission—see below), the mass derived for the black hole in IC 1459 from gas motions is significantly lower than that derived from stellar motions. It will be very useful to carry out such comparisons for other nuclei, as this is the only way to put the masses determined by different methods onto the same scale, a prerequisite for an unbiased analysis of the black hole demography. It seems clear that the kinematic model of circular rotation for the gas is nearly always too simple. In many cases the gas velocity dispersion increases strongly inward. This is evidence for "turbulence," presumably caused by nongravitational motions, which probably include in/outflows, winds, and genuine turbulence in the gas near the central monster (similar to what is seen in the simulations reported by Wada). An improved (hydrodynamic) standard model for analysis of the gas motions would be very useful.

29.2.5   *Masers in Seyferts and LINERs*

VLBI measurements of H_2O maser emission allow probing the gas kinematics to very small scales, and have provided the cleanest extragalactic black hole mass determination, for NGC 4258. This method achieves the highest spatial resolution to date, but is possible only when the circumnuclear disk is nearly edge-on. A search of $\sim$700 nuclei has revealed maser emission in 21 additional cases, and regular kinematics on VLBI scales in four of these, to distances of up to 70 Mpc. The inferred black hole masses range from $(1-40) \times 10^6 M_\odot$. This mass range is currently very difficult to probe otherwise.

29.2.6   *Reverberation Mapping*

The observed time-variation of broad emission lines (e.g., $H\beta$) relative to the continuum in nearby Seyfert 1 nuclei and quasars can be used to estimate the radius of the spatially unresolved broad-line region by means of the light travel-time argument. In combination with simple kinematic models for the motion of the broad-line clouds, the observed width of the lines then provides a typical dispersion, so that an estimate of the mass of the central object responsible for these motions follows (Green, Peterson).

This is a powerful method in principle, as it allows probing nuclei out to significant distances. Early mass determinations appeared to be statistically systematically lower than those obtained by other means, but this discrepancy has now disappeared. However, there is not yet a truly independent calibration of the mass scale. The nearest well-studied case is too faint for obtaining, e.g., the resolved stellar kinematics of the nucleus, and for this reason the simple kinematic model is calibrated by requiring results to fit the local $M_\bullet - \sigma$ relation, so that the masses agree by definition. This makes reverberation mapping effectively a "secondary" mass indicator.

29.2.7   *Line Widths*

A number of contributions (e.g., Vestergaard, Kukula, Jarvis & McClure) advocated using the widths of the Hβ, Mg II, or C IV emission lines to estimate a velocity dispersion σ, and to infer a typical radius R from a locally observed correlation between luminosity and broad-line region size, so that a black hole mass follows. These authors set the overall scale by calibrating versus reverberation mapping masses, which in turn is calibrated by the local $M_\bullet - \sigma$ relation, making this a tertiary mass indicator, which may still contain significant systematic uncertainties. This approach is potentially very useful, as it can be applied even to the highest redshift quasars. The method suggests that some of the $z > 6$ quasars may contain $5 \times 10^9 M_\odot$ black holes. If true, this would require very rapid growth of any seed black hole to acquire so much mass at such early times. It will be very useful to calibrate this method more directly.

29.2.8   *X-ray Emission*

Recently, it has become possible to measure the profile of the Fe Kα line at 6.4 keV in the nuclei of Seyfert galaxies such as MCG-6-30-15 and NGC 3516. The early *ASCA* results have now been superseded by *XMM-Newton* spectra (Fabian et al. 2002). The data reduction effort is significant, but the asymmetric line profile shows a width of about 10^5 km s^{-1}, a sure sign that the emission must originate near 10 Schwarzschild radii of a (rotating) relativistic object. The radius where the emission originates is not measured independently, so these observations do not provide the mass of the black hole.

29.3   Black Hole Demography

29.3.1   *$M_\bullet - \sigma$ Relation*

Early attempts to relate measured black hole masses to global properties of the host galaxy focused on $M_\bullet$ versus bulge luminosity L_B (e.g., Kormendy & Richstone 1995). Following a suggestion by Avi Loeb, two groups showed that there is a tighter correlation between $M_\bullet$ and host galaxy velocity dispersion σ (Ferrarese & Merritt 2000; Gebhardt et al. 2000). This generated many follow-up papers either explaining the relation, or "predicting" it afterwards, and also sparked a remarkable (and sometimes even amusing) debate about the slope of the correlation, and the proper way to measure it (see Tremaine et al. 2002 for a recent summary, and contributions by Gebhardt, Kormendy, and Richstone).

We have seen that the black hole masses that form the foundation of the $M_\bullet - \sigma$ relation may still contain significant errors. The black hole masses for high-σ galaxies (giant ellipticals) generally rely on gas kinematics, and these should be treated with caution until we have independent measurements derived from stellar kinematics. Many of the available stellar dynamical measurements in E/S0 galaxies cover black hole masses in the range 5×10^7 to $5 \times 10^8 M_\odot$, but rely mostly on edge-on axisymmetric dynamical models. Many of the host galaxies show signs of triaxiality, which must cause considerable changes in the anisotropy of the orbital structure relative to that in an axisymmetric model. Depending on the direction of viewing, one might observe more or less of this anisotropy in the line-of-sight kinematics, and accordingly ascribe less or more of a higher central velocity dispersion to the presence of a black hole. This may be a significant effect for any given galaxy (Gerhard 1988). While the resulting shifts in black hole mass may tend to average out when observing a sample of galaxies with random orientations, there may still be residual systematic errors in $M_{\bullet}$ when derived with an axisymmetric model. If the distribution of intrinsic shapes of early-type galaxies varies with luminosity (or σ) then relying on axisymmetric models may introduce a bias in the slope of the $M_{\bullet} - \sigma$ relation.

The number of reliable black hole masses below $4 \times 10^7 M_{\odot}$ is still modest, especially for spiral galaxies. The expectation is that the spirals with classical spheroidal bulges should all contain a black hole, in agreement with the fact that a very large fraction of them also contains a low-luminosity active nucleus (Heckman, Ho, Sadler). The smallest reported black hole masses for spheroidal systems are those for the Galactic globular cluster M15 and for the M31 companion cluster G1. While both are consistent with an extrapolation of the $M_{\bullet} - \sigma$ relation to small σ, the mass determinations are not yet secure, in particular for M15.

In view of this, it seems to me premature to consider the $M_{\bullet} - \sigma$ relation ironclad over the entire range of $10^3 - 3 \times 10^9 M_{\odot}$, *and* valid for all Hubble types to all redshifts, as some speakers assumed, and as appears to have become almost universally accepted outside the immediate field. Rather than debating the details of finding the slope of the correlation, it seems more productive to measure more reliable black hole masses, and indeed to work out why this correlation should be there, especially if it were to extend all the way from giant ellipticals to globular clusters.

29.3.2 Galaxies with Pseudo-bulges

Many spiral galaxies do not have a classical $R^{1/4}$ spheroid, but instead contain a separate central component that has many properties resembling those of disks (Kormendy 1993; Carollo 1999), and is commonly referred to as a pseudo-bulge. Carollo reviewed recent work on these objects, based on large imaging surveys with *HST*. She showed that all pseudo-bulges in her sample contain nuclear star clusters that are barely resolved at *HST* resolution (e.g., Carollo et al. 2002). The example of M33 mentioned in the above suggests that these galaxies may not have a central black hole, consistent with the almost complete absence of emission-line signatures of activity (Heckman, Ho). The sole exception appears to be NGC 4945 which has a black hole mass based on maser emission. It will be very interesting to establish what the demography of black holes is in galaxies with pseudo-bulges. The presence of the tight central cluster will make it very difficult to measure central black hole masses by stellar dynamical methods.

29.3.3 Quasars

We heard much about recent work on quasars, driven by new and large samples, notably that provided by the Sloan Digital Sky Survey. This has produced accurate and detailed luminosity functions (Osmer, Fan), and also tightened the limits on the typical lifetime of the quasar phenomenon derived from the spectacular decline in numbers at redshifts below two, to $\sim 4 \times 10^7$ yr (Martini).

The $z > 6$ quasars seen in the Sloan Survey are interesting objects for a number of reasons. The most luminous of these demonstrate that the central black hole responsible for the observed activity must have grown about as fast as possible. The spectrum shortward of Lyα

also provides evidence for a Gunn-Peterson trough, suggesting the tail end of reionization occurred a little above a redshift of six.

Many apparently conflicting opinions were stated on the nature of the host galaxies of quasars. Lacy, Kukula and Haehnelt argued that the high-z quasars are located in (boxy) ellipticals. Peng reported a mix of host galaxy types at $z \approx 1$, Heckman argued for "very luminous galaxies," and Scoville presented evidence that many of the nearest low-luminosity quasars reside in disks. This underlines the importance of careful sample definition and nomenclature before drawing strong conclusions on host-galaxy evolution with redshift.

A number of speakers revisited Sołtan's (1982) line of reasoning, and used recent data to work out the mean density in black holes versus the mean density in extragalactic background light. Fall derived that the quasar contribution is 1% of that of stars. The consensus value for the mean density of black holes is $\sim 2.5 \times 10^5 M_\odot$ Mpc^{-3}. This assumes an efficiency of about 0.1, and agrees with the local census in average value. However, the comparison of the quasar luminosity function and the current estimate of the local black hole mass function is less straightforward, and requires more work (Yu, Richstone).

Fabian addressed the possibility that the recent measurements of the X-ray background with *Chandra* and *XMM-Newton* indicate a black hole density a factor of 4 larger than the above estimate. This would either require a much-increased accretion efficiency (which seems hardly possible), or over two-thirds of the active nuclei being obscured. He however also presented evidence that the X-ray objects evolve differently with redshift than do the quasars, and in the end concluded (or, as some pointed out, recanted) that any discrepancy might be less than a factor of 2. In my primitive understanding of this issue this means that we are close to having the correct census of black holes in the Universe.

29.4   Black Hole Formation and Growth

Rees (1984) published a celebrated diagram illustrating that the formation of (seed) black holes early on during galaxy formation is essentially inevitable. Many of the processes he discussed were considered also during this conference.

29.4.1   *Formation*

Recent work on the formation of the first objects in the Universe suggests that the initial collapses occurred as early as $z \approx 20$. Whether these lead to mini-galaxies or mini-quasars is less clear (Haiman), but either way, the reionization of the Universe must have started about this time. The theorists seemed in agreement that most of the mass in the black holes that power the active nuclei must have grown from seeds formed around $z \approx 15$ (Phinney). Clarke argued for competitive accretion (by analogy with the normal star formation process), while others considered the collapse of a supermassive star to a Kerr black hole (Shapiro) and the dynamical evolution of dense stellar clusters (Freitag, Rasio). It is clear from the existence of the $z \approx 6$ quasars that whatever happened, the process was able to build black holes with masses larger than $10^9 M_\odot$ very quickly.

29.4.2   *Adiabatic Growth*

A number of speakers considered processes that cause (secular) growth and evolution of a central black hole, with as ultimate aim to deduce (or at least constrain) the formation history of the black hole from the observed morphology and kinematics of the host nucleus. Sigurdsson considered the problem of adiabatic growth through accretion of stars.

This process is well understood in spherical geometry, and produces a power-law density profile with a range of slopes, as well as tangential orbital anisotropy near the black hole. However, the same observed properties can be produced by other formation mechanisms, such as violent relaxation around a pre-existing black hole (Stiavelli 1998). The problem of adiabatic growth in axisymmetric or triaxial geometry deserves more attention. There is a fairly wide-spread expectation of inside-out evolution of the shape toward a more nearly spherical geometry, but recent numerical work by, e.g., Holley-Bockelmann et al. (2002) suggests that this may not occur on interesting time scales. Construction of dynamical models of galaxies with measured black hole masses can provide significant constraints on the intrinsic shapes and orbital structure, and hence may shed further light on this problem.

29.4.3   *Mergers, Binary Black Holes, and Gravitational Waves*

In the currently popular scenario of hierarchical galaxy formation, the giant ellipticals with extended cores had their last major merger many Gyr ago, and since then accreted mostly small lumps. An infalling galaxy will lose its steep central cusp if the host contains a massive black hole. This tidally disrupts the cusp, and transforms it into a nuclear stellar disk of the kind seen by *HST* in nearby galaxies. The core itself was formed during the last major merger that presumably involved two cusped systems, each containing a black hole. Dynamical friction causes the black holes to form a binary, after which star-binary interactions remove many of the stars from the central region, generating an extended low-surface brightness core, or even a declining central luminosity profile (Lauer). The dynamical interactions harden the black hole binary, which eventually may reach the stage where gravitational radiation becomes effective so that rapid black hole coalescence follows (e.g., Begelman, Blandford & Rees 1980). Overall triaxiality of the host galaxy increases the black hole merging rate because a large fraction of stars are on orbits that bring them close to the center so that the binary evolution speeds up (e.g., Yu 2002).

Merritt and Milosavljević reported on recent progress in this area. No reliable *N*-body simulations of the complete process of formation of a core-galaxy via a binary black hole merger of two cusped systems are available yet, and not much is known about the expected shape of the resulting core, but this situation should change in the next few years. To date, it remains unclear whether black hole coalescence goes to completion in less than a Hubble time, or whether the binary stalls. If the process is slow, then binary black holes might be detectable with high spatial resolution spectroscopy in nearby systems such as Centaurus A, and they would provide a natural explanation for the twisted jets seen in some radio galaxies (Merritt & Ekers 2002). In this case repeated mergers will cause three-body interactions of the black holes, which may eject one of them from the galaxy altogether, and hence generate a population of free-floating massive black holes hurtling through space. If, on the other hand, black hole coalescence is fast, then giant ellipticals will contain a slowly rotating single black hole (Hughes & Blandford 2003), and many binary black hole mergers must have occurred in the past. A number of speakers discussed the exciting possibility of detecting the gravitational wave signature of these black hole mergers with LISA, the planned ESA/NASA space observatory to be launched in the next decade (Armitage, Backer, P. Bender, Phinney).

29.4.4 *Fueling*

The bulk of the fuel for the central engine that powers an active nucleus must be in gaseous form. While transporting gas inward to about 100 pc from the nucleus is fairly straightforward, it is not at all clear how to get it inside 1 pc. Shloshman, Frank & Begelman's (1990) bar-in-bar scenario has been updated, but observational evidence suggests that large-scale bars may not be very important for fueling. However, Emsellem presented intriguing kinematic evidence for inner density waves, perhaps related to central bars, which may hold the key to efficient gas transport into the nucleus.

Three-dimensional simulations of the structure of nuclear gas continue to improve. Wada showed a beautiful example in which the gas is turbulent and filamentary, but is in ordered motion with a significant velocity dispersion, in qualitative agreement with observations. The morphology of his simulated disk resembles that of the celebrated $H\alpha$ disk in M87 (Ford et al. 1994). It will be interesting to see how well the kinematic properties compare. The turbulent filamentary structure in the simulated disk is caused by local energy input which corresponds to an assumed supernova rate of 1 per year. While this may be appropriate for the peak of activity in high-luminosity quasars, it is less evident that this would apply to, say, M87.

29.4.5 *Accretion Physics and Feedback*

Blandford reviewed recent progress in our understanding of accretion physics. The basic paradigm for an active nucleus—an accretion disk surrounding a supermassive black hole—has been agreed for over 30 years, but detailed models have been fiendishly difficult to construct as they require inclusion of a remarkable range of physical phenomena in the context of general relativistic magneto-hydrodynamics. The goal is to model the morphology and spectra of individual objects in detail, and to reproduce the full zoo of the AGN taxonomy, where inclination of the accretion disk to the line of sight, but also age and accretion mode are parameters that can be varied (e.g., Umemura). Theory has identified three viable modes of accretion. In the "low" mode, accretion is adiabatic. This is likely to apply to "inactive" nuclei (low-luminosity AGNs), including Sgr A* in our own Galaxy. The "intermediate" mode may apply to Seyferts (Heckman) and corresponds to the classical thin disk (Shakura & Sunyaev 1973), while the "high" mode is again adiabatic, and may be most relevant for obscured quasars. Amongst the adiabatic models much work has been done on advection-dominated accretion flows (Narayan & Yi 1994), which are radiatively inefficient. There is disagreement as to whether this ADAF model fits the spectrum of, e.g., Sgr A*. The theory of the more recent convection-dominated accretion flow (CDAF) variant is less well developed, and in particular the magneto-hydrodynamic flow in these models has not been worked out in full. The adiabatic inflow-outflow solution (ADIOS) of Blandford & Begelman (1999) was developed to overcome some of the difficulties encountered by the ADAF models, but more work on this is needed as well.

Begelman reminded us that much of the output of an active nucleus is in kinetic form (outflows, jets), which may inhibit accretion and affect the surroundings. Some of this feedback may be observable in radio galaxies, as episodic activity that he argued is buoyancy-driven, with the energy distributed by effervescent heating. It will be interesting to work out in more detail how feedback operates in the formation of the luminous high-z quasars. We have seen that these require rapid and efficient growth of their central supermassive black hole. It is likely that efficient cooling is required to bring in enough gas to grow the black hole efficiently. Statistical evidence from the Sloan Survey suggests that powerful active nuclei may be accompanied by a significant population of young and intermediate-age stars, and that the most powerful ones may well reside in massive young galaxies (Heckman, Kauffmann). The combined effects of feedback by the growing nucleus, and of the massive accompanying starburst, may well be reponsible for lifting the obscuration provided by the dust and gas revealing the quasar for all the Universe to see.

29.5   Black Holes and Galaxy Formation

It has been clear for some time that nuclear and global properties of galaxies correlate. This includes the correlation of cusp slope with total luminosity, the $M_\bullet - \sigma$ relation, and also the relation between $M_\bullet$ and the Sersic index n that characterizes the overall (rather than the central) luminosity profile (Graham, Erwin). Understanding these correlations, and establishing which of these is fundamental, is a key challenge for galaxy formation theories.

The current galaxy formation paradigm is based upon hierarchical merging, where the dark matter clumps through gravitational instability in the early Universe, after which the primordial gas settles in the resulting potential well, in the form of a disk, and starts forming stars. These produce heavier elements that they return to the gas at the end of their lives, so that subsequent generations of stars are increasingly metal rich. The proto-galactic disks experience frequent interactions, ranging from infall of smaller clumps that hardly disturb the disk to equal-mass mergers that result in a spheroidal galaxy, which may reacquire a stellar disk by further gas infall.

Somerville and Haehnelt suggested that during the first collapse, the low angular momentum material may already form a bulge, containing a black hole, and that its properties are related to those of the dark halo. In this view, the main disk would form/accrete later, and its properties may have little to do with the black hole, so that correlations between nuclear and global properties arise naturally for galaxies with classical bulges. The theorists have various scenarios to reproduce the $M_\bullet - \sigma$ relation. The models also predict the evolution of the black hole mass function, and could in principle provide the expected LISA signal for merging massive black holes once the vexing details of black hole binary coalescence have been sorted out.

The observed range of central cusp slopes in ellipticals, lenticulars, and classical spiral bulges is larger than predicted by simple adiabatic growth, but qualitatively consistent with successive merging of smaller cusped systems with their own black holes. It will be very interesting to compare detailed numerical simulations of this merging history with the internal orbital structure derived from dynamical modeling for individual galaxies. Burkert reported that the observed tangential orbital anisotropy is not (yet?) seen in his merger simulations.

This leaves the disk galaxies with pseudo-bulges, which show little evidence for nuclear activity. Could they have formed late, as a stable disk in an isolated dark halo, and without a central black hole? It has been argued that the pseudo-bulges might be the result of bar dissolution, but despite earlier reports to the contrary, stellar bars appear to be robust against dynamical evolution caused by a central "point mass," as long as its mass is as modest as that of the observed central star clusters (Carollo, Sellwood, Shen). Understanding the formation and subsequent internal evolution of these disks is an area that deserves much more attention.

29.6   Challenges

Rather than summarize this summary, or repeat suggestions for future work made in the above, I conclude by mentioning two areas where progress is most needed.

It will be very useful to measure more black hole masses, to (i) establish the local demography over the full mass range from globular clusters to giant ellipticals, (ii) settle whether or not the galaxies with pseudo-bulges contain central black holes, and (iii) calibrate the secondary and tertiary mass estimators that can then hopefully be used to estimate reliable black hole masses to high redshift. On the theoretical side, it will be interesting to see if the measured black hole masses can help pin down the models for the various active nuclei, e.g., to establish accretion mode and age.

Pressing projects related to galaxy formation include (i) establishing what are the fundamental parameters underlying the various correlations between global and nuclear properties of galaxies, (ii) understanding the evolution of the black hole mass function, from the formation of the first seed black holes to the present day, including the role of binaries, and (iii) working out where the galaxies with pseudo-bulges fit in.

Acknowledgements. It is a pleasure to acknowledge stimulating conversations with many of the participants at this Symposium, to thank Michele Cappellari for expert assistance during the preparation of the summary talk, and to thank him, Peter Barthel, Marcella Carollo, Eric Emsellem, Karl Gebhardt, Luis Ho, and Gijs Verdoes Kleijn for constructive comments on an earlier version of this manuscript.

References

Bacon, R., Emsellem, E., Combes, F., Copin, Y., Monnet, G., & Martin, P. 2001, A&A, 371, 409
Barth, A. J., Sarzi, M., Rix, H.-W., Ho, L. C., Filippenko, A. V., & Sargent, W. L. W. 2001, ApJ, 555, 685
Begelman, M. C., Blandford, R. D., & Rees, M. J. 1980, Nature, 287, 307
Blandford, R. D., & Begelman, M. C. 1999, MNRAS, 303, L1
Carollo, C. M. 1999, ApJ, 523, 566
Carollo, C. M., Stiavelli, M., Seigar, M., de Zeeuw, P. T., & Dejonghe, H. B. 2002, AJ, 123, 159
Fabian, A. C., et al. 2002, MNRAS, 335, L1
Ferrarese, L., & Merritt, D. R. 2000, ApJ, 529, L9
Ford, H. C., et al. 1994, ApJ, 435, L27
Gebhardt, K., et al. 2000, ApJ, 529, L13
——. 2001, AJ, 122, 2469
——. 2003, ApJ, 583, 92
Gerhard, O. E. 1988, MNRAS, 232, 13P
Holley-Bockelman, K., Mihos, C. J., Sigurdsson, S., Hernquist, L., & Norman, C. A. 2002, ApJ, 567, 817
Hughes, S. A., & Blandford, R. D. 2003, ApJ, 585, L101
King, I. R., & Anderson, J. 2002, in Omega Centauri, A Unique Window into Astrophysics, ed. F. van Leeuwen, J. D. Hughes, & G. Piotti (San Francisco: ASP), 21
Kormendy, J. 1993, in Galactic Bulges, ed. H. Dejonghe & H. J. Habing (Dordrecht: Kluwer), 209
Kormendy, J., & Richstone, D. O. 1995, ARA&A, 33, 581
Light, E. S., Danielson, R. E., & Schwarzschild, M., 1974, ApJ, 194, 257
Merritt, D. R., & Ekers, R. D. 2002, Science, 297, 1310
Narayan, R., & Yi, I., 194, ApJ, 428, L13
Rees, M. J. 1984, ARA&A, 22, 471
Rix, H.-W., de Zeeuw, P. T., Cretton, N. C., van der Marel, R. P., & Carollo, C. M. 1997, ApJ, 488, 702
Schwarzschild, M., 1979, ApJ, 232, 236
Shakura, N. I., & Sunyaev, R. A., 1973, A&A, 24, 337
Shlosman, I., Frank, J., & Begelman, M. C. 1989, Nature, 338, 45
Sołtan, A. 1982, MNRAS, 200, 115
Stiavelli, M. 1998, ApJ, 495, L1

Tremaine, S. D., et al. 2002, ApJ, 574, 740

van der Marel, R. P., Cretton, N. C., de Zeeuw, P. T., & Rix, H.-W., 1998, ApJ, 493, 613

Verolme, E. K., et al. 2002, MNRAS, 335, 517

Verolme, E. K., Cappellari, M., Emsellem, E., van de Ven, G., & de Zeeuw, P. T. 2003, MNRAS, submitted

Yu, Q. 2002, MNRAS, 331, 935

Credits

The following figures in this volume were reproduced with permission from the original author and publisher.

Figure 1.5: Bender, R., et al. (2004), ApJ, submitted. Reproduced with permission from the American Astronomical Society.

Figure 4.4: Ghez, A. M., et al. 2003, ApJ, 586, L127, "The First Measurement of Spectral Lines in a Short-Period Star Bound to the Galaxy's Central Black Hole: A Paradox of Youth." Reproduced with permission from the American Astronomical Society.

Figure 6.1: Quinlan, G. D., Hernquist, L., & Sigurdsson, S. 1995, ApJ, 440, 554, "Models of Galaxies with Central Black Holes: Adiabatic Growth in Spherical Galaxies." Reproduced with permission from the American Astronomical Society.

Figure 6.3: Holley-Bockelmann, K., et al. 2002, ApJ, 567, 817, "The Evolution of Cuspy Triaxial Galaxies Harboring Central Black Holes." Reproduced with permission from the American Astronomical Society.

Figure 7.3: Shibata, M., & Shapiro, S. L. 2002, ApJ, 572, L39, "Collapse of a Rotating Supermassive Star to a Supermassive Black Hole: Fully General Relativistic Simulation." Reproduced with permission from the American Astronomical Society.

Figure 7.5–7.6: Shapiro, S. L., & Teukolsky, S. A. 1986, ApJ, 307, 575, "Relativistic Stellar Dynamics on the Computer. IV. Collapse of a Star Cluster to a Black Hole." Reproduced with permission from the American Astronomical Society.

Figure 7.7–7.8: Balberg, S., Shapiro, S. L., & Inagaki, S. 2002, ApJ, 568, 475, "Self-Interacting Dark Matter Halos and the Gravothermal Catastrophe." Reproduced with permission from the American Astronomical Society.

Figure 8.3: Clarke, C. J., & Bromm, V. 2003, MNRAS, 343, 1224, "The Characteristic Stellar Mass as a Function of Redshift." Reproduced with permission from Blackwell Publishing.

Figure 8.4: Bonnell, I., et al. 2001, MNRAS, 324, 573, "Accretion in Stellar Clusters and the IMF." Reproduced with permission from Blackwell Publishing.

Figure 11.1–11.2: Yu, Q., & Tremaine, S. 2002, MNRAS, 335, 965, "Observational

Constraints on Growth of Massive Black Holes." Reproduced with permission from Blackwell Publishing.

Figure 11.3–11.4: Martini, P., & Weinberg, D. H. 2001, ApJ, 547, 12, "Quasar Clustering and the Lifetime of Quasars." Reproduced with permission from the American Astronomical Society.

Figure 12.2: Wada, K. 2001, ApJ, 559, L41, "The Three-dimensional Structure of a Massive Gas Disk in the Galactic Central Region." Reproduced with permission from the American Astronomical Society.

Figure 12.5: Wada, K., & Norman, C. A. 2002, ApJ, 566, L21, "Obscuring Material around Seyfert Nuclei with Starbursts." Reproduced with permission from the American Astronomical Society.

Figure 14.2: Faber, S. M., et al. 1997, AJ, 114, 1771, "The Centers of Early-type Galaxies with HST. IV. Central Parameter Relations." Reproduced with permission from the American Astronomical Society.

Figure 14.4: Ravindranath, S., et al. 2001, AJ, 122, 653, "Central Structural Parameters of Early-Type Galaxies as Viewed with HST/NICMOS." Reproduced with permission from the American Astronomical Society.

Figure 15.4: Davies, R. L., & Illingworth, G. D. 1983, ApJ, 266, 516, "Dynamics of Yet More Ellipticals and Bulges." Reproduced with permission from the American Astronomical Society.

Figure 20.1: Boyle, B. J., et al. 2000, MNRAS, 317, 1014, "The 2dF QSO Redshift Survey - I. The Optical QSO Luminosity Function." Reproduced with permission from Blackwell Publishing.

Figure 20.2: Boyle, B. J., et al. 2000, MNRAS, 317, 1014, "The 2dF QSO Redshift Survey - I. The Optical QSO Luminosity Function." Reproduced with permission from Blackwell Publishing. Fan, X., et al. 1999, AJ, 118, 1, "High-redshift Quasars Found in SDSS Commissioning Data." Reproduced with permission from the American Astronomical Society.

Figure 20.3: Fan, X., et al. 2001, AJ, 121, 54, "High-redshift Quasars Found in SDSS Commissioning Data IV: Luminosity Function from the Fall Equatorial Stripe Sample." Fan, X., et al. 2001, AJ, 122, 2833, "A Survey of $z > 5.8$ Quasars in the SDSS. I. Discovery of Three New Quasars and the Spatial Density of Luminous Quasars at z 6." Reproduced with permission from the American Astronomical Society.

Figure 20.4: Rosati, P., et al. 2002, ApJ, 566, 667, "The Chandra Deep Field South: The 1 Million Second Exposure." Reproduced with permission from the American Astronomical Society.

Figure 20.5: Hasinger, G., 2003, in IAU Symp. 214, High Energy Processes and Phenomena in Astrophysics, ed. X. Li, Z. Wang, & V. Trimble (San Francisco: ASP), in press. Reproduced with permission from The Astronomical Society of the Pacific. Cowie, L. L., et al. 2003, ApJ, 584, L57, "The Redshift Evolution of the 2-8 keV X-Ray Luminosity Function." Reproduced with permission from the American Astronomical Society.

Figure 21.1–21.3: Dunlop et al. 2003, MNRAS, 340, 1095, "Quasars, their Host Galaxies, and their Central Black Holes." Reproduced with permission from Blackwell Publishing.

Figure 22.1: Kauffmann, G., et al. 2003, MNRAS, 341, 33, "Stellar Masses and Star Formation Histories for 10^5 Galaxies from the SDSS." Reproduced with permission from Blackwell Publishing.

Figure 22.2–22.3: Kauffmann, G., et al. 2003, MNRAS, 346, 1055, "The Host Galaxies of AGN." Reproduced with permission from Blackwell Publishing.

Figure 23.1: Fabian, A. C., et al. 2000, MNRAS, 318, L65, "Chandra Imaging of the Complex X-ray Core of the Perseus Cluster." Reproduced with permission from Blackwell Publishing.

Figure 23.2: Ruszkowski, M., & Begelman, M. C. 2002, ApJ, 581, 223, "Heating, Conduction, and Minimum Temperatures in Cooling Flows." Reproduced with permission from the American Astronomical Society.

Figure 24.1: Vitvitska, M., et al. 2002, ApJ, 581, 799, "The Origin of Angular Momentum in Dark Matter Halos." Reproduced with permission from the American Astronomical Society.

Figure 25.2: Haehnelt, M. G., & Rees, M. J. 1993, MNRAS, 263, 168, "The Formation of Nuclei in Newly Formed Galaxies and the Evolution of the Quasar Population." Reproduced with permission from Blackwell Publishing.

Figure 25.5: Haehnelt, M. G. 2002, in A New Era in Cosmology, ed. N. Metcalfe & T. Shanks (San Francisco: ASP), 95, "Growing Supermassive Black Holes at the Centre of Galaxies." Reproduced with permission from The Astronomical Society of the Pacific.

Figure 25.7–25.8: Haehnelt, M. G. 2003, Classical and Quantum Gravity, 20, S31, "Hierarchical Build-up of Galactic Bulges and the Merging Rate of Supermassive Binary Black Holes." Reproduced with permission from the Institute of Physics Publishing Limited.

Figure 28.1: Vignati, P., et al. 1999, A&A, 349, L57, "BeppoSAX Unveils the Nuclear Component in NGC 6240." Reproduced with permission from *Astronomy and Astrophysics*. Iwasawa, K., Fabian, A. C., & Ettori, S. 2001, MNRAS, 321, L15, "Chandra Detection of Reflected X-ray Emission from the Type 2 QSO in IRAS 09104+4109." Reproduced with permission from Blackwell Publishing. Gandhi, P. 2002, Ph.D. Thesis, Cambridge University.